Vascular Engineering

Kazuo Tanishita • Kimiko Yamamoto
Editors

Vascular Engineering

New Prospects of Vascular Medicine and Biology with a Multidiscipline Approach

Springer

Editors
Kazuo Tanishita
Waseda University
Tokyo
Japan

Kimiko Yamamoto
The University of Tokyo
Tokyo
Japan

ISBN 978-4-431-54800-3 ISBN 978-4-431-54801-0 (eBook)
DOI 10.1007/978-4-431-54801-0

Library of Congress Control Number: 2016931275

Springer Tokyo Heidelberg New York Dordrecht London

Printed on acid-free paper

Springer Japan KK is part of Springer Science+Business Media (www.springer.com)

Contents

Introduction

1

Kazuo Tanishita and Kimiko Yamamoto

Abstract

The vascular system maintains the homeostasis by coordinating various transport phenomena- momentum, mass and the heat transfer- in the circulatory system. To understand the integrated view of dynamic equilibrium state, we need all the conventional macroscopic engineering, cellular and molecular approach. The integrated view of vascular system at multiple scales based on the multi-disciplinary approaches is referred to as vascular engineering. The vascular engineering aims to construct an integrated view, which comprises analysis, synthesis and medical applications. This book focuses on these three aspects of vascular engineering.

Keywords

Vascular engineering • Transport phenomena • Vascular system • Integrated view • Multi-disciplinary approaches

1.1 What Is Vascular Engineering?

The heart supplies blood to the vascular system to provide essential substances (nutrients, oxygen, hormones) to tissue cells and to remove biological waste. The vascular system, which consists of arterial and venous conduits and peripheral beds, generates flow resistance, being the mechanical load due to the pumping function of the heart. Because cells are in the range of several tens of microns in size, the

---

K. Tanishita (✉)
Waseda University Research Organization for Nano and Life Innovation,
513 Wasedatsurumakicho, Tokyo, Shinjuku-ku 162-0041, Japan
e-mail: tanishita@aoni.waseda.jp

K. Yamamoto
Department of Biomedical Engineering, The University of Tokyo, Tokyo, Japan

K. Tanishita, K. Yamamoto (eds.), *Vascular Engineering*,
DOI 10.1007/978-4-431-54801-0_1

capillary vessels in the peripheral beds are of 6-μm diameter, to facilitate mass and heat transport between the bloodstream and cells. In contrast, the aortic vessel, at which the distribution of the bifurcating blood-vessel network begins, is almost 20 mm in diameter. Thus, the structure of blood vessels differs significantly depending on their position within the circulatory system. The wall of the large aortic vessel is characterized by large quantities of elastic fibers; in contrast, arterioles have thick walls composed of smooth muscle fibers.

Vascular blood flow mediates the transport of substances of various molecular sizes to maintain a state of dynamic equilibrium in the tissue. This function of the vascular system is mediated by a series of dynamic processes (Lee 2004). That is, the vascular system maintains homeostasis by coordinating various transport phenomena—momentum (blood flow), mass, and heat transfer—in the circulatory system. These transport phenomena are described most frequently using a conventional engineering approach based on macroscopic fluid and solid mechanics (Fung 1990, 1997, 2010; Pedley 1980; Caro et al. 1978; Fournier 2012; Probstein 1989; Berger et al. 1996). However, analysis at the cellular and molecular levels is necessary for detailed investigation of mass transport in tissue (Probstein 1989). Indeed, mass transport at the cellular and molecular levels is frequently the focus of studies in the fields of biology and medicine (Friedman 2008; Hall 2011; Kierszenbaum and Tres 2012; Alberts et al. 1983). Therefore, investigation of transport phenomena in the vascular system must be conducted at multiple scales, which is facilitated by the use of multi-disciplinary approaches, including engineering, biology, and medicine (De et al. 2015). The field into the vascular (circulatory) system at multiple scales based on multi-disciplinary approaches is referred to as vascular engineering. Thus, vascular engineering addresses the coordinated dynamics of vessel functions, whereas vascular biology deals with the coordinated state of vessel structure, as shown in Fig. 1.1

A multi-scale view of the vascular system includes a range of scales from the whole body to the level of the individual molecule, as shown in Fig. 1.2. Vascular biology (Hall 2011; Kierszenbaum and Tres 2012) deals primarily with the molecular and cellular scales, while vascular engineering aims to explore the dynamical process of vessel function at a wider range of scales. Cardiovascular physiology (Klabunde 2012) involves investigation of the structure and function of the

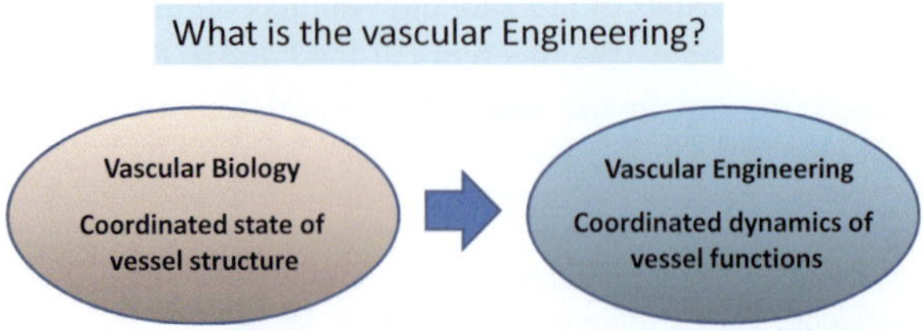

Fig. 1.1 The function of the vascular system is mediated by a series of dynamic processes. Vascular engineering involves investigation of the vascular system at multiple scales based on a multi-disciplinary approach. Thus, vascular engineering focuses on the coordinated dynamics of vessel functions, whereas vascular biology addresses the coordinated state of vessel structure

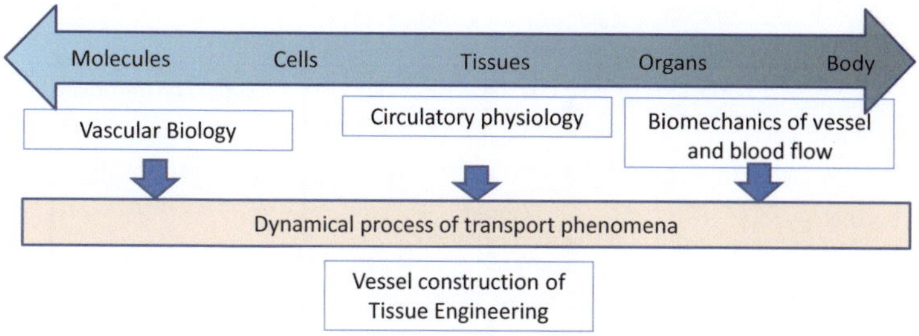

Fig. 1.2 The concept of vascular engineering involves the molecular to macroscopic scales and involves investigation of dynamic transport processes within the vascular system

cardiovascular network at a wide range of scales. Importantly, the detailed mechanical properties of the vascular system at scales larger than the capillaries can be investigated using fluid and solid mechanics together with mechanical analysis. Unfortunately, this macroscopic approach does not provide information on molecular-scale events; therefore, investigations at multiple scales are required to understand the roles of macro- and molecular-scale events in the vascular system.

In the vascular system, endothelial cells play an important role as the interface between mechanical and biological events (Davies 1995; Ando and Yamamoto 2011; Frangos 1993; Gefen 2011; Jacobs et al. 2012; Nagatomi 2011; Aranda-Espinoza 2015; Kiseleva and Kamkin 2010; Engler and Kumar 2014). Vascular endothelial cells—which comprise the inner lining of blood vessels—are exposed directly to the mechanical stimuli generated by blood flow and cyclic stretch due to left ventricle contraction. The response to these mechanical stimuli serves important homeostatic functions, including vascular remodeling. In other words, endothelial cells transduce mechanical signals into biochemical signals in the blood vessel. This phenomenon is designated mechano-transduction and is a rapidly growing area of interest in several fields (Kiseleva and Kamkin 2010; Engler and Kumar 2014).

One of the most difficult aspects of vascular engineering is achieving an integrated multi-scale view of events in the vascular system. Figure 1.3 illustrates the structural hierarchy of the vascular system at the molecular, organ, and whole-body scales; vascular engineering requires an integrated view of this hierarchical structure. Although the discipline necessary to fuse the molecular and macroscopic approaches is not yet available, vascular engineering aims to construct an integrated view of the multi-scale dynamic processes of the vascular system. The integrated view comprises analysis, synthesis, and medical applications (development of therapies). Modern biology, such as molecular biology, uses the analytical approach (Alberts 1983), whereas engineering uses the synthesis approach. The function of the vascular system can be understood using the synthesis approach, which is referred to as the constructive approach, by means of tissue engineering (Atala et al. 2008; Mofrad and Kamm 2009) and computer simulation (Rakocevic et al. 2013; Geris 2012). Both the analytical and synthesis approaches contribute to the construction of models of the vascular system; therefore, these approaches are

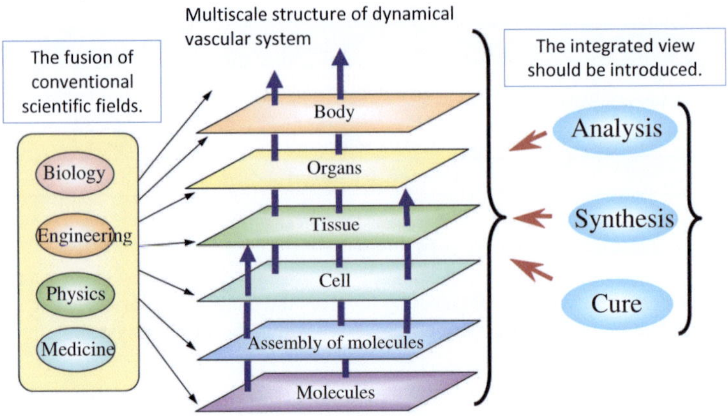

Fig. 1.3 Hierarchical structure of the vascular system at the molecular, organ and whole-body scales. Vascular engineering requires an integrated view of this hierarchical structure. This dynamic system can be understood only using a multi-scale and multi-disciplinary approach. Although the technology necessary to fuse the molecular and macroscopic approaches is not yet available, vascular engineering aims to construct an integrated view of the multi-scale dynamic processes of the vascular system. The integrated view comprises analysis, synthesis, and medical applications (development of therapies)

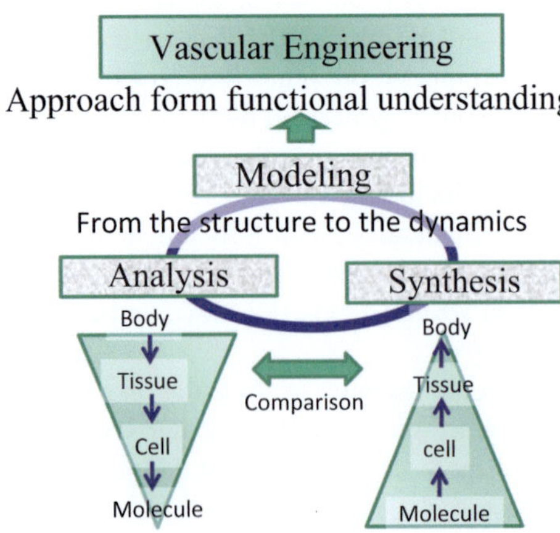

Fig. 1.4 Both analytical and synthesis approaches contribute to the construction of models of the vascular system; therefore, these approaches are utilized for medical applications (development of therapies)

utilized for medical applications (development of therapies), as shown in Fig. 1.4. The macroscopic view of the vascular system provides information on physico-chemical processes based on continuum mechanics, such as fluid and solid mechanics. Functional biological processes occur in the peripheral vascular network and at the cellular level in the tissue. Modern fields of biology, such as molecular biology, focus on the behavior of individual molecules, which are involved in maintaining the function of cells and tissues (Alberts 1983).

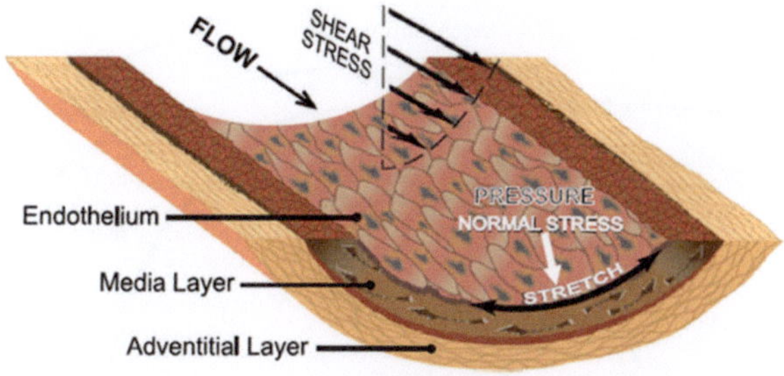

Fig. 1.5 The inner structure of an arterial blood vessel (Chiu and Chien (2011) The walls of arterial blood vessels consist of elastin, collagen and smooth muscle embedded in a mucopolysaccharide ground substance. Arterial vessel walls consist of three layers. The mechanical effects of blood flow are mediated by two types of stress; namely tangential stress (shear stress) and normal stress (pressure)

Mathematical models of blood flow and the mechanical properties of blood vessels have been well established and documented (Lee (2004), Fung (1990, 1997, 2010), Pedley (1980)) based on fluid and solid mechanics. On the other hand, experimental measurements provide quantitative information on the biological processes that occur within blood vessels. For detailed investigation of how the blood flow is coordinated with the blood vessel, knowledge of the inner structure of arterial blood vessels is required, as shown in Fig. 1.5 (Geris 2012). Arterial blood vessels consist of elastin, collagen, and smooth muscle embedded in a mucopolysaccharide ground substance. The wall of arterial vessels consists of three layers. The innermost layer is the tunica intima, which comprises endothelial cells, connective tissue, and basement membrane. The next layer is the thick tunica media, which contains elastin, smooth muscle, and collagen. The outermost layer is the adventitia, which comprises stiff collagen fibers.

Blood flow induces mechanical effects in terms of two types of stress, namely tangential stress (shear stress) and normal stress (pressure). The magnitude of tangential stress is proportional to the viscosity of the blood and the velocity gradient; this is commonly referred to as Newton's law. Tangential stress affects endothelial cells, and this mechanical stimulus has many biological and functional consequences that influence vessel dynamics. In contrast, pressure gives rise to cyclic stretch in vessel walls, which influences multiple biological processes. Furthermore, the various hemodynamic forces induce endothelial dysfunction by unfavorably moderating EC signaling and gene expression, resulting in the development of vascular pathologies (Chiu and Chien 2011).

1.2 Book Contents (Fig. 1.6)

This book focuses on three aspects of vascular engineering (Fig. 1.6); namely, analysis, synthesis, and medical applications (development of therapies). The contribution of each of these to the concept of vascular engineering is discussed.

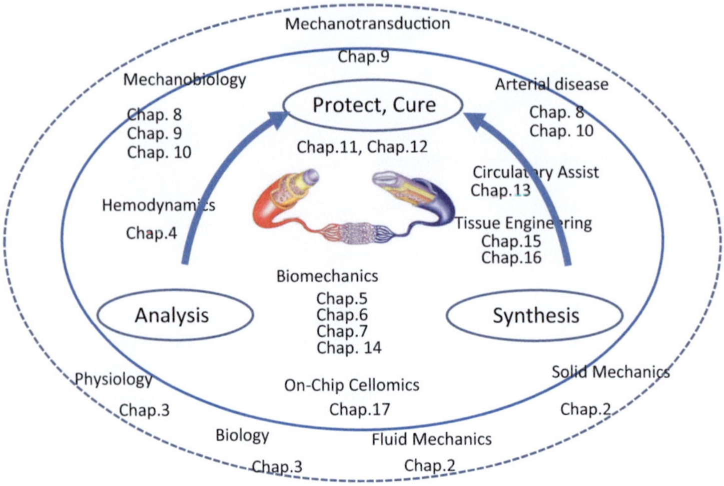

Fig. 1.6 This book focuses on three major aspects of blood vessel and flow dynamics

Chapter 2 deals with the mechanical viewpoint, which provides the basis for understanding various physiological and pathophysiological phenomena. The fundamentals of fluid and solid mechanics in relation to the circulatory system are introduced. In the first part, focusing on fluid mechanics, the concepts of Newtonian and non-Newtonian fluids are described. The rheological properties of blood are described and the universal mechanical law for flow through cylindrical tubes is explained. Based on this law, the characteristics of tube flow for Newtonian and non-Newtonian fluids are described and several mathematical models of blood flow through vessels presented. These models are closely related to important physiological phenomena. In the second part, the concept of continuum mechanics for a large deformation of the vascular wall is first discussed. Then, we introduce passive hyperelastic models, an active smooth muscle model, and discuss incorporation of residual strain and smooth muscle contractions. We demonstrate typical axisymmetric solutions of arterial wall stress for a tube model under physiological loading conditions; i.e., longitudinal stretch and intraluminal pressure. We also present several approaches for arterial diseases, such as atherosclerosis, aortic aneurysm, and dissection.

In Chap. 3, Dr. Aoki describes the basic mechanisms of homeostatic control in the cardiovascular system from the biological, physiological, and pathological point of view. How hemodynamic forces, hormones, cytokines, and neurotransmitters affect the control of vascular functions are stated at both the cellular and molecular levels. The mechanisms of some of the pathological conditions that develop from the breakdown of the proper regulation of the cardiovascular system, such as circulatory shock, hypertension, heart failure, and coronary artery disease, are also stated. In particular, new insights regarding the molecular mechanisms controlling the development of vascular diseases causing cerebral hemorrhage or cardiac infarction, including atherosclerosis and aneurysm, are described comparing the different influences from either hemodynamic forces or chemical stimuli.

In Chap. 4, hemodynamics, being a part of cardiovascular physiology, deals with the forces that drive the blood circulation in mammalian cardiovascular systems. Pressure generated in the heart propels blood through the system continuously. In this chapter, the concepts of basic hemodynamics, essential to the interpretation of arterial disease with regard to bio-fluid mechanics, are introduced. A study of hemodynamics contributes to a better understanding of clinical and pathological observations and, in connection with mathematical models, facilitates the development of new methods for diagnosis. In particular, hemodynamic factors, such as wall shear stress and oscillatory shear index, correlate substantially with the generation and progression of arterial disease, including intimal thickening and atherosclerosis. The main aim of this chapter was to introduce hemodynamic applications of mathematical modeling of fluid mechanics. Mathematical models of fluid mechanics are used to quantify hemodynamic factors and their relationship to vascular disease. The majority of cardiovascular diseases and disorders are related to systemic hemodynamic dysfunction.

In Chap. 5, Dr. Kaunas and Dr. Deguchi focus on the reorganization of stress fibers in endothelial cells (ECs) in response to cyclic stretch. Cyclic stretch is caused by blood pressure acting on vascular walls in a perpendicular fashion based on the pulsatile flow and elongates endothelial cells and smooth muscle cells to the circumferential direction of the vascular walls. Vascular ECs are cultured on elastic stretch chambers to apply cyclic stretch; cells are then elongated and oriented perpendicularly to the stretch with reorganization of the stress fibers. They discuss the details of the biological characteristics of stress fibers in the first section, with the roles and molecular mechanisms of stress fibers in the cyclic stretch–induced signal transduction described in the latter section.

In Chap. 6, Dr. Ohashi focuses on the mechanical characteristics of vascular ECs in response to fluid shear stress. Fluid shear stress arises in ECs when blood flow disturbs the vascular endothelium. Shear stress in the human aorta is 10–20 dynes/ cm^2, whereas shear stress on the walls of veins is 1–6 dynes/cm^2 under physiological conditions. Since blood flow changes in a pulsatile manner with each heartbeat, shear stress also regularly changes. ECs covering the inner surface of blood vessels sense shear stress generated by flowing blood and transmit the signal into the interior of the cell, which evokes a cellular response. Here Dr. Ohashi introduces some topics of cellular responses and mechanical properties, which were measured using flow-loading devices developed by his group members.

Chapter 7 describes how artery walls change their dimensions and mechanical properties adaptively in response to mechanical stimulation. These responses are mediated by the vascular smooth muscle cells (VSMCs) in the tunica media, and so detailed knowledge of the mechanical environment of the VSMCs is indispensable for understanding the mechanism underlying the adaptation. In this chapter, we first introduce the experimental techniques used for tensile testing of tissues and cells at a microscopic scale, and review in detail the tensile properties of VSMCs, followed by those of elastin and collagen fibers. In contrast to elastin and collagen fibers, which are simple, passive materials, VSMCs are living entities with highly complex mechanical properties. Their mechanical properties are reviewed from the viewpoints of smooth muscle contraction, anisotropy in cytoskeletal structure, and viscoelasticity.

Chapter 8 describes the formation of atherosclerosis in the arterial vessel. Atherosclerosis is a serious disease that causes cardiovascular diseases such as cerebral infarction and myocardial infarction. Endothelial injury is the first step in atherogenesis by inducing increases in production of chemoattractant protein and adhesion molecules to leukocytes. One of the key events in atherogenesis is the recruitment of blood leukocytes, especially monocytes, to proatherogenic vascular regions and their subsequent transmigration across endothelial cells. Interactions between leukocytes and endothelial cells involve multi-step processes, including rolling, adhesion, locomotion, and transmigration. Therefore, mechanics and dynamics of endothelial cells and monocytes are important matters to understand the whole process of atherogenesis. Thus, Chap. 8 shows and discusses the mechanobiology of endothelial cells related to the formation of arterial disease.

In Chap. 9, Dr. Butler describes details concerning the molecular mechanisms of the mechanotransduction of fluid shear stress in vascular ECs. When shear stress acts on ECs, it can be converted into biological signals through various membrane-associated molecules and transmitted into the cell interior. The EC response to shear stress is closely linked to the regulation of vascular tone, blood coagulation as well as fibrinolysis, angiogenesis, and vascular remodeling, and plays an important role in maintaining the homoeostasis of the circulatory system. Multiple downstream pathways are involved in shear stress signaling, and they lead to changes in gene expression through the activation of a variety of transcription factors, which results in alterations in EC functions. However, the mechanisms and sensors by which ECs initially recognize shear stress have yet to be identified. Here, Dr. Butler will describe the candidates for shear stress sensors. Understanding the shear stress mechanotransduction, pathways will provide methods of promoting vascular health while also predicting, diagnosing, treating, and preventing vascular disease.

Chapter 10 describes bio-transport in endothelial cells, taking into consideration the effect of shear stress. Arteriosclerosis occurs preferentially at the inner side of the curvature of blood vessels, where shear stress induced by blood flow is low. It has been reported that macromolecules accumulate in these regions. Endothelial cells line the inner surface of blood vessels. A variety of substances are transported from blood vessels to tissues through endothelial cells. Many functions of endothelial cells are known to be affected by fluid shear stress, and dysfunction of endothelial transport is thought to be related to arteriogenesis. Endothelial transport is divided into two main types: passive paracellular transport and active transcellular transport. In this chapter, we introduce these two endothelial transport processes and describe the effects on them in various shear stress conditions.

Chapter 11 describes blood coagulation associated with platelet function. Platelet aggregation occurs at sites of vascular injury due to the high shear rate, and platelet behavior is sensitive to blood flow, which represents a mechanism for maintenance of hemostasis. Thus blood coagulation and platelets play a crucial role in the maintenance of homeostasis in the vascular system. The identity of the majority of major players and the details of their roles in hemostasis and thrombus formation are known, but in most cases only qualitatively. In future, quantitative biological experiments and new computer simulation technology will contribute to advancement of the novel field of vascular engineering.

Chapter 12 focuses on novel technology for application of blood-compatible surfaces to vascular grafts. Despite much effort in the past several decades, no artificial graft has provided a satisfactory patency rate following application to small-caliber vessels (less than 5-mm diameter) in clinical applications. Owing to thrombus formation at the acute stage and intimal thickening caused by compliance mismatch, the long-term patency of these small-caliber vascular grafts is disappointing. Endothelial cell seeding has been proposed to improve the blood compatibility of small-diameter vascular grafts by creating an inner lining with non-thrombogenic surface characteristics similar to those of native blood vessels. We review a surface modification technique that enables the production of consistent and firm endothelial cell linings for hybrid vascular grafts—ion beam implantation.

Chapter 13 describes the vascular engineering of circulatory assist devices. Particular emphasis is placed on recent progress in ventricular assist devices and cardiopulmonary bypass pumps. These important medical devices assist human circulation during either the chronic or the acute phase. Because these devices are derived from an industrial pump, a large number of studies have been conducted from not only medical but also industrial and engineering perspectives. In this chapter, current ventricular assist devices and cardiopulmonary bypass pumps, their specifications, their classifications, and methods for their design and evaluation are presented, including analysis of flow within the pump to optimize the geometry.

In Chap. 14, Dr. Bonakdar and Dr. Goldmann et al. introduce several specific methods for analyzing cellular and molecular responses for different types of mechanical forces, such as shear stress, stretch, adhesion, and pressure, which are applied to several kinds of adherent cells. They review the following different methods for either measuring mechanical properties or applying mechanical forces: (i) a nano-scale particle tracking system, (ii) a magnetic tweezer system, (iii) a rotation disc rheometer, (iv) a magnetic twisting cytometry with optical detection system, (v) a cell poking system, and (vi) a traction force microscopy system. The latter part deals with the response of smooth muscle cells to these systems.

In Chap. 15, Dr. Takahashi and Dr. Naruse discuss how vascular ECs and smooth muscle cells (SMCs) can sense mechanical forces, including stretch elongation or fluid shear stress, particularly vascular SMCs when cyclic stretch is applied by pulsatile blood flow and vasoconstriction. Cultured SMCs are perpendicularly oriented with response to the direction of stretch elongation through the activation Notch signaling. They also mention some of the possibilities for applying knowledge from mechanobiology into "mechano-medicine," including regenerative medicine. The pathological conditions due to vascular mechanotransduction pathways, for instance, arteriosclerosis and in-stent restenosis are also described. Appropriate methods for inducing vasculogenesis have been recently developed using three-dimensional culture of ECs and SMCs with extracellular matrix protein, helping to predict the prognosis of artery disease.

In Chap. 16, the authors describe the culture methods to construct microvascular networks as well as approaches to integrating capillary networks with 3D epithelial tissue-engineered constructs. First, culture models of microvascular networks, such as *in vitro* angiogenesis and vasculogenesis models, are introduced. Using these culture models, the roles of endothelial cells (ECs), such as endothelial tip, stalk, and phalanx cells, are demonstrated. Additionally, regulatory factors, including both biochemical and biophysical factors, are discussed in the context of 3D capillary formation, including the process of vascular development, growth, and maturation. Next, we focus on the use of microfluidics technologies for investigating capillary morphogenesis. Examples of 3D capillary formation assays with growth factor gradients and different extracellular matrix materials are described. Cocultures of ECs and the other cell types in microfluidic devices are also introduced to show the potential of microfluidic vascular formation models. The vascularization of constructed tissues is discussed from the viewpoints of horizontal and vertical approaches for combining capillary structures and epithelial tissues *in vitro*. Finally, the concept of integrated vascular engineering and future perspectives are discussed.

In Chap 17, Dr. Yasuda introduces the unique technology of "on-chip cellomics" combining on-chip technology with the cellomics one. The on-chip cellomics system was originally developed by his group for analyzing epigenetic and genetic information or for use in the drug screening and the regenerative medicine. He describes detailed information regarding the mathematical model of the basic ideas and the specific methods of their systems. The single cell behavior of neurons or cardiomyocytes was then analyzed in order to apply the system to drug screening or tissue regeneration.

References

Alberts B et al (1983) Molecular biology of the cell. Garland, New York

Ando J, Yamamoto K (2011) Effects of shear stress and stretch on endothelial function. Antioxid Redox Signal 15:1389–1403

Aranda-Espinoza H (2015) Mechanobiology of the endothelium. CRC Press, Boca Raton

Atala A, Lanza R, Thomson JA, Nerem RM (2008) Principles of regenerative medicine. Elsevier, Amsterdam

Berger SA et al (eds) (1996) Introduction to bioengineering. Oxford University Press, New York

Caro CG, Pedley TJ, Schroter RC, Seed WA (1978) The mechanics of the circulation. Oxford University Press, Oxford

Chiu JJ, Chien S (2011) Effects of disturbed flow on vascular endothelium: pathophysiological basis and clinical perspectives. Physiol Rev 91:327–387

Davies PF (1995) Flow-mediated endothelial mechanotransduction. Physiol Rev 75:519–560

De S, Hwang W, Kuhl E (2015) Multiscale modeling in biomechanics and mechanobiology. Springer, London

Engler AJ, Kumar S (2014) Mechanotransduction. Academic, Amsterdam

Fournier R (2012) Basic transport phenomena in biomedical engineering, 3rd edn. CRC Press, Baca Raton
Frangos JA (ed) (1993) Physical forces and the mammalian cell. Academic, San Diego
Friedman MH (2008) Principles and models of biological transport. Springer, New York
Fung YC (1990) Biomechanics, motion, flow, stress and growth. Springer, New York
Fung YC (1997) Biomechanics: circulation. Springer, New York
Fung YC (2010) Biomechanics: mechanical properties of living tissues. Springer, New York
Gefen A (2011) Cellular and biomolecular mechanics and mechanobiology. Springer, Berlin
Geris L (2012) Computational modeling in tissue engineering. Springer, Berlin/Heidelberg
Hall JE (2011) Textbook of medical physiology, 12th edn. Elsevier, Philadelphia
Jacobs CR, Huang H, Kwon RY (2012) Introduction to cell mechanics and mechanobiology. Garland Science, New York
Kierszenbaum AL, Tres L (2012) Histology and cell biology. Elsevier, Philadelphia
Kiseleva I, Kamkin A (2010) Mechanosensitivity and mechanotransduction. Springer, Dordrecht
Klabunde RE (2012) Cardiovascular physiology concepts. Lippincott Williams & Wilkins, Baltimore
Lee JKJ (2004) Dynamics of the vascular system. World Scientific Publishing Co. Pte. Ltd, River Edge, Singapore
Mofrad MRK, Kamm RD (2009) Cellular mechanotransduction: diverse perspectives from molecules to tissues. Cambridge University Press, Cambridge
Nagatomi J (2011) Mechanobiology handbook. CRC Press, Boca Raton
Pedley TJ (1980) Fluid mechanics of large blood vessels. Cambridge University Press, New York
Probstein RF (1989) Physicochemical hydrodynamics. Butterworths, Boston
Rakocevic G, Djukic T, Filipovic N, Milutinovic V (2013) Computational medicine in data mining and modeling. Springer, New York

Fundamentals of Vascular Bio-fluid and Solid Mechanics

2

Masako Sugihara-Seki and Hiroshi Yamada

Abstract

While considering the circulatory system, a mechanical view point provides the basis for understanding various physiological and pathophysiological phenomena. In this section, we introduce the fundamentals of fluid and solid mechanics in relation to the circulatory system. In the first part on fluid mechanics, we start by introducing the concepts of Newtonian and non-Newtonian fluids, viscosity, and apparent viscosity. Then, the rheological properties of blood are described, and the universal mechanical law for flow through cylindrical tubes is derived. Based on that law, we consider the characteristics of tube flow for Newtonian and non-Newtonian fluids and present several representative examples of mathematical models for blood flow through vessels. These models have a close relationship with important physiological phenomena. In the second part, we first discuss the concept of continuum mechanics for a large deformation of the vascular wall. Then, we introduce passive hyperelastic models, an active smooth muscle model, and incorporations of residual strain and smooth muscle contractions. We demonstrate typical axisymmetric solutions of arterial wall stress for a tube model under physiological loading conditions, i.e., longitudinal stretch and intraluminal pressure. We also show some approaches for arterial diseases such as atherosclerosis, aortic aneurysm, and aortic dissection.

Keywords

Apparent viscosity • Blood flow • Boundary-value problem • Continuum mechanics • Constitutive equation

M. Sugihara-Seki (✉)
Department of Pure and Applied Physics, Kansai University, Suita, Japan
e-mail: sekim@kansai-u.ac.jp

H. Yamada
Department of Biological Functions Engineering, Graduate School of Life Science and Systems Engineering, Kyushu Institute of Technology, Kitakyushu, Japan

K. Tanishita, K. Yamamoto (eds.), *Vascular Engineering*,
DOI 10.1007/978-4-431-54801-0_2

2.1 Vascular Fluid Mechanics

In this section, we introduce the basic concepts of fluid mechanics to understand the circulatory system from a mechanics standpoint. Focusing on steady flow through circular cylindrical tubes, we describe the rheological behaviors of Newtonian and non-Newtonian fluids, and consider several distinctive features of blood flow through vascular vessels by utilizing simple mathematical models.

2.1.1 Newtonian and Non-Newtonian Fluids

First, the definition of "stress" in a fluid will be introduced. Suppose a small surface, S, is inside the fluid, and a unit vector normal to this surface is denoted as $\boldsymbol{n}$. As shown in Fig. 2.1, the fluid on the side of the surface, to which the normal vector points, exerts a force, $\boldsymbol{F}$, on this surface. In mechanics, the term "stress" is used to express force per unit area. Thus, the stress acting on this side of the surface, $\boldsymbol{\sigma}_n$, is defined as $\boldsymbol{F}/S$ as the limit of S approaches 0. The stress acting on the other side of the surface is expressed as $-\boldsymbol{F}/S$ in the same limit, as may be understood by Newton's third law. The stress has three components in space that depend on the location and orientation of the normal vector $\boldsymbol{n}$. In this sense, the stress can be specified by the so-called stress tensor of rank 2 (For details, refer to a text book of fluid mechanics, e.g., Farber 1995).

It is usually convenient to decompose stress, $\boldsymbol{\sigma}_n$, into a normal stress, normal to the surface or parallel to the vector $\boldsymbol{n}$, and a tangential stress, tangent to the surface, as shown in Fig. 2.1. The normal stress is sometimes called the tension if it works in the direction of $\boldsymbol{n}$, whereas it is called the pressure if it works in the opposite direction of $\boldsymbol{n}$.

2.1.1.1 Viscosity and Apparent Viscosity

As may be expected, only pressure acts on a fluid at rest. When the fluid flows, tangential stress is also generated. With regard to this issue, Newton first introduced the concept of "viscosity" of a fluid, utilizing a simple flow configuration sketched in Fig. 2.2.

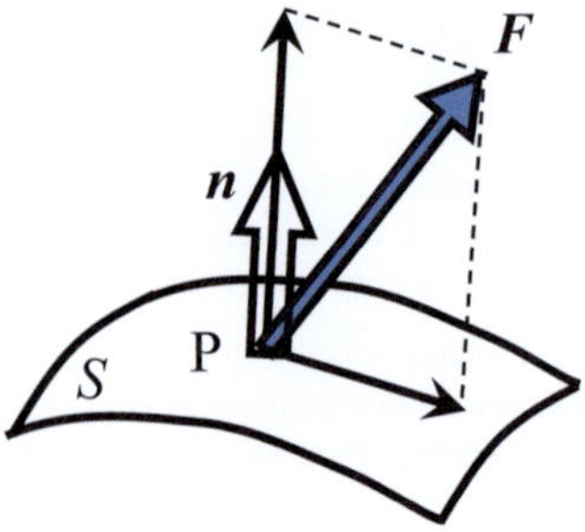

Fig. 2.1 Force $\boldsymbol{F}$ exerted on a surface, S, in a fluid

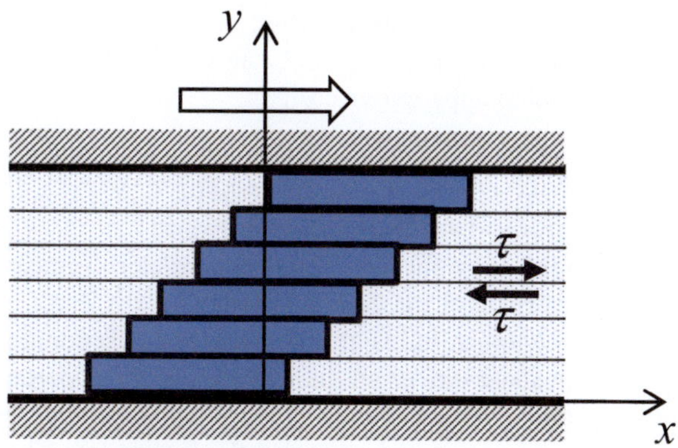

Fig. 2.2 Couette flow between two parallel plates

Suppose that a fluid is confined between two parallel plates, an upper plate that moves with a constant velocity and a lower plate that is stationary. Following the movement of the upper plate, planar laminae of the fluid lying parallel to the plates move steadily and slide over one another creating a uniform velocity gradient (or shear rate), $\dot{\gamma}$. Newton postulated that in such circumstances, a frictional shear stress (tangential stress) arises between adjacent laminae, the magnitude of which is proportional to the velocity gradient:

$$\tau = \eta\dot{\gamma}. \tag{2.1}$$

The proportional constant, η, is called the viscosity or shear viscosity. For homogeneous fluids such as water or saline, η is a material constant if temperature and pressure are specified. These fluids are referred to as Newtonian. Equation (2.1) is called the Newton formula. More rigorously, a Newtonian fluid is one in which the stress tensor in the fluid is a linear function of the rate of strain tensor, the components of which are expressed in terms of velocity gradients (e.g., Farber 1995).

In Fig. 2.2, the lamina next to the plate exerts the shear stress expressed by Eq. (2.1) on the plate. Thus, to keep the movement of the upper plate steady, we need to apply an external force on the plate that balances such a tangential force over the plate. Accordingly, if the force applied to the upper plate is measured, the fluid viscosity, η, can be estimated using Eq. (2.1). This is the principle used by rotational viscometers to measure viscosity of a fluid.

A fluid not obeying the Newton formula (2.1) is referred to as non-Newtonian. Blood plasma is generally regarded as a Newtonian fluid. Blood, however, behaves as a non-Newtonian fluid; the shear stress, τ, is a nonlinear function of $\dot{\gamma}$. Before studying blood viscosity, let us consider some representative characteristics of non-Newtonian fluids.

2.1.1.2 Representative Types of Non-Newtonian Fluids

A number of complex fluids, including whole blood, suspensions and slurries of rigid and deformable solid matter, and pastes and emulsions, are non-Newtonian. If measurements of the shear stress, τ, and shear rate, $\dot{\gamma}$, are made for the flow configuration shown in Fig. 2.2, the ratio of these quantities defines a function, which is called apparent viscosity:

$$\eta_a(\dot{\gamma}) = \frac{\tau}{\dot{\gamma}}. \tag{2.2}$$

Although the apparent viscosity is not a material constant, it represents a rheological characteristic of the complex fluid.

Non-Newtonian fluids may be classified depending on the variation of the shear stress with shear rate or on the variation of the apparent viscosity with shear stress or shear rate. Figure 2.3 shows several representative behaviors of non-Newtonian fluids as well as that of Newtonian fluids (line (i)). The materials corresponding to lines (ii), (iii), and (iv) in Fig. 2.3 are called "Bingham plastic," "pseudoplastic," and "dilatant," respectively. Bingham plastic does not deform at all when subjected to a shear stress smaller than a certain value. This is mostly due to the presence of microstructures or clusters of molecules, and a certain stress is required to break them. This critical value is called the "yield stress." For stresses above the yield stress, the Bingham plastic flows with the shear stress as a linear function of shear rate. The pseudoplastic (line (iii)) exhibits a decrease in apparent viscosity as the shear rate increases. This phenomenon is sometimes called "shear-thinning." Many materials, such as various polymer solutions, show pseudoplastic behavior in varying degrees. There are some materials that appear to expand when subjected to shear stresses; the dilatant (line (iv)) exhibits an increase in apparent viscosity as the shear rate increases because of this dilation. This phenomenon is sometimes called "shear-thickening," and some concentrated suspensions of solids, slurries, and proteins show dilatant behavior.

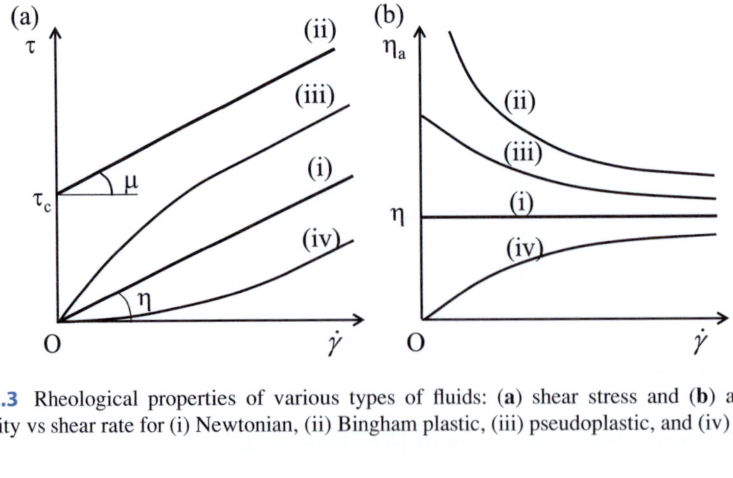

Fig. 2.3 Rheological properties of various types of fluids: (**a**) shear stress and (**b**) apparent viscosity vs shear rate for (i) Newtonian, (ii) Bingham plastic, (iii) pseudoplastic, and (iv) dilatant

It is evident that the straight line (i) in Fig. 2.3a for a Newtonian fluid is described by Eq. (2.1). The slope of this line corresponds to the viscosity, η. The line (ii) for Bingham plastic is expressed by

$$\begin{cases} \tau = \tau_c + \mu\dot{\gamma} & (\tau \geq \tau_c) \\ \dot{\gamma} = 0 & (\tau < \tau_c) \end{cases}, \tag{2.3}$$

where τ_c represents the yield stress, and μ is the plastic viscosity. The shear-thinning and shear-thickening may be ideally expressed by the so-called power-law fluid:

$$\tau = k\dot{\gamma}^n, \tag{2.4}$$

where n and k are constant, and $n < 1$ and $n > 1$ represent the shear-thinning and shear-thickening, respectively. $n = 1$ represents the Newtonian fluid. Equations (2.1), (2.3), and (2.4) specify the properties of the materials and are collectively called the constitutive equation.

2.1.2 Blood Viscosity

Blood is a suspension of red blood cells (erythrocytes), white blood cells (leukocytes), and platelets in plasma. Plasma is about 90 % water by weight, 7 % plasma protein, 1 % inorganic substances, and 1 % other organic substances. Normally, plasma is an incompressible Newtonian fluid with a viscosity of about 1.2 cP at 37 °C. The red cells constitute the largest percentage of cells with a hematocrit (volume fraction of red cells) of about 40–45 % in normal blood. The white cells of various categories make up less than 1/600th of the total cellular volume, and the platelets occupy less than 1/800th of the cellular volume (Fung 1993). Thus, blood can be regarded as a highly concentrated suspension of red cells, and its rheological properties are strongly influenced by these red cells.

The viscosity of blood at a given temperature is a function of shear rate, hematocrit, plasma viscosity, physical properties of the red cells, and cell aggregation. At a normal state of 45 % hematocrit, the apparent viscosity of whole blood is about 4 cP at relatively high shear rates. The relationship between the apparent viscosity, η_a, and hematocrit is nonlinear, with η_a rising at an increasing rate as the hematocrit increases.

Human red cells can form aggregates under low-flow conditions. They can deform when subjected to shear. Figure 2.4 is a typical diagram used to show the importance of red cell deformation and aggregation (Chien 1970). The apparent viscosities relative to the medium viscosity are shown for three different red cell suspensions, NP, NA, and HA. The line NP refers to normal blood. When the shear rate increases, blood aggregates tend to be broken up, and the viscosity of blood decreases. As the shear rate further increases, the deformation of the red cells becomes more evident, and the red cells are elongated and aligned with the flow

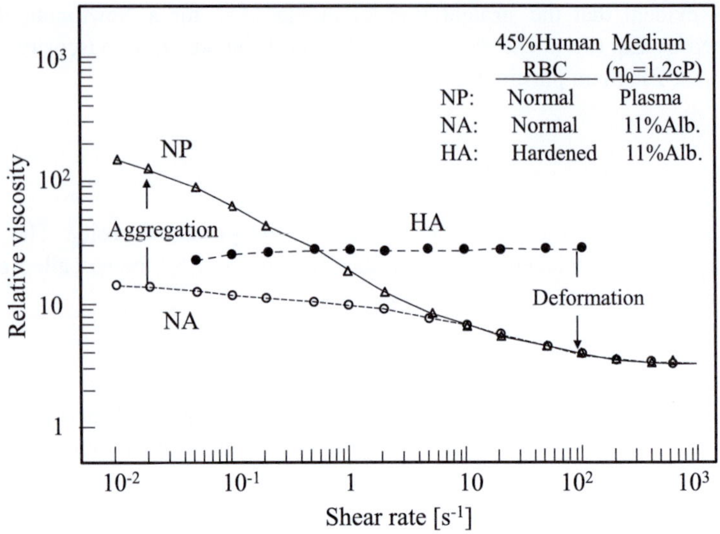

Fig. 2.4 Logarithmic relation between viscosity and shear rate in three types of suspensions at 37 °C, each containing 45 % human RBC by volume (Modified from Chien 1970)

direction, which reduces blood viscosity. The line NA refers to a suspension of normal red cells in isotonic saline containing 11 % albumin, and they show deformation but no aggregation. The third line, HA, refers to a suspension of hardened red cells in the same saline solution, and they show neither deformation nor aggregation. At low shear rates, the apparent viscosity relative to the medium viscosity is higher for NP than for NA because of red cell aggregation, and at high shear rates it is lower for NP than for HA because of red cell deformation.

It is convenient to have analytical expressions representing the blood apparent viscosity as a function of rheological factors such as hematocrit and shear rate. A number of equations have been proposed to describe the rheological behavior of blood. Among them, Scott-Blair (1959) used Casson equation derived by Casson (1959) based on a mechanical model of suspension flow:

$$\sqrt{\tau} = \sqrt{\tau_c} + \sqrt{\eta_c}\sqrt{\dot{\gamma}}, \tag{2.5}$$

where η_c and τ_c are called the Casson viscosity and Casson yield stress, respectively. In this model, the suspension was assumed to contain particles that can aggregate at low shear rates to form rod-like aggregates, which increase in length as the shear rate decreases. Scott Blair plotted $\tau^{1/2}$ against $\dot{\gamma}^{1/2}$ for whole blood, plasma, and serums of several species and showed that they produced excellent straight lines. Cokelet et al. (1963) showed that for small shear rates, $\dot{\gamma} < 10\ \mathrm{s}^{-1}$, and for a hematocrit less than 40 %, the blood viscosity is well described by the Casson Eq. (2.5).

It is generally accepted that blood does not flow unless the shear stress exceeds yield stress, τ_c, the value of which is markedly influenced by several factors such as hematocrit and macromolecular composition. Since τ_c is very small, ~ 0.05 dyn/cm^2 (Merrill et al. 1963), experimental determination of yield stress is complicated. It is known that a suspension of red cells in plasma containing fibrinogen has a finite yield stress, but a suspension of red cells in saline plus albumin has zero yield stress. At high shear rates, whole blood behaves like a Newtonian fluid with a constant viscosity.

2.1.3 Steady Flow Through Circular Cylindrical Tubes

As a basis to understand blood flow through vascular vessels, we consider the steady laminar flow of a homogeneous fluid through a straight circular cylindrical tube of radius R and length L under a pressure difference, Δp. Since the tube radius is uniform, it is reasonable to assume that the flow is parallel to the cylindrical axis, the x-axis. In this section, we derive an expression for shear stress in the fluid using the property that the net force acting on any fluid element vanishes because of steady flow.

We consider a fluid element of circular cylindrical annulus of radius r, thickness Δr and length Δx, the center of which lies on the centerline of the tube, as shown in Fig. 2.5. If shear stresses acting on the circumferential surfaces of the annulus at $r = r$ and $r + \Delta r$ are expressed as $\tau(r)$ and $\tau(r + \Delta r)$, respectively, then the total force exerted on the circumferential surfaces becomes

$$\tau(r + \Delta r) \times 2\pi(r + \Delta r)\Delta x - \tau(r) \times 2\pi r \Delta x \quad (2.6)$$

in the x-direction. This force is balanced with the pressure difference exerted at both ends of the cylindrical annulus at $x = x$ and $x + \Delta x$. This condition of force balance yields

$$[(r + \Delta r)\tau(r + \Delta r) - r\tau(r)]\, 2\pi\Delta x = [p(x + \Delta x) - p(x)]\, 2\pi r \Delta r. \quad (2.7)$$

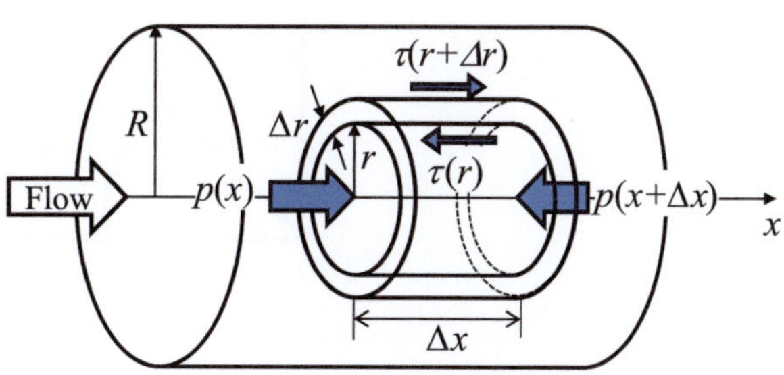

Fig. 2.5 A fluid element of circular cylindrical annulus of radius r, thickness Δr, and length Δx

Using Taylor expansions, we obtain

$$-\frac{\mathrm{d}p}{\mathrm{d}x}+\frac{1}{r}\frac{\mathrm{d}}{\mathrm{d}r}(r\tau)=0. \tag{2.8}$$

Integration of this equation in the r-direction leads to

$$\tau=-\frac{\alpha}{2}r+\frac{C}{r}, \tag{2.9}$$

where $\alpha=-\mathrm{d}p/\mathrm{d}x=\Delta p/L$ represents the pressure gradient, and C is a constant. Since the shear stress is finite at $r=0$, C is determined to be 0, and the shear stress at the radial position r is obtained as

$$\tau=-\frac{\alpha}{2}r. \tag{2.10}$$

The magnitude of shear stress is illustrated in Fig. 2.6. It vanishes at the centerline of the tube and increases linearly with r up to a maximum value, τ_w (= $\alpha R/2$), on the tube wall. Equation (2.10) is valid for both Newtonian and non-Newtonian fluids. Next, we obtain a velocity profile for several types of fluids based on this relationship (2.10).

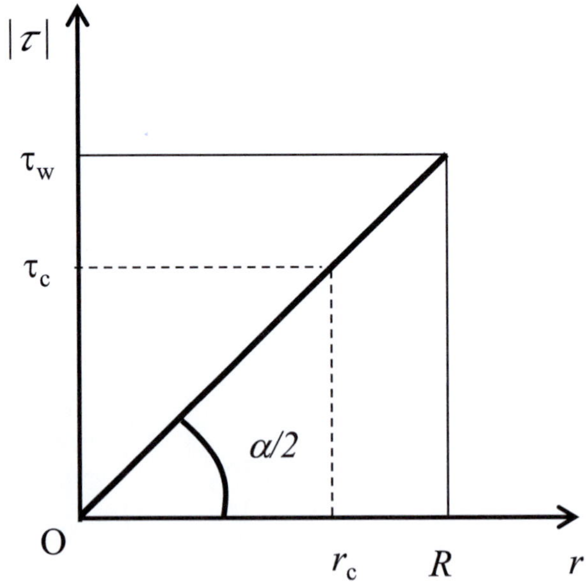

Fig. 2.6 Magnitude of shear stress in a steady tube flow. The shear stress reaches the yield stress τ_c at radial distance r_c

2.1.3.1 Flow of Newtonian Fluids

We consider the flow of a Newtonian fluid with viscosity η through a cylindrical tube treated above. Since velocity gradient in this case is expressed as du/dr, Eq. (2.1) is rewritten as

$$\tau = \eta \frac{\mathrm{d}u}{\mathrm{d}r}. \tag{2.11}$$

Thus, substituting Eq. (2.11) in Eq. (2.10) and integrating under the no-slip boundary condition on the tube wall yields the velocity profile in the tube:

$$u = \frac{\alpha}{4\eta}\left(R^2 - r^2\right). \tag{2.12}$$

Equation (2.12) represents a parabolic velocity profile, called the Poiseuille flow. The flow rate per unit time is easily obtained as

$$Q = \int_0^R u 2\pi r dr = \frac{\pi R^4}{8\eta}\alpha. \tag{2.13}$$

For a given pressure gradient, $\alpha(=\Delta p/L)$, the flow rate Q is proportional to R^4, indicating that it depends substantially on the tube radius. In other words, a small change in the tube radius drastically affects the flow rate through the tube. This may reflect effective control of blood flow in arterioles by changing their radii.

2.1.3.2 Flow of Casson Fluids

Next, let us consider the steady flow of a Casson fluid through a tube under the same condition. The constitutive Eq. (2.5) can be used if $\tau > \tau_c$. If not, the velocity gradient vanishes, i.e.

$$\dot{\gamma} = \frac{\mathrm{d}u}{\mathrm{d}r} = 0. \tag{2.14}$$

Thus, we notice from Fig. 2.6 that if $\tau_w < \tau_c$, then the magnitude of the shear stress is smaller than τ_c in the entire region of the tube. In this case, Eq. (2.14) together with the no-slip boundary condition on the tube wall leads to

$$u = 0 \quad (\tau_w < \tau_c). \tag{2.15}$$

The condition of $\tau_w = \tau_c$ provides a critical pressure difference $(\Delta p)_c$ to generate fluid flow through the tube:

$$(\Delta p)_c = \frac{2\tau_c}{R}L. \tag{2.16}$$

If the pressure difference across the tube is smaller than $(\Delta p)_c$, flow does not occur.

For the case of $\tau_w > \tau_c$, we define the critical radius r_c by

$$\tau_c = \frac{\alpha}{2} r_c. \tag{2.17}$$

It can be seen from Fig. 2.6 that for $r > r_c$, the magnitude of shear stress is larger than τ_c for cases where Eq. (2.5) is satisfied. For $r < r_c$, the magnitude of shear stress is smaller than τ_c for cases where Eq. (2.14) is satisfied. Combining these equations and Eq. (2.10) yields

$$\begin{cases} \sqrt{\frac{\alpha}{2} r} = \sqrt{\tau_c} + \sqrt{\eta_c}\sqrt{-\frac{du}{dr}} & (r_c \le r \le R) \\ \frac{du}{dr} = 0 & (r < r_c) \end{cases}. \tag{2.18}$$

The second equation implies that the velocity profile is flat for $r < r_c$. From the first equation, we obtain

$$-\frac{du}{dr} = \frac{1}{\eta_c}\left(\sqrt{\frac{\alpha}{2} r} - \sqrt{\tau_c}\right)^2 \quad (r_c \le r \le R). \tag{2.19}$$

Integrating this equation and using the no-slip boundary condition on the tube wall at $r = R$, we have

$$u = \frac{\alpha}{4\eta_c}\left[R^2 - r^2 - \frac{8}{3} r_c{}^{1/2}\left(R^{3/2} - r^{3/2}\right) + 2r_c(R - r)\right] \quad (r_c \le r \le R). \tag{2.20}$$

Inserting $r = r_c$ into Eq. (2.20), we obtain the velocity u_c, which represents the constant velocity in the region $r < r_c$:

$$u_c = \frac{\alpha}{4\eta_c}\left(\sqrt{R} - \sqrt{r_c}\right)^3\left(\sqrt{R} + \frac{1}{3}\sqrt{r_c}\right) \quad (r < r_c). \tag{2.21}$$

Equations (2.20) and (2.21) provide a blunt velocity profile as shown in Fig. 2.7. The flow rate is easily obtained as

$$Q = \frac{\pi R^4 \alpha}{8\eta_c} F(\xi) \quad (\tau_w > \tau_c), \tag{2.22}$$

where $F(\xi)$ was formulated by Oka (1981) as

$$F(\xi) = 1 - \frac{16}{7}\xi^{1/2} + \frac{4}{3}\xi - \frac{1}{21}\xi^4, \quad \xi = \frac{r_c}{R}. \tag{2.23}$$

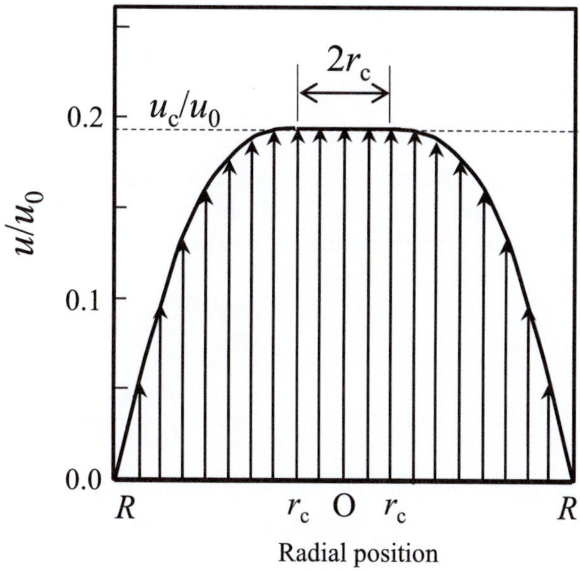

Fig. 2.7 Velocity profile of a Casson fluid in a tube ($r_c/R = 0.2$). $u_0 = \alpha R^2/(4\eta_c)$

In the case of $r_c = 0$ or $\tau_c = 0$, as expected, Eqs. (2.20) and (2.22) reduce to Eqs. (2.12) and (2.13), respectively. In the other cases, since r_c is related to α (Eq. 2.17), the velocity profile and flow rate depend nonlinearly on pressure gradient.

Here, if we define $\eta_a = \eta_c/F(\xi)$, then Eq. (2.22) can be rewritten as

$$Q = \frac{\pi R^4}{8\eta_a}\alpha. \tag{2.24}$$

This equation has the same form as Eq. (2.13) for Poiseuille flow. The quantity η_a in Eq. (2.24) is also called apparent viscosity. From Eq. (2.23), we note that apparent viscosity for a Casson fluid increases rapidly with increasing ξ.

2.1.3.3 Flows of Bingham Plastic and Power-Law Fluids

It may be interesting to consider the flow of other types of non-Newtonian fluids in a tube. Here, we briefly treat a Bingham plastic and power-law fluid as representative examples.

For a Bingham plastic fluid with the constitutive equation given by Eq. (2.3), a straightforward calculation similar to Sect. 2.1.3.2 for a Casson fluid leads to the velocity profile and apparent viscosity under the condition of $\tau_c < \tau_w$:

$$u = \begin{cases} \frac{\alpha}{4\mu}\left[\left(R^2 - r^2\right) - 2r_c(R - r)\right] & (r_c \le r \le R) \\ \frac{\alpha}{4\mu}(R - r_c)^2 & (r < r_c) \end{cases}, \tag{2.25}$$

and

$$\eta_a = \mu\left[1 - \frac{4}{3}\xi + \frac{\xi^4}{3}\right]^{-1}, \quad \xi = \frac{r_c}{R}, \tag{2.26}$$

where r_c is defined by Eq. (2.17). Under the condition of $\tau_c > \tau_w$, there is no flow.

For a power-law fluid with the constitutive equation given by Eq. (2.4), we obtain the velocity profile and apparent viscosity:

$$u = u_0\left(1 - \left(\frac{r}{R}\right)^{\frac{n+1}{n}}\right), \quad u_0 = \left(\frac{\alpha}{2k}\right)^{\frac{1}{n}}\frac{n}{n+1}R^{\frac{n+1}{n}}, \tag{2.27}$$

$$\eta_a = \frac{R^2\alpha}{8u_0}\frac{3n+1}{n+1}. \tag{2.28}$$

Except for the case of $n = 1$, corresponding to a Newtonian fluid, the velocity profile as well as flow rate depend on the pressure difference in a non-linear fashion as a power of $1/n$.

2.1.4 Application to Various Blood Flows

The basic ideas introduced above are helpful to understand several important and unique characteristics of blood flow through blood vessels. Here, we present some typical examples. In each case, an idealized model describes the distinct features of blood flow observed *in vivo* and predicts its physiological impact.

2.1.4.1 "Stacked-Coins" Model

The apparent viscosity of blood flowing through small vessels is known to be much smaller than the bulk viscosity of blood (Fahraeus-Lindqvist effect) (Fahraeus and Lindqvist 1931). Fluid mechanical investigations on this phenomenon are rather complicated, since the continuum approach considered so far may be inappropriate because of the comparable sizes of blood cells and vessel radii. Instead, we need to analyze the individual motion of the blood cells that interact with the surrounding plasma. However, the idea introduced at the beginning of this section is useful for conducting a simple analysis that accounts for the phenomena in small capillaries.

In a vessel with a diameter smaller than ~6 μm, red cells are most likely to flow in a single file, and a layer of plasma exists between the red cells and vessel wall. When the distance between neighboring cells is much smaller than vessel diameter, this flow situation may be described as the first approximation by the "axial-train" model developed by Whitmore (1968), depicted in Fig. 2.8, and also referred to as the "stacked-coins" model. In this model, the red cells are confined in a cylindrical core of radius λR that is concentric with the vessel and moves as a rigid body. The spacing between the core and vessel wall is assumed to be filled with plasma, a Newtonian fluid of viscosity η.

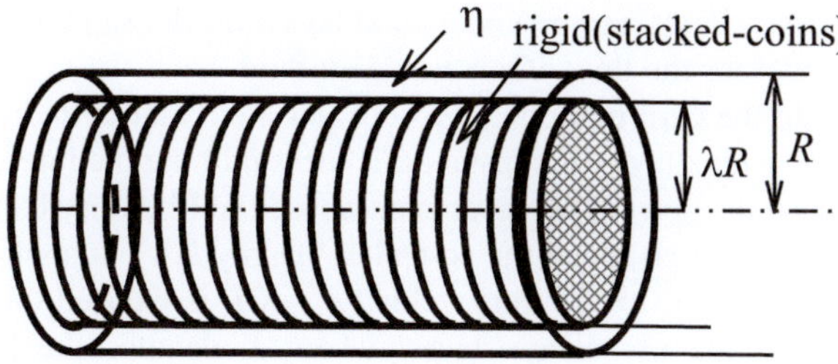

Fig. 2.8 Stacked-coins model

If we follow procedure similar to Sect. 2.1.3.1 for the plasma in the region of $\lambda R \leq r \leq R$, then we obtain Eq. (2.12) for the velocity profile of the plasma in the gap region. From the no-slip condition at the interface between red cells and plasma, putting $r = \lambda R$ in Eq. (2.12) provides the velocity of the stacked coins (red cell core):

$$u_c = \frac{R^2 \alpha}{4\eta} \left(1 - \lambda^2\right). \tag{2.29}$$

The flow rate of the whole blood including red cells and plasma is given by

$$Q = \frac{\pi R^4 \alpha}{8\eta} \left(1 - \lambda^4\right). \tag{2.30}$$

Thus, apparent viscosity of the stacked-coins model is given in terms of plasma viscosity, η, and radius ratio, λ, as

$$\eta_a = \frac{\eta}{1 - \lambda^4}. \tag{2.31}$$

This model may give some insight into the low viscosity observed in small capillaries. Equation (2.31) indicates that the presence of even a relatively thin layer of plasma in a capillary is sufficient to keep apparent viscosity at a reasonably low level. For instance, a plasma layer of 0.3 μm thickness in a capillary of diameter 6 μm gives $\lambda = 0.9$ and $\eta_a = 2.9\eta$, which is lower than the bulk viscosity of whole blood.

2.1.4.2 Capillary Flow in the Presence of a Glycocalyx Layer

Although it has long been known that the luminal surface of the vascular endothelium is lined with a layer of glycocalyx, it was not known until recently that this glycocalyx layer plays important roles in multifaceted endothelial functions such as microvessel permeability, regulation of blood flow, and blood cell interactions (Pries et al. 2000; Weinbaum et al. 2007). If the glycocalyx layer impedes the flow, the presence of the layer decreases the effective cross-sectional area of the vessel available for plasma and red cell motion, so it could possibly cause an increase in apparent viscosity as well as a decrease in hematocrit in small vessels.

These effects can be understood based on the same idea by using a mathematical model for the flow of a Newtonian fluid.

In the current model, we consider a circular cylindrical tube of radius R with a glycocalyx layer of constant thickness h lining the interior surface of the tube wall. Since the glycocalyx layer is composed of fiber matrixes, it can be assumed to be a porous material and the so-called "Darcy law" is applied. Thus, a frictional force proportional to the fluid velocity, $-Ku$, is exerted on the fluid per unit volume within thc glycolcayx layer, where K denotes the hydraulic resistivity of the glycocalyx layer. Adding this frictional force term, $-Ku \cdot 2\pi r \Delta r \Delta x$, to the left-hand side of Eq. (2.7), we obtain, in place of Eq. (2.8), for the steady flow of a Newtonian fluid within the glycocalyx layer ($R - h \leq r \leq R$):

$$-\frac{\mathrm{d}p}{\mathrm{d}x} + \frac{1}{r}\frac{\mathrm{d}}{\mathrm{d}r}(r\tau) - Ku = 0 \quad (R - h \leq r \leq R). \tag{2.32}$$

Applying Eq. (2.32) for $R - h \leq r \leq R$ and Eq. (2.8) for $r < R-h$ together with the constitutive Eq. (2.1) for the Newtonian fluid, we obtain the velocity profile:

$$u = \begin{cases} \dfrac{\alpha}{K}\left[AI_0\left(\dfrac{r}{\delta}\right) + BK_0\left(\dfrac{r}{\delta}\right) + 1\right] & (R - h \leq r \leq R) \\ \dfrac{\alpha}{4\eta}\left(-r^2 + C\right) & (r < R - h) \end{cases}, \tag{2.33}$$

where I_0 and K_0 are modified Bessel functions and $\delta = \sqrt{\eta/K}$. A, B, and C are constants expressed in terms of R, h, and δ (for details see Sugihara-Seki and Fu 2005). The dimension of δ is the length, and it represents a typical thickness to which the flow outside the layer penetrates within the layer.

Figure 2.9 illustrates the velocity profiles of Eq. (2.33) for $\delta = 0.02$, 0.05, 0.1, and 0.2 μm, when the pressure gradient α is kept constant at $h = 0.5$ μm and $R = 3$ μm. In this figure, the velocity is normalized by the centerline velocity of a Poiseuille flow at $K = 0$.

No direct measurements of the mechanical properties of the glycocalyx layer have been reported. However, observations of its functional behavior and theoretical considerations have given estimates that the value of K for the vascular endothelial glycocalyx layer is at least 10^{11} N s/m^4 or 10^{14} N s/m^4, leading to $\delta \approx 0.01 - 0.1$ μm (Pries et al. 2000; Weinbaum et al. 2003; Sugihara-Seki 2006). Figure 2.9 indicates that the fluid is almost stationary within the glycocalyx layer, so the presence of the glycocalyx layer is nearly equivalent to a decrease in the vessel radius from R to $R - h$. This result implies that the flow of Newtonian fluids exhibits an increase in apparent viscosity by $R^4/(R - h)^4$ times. For instance, a 0.5-μm thick glycocalyx in a 6-μm diameter capillary gives $R/(R–h) = 1.2$ and $R^4/(R - h)^4 \approx 2$, which is a significant increase in apparent viscosity. In addition, the presence of the glycocalyx layer causes a considerable decrease in the hematocrit in small vessels since red cells do not invade the glycocalyx layer under normal conditions.

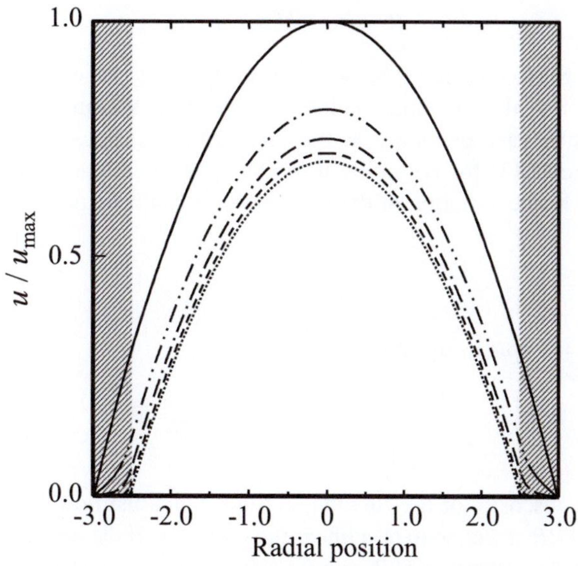

Fig. 2.9 Velocity profiles of a Newtonian fluid through a tube in the presence of a glycocalyx layer at its luminal surface ($h = 0.5$ μm, $R = 3.0$ μm). *Dotted curve*, $\delta = 0.02$ μm; *dashed curve*, $\delta = 0.05$ μm; *dash-dotted curve*, $\delta = 0.1$ μm; *dash-double dotted curve*, $\delta = 0.2$ μm. The *solid curve* represents the velocity profile of Poiseuille flow with the same pressure gradient α (Modified from Sugihara-Seki and Fu 2005)

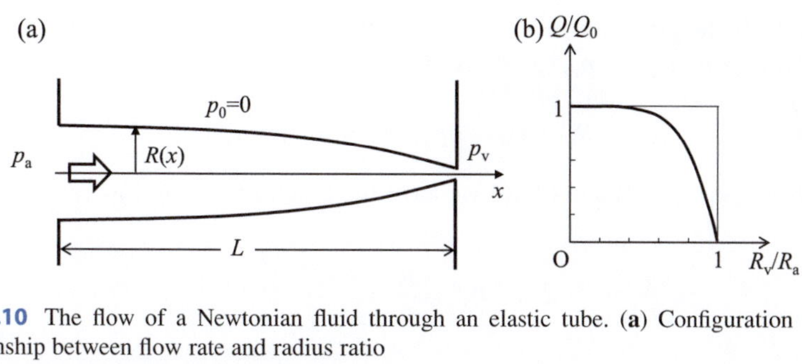

Fig. 2.10 The flow of a Newtonian fluid through an elastic tube. (**a**) Configuration and (**b**) relationship between flow rate and radius ratio

2.1.4.3 Steady Flow in an Elastic Tube

Another type of flow in circular cylindrical tubes to consider is the steady flow of a Newtonian fluid in an elastic tube when the upstream and downstream pressures are prescribed, as shown in Fig. 2.10a. The flow through distensible vessels was first analyzed in a model study for sheet flow in pulmonary alveolar blood vessels (Fung and Sobin 1972).

We assume that the cross section of the tube is maintained in a circular shape, and the radius of the tube is governed by the pressure difference between the inside and outside of the tube, $p - p_0$, in such a way that

$$R = R_0 + k(p - p_0), \tag{2.34}$$

where R_0 and k are constants. The pressure p_0 represents the external pressure; we assume $p_0 = 0$ hereafter, without the loss of generality. We further assume that the radii of the tube vary gradually in the flow direction along the x-axis, so that Eqs. (2.12) and (2.13) for Poiseuille flow can apply locally at each cross section. From Eq. (2.34), the variation of the tube radius R can be expressed as

$$\frac{\mathrm{d}R}{\mathrm{d}x} = k\frac{\mathrm{d}p}{\mathrm{d}x}. \tag{2.35}$$

Then, since $\alpha = -\mathrm{d}p/\mathrm{d}x$, Eq. (2.13) reduces to

$$Q = \frac{\pi R^4}{8\eta}\frac{-\mathrm{d}R/\mathrm{d}x}{k} = -\frac{\pi}{8\eta}\frac{\mathrm{d}R^5/\mathrm{d}x}{5k}, \tag{2.36}$$

where η is the viscosity of the Newtonian fluid. Since the flow rate Q is constant along the x-axis, R^5 is found to be a linear function of x. If the radii of the tube at the inlet and outlet are expressed as R_a and R_v, respectively, and the disturbances due to entry and exit are assumed to be negligible, then the radius of the tube, $R(x)$, can be expressed as

$$R^5 = {R_a}^5 - \frac{{R_a}^5 - {R_v}^5}{L}x, \tag{2.37}$$

where L is the length of the tube, and R_a and R_v are related to the pressures upstream and downstream, p_a and p_v, respectively, as

$$R_a = R_0 + kp_a, \quad R_v = R_0 + kp_v. \tag{2.38}$$

Then, from Eqs. (2.37) and (2.36) we have the tube radius and flow rate:

$$R = \left[{R_a}^5 - \frac{{R_a}^5 - {R_v}^5}{L}x\right]^{\frac{1}{5}}, \tag{2.39}$$

$$Q = Q_0\left(1 - \left(\frac{R_v}{R_a}\right)^5\right), \quad Q_0 = \frac{\pi {R_a}^5}{40\eta k L}. \tag{2.40}$$

Figure 2.10b shows Q/Q_0 versus R_v/R_a. It is seen that $Q/Q_0 \approx 1$ for a rather wide range of R_v/R_a from 0 to ~0.5. Since Q_0 is independent of the downstream condition, this implies that the flow rate is mainly determined by the upstream condition, unless $R_v/R_a \approx 1$. This feature exhibits a sharp contrast to the flow through a rigid tube, where the flow rate is proportional to the pressure difference, $\Delta p \; (= p_a - p_v)$.

The solutions (2.39) and (2.40) are obtained based on the linear relationship between tube radius and pressure difference (2.34). Instead of this assumption, we may adopt Hooke's law for the dilatation of the tube (Fung 1996).

Noting that the circumferential stress in the tube wall of thickness h is expressed as pR/h by Laplace's law, we have an expression for Hooke's law in terms of Young's modulus E:

$$\frac{pR}{h} = E\left(\frac{R}{R_0} - 1\right). \tag{2.41}$$

Here, we assume E, h and R_0 are constant. Inserting this expression into Eq. (2.13) yields

$$-\frac{\mathrm{d}p}{\mathrm{d}x}\left(1 - \frac{R_0}{hE}p\right)^{-4} = \frac{8\eta}{\pi {R_0}^4}Q. \tag{2.42}$$

Since all quantities except $p(x)$ are constant, Eq. (2.42) can be integrated to obtain the solution in terms of tube radii at the inlet and outlet, R_a and R_v, respectively,

$$Q = Q_0\left(1 - \left(\frac{R_v}{R_a}\right)^3\right), \quad Q_0 = \frac{\pi h E {R_a}^3}{24\eta L}. \tag{2.43}$$

This equation is similar to Eq. (2.40) and has a similar feature, i.e., the flow rate does not depend linearly on pressure difference, Δp, and instead depends mostly on the upstream condition.

2.2 Vascular Solid Mechanics

Mechanical stress is one of the factors that regulate the behavior of blood vessels (Hayashi et al. 2001; Humphrey 2002). It affects the physiological and pathological responses of the vascular wall. Macroscopically, mechanical stress can cause a vascular wall to rupture. Microscopically, it can trigger responses in the mechanoreceptors of vascular cells. This section focuses on the macroscopic effects of mechanical stress on the vascular wall.

2.2.1 Fundamentals of Solid Mechanics

There are two approaches to studying the mechanics of the vascular wall: one involves formulating an equation to balance force to obtain the stress, and the other formulates a stress-strain relationship for the vascular wall and solves a boundary-value problem for a large deformation of the blood vessel.

Considering the first approach, Laplace's law can be applied to a membrane, which has negligible resistance to deformation with bending (Fung 1993). Denoting the principal curvatures and tensions (force per unit length) as κ_i and T_i ($i = 1, 2$), respectively, and the pressures on the inner and outer wall as p_i and p_0, Laplace's equation is expressed as

$$T_1\kappa_1 + T_2\kappa_2 = p_i - p_0 \tag{2.44}$$

where κ_i is related to the principal radius of curvature r_i as

$$r_1 = 1/\kappa_1, \quad r_2 = 1/\kappa_2. \tag{2.45}$$

For a thin cylindrical tube with inner diameter d_i and wall thickness t subjected to intraluminal pressure p ($p_0 = 0$), the relations

$$r_1 = d_i/2, \quad r_2 = \infty \tag{2.46}$$

hold. Therefore, Laplace's equation can be written as

$$2T_1/d_i = 2\sigma t/d_i = p \tag{2.47}$$

where σ is the average of the circumferential stress (hoop stress) distributed over the wall thickness.

The other approach to revealing the mechanics of the vascular wall involves formulating a stress-strain relationship for the vascular wall and solves a boundary-value problem for a large deformation of the blood vessel. To solve the problem, finite deformation theory is introduced from continuum mechanics (Holzapfel 2000).

Consider the Cartesian coordinate system O-$x_1x_2x_3$. A particle at an initial position $\boldsymbol{X}$ in a body at time $t = 0$ moves to its current position $\boldsymbol{x}$ at time $t = t$ with deformation of the body. Using basis vectors $\boldsymbol{e}_i$ and components X_i and x_i ($i = 1$, 2, 3), the initial and current positions $\boldsymbol{x}$ and $\boldsymbol{X}$ are expressed as

$$\boldsymbol{X} = X_i\boldsymbol{e}_i = X_1\boldsymbol{e}_1 + X_2\boldsymbol{e}_2 + X_3\boldsymbol{e}_3, \quad \boldsymbol{x} = x_i\boldsymbol{e}_i = x_1\boldsymbol{e}_1 + x_2\boldsymbol{e}_2 + x_3\boldsymbol{e}_3 \tag{2.48}$$

where the repeat of the dummy index i obeys the summation convention.

We assume that the relations

$$\boldsymbol{x} = \boldsymbol{x}(\boldsymbol{X}, t), \quad \boldsymbol{X} = \boldsymbol{X}(\boldsymbol{x}, t) \tag{2.49}$$

hold. The deformation gradient $\boldsymbol{F}$, a two-point tensor, is defined as

$$\boldsymbol{F} = F_{ij}\boldsymbol{e}_i \otimes \boldsymbol{e}_j = \frac{\partial \boldsymbol{x}}{\partial \boldsymbol{X}} = \frac{\partial x_i}{\partial X_j}\boldsymbol{e}_i \otimes \boldsymbol{e}_j \tag{2.50}$$

where the symbol $\otimes$ indicates a tensor product.

The inverse of $\boldsymbol{F}$ is obtained as

$$\boldsymbol{F}^{-1} = \frac{1}{\det\boldsymbol{F}}\left(\frac{1}{2}e_{jkl}e_{imn}F_{km}F_{ln}\right)\boldsymbol{e}_i \otimes \boldsymbol{e}_j \tag{2.51}$$

where e_{ijk} is a permutation symbol defined as

$$e_{ijk} = \begin{cases} 1, & (i,j,k) = (1,2,3),\ (2,3,1),\ (3,1,2) \\ -1, & (i,j,k) = (3,2,1),\ (2,1,3),\ (1,3,2) \\ 0, & i = j \quad \text{or} \quad j = k \quad \text{or} \quad k = i \end{cases} \tag{2.52}$$

and the determinant of $\boldsymbol{F}$ is calculated as

$$\det\boldsymbol{F} = e_{ijk}F_{1i}F_{2j}F_{3k}. \tag{2.53}$$

A transpose of $\boldsymbol{F}$ is expressed as

$$\boldsymbol{F}^T = F^T{}_{ij}\boldsymbol{e}_i \otimes \boldsymbol{e}_j = F_{ij}\boldsymbol{e}_j \otimes \boldsymbol{e}_i = \frac{\partial x_i}{\partial X_j}\boldsymbol{e}_j \otimes \boldsymbol{e}_i = \frac{\partial x_j}{\partial X_i}\boldsymbol{e}_i \otimes \boldsymbol{e}_j. \tag{2.54}$$

The right Cauchy-Green deformation tensor $\boldsymbol{C}$ and the left Cauchy-Green deformation tensor $\boldsymbol{B}$ are defined as

$$\boldsymbol{C} = \boldsymbol{F}^{\mathrm{T}}\boldsymbol{F}, \quad \boldsymbol{B} = \boldsymbol{F}\boldsymbol{F}^{\mathrm{T}} \tag{2.55}$$

Therefore, the components of the above deformation tensors can be written as

$$\boldsymbol{C} = \left(\frac{\partial x_k}{\partial X_i}\frac{\partial x_k}{\partial X_j}\right)\boldsymbol{e}_i \otimes \boldsymbol{e}_j, \quad \boldsymbol{B} = \left(\frac{\partial x_i}{\partial X_k}\frac{\partial x_j}{\partial X_k}\right)\boldsymbol{e}_i \otimes \boldsymbol{e}_j. \tag{2.56}$$

Green's strain tensor $\boldsymbol{E}$ is defined in the reference configuration as

$$\boldsymbol{E} = E_{ij}\boldsymbol{e}_i \otimes \boldsymbol{e}_j = \frac{1}{2}(\boldsymbol{C} - \boldsymbol{I}) = \frac{1}{2}\left(C_{ij} - \delta_{ij}\right)\boldsymbol{e}_i \otimes \boldsymbol{e}_j \tag{2.57}$$

where $\boldsymbol{I}$ is a second-order unit tensor and δ_{ij} is Kronecker delta defined as

$$\boldsymbol{I} = \delta_{ij}\boldsymbol{e}_i \otimes \boldsymbol{e}_j \quad \left(\delta_{ij} = 1 \quad \text{if} \quad i = j \quad \text{and} \quad \delta_{ij} = 0 \quad \text{if} \quad i \neq j\right). \tag{2.58}$$

The first, second, and third invariants of the right Cauchy-Green deformation tensor $\boldsymbol{C}$ are defined as

$$I_1 = \mathrm{tr}\boldsymbol{C} = C_{ii} \tag{2.59}$$

$$I_2 = \frac{1}{2}\left(\mathrm{tr}^2\boldsymbol{C} - \mathrm{tr}\boldsymbol{C}^2\right) = \frac{1}{2}\left(C_{ii}C_{jj} - C_{ij}C_{ji}\right) \tag{2.60}$$

$$I_3 = \det\boldsymbol{C} \tag{2.61}$$

If $\boldsymbol{v}$ is a velocity vector and v_i is its component, the material time derivative of variable $\boldsymbol{A}$ is defined as

$$\dot{\boldsymbol{A}}(x,\ t) = \frac{\partial \boldsymbol{A}}{\partial t} + \frac{\partial \boldsymbol{A}}{\partial x_i} v_i. \tag{2.62}$$

The deformation rate tensor $\boldsymbol{D}$ and spin tensor $\boldsymbol{W}$ are defined as

$$\begin{aligned} \boldsymbol{D} &= \frac{1}{2}\left\{\frac{\partial \boldsymbol{v}}{\partial \boldsymbol{x}} + \left(\frac{\partial \boldsymbol{v}}{\partial \boldsymbol{x}}\right)^T\right\} = \frac{1}{2}\left\{\dot{\boldsymbol{F}}\boldsymbol{F}^{-1} + (\dot{\boldsymbol{F}}\boldsymbol{F}^{-1})^T\right\} \\ &= \frac{1}{2}\left(\frac{\partial v_i}{\partial x_j} + \frac{\partial v_j}{\partial x_i}\right)\boldsymbol{e}_i \otimes \boldsymbol{e}_j \end{aligned} \tag{2.63}$$

$$\begin{aligned} \boldsymbol{W} &= \frac{1}{2}\left\{\frac{\partial \boldsymbol{v}}{\partial \boldsymbol{x}} - \left(\frac{\partial \boldsymbol{v}}{\partial \boldsymbol{x}}\right)^T\right\} = \frac{1}{2}\left\{\dot{\boldsymbol{F}}\boldsymbol{F}^{-1} - (\dot{\boldsymbol{F}}\boldsymbol{F}^{-1})^T\right\} \\ &= \frac{1}{2}\left(\frac{\partial v_i}{\partial x_j} - \frac{\partial v_j}{\partial x_i}\right)\boldsymbol{e}_i \otimes \boldsymbol{e}_j. \end{aligned} \tag{2.64}$$

2.2.2 Structure and Components of Vascular Walls

In the order of the blood circulation, blood vessels are classified as the aorta (main artery), arteries, arterioles, capillaries, venules, veins, and the vena cava (Burton 1954; Rhodin 1979). Typical dimensions and intraluminal pressures for these vessels are listed in Table 2.1 (Mashima 1986; Berne and Levy 1993).

Blood vessels consist of fiber components; *i.e*., elastic and collagen fibers, ground substances, and cellular components such as endothelial and smooth muscle cells (SMCs). The elastic and collagen fibers are load-bearing components. Elastic fibers are extensible, while collagen fibers are stiff. SMCs generate a contractile force to shrink muscular arteries.

The aorta, arteries, veins, and the vena cava have load-bearing structures. The vascular walls of these vessels consist of concentric layers called the intima, media, and adventitia. The intima is a thin layer, which comprises endothelial cells at the

Table 2.1 Typical dimensions and pressure of blood vessels (Burton 1954; Mashima 1986; Berne and Levy 1993)

Vessel	Diameter	Wall thickness	Pressure (mmHg)
Aorta	25 mm	2 mm	120–80
Artery	4 mm	1 mm	120–80
Arteriole	30 μm	6 μm (*)	60–40
Capillary	8 μm	0.5 μm (*)	30(−15)
Venule	20 μm	1 μm (*)	15(−10)
Vein	5 mm	0.5 mm	10(−5)
Vena cava	30 mm	1.5 mm	5–0

100 mmHg = 13.3 kPa. (*) The listed values (Berne and Levy 1993) differ from those in Burton (1954)

luminal surface and an internal elastic lamina at the boundary with the media. The media is the major region of the vascular wall. The media of the aorta and elastic arteries has a multilayer structure with an elastic lamina. Muscular arteries have a large number of smooth muscle cells, which generate force actively, mainly in the circumferential direction. The adventitia is the outermost layer of the vascular wall and connects the vascular wall to the surrounding tissues.

2.2.3 Mechanics of Normal Arterial Walls

We focus on the mechanics of the aorta and arteries. These vascular walls are subjected to blood pressure and longitudinal stretch. The luminal surface of the blood vessels are subjected to the blood pressures listed in Table 2.1 and the pressure of the blood column, for which the reference level is located at the height of the heart. Posture primarily affects the blood pressure in the lower limbs (Thubrikar 2007). If a blood vessel is located at heart level, the intraluminal pressure equals the value in Table 2.1.

2.2.3.1 Structure

Two types of multilayer wall model have been proposed (von Maltzahn et al. 1981; Yamada et al. 1999; Holzapfel et al. 2000; Matsumoto et al. 2004). One type of model includes the intima, media, and adventitia. The intima of the aorta/arteries is thickened physiologically in elderly people, while it is very thin in the young. The other type of model comprises alternate elastic and SMC layers in the media. A homogeneous single-layer wall model is the simplest and most commonly used model.

2.2.3.2 Geometry

The arterial tree includes straight, curved, and branched portions. The transverse cross-section of the vascular wall is almost circular and the wall thickness can vary along the circumference. Such variation occurs as a result of the remodeling of vascular tissues in a variety of environments.

The simplest geometrical model is a hollow cylinder. Realistic patient-specific models can also be built using image slices obtained with magnetic resonance imaging (MRI) or computed tomography (CT) (Tang et al. 2009). A model is subject-specific if the dimensions of a specimen are measured from an actual body using any technique.

2.2.3.3 Incompressibility

The vascular wall is nearly incompressible because of its high water content (Carew et al. 1968; Chuong and Fung 1984). The incompressibility condition is expressed as

$$I_3 = \det \boldsymbol{C} = 1 \quad \text{or} \quad J = \det \boldsymbol{F} = 1. \tag{2.65}$$

This incompressibility constraint is added to an equation for an incompressible strain energy density function as $H/2 \cdot (I_3 - 1)$, where H is a Lagrange multiplier or an indeterminate pressure, which is determined by the boundary conditions. An example showing how to calculate H is given in Sect. 2.2.3.5.

2.2.3.4 Passive and Active Mechanical Properties

The stress-strain relationship of the arteries depends on the direction of deformation. It tends to be isotropic in the portion near the heart and anisotropic (orthotropic or transversely isotropic) peripherally (Azuma and Hasegawa 1971).

Elastic and collagen fibers are passive components, while smooth muscle is an active component. In a passive state, the vascular wall behaves as a nonlinear viscoelastic material. In an active state, generation of a contractile force in the smooth muscle reduces the vascular diameter (Murphy 1980). The effect of smooth muscle contraction is strong for muscular arteries; *e.g.*, the renal and femoral arteries, and weak for elastic arteries; *e.g.*, the carotid artery (Cox 1978). The maximum contractile force depends on the length of the smooth muscle (Murphy 1976).

2.2.3.5 Constitutive Equations

Hyperelastic, viscoelastic, and poroelastic models have been proposed (Vito and Dixon 2003). Of these, incompressible hyperelastic models are used most frequently. One postulates a strain energy density function in a stress-free configuration. Unloaded and stress-free configurations should be distinguished because a vascular wall can change its geometry by releasing any residual stress (See Fig. 2.11) (Fung 1993).

An isotropic, transversely isotropic, or orthotropic incompressible strain energy function is introduced and applied to a hollow cylinder representing a blood vessel. Stresses are calculated under physiological loads; *i.e.*, a constant axial stretch, λ_z, and a uniform intraluminal pressure, p. The residual strain component in the

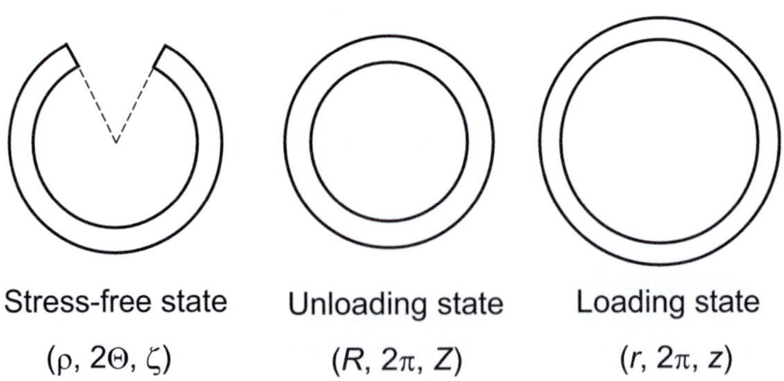

Fig. 2.11 Transverse cross-sections of a hollow cylindrical vessel in the stress-free, unloaded, and loaded states. The corresponding radius, central angle, and longitudinal-axis coordinates are shown in each state (Note that the stress-free geometry is an assumption)

circumferential direction is assumed as a function of the position along the wall thickness.

Under axisymmetric deformation of a hollow cylinder, the radial, circumferential, and longitudinal directions are the principal directions. Then, the principal stretches in these three directions are defined respectively as

$$\lambda_r = \frac{dR}{d\rho}, \quad \lambda_\theta = \frac{\pi R}{\Theta \rho}, \quad \lambda_z = \frac{Z}{\zeta} \qquad \text{(unloaded state)} \tag{2.66}$$

$$\lambda_r = \frac{dr}{d\rho}, \quad \lambda_\theta = \frac{\pi r}{\Theta \rho}, \quad \lambda_z = \frac{z}{\zeta} \qquad \text{(loaded state)}. \tag{2.67}$$

If there is no residual stress, the relationships

$$\Theta = \pi, \quad \rho = R, \quad \zeta = Z \tag{2.68}$$

hold. Therefore, the stretches can be written as

$$\lambda_r = \frac{dr}{dR}, \quad \lambda_\theta = \frac{r}{R}, \quad \lambda_z = \frac{z}{Z} \qquad \text{(loaded state, no residual stress)}. \tag{2.69}$$

Based on a vascular wall with a constant volume, the radius in a loaded state is expressed as

$$r = \sqrt{{r_i}^2 + \frac{\Theta(\rho^2 - {\rho_i}^2)}{\pi \lambda_z}} \quad (\rho_i < \rho_0), \qquad r = \sqrt{{r_i}^2 + \frac{\Theta({\rho_i}^2 - \rho^2)}{\pi \lambda_z}} \quad (\rho_i > \rho_0). \tag{2.70}$$

where the subscripts i and o denotes the inner (radius of the inner surface of the intima) and outer (radius of the outer surface of the adventitia) radius, respectively (Yamada 2012).

Since the radial, circumferential, and longitudinal directions are the principal directions, the incompressibility condition (Eq. 2.65) is written as

$$\lambda_r \lambda_\theta \lambda_z = 1. \tag{2.71}$$

An equilibrium equation in the radial direction is expressed as

$$\frac{d\sigma_r}{dr} + \frac{\sigma_r - \sigma_\theta}{r} = 0 \tag{2.72}$$

where σ_r and σ_θ are the Cauchy stress in the radial and circumferential directions, respectively. The boundary conditions are written as

$$\sigma_r = \begin{cases} -p & (r = r_i) \\ 0 & (r = r_o) \end{cases} \tag{2.73}$$

Then, the radial component of the Cauchy stress is calculated using

$$\sigma_r(r) = \int_{r_i}^{r} \frac{\sigma_\theta - \sigma_r}{r} dr - p. \tag{2.74}$$

The first and second invariants of $\boldsymbol{C}$ are written as

$$I_1 = \lambda_r{}^2 + \lambda_\theta{}^2 + \lambda_z{}^2, \quad I_2 = \lambda_r{}^2\lambda_\theta{}^2 + \lambda_\theta{}^2\lambda_z{}^2 + \lambda_z{}^2\lambda_r{}^2. \tag{2.75}$$

For an incompressible isotropic hyperelastic model, a strain energy density function is a function of I_1 and I_2 and the Cauchy stress components are written as

$$\begin{aligned} \sigma_r &= 2\lambda_r{}^2\{W_1 + (\lambda_\theta{}^2 + \lambda_z{}^2)W_2\} + H \\ \sigma_\theta &= 2\lambda_\theta{}^2\{W_1 + (\lambda_z{}^2 + \lambda_r{}^2)W_2\} + H \\ \sigma_z &= 2\lambda_z{}^2\{W_1 + (\lambda_r{}^2 + \lambda_\theta{}^2)W_2\} + H \end{aligned} \tag{2.76}$$

where W_1 and W_2 are defined as

$$W_1 = \frac{\partial W(I_1, \ I_2)}{\partial I_1}, \quad W_2 = \frac{\partial W(I_1, \ I_2)}{\partial I_2}. \tag{2.77}$$

Vito and Hickey (1980) postulated that

$$W = b_1(I_1 - 3) + b_2(I_1 - 3)^2 + b_3(I_1 - 3)^3 + b_4(I_1 - 3)^4 \tag{2.78}$$

and Yamada et al. (1999) postulated that

$$W = C_1(I_1 - 3) + C_2(I_2 - 3) + C_3(I_2 - 3)^2 + C_4(I_2 - 3)^3 \tag{2.79}$$

as polynomials of the first and second invariants of $\boldsymbol{C}$ where b_i and C_i $(i = 1, 2, 3, 4)$ are material constants.

Substituting Eq. (2.76) into Eq. (2.74), the radial component of the Cauchy stress is expressed as

$$\sigma_r(r) = -\int_{r}^{r_o} \frac{2(\lambda_\theta{}^2 - \lambda_r{}^2)(W_1 + \lambda_z{}^2 W_2)}{r} \mathrm{d}r = -L(r). \tag{2.80}$$

To optimize the set of material constants in a strain energy density function, one minimizes the error between the experimental (data points) and estimated (calculated points) curves. For each given pressure p, a value of the outer radius (or inner radius using Eq. (2.70)) should be determined to satisfy the relation

$$L(r_i) = p. \tag{2.81}$$

If an outer (inner) radius is given, the pressure p is calculated using Eq. (2.80). The stress components are obtained by

$$\begin{aligned}
\sigma_r &= -L(r) \\
\sigma_\theta &= 2\left(\lambda_\theta{}^2 - \lambda_r{}^2\right)\left(W_1 + \lambda_z{}^2 W_2\right) - L(r) \\
\sigma_z &= 2\left(\lambda_z{}^2 - \lambda_r{}^2\right)\left(W_1 + \lambda_\theta{}^2 W_2\right) - L(r)
\end{aligned} \tag{2.82}$$

In the longitudinal direction, the equation for the balance of force

$$F_z + \pi r_i{}^2 p = 2\pi \int_{r_i}^{r_o} \sigma_z r \mathrm{d}r \tag{2.83}$$

holds, where F_z is the force in the longitudinal direction. For an incompressible isotropic hyperelastic model, the longitudinal force can be calculated as

$$F_z = 2\pi \int_{r_i}^{r_o} \left\{2\left(\lambda_z{}^2 - \lambda_r{}^2\right)\left(W_1 + \lambda_\theta{}^2 W_2\right) - \left(\lambda_\theta{}^2 - \lambda_r{}^2\right)\left(W_1 + \lambda_z{}^2 W_2\right)\right\} r \mathrm{d}r. \tag{2.84}$$

As an incompressible orthotropic hyperelastic model, Chuong and Fung (1983) formulated an exponential type of strain energy density function

$$\begin{aligned}
W &= \frac{C}{2} \exp Q \\
Q &= b_1 E_\theta{}^2 + b_2 E_z{}^2 + b_3 E_r{}^2 + 2b_4 E_\theta E_z + 2b_5 E_z E_r + 2b_6 E_r E_\theta
\end{aligned} \tag{2.85}$$

in terms of the components of Green's strain tensor, which are expressed by the principal stretches as

$$E_r = \frac{1}{2}\left(\lambda_r{}^2 - 1\right), \quad E_\theta = \frac{1}{2}\left(\lambda_\theta{}^2 - 1\right), \quad E_z = \frac{1}{2}\left(\lambda_z{}^2 - 1\right). \tag{2.86}$$

The stress components and longitudinal force are expressed as

$$\begin{aligned}
\sigma_r &= -L(r) \\
\sigma_\theta &= C e^Q \left[(1 + 2E_\theta)\left\{b_1 E_\theta + b_4\left(\lambda_z{}^2 - 1\right)/2 + b_6 E_r\right\}\right. \\
&\quad \left. - (1 + 2E_r)\left\{b_3 E_r + b_5\left(\lambda_z{}^2 - 1\right)/2 + b_6 E_\theta\right\}\right] - L(r) \\
\sigma_z &= C e^Q \left[\lambda_z{}^2\left\{b_2\left(\lambda_z{}^2 - 1\right)/2 + b_4 E_\theta + b_5 E_r\right\}\right. \\
&\quad \left. - (1 + 2E_r)\left\{b_3 E_r + b_5\left(\lambda_z{}^2 - 1\right)/2 + b_6 E_\theta\right\}\right] - L(r)
\end{aligned} \tag{2.87}$$

$$L(r) = C\int_{r}^{r_o} \left[(1+2E_\theta)\left\{b_1E_\theta + b_4\left(\lambda_z^2-1\right)/2 + b_6E_r\right\} \right.$$
$$\left. -(1+2E_r)\left\{b_3E_r + b_5\left(\lambda_z^2-1\right)/2 + b_6E_\theta\right\}\right]\left(e^Q/r\right)\mathrm{d}r \tag{2.88}$$

$$F_z = 2\pi C\int_{r_i}^{r_o} re^Q\left[\lambda_z^2\left\{b_2\left(\lambda_z^2-1\right)/2 + b_4E_\theta + b_5E_r\right\}\right.$$
$$-(1+2E_r)\left\{b_3E_r + b_5\left(\lambda_z^2-1\right)/2 + b_6E_\theta\right\}/2$$
$$\left.-(1+2E_\theta)\left\{b_1E_\theta + b_4\left(\lambda_z^2-1\right)/2 + b_6E_r\right\}/2\right]\mathrm{d}r \tag{2.89}$$

In a loaded state, the resultant force of the circumferential Cauchy stress is balanced by the resultant intraluminal pressure as

$$\int_{r_i}^{r_o} \sigma_\theta \mathrm{d}r = pr_i. \tag{2.90}$$

A comparison of the above equation with Laplace's equation (Eq. 2.47) shows that the circumferential stress in Eq. (2.47) is equal to an average of the circumferential stress over the wall thickness for a deformed blood vessel as

$$\sigma = \int_{r_i}^{r_o} \sigma_\theta \mathrm{d}r/t. \tag{2.91}$$

Yamada et al. (2007) demonstrated the effect of residual strain on the circumferential stress over the wall thickness. They used the following strain energy density function

$$W = C_1(I_1-3) + C_2(I_1-3)^2 + C_3(I_1-3)^3. \tag{2.92}$$

They determined the material constants C_1, C_2, and C_3, and the inner or outer radius in the stress-free state to reproduce the same pressure-diameter relationship for various cases of residual strain.

In the model, the opening angle Ψ is defined as

Table 2.2 Dimensions of an opened hollow cylindrical wall in the stress-free state and the material constants used to reproduce the same pressure-diameter relationship under a constant longitudinal stretch of 1.3 and intraluminal pressure (Yamada et al. 2007)

Ψ (degree)	r_i (mm)	r_0 (mm)	C_1 (kPa)	C_2 (kPa)	C_3 (kPa)
0	2.6	3.5	12.0	8.0	37.5
60	5.0	4.1	9.9	11.3	38.1
100	7.3	6.4	8.3	12.8	37.2

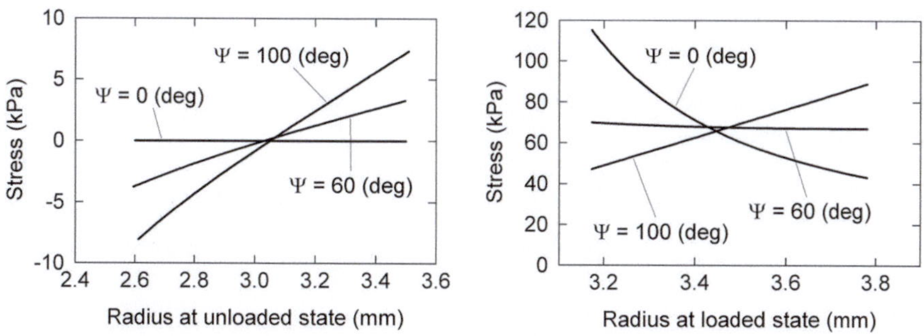

Fig. 2.12 Distributions of the circumferential stress in the unloaded stage (*left*) and loaded state under a longitudinal stretch of 1.3 and intraluminal pressure of 13 kPa (Modified from Yamada et al. 2007)

$$\Theta + \Psi = \pi. \tag{2.93}$$

The dimensions and material constants of the model were determined as shown in Table 2.2. Figure 2.12 shows the distributions of the circumferential component of the Cauchy stress in the unloaded and loaded states under the boundary conditions of a longitudinal stretch of 1.3 and an intraluminal pressure of 13 kPa (98 mmHg). The results indicate that an increase in the opening angle decreases the concentration of the circumferential stress at the intimal side of the wall, and a further increase causes a negative stress gradient across the wall thickness.

It is important to model the mechanical behavior of the active components; *i.e.*, the smooth muscles, to describe the mechanical behavior of blood vessels. The active force of skeletal muscles is modeled as a function of the length, velocity, and activation level. The following model of smooth muscles considers their length and activation (Yamada 1999).

The total Cauchy stress tensor of an active component is assumed to be the sum of the passive and active stresses or

$$\boldsymbol{\sigma}_t = \boldsymbol{\sigma}_p + \boldsymbol{\sigma}_a. \tag{2.94}$$

The stress-strain relationship of the passive stress is formulated as

$$\hat{\boldsymbol{\sigma}}_p = \kappa(\mathrm{tr}\boldsymbol{D})\boldsymbol{I} + 2\mu\boldsymbol{D} \tag{2.95}$$

where κ and μ are Lame constants. These constants are related to Young's modulus E and Poisson's ratio v as

$$\kappa = \frac{Ev}{(1+v)(1-2v)}, \quad \mu = \frac{E}{2(1+v)}. \tag{2.96}$$

The Jaumann time derivative of the passive stress is defined as

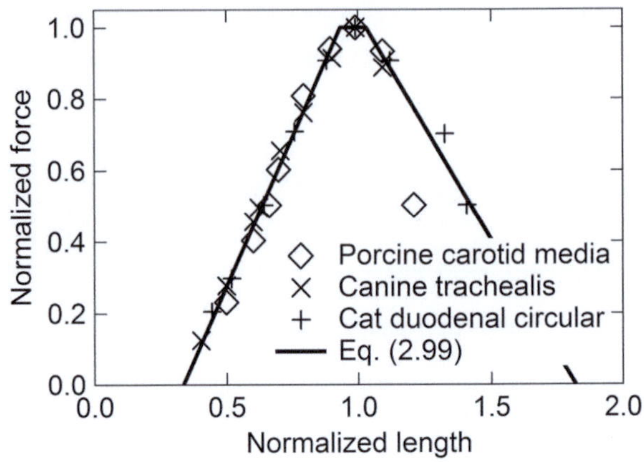

Fig. 2.13 Length dependence of the isometric maximum force of various smooth muscles (Murphy 1976) and a mathematical model (Eq. 2.99) formulated by Yamada (1999) (Modified from Yamada 1999)

$$\hat{\boldsymbol{\sigma}}_p = \dot{\boldsymbol{\sigma}}_p + \boldsymbol{\sigma}_p \boldsymbol{W} - \boldsymbol{W}\boldsymbol{\sigma}_p. \tag{2.97}$$

Active stress of smooth muscle that aligns in the direction of a unit vector $\boldsymbol{m}$ and has a stretch of λ is formulated as a product of the activation level ξ and the length-dependent maximum stress $\tau_{\max}$ as

$$\boldsymbol{\sigma}_a = \xi \tau_{\max}(\lambda) \boldsymbol{m} \otimes \boldsymbol{m} \quad (0 \le \xi \le 1) \tag{2.98}$$

where the first unit vector $\boldsymbol{m}$ shows the direction of the stress component and the second $\boldsymbol{m}$ shows the normal direction of the face on which the stress acts.

Based on Murphy's report (1976), Yamada (1999) determined the length-dependent maximum stress for smooth muscles as (Fig. 2.13)

$$\frac{\tau_{\max}(\lambda)}{\tau_{\max}(\lambda_{\max})} = \begin{cases} 1.67(\lambda/\lambda_{\max}) - 0.56, & 0.34 \le \lambda/\lambda_{\max} \le 0.93 \\ 1, & 0.93 < \lambda/\lambda_{\max} < 1.03 \\ -1.27(\lambda/\lambda_{\max}) + 2.31, & 1.03 \le \lambda/\lambda_{\max} \le 1.82 \\ 0, & \lambda/\lambda_{\max} < 0.34, \ \lambda/\lambda_{\max} > 1.82 \end{cases} \tag{2.99}$$

A finite element analysis was conducted using a model with nine concentric layers, including five passive and four active layers alternately, with the same thickness (Yamada et al. 1999). Figure 2.14 shows the stress distributions in the circumferential direction in the passive and active states and the change in the opening angle as an index of the residual strain with an increase in the activation level.

The simulation results indicate that activation decreases the circumferential stress of the passive components; *i.e.*, elastic and collagen fibers, and increases the stress of the active component; *i.e.*, smooth muscles. A comparison of Figs. 2.12 and 2.14 indicates that both the opening angle and contraction of smooth muscles change the gradient of circumferential stress of a homogeneous model in the

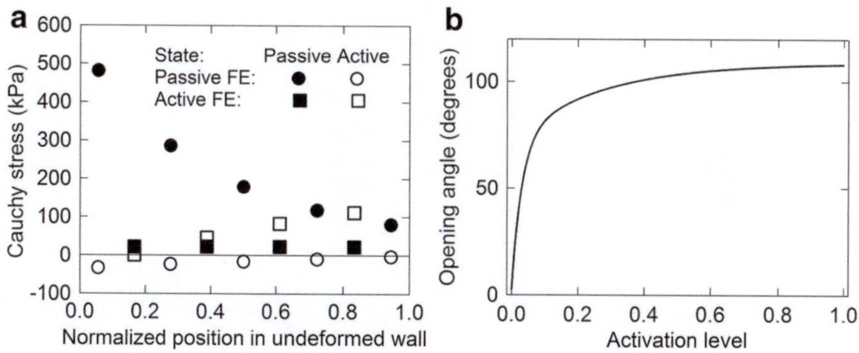

Fig. 2.14 (**a**) Distributions of the Cauchy stress in the circumferential direction in the passive and active states (activation level $\xi = 0$ and 1, respectively) for a nine-layer finite element model under a longitudinal stretch of 1.7 and an intraluminal pressure of 13 kPa. Residual stress is not considered. (**b**) Change in the opening angle, Ψ, which is defined at a stress-free vascular wall as an index of residual strain, with an increase in the activation level of ξ (Modified from Yamada et al. 1999)

negative direction. In other words, they cause the stress to decrease on the inner side and increase on the outer side.

2.2.4 Mechanics of Arterial Walls in Diseases

The structure, components, geometry, and accompanying mechanical properties of arterial walls change with age (Hayashi 1993; Nichols et al. 2011). Elderly people frequently develop arterial diseases such as atherosclerosis, aneurysms, and aortic dissection as a result of degeneration with aging.

2.2.4.1 Atherosclerosis

Atherosclerosis involves the accumulation of lipoproteins in a deep intimal layer and the development of a fibrous cap. The rupture of a plaque causes the formation of a thrombus, which can lead to ischemia or infarction. Many fluid and solid mechanics studies have examined atherosclerosis.

The earliest studies of atherosclerosis focused on the hemodynamic conditions that accelerate the invasion of low-density lipoproteins in the blood into the intima. In recent years, many researchers have investigated the stress state in plaque regions using the finite element method. They evaluated the possibility of plaque rupture or compared the stress between vulnerable and stable plaques or between ruptured and un-ruptured ones.

Yamada et al. (2010) conducted a finite element analysis of ring-like specimens of the common carotid artery with a plaque by releasing the stress by cutting the wall, as shown in Fig. 2.15. They demonstrated the deformation and stress distribution of the arterial wall under a cyclic intraluminal pressure of 0–13 kPa and a

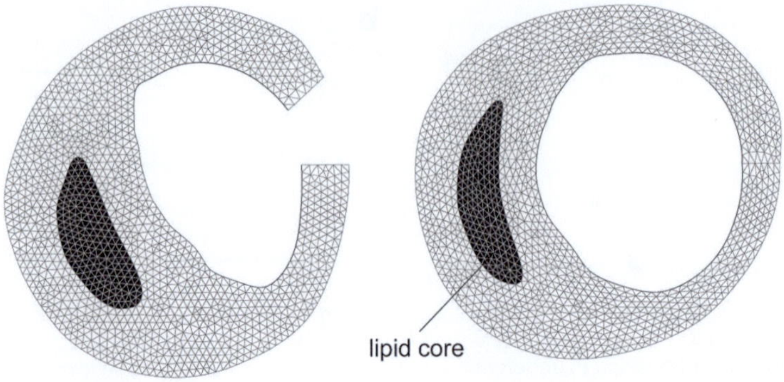

Fig. 2.15 Geometries of ring-like specimens of the common carotid artery in a stress-released state and under a physiological load with a longitudinal stretch of 1.07 and an intraluminal pressure of 13 kPa (Modified from Yamada and Sakata 2013)

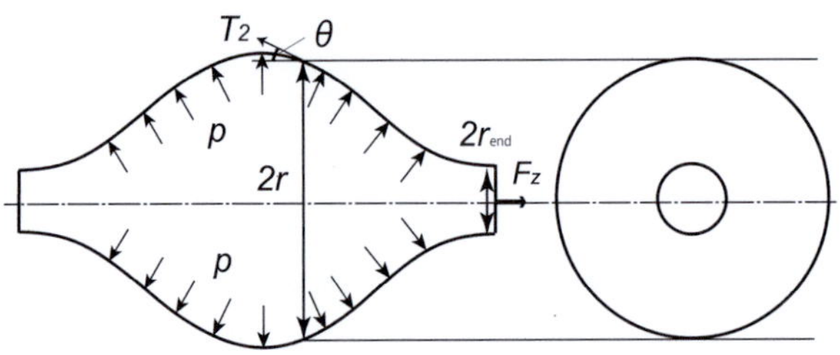

Fig. 2.16 The axisymmetric AAA model considered by Elger et al. (1996) (Modified from Yamada and Hasegawa 2007)

constant longitudinal stretch of 1.07. Yamada and Sakata (2013) observed a fluid-like lipid core. They modeled the lipid core using the bulk modulus of water and simulated a small change in pressure in the lipid core.

2.2.4.2 Aortic Aneurysms and Dissection

Abdominal aortic aneurysms (AAA) have frequently been investigated using solid mechanics. Two approaches can be used to estimate the stress in the aneurysm wall: one estimates the average stress based on the equilibrium of force, assuming a membrane-like aneurysm, and the other estimates the stress distribution using the equilibrium of stress and a constitutive equation, utilizing a finite element method.

Elger et al. (1996) assumed an axisymmetric spindle-like geometry for an AAA, as shown in Fig. 2.16, and formulated two equations for the equilibrium of force; *i. e.*, Laplace's Eq. (2.44), and, in the longitudinal direction,

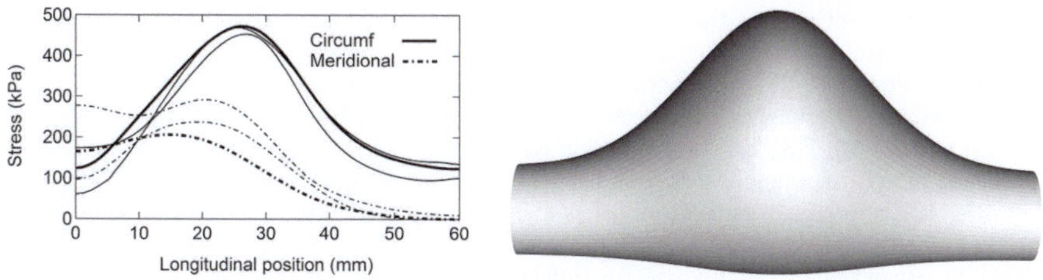

Fig. 2.17 Comparisons of the circumferential (*solid curve*) and meridional (*dot dashed curve*) components of the Cauchy stress using the results of equations for the force equilibrium (*thick lines*) and finite element analysis (*thin lines*: maximum and minimum values in the cross-sectional wall) along the median line on the side with the large bulge. The model has a maximum bulge with an inner diameter of 60 mm in the unloaded state at the origin (0 mm) of the longitudinal coordinate and ends at 60 mm (Modified from Yamada and Hasegawa 2007)

$$2\pi r T_2 \cos\theta = \pi\left(r^2 - r_{\mathrm{end}}^2\right)p + F_z. \qquad (2.100)$$

Yamada and Hasegawa (2007) extended this model to bilaterally symmetric AAA models to estimate the peak stress in the wall. They estimated the circumferential and meridional stresses along the anterior median line of the symmetric model using the two equations of force equilibrium and compared them with the results of a finite element analysis. This method provides a reasonable value and gives the location of the peak stress to evaluate the risk of wall rupture without utilizing the stress-free geometry and a constitutive equation (Fig. 2.17).

Aortic dissection follows an intimal tear and the penetration of blood into the delaminated space of the media in the longitudinal direction. An aortic dissection extends from the ascending to the descending thoracic aorta. This portion is subjected to pulsatile flow from the left ventricle. From the viewpoint of solid mechanics, the stress-strain relationship and mechanical strength have been investigated using the tensile loading and peeling tests (Okamoto et al. 2003; Holzapfel 2009; Pasta et al. 2012).

References

Azuma T, Hasegawa M (1971) A rheological approach to the architecture of arterial walls. Jpn J Physiol 21(1):27–47

Berne RM, Levy MN (1993) Physiology, 3rd edn. Mosby, St. Louis, p 362

Burton AC (1954) Relation of structure to function of the tissues of the wall of blood vessels. Physiol Rev 34(4):619–642

Carew TE, Vaishnav RN, Patel DJ (1968) Compressibility of the arterial wall. Circ Res 23 (1):61–68

Casson N (1959) A flow equation for pigment-oil suspensions of the printing ink type. In: Mill CC (ed) Rheology of disperse systems. Pergamon Press, Oxford, p 84

Chien S (1970) Shear dependence of effective cell volume as a determinant of blood viscosity. Science 168:977–979

Chuong CJ, Fung YC (1983) Three-dimensional stress distribution in arteries. J Biomech Eng 105 (3):268–274

Chuong CJ, Fung YC (1984) Compressibility and constitutive equation of arterial wall in radial compression experiments. J Biomech 17(1):35–40. doi:10.1016/0021-9290(84)90077-0

Cokelet GR, Merrill EW, Gilliland ER, Shin H, Britten A, Wells RE Jr (1963) The rheology of human blood – measurement near and at zero shear rate. Trans Soc Rheol 7:303–317

Cox RH (1978) Regional variation of series elasticity in canine arterial smooth muscles. Am J Physiol 234(5):H542–H551

Elger DF, Blackketter DM, Budwig RS, Johansen KH (1996) The influence of shape on the stresses in model abdominal aortic aneurysms. J Biomech Eng 118(3):326–332. doi:10.1115/1.2796014

Fahraeus R, Lindqvist T (1931) The viscosity of the blood in narrow capillary tubes. Am J Physiol 96:562–568

Farber TE (1995) Fluid dynamics for physicists. Cambridge University Press, Cambridge

Fung YC (1993) Biomechanics: mechanical properties of living tissues, 2nd edn. Springer, New York

Fung YC (1996) Biomechanics: circulation, 2nd edn. Springer, New York

Fung YC, Sobin SS (1972) Pulmonary alveolar blood flow. Circ Res 30:470–490

Hayashi K (1993) Experimental approaches on measuring the mechanical properties and constitutive laws of arterial walls. J Biomech Eng 115(4):481–488. doi:10.1115/1.2895528

Hayashi K, Stergiopulos N, Meister J-J, Greenwald SE, Rachev A (2001) Techniques in the determination of the mechanical properties and constitutive laws of arterial walls. In: Leondes C (ed) Biomechanical systems: techniques and applications, vol. II, Cardiovascular techniques. CRC Press, Boca Raton

Holzapfel GA (2000) Nonlinear solid mechanics: a continuum approach for engineering. Wiley, Chichester

Holzapfel GA (2009) Arterial tissue in health and disease: experimental data, collagen-based modeling and simulation, including aortic dissection. In: Holzapfel G, Ogden R (eds) Biomechanical modelling at the molecular, cellular and tissue levels, vol 508. CISM International Centre for Mechanical Sciences. Springer, Vienna, pp 259–344. doi:10.1007/978-3-211-95875-9_4

Holzapfel GA, Gasser TC, Ogden RW (2000) A new constitutive framework for arterial wall mechanics and a comparative study of material models. J Elast 61(1–3):1–48. doi:10.1023/A:1010835316564

Humphrey JD (2002) Cardiovascular solid mechanics: cells, tissues, and organs. Springer, New York

Mashima H (1986) Physiology, 18th edn. Bunkodo, Tokyo, pp 396–397, in Jap

Matsumoto T, Goto T, Sato M (2004) Microscopic residual stress caused by the mechanical heterogeneity in the lamellar unit of the porcine thoracic aortic wall. JSME Int J, Ser A 47 (3):341–348. doi:10.1299/Jsmea.47.341

Merrill EW, Cokelet GC, Britten A, Wells RE Jr (1963) Non-Newtonian rheology of human blood -effect of fibrinogen deduced by "subtraction". Circ Res 13:48–55

Murphy RA (1976) Contractile system function in mammalian smooth muscle. Blood Vessels 13 (1–2):1–23

Murphy RA (1980) Mechanics of vascular smooth muscle. In: Handbook of physiology. American Physiological Society, Bethesda, pp 325–351

Nichols WW, O'Rourke MF, Vlachopoulos C (2011) McDonald's blood flow in arteries: theoretical, experimental and clinical principles, 6th edn. Hodder Arnold, London, pp 411–446

Oka S (1981) Cardiovascular hemorheology. Cambridge University Press, Cambridge

Okamoto RJ, Xu HD, Kouchoukos NT, Moon MR, Sundt TM (2003) The influence of mechanical properties on wall stress and distensibility of the dilated ascending aorta. J Thorac Cardiovasc Surg 126(3):842–850. doi:10.1016/S0022-5223(03)00728-1

Pasta S, Phillippi JA, Gleason TG, Vorp DA (2012) Effect of aneurysm on the mechanical dissection properties of the human ascending thoracic aorta. J Thorac Cardiovasc Surg 143 (2):460–467. doi:10.1016/j.jtcvs.2011.07.058

Pries AR, Secomb TW, Gaehtgens P (2000) The endothelial surface layer. Pflugers Arch - Eur J Physiol 440:653–660

Rhodin JAG (1979) Architecture of the vessel wall. In: Burne RM (ed) Handbook of physiology, sec 2, vol 2. American Physiological Society, Bethesda, pp 1–31

Scott Blair GW (1959) An equation for the flow of blood, plasma and serum through glass capillaries. Nature 183:613–614

Sugihara-Seki M (2006) Transport of spheres suspended in the fluid flowing between hexagonally arranged cylinders. J Fluid Mech 55:309–321

Sugihara-Seki M, Fu BM (2005) Blood flow and permeability in microvessels. Fluid Dyn Res 37:82–132

Tang DL, Teng ZZ, Canton G, Yang C, Ferguson M, Huang XY, Zheng J, Woodard PK, Yuan C (2009) Sites of rupture in human atherosclerotic carotid plaques are associated with high structural stresses – an in vivo MRI-based 3D fluid-structure interaction study. Stroke 40 (10):3258–3263. doi:10.1161/Strokeaha.109.558676

Thubrikar MJ (2007) Vascular mechanics and pathology. Springer, New York

Vito RP, Dixon SA (2003) Blood vessel constitutive models-1995-2002. Annu Rev Biomed Eng 5:413–439. doi:10.1146/annurev.bioeng.5.011303.120719

Vito RP, Hickey J (1980) The mechanical properties of soft tissues.2. The elastic response of arterial segments. J Biomech 13(11):951–957. doi:10.1016/0021-9290(80)90166-9

von Maltzahn WW, Besdo D, Wiemer W (1981) Elastic properties of arteries – a nonlinear two-layer cylindrical model. J Biomech 14(6):389–397. doi:10.1016/0021-9290(81)90056-7

Weinbaum S, Zhang X, Han Y, Vink H, Cowin SC (2003) Mechanotransduction and flow across the endothelial glycocalyx. PNAS 100:7988–7995

Weinbaum S, Tarbell JM, Damiano ER (2007) The structure and function of the endothelial glycocalyx layer. Annu Rev Biomed Eng 9:121–167

Whitmore RL (1968) Rheology of the circulation. Pergamon, Oxford

Yamada H (1999) A mathematical model of arteries in the active state (Incorporation of active stress and activation parameter). JSME Int J, Ser C 42(3):545–551

Yamada H (2012) Fundamental mechanics and biomechanics. Corona Publishing, Tokyo, in Jap

Yamada H, Hasegawa Y (2007) A simple method of estimating the stress acting on a bilaterally symmetric abdominal aortic aneurysm. Comput Methods Biomech Biomed Engin 10 (1):53–61. doi:10.1080/10255840601086531

Yamada H, Sakata N (2013) Low pressure condition of a lipid core in an eccentrically developed carotid atheromatous plaque: a static finite element analysis. J Biorheol 27(1):9–17. doi:10.1007/s12573-012-0051-x

Yamada H, Shinoda T, Tanaka E, Yamamoto S (1999) Finite element modeling and numerical simulation of the artery in active state. JSME Int J, Ser C 42(3):501–507

Yamada H, Yoshitake Y, Iwata N (2007) Comparisons of the finite element analysis solutions and the analytical ones for various opening-angled arterial walls. In: Proc Mech Eng Congress, 2007 Japan, No. 07–1, vol. 5, pp 163–164, in Jap

Yamada H, Yuri K, Sakata N (2010) Correlation between stress/strain and the retention of lipoproteins and rupture in atheromatous plaque of the human carotid artery: a finite element study. J Biomech Sci Eng 5(4):291–302

Fundamentals of Physiology and Biology of Vascular System

3

Tomohiro Aoki and Kimiko Yamamoto

Abstract

Vascular system supports the homeostasis of our body through supplying adequate oxygen and nutrition to organs and retrieving wastes. This system is fundamental to maintain the physiological function of each organ connected to parallel fashion. Once vascular system becomes malfunction, the severe pathological condition threatening our lives occurs in our body. This is why vascular system finely adjusts the blood flow in corresponding to the demand of each organ. In this step, several regulatory mechanisms present in our body and a change of radius of arteries supplying to each organ is a major one. Radius of arteries is regulated by the dilation or contraction of vascular walls and the latter mainly contribute the regulation of blood flow. This contractile status of vascular system, regulated by the power stroke of heads of myosin along actin fibers driven by the increase of intracellular Ca^{2+} level, is termed as vascular tone. Because the disturbance of vascular tone can cause the pathological consequence in organs and our body such as shock, vascular tone is strictly regulated by many factors including both intrinsic and extrinsic ones such as renin-angiotensin system and sympathetic nerve system. In this chapter, we summarize the fundamentals of the physiology and also the biology of vascular system.

Keywords

Vascular tone • Sympathetic nerve • Renin-angiotensin system • Physiology • Ca^{2+}

T. Aoki (✉)
Innovation Center for Immunoregulation Technologies and Drugs (AK project), Kyoto University Graduate School of Medicine, Konoe-cho Yoshida, Sakyo-ku, Kyoto City, Kyoto 606-8501, Japan
e-mail: tomoaki@kuhp.kyoto-u.ac.jp

K. Yamamoto
Department of Biomedical Engineering, The University of Tokyo, Tokyo, Japan

K. Tanishita, K. Yamamoto (eds.), *Vascular Engineering*,
DOI 10.1007/978-4-431-54801-0_3

3.1 Fundamental Role of Vascular System

Vascular system plays a crucial role to maintain the condition and keep homeostasis of our body through transporting the adequate oxygen and nutrition to demanding tissues and retrieving waste products. Watery fluid takes up about 60 % of our body and is distributed in intra-cellular, intestinal and intra-vascular spaces. Among them, watery fluid is most abundantly present in intra-cellular space, up to about two-thirds of total body water, and the remains intestinally (about one-fourth of total body water) or intra-vascularly (less than 10 % of total body water) distribute. Cells in various tissues and organs directly communicate with surrounding intestinal fluids, not with the vascular system itself, to exchange nutrition and waste for keeping cell conditions constant. However, because intestinal space is not large enough to supply and store adequate nutrition as required by cells, the continuous supply of oxygen and nutrition from capillaries are essential to maintain the appropriate function of cells, tissues and organs. In this sense, vascular system is critical for keeping the homeostasis of our body.

Basic composition of vascular system is heart, as a pump to generate output, and vessels, as a tube to circulate blood among heart and organs. Heart functionally consists of two elements, left heart pump to deliver blood to organs, so called *systemic circulation*, and right heart pump to deliver blood to lung to ingest oxygen and egest carbon dioxide, so called *pulmonary circulation* (Fig. 3.1) (Klabunde 2012; Mohrman and Heller 2010). Here, heart and organs, except lung, are connected in a parallel fashion not in series (Fig. 3.1). Blood pumped out from left ventricle of heart flows through arteries and reaches various organs. In this process, the supply of blood flow in a parallel fashion enables to deliver the blood with identical compositions to organs. Further, a parallel fashion also makes it

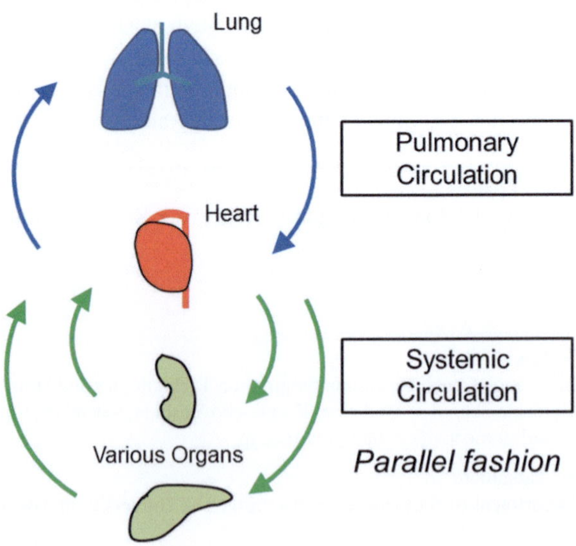

Fig. 3.1 Global structure of vascular system (Note that the lung is connected in series with heart (pulmonary circulation) and other organs are connected in a parallel fashion (systemic circulation))

easier to regulate blood flow independently according to the demand of organs. For example, when blood volume in our body is urgently decreased, the distribution of blood flow is rapidly rearranged resulting in the maintenance of blood flow in vital organs, such as brain and myocardium, accompanied with the decrease of blood supply in non-vital organs such as skin. In other point of view, the tolerance to rapid reduction in the supply of blood flow is relatively low in brain and myocardium because these two organs need a lot of oxygen and glucose to perform their appropriate function and blood flow does not prepare enough allowance. Indeed, we fall in unconsciousness at only the short period of blood flow arrest to brain. Furthermore, the infarction, which means the death of neuronsin brain or myocardium in heart, can easily develop after the stoppage of blood flow even in only the short period.

On the contrary, heart and lung are connected in series (Fig. 3.1) (Klabunde 2012; Mohrman and Heller 2010). Blood returned from systemic organs through venous system is pumped out from right ventricle of heart to lung through pulmonary artery, in which blood has a lowest oxygen concentration in our body. In lung, gas composition in blood is immediately re-conditioned through the rapid egestion of carbon dioxide and the ingestion of oxygen. Then, blood with enough oxygen returns to left atrium of heart through the pulmonary vein, in which blood has a highest oxygen concentration in our body, and again pumped out from left ventricle to organs.

3.2 Fundamental Physiology of Vascular System

3.2.1 Physics of Vascular System

In our body, about 5 liter per minute of blood is delivered to organs in the resting state. However, once we do our daily lives or exercise, the demand of oxygen and nutrition in our body remarkably increases. In these situations, cardiac output is increased and, at the same time, the distribution of blood flow to each organ is rearranged to supply adequate blood to demanding organs.

Fluid flow in vessels, which can be considered as a tube, is influenced and regulated by several physical factors. First, the flow rate in a tube is determined by the pressure difference between inlet and outlet, which makes a driving force of blood flow, and the resistance of vascular walls (please also see Chap. 2).

$$Q(\text{flow rate}) = (P_{\text{inlet}} - P_{\text{outlet}})/R\ (\text{vascular resistance})$$

Resistance to flow depends on the internal radius of tube (r), the length of tube (L) and the viscosity of fluid (η), in this case blood.

$$R = 8\,L\eta/\pi r^4$$

These equations can be combined into one equation, well known as the *Poiseuille* Equation (please also see Chap. 2).

$$Q = (P_{\text{inlet}} - P_{\text{outlet}})\pi r^4/8\,L\eta$$

Note that the difference of pressure ($P_{\text{inlet}} - P_{\text{outlet}}$) and radius of arteries (r) are major regulatory factors of blood flow to organs because the length of vessels and the viscosity of blood are not greatly changes. Further, considered the fact that the difference of pressure ($P_{\text{inlet}} - P_{\text{outlet}}$) is generated by cardiac output and that blood pressure to organs is basically the same independent of the kind of organs, the radius of arteries mainly regulates the blood supply to each organ.

3.2.2 Control of Blood Flow to Organs

As described in the previous section, the regulation of the radius of vascular system is fundamental for the control of blood flow to various organs. Here, because the radius of vascular system is mainly regulated by the contraction and the dilatation of vessels, we should understand mechanisms underlying them (Klabunde 2012; Mohrman and Heller 2010).

3.2.2.1 Mechanisms for the Contraction

The radius of vessels is basically regulated by the contractile activity of vascular smooth muscle cell layer. Because almost all vessels equips the smooth muscle cells in their walls except for capillaries, the contribution of vascular smooth muscle cells to the regulation of vascular diameter can be widely adapted in vascular system (Klabunde 2012; Mohrman and Heller 2010).

Vascular smooth muscle cell is a spindle-shaped and circumferentially aligned cell in media of vascular walls. Because vascular smooth muscle cells are electronically connected with each other by forming gap junctions among neighborhood cells, these cells can coordinately contract according to stimuli to decrease the vascular diameter which leads to the remarkable reduction of blood flow as described in previous section. Contraction of vascular smooth muscle cells is performed through the interaction of contractile filaments, actin and myosin. In resting state, heads of myosin tightly interact with actin fibers. Under the presence of ATP, ATP binds to the head of myosin and, then, as a result of the little conformational change, the head of myosin is dissociated from the actin fiber. Subsequently, with the hydrolysis of ATP, the head of myosin is displaced about 5 nm length along the actin fiber. Finally, the head of myosin moves the actin fibers, known as 'power-stroke', and loses ADP, hydrolytic product of ATP (Fig. 3.2). Here, not like the skeletal muscle or myocardium, smooth muscle cells lack the striations consisting of regularly and repeatedly arranged contractile fibers. In smooth muscle cells, actin fibers attach to the so-called *dense bodies*, instead of

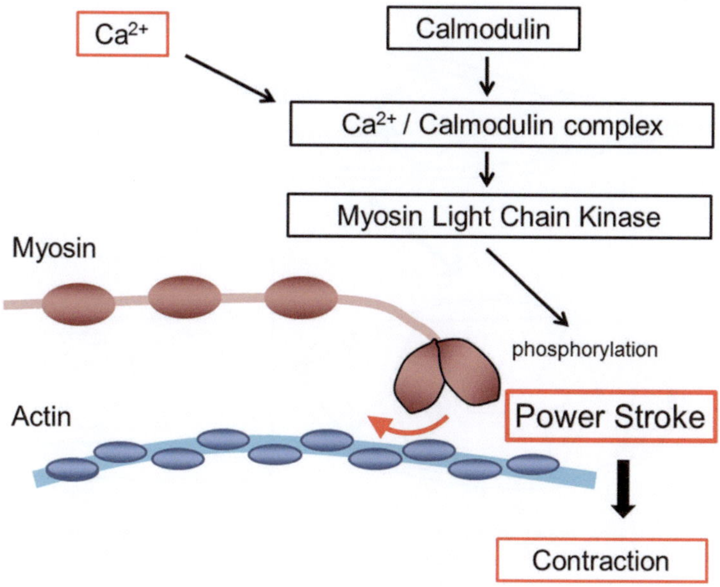

Fig. 3.2 Mechanisms of Ca^{2+} induced contraction of actomyosin fiber

Z-line observed in other type of muscle cells (Fig. 3.3). Therefore, as a natural consequence, the arrangement of contractile fibers in smooth muscle cells and other type of cells is completely different. In smooth muscle cells, contractile fibers are arranged as a net-like structure not as an alignment along the long axis of cells observed in other type of muscle cells (Fig. 3.3). The net-like structure of contractile fibers and circumferential alignment of cells probably enable to effectively contractile the vessels and generate tension.

Contractile of smooth muscle cells is regulated by the increase of free intracellular Ca^{2+} content. In smooth muscle cells, the free intracellular Ca^{2+} content is increased via two major pathways, *electromechanical coupling* and *pharmacomechanical coupling* (Fig. 3.4) (Klabunde 2012; Mohrman and Heller 2010). In electromechanical coupling, depolarization of membrane in response to stimuli, which means the activation of smooth muscle cells, increases the opening of voltage-operated calcium channel located in the surface membrane of smooth muscle cells and, as a result, Ca^{2+} influx increases (Fig. 3.4). In pharmacomechanical coupling, once some agonists with vasoconstrictor activity such as norepinephrine bind to their own receptors such as the α1-adrenergic receptor in the case of norepinephrine, free intracellular Ca^{2+} content is increased mainly via two mechanisms (Fig. 3.4). First, some receptors functionally bind to receptor operated calcium channels and the activation of receptors opens the coupled calcium channel resulting in the increase of Ca^{2+} influx (Fig. 3.4). Second, some receptors activate phospholipase C and increases inositol triphosphate (IP_3) through the hydrolysis of phosphatidylinositol 4,5-bisphosphate (PIP_2) located in plasma membrane. IP_3 is

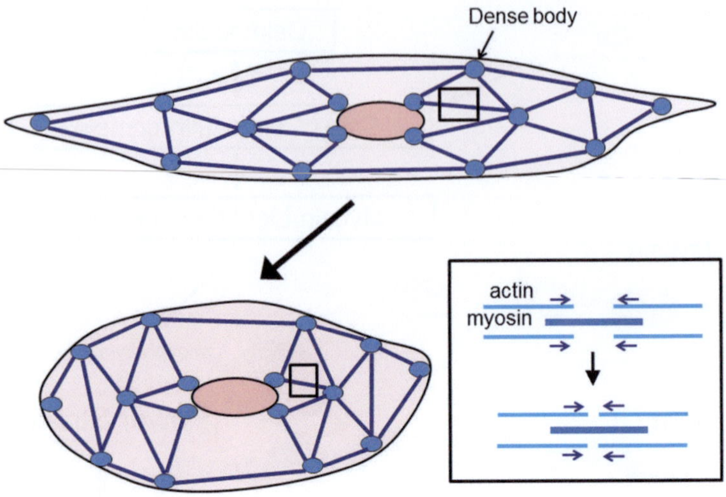

Fig. 3.3 Net-like structure of actomyosin bands in smooth muscle cells (Note that each actomyosin band is connected with dense body in cells)

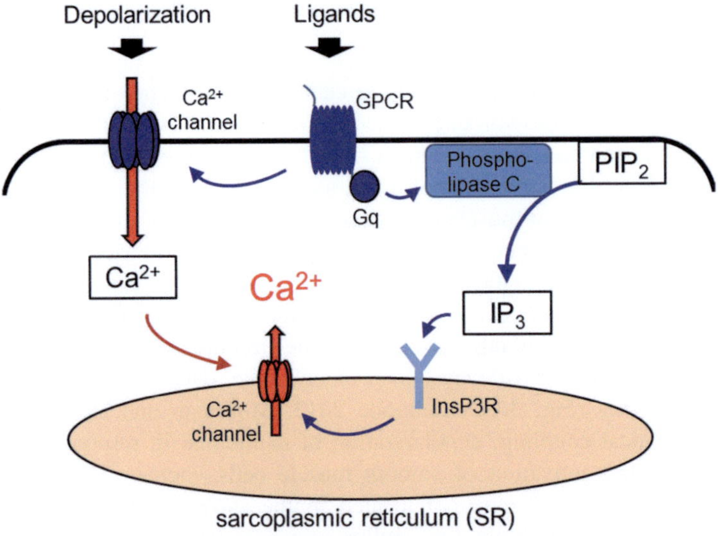

Fig. 3.4 Mechanisms underlying the increase of intracellular Ca^{2+} content under the depolarization (electromechanical coupling) and the stimuli by ligands (pharmacomechanical coupling). *GPCR* G-protein coupled receptor, *IP_3* inositol triphosphate, *PIP_2* phosphatidylinositol 4,5-bisphosphate, *InsP3R* inositol triphosphate receptor

diffusing to endoplasmic reticulum (ER) or sarcoplasmic reticulum (SR) in the case of muscle cells and binds to its receptor, inositol triphosphate receptor (InsP3R) on the surface. The binding of IP_3 to InsP3R, then, opens the calcium channel located in the membrane of ER or SR and, as a result, Ca^{2+} is released from ER or SR. This step is well known as calcium-induced calcium release (Fig. 3.4).

In smooth muscle cells, free intracellular Ca^{2+} binds to the calcium-binding protein, *Calmodulin* and forms complex. Ca^{2+}-Calmodulin complex, then, activates myosin light chain kinase which phosphorylates myosin light chain and enables the power-stroke by using the energy of ATP (Fig. 3.2).

3.2.2.2 Mechanisms for the Dilatation

Epinephrine, histamine and vasoactive intestinal peptide (VIP) plays the vasodilator function via the increase of intracellular cAMP level (Klabunde 2012; Mohrman and Heller 2010). Increase of cAMP in cells activates protein kinase A (PKA) and, then, PKA phosphorylates various bioactive molecules resulting in the increase of Ca^{2+} efflux, the hyperpolarization of membrane potential and so on. In addition, cGMP mediates a vasodilator function of nitric oxide, well-known vasodilator, through the activation of cGMP-dependent protein kinase.

3.2.3 Regulation of Vascular Tone

The contractile state of vascular system, which is generated by the contraction of vascular smooth muscle cells, is represented by the term, 'vascular tone' (Klabunde 2012; Mohrman and Heller 2010). Here, because blood flow to each organ mainly depends on the diameter of efferent arteries to organs, vascular tone of arteries is the key regulator of the blood flow. In the resting state, arteries are partially contracted. And the vascular tone in this partially constricted state of arteries without any external stimuli is known as the '*basal tone*' of vascular system. In addition to basal tone, vascular system increases or decreases its tone through the further contraction or the relaxation of vascular smooth muscle cells to regulate vascular diameter and resistance and resultant blood flow to organs. Several factors, including both extrinsic factors and intrinsic ones, regulate this step (Fig. 3.5).

3.2.3.1 Extrinsic Factors Regulating Vascular Tone

Major extrinsic factors regulating vascular tone are sympathetic nerve system and circulating hormones (Fig. 3.5) (Klabunde 2012; Mohrman and Heller 2010).

Sympathetic Nerve System

Arteries in all organs are innervated by sympathetic nerve fibers. The control of sympathetic nerve system is regulated in the *rostral ventrolateral medulla* (RVLM) preset in brain stem. In RVLM, sympathetic neurons spontaneously generate action potentials and consequently keep the arteries in the relative contractive status. In other word, the sympathetic nerve system plays the crucial role to keep basal tone of vascular system. Therefore, if the sympathetic nerve activity is suddenly lost,

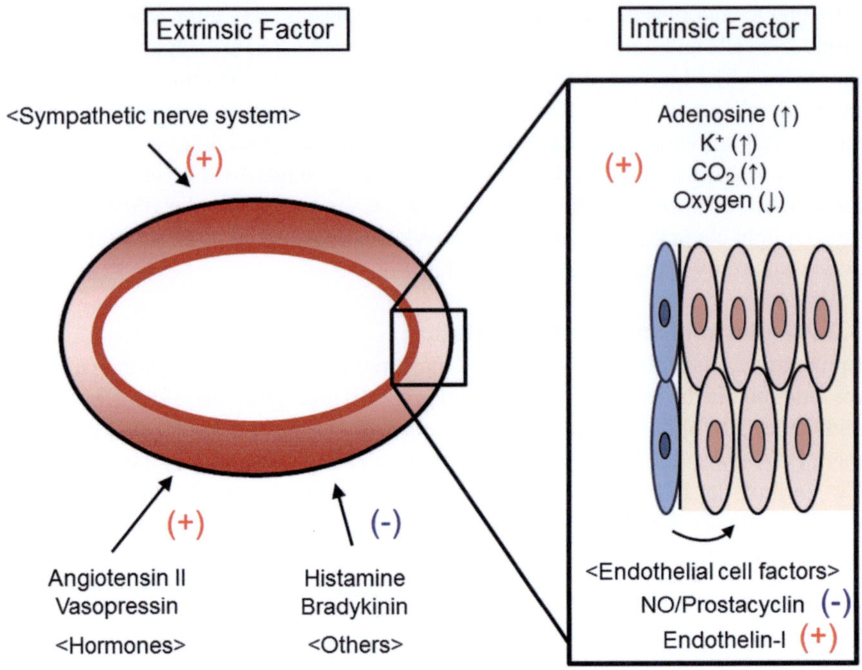

Fig. 3.5 Extrinsic or intrinsic factors regulating the vascular tone. Plus (+) or minus (−) represents the factors which increase or decrease the vascular tone, respectively

arteries lose the basal tone and are markedly dilated. Axons of sympathetic nerve in RVLM descend into spinal cord and, in the intermediolateral column of spine, transmit to next neurons by synapse formation. Sympathetic nerves, then, exit from the spinal cord at the level of Th1 to L2 depending on the location of arteries they regulate. After then through synapse formation in paravertebral ganglions or prevertebral ganglions, sympathetic nerve terminals reach the adventitia layer of governing arteries and innervate them. Sympathetic nerves release norepinephrine from their terminals depending on the extent of the activation status of sympathetic nerve system. Norepinephrine, then, binds to one of its receptors, α1-adrenergic receptor on vascular smooth muscle cells. Because α1-adrenergic receptor couples with phospholipase C via Gq, the activation of the receptor results in the increase of intracellular Ca^{2+} content leading the contraction of vascular smooth muscle cells in a pharmacomechanical fashion as described previously (Fig. 3.4).

In contrast to the sympathetic nerve system, the contribution of parasympathetic nerve system to vascular tone is quite limited. Here note that in heart, the contribution of parasympathetic nerve system to the regulation of heart rate is significant and cannot be neglected.

Hormones

Hormones released in blood flow from several endocrine organs or tissues also regulate vascular tone (Klabunde 2012; Mohrman and Heller 2010). As a result of the activation of sympathetic nerves, catecholamines, representatively epinephrine and norepinephrine, are released from adrenal gland. Catecholamines, then, activate β1-adrenergic receptor in pace-maker cells heart and increases heart rate. In addition, catecholamines activate α1-adrenergic receptor located in vascular smooth muscle cells resulting in the vasoconstriction accompanied and synergistically with the vasoconstriction through the activation of sympathetic nerve itself.

Angiotensin II is a derivative of renin-angiotensin system. Angiotensin II is a peptide hormone derived from angiotensinogen through sequential actions of renin and angiotensin-converting enzyme (ACE) and exerts its role via acting on AT1 receptor. Because renin is secreted under the decrease of renal blood flow, angiotensin II is generally formed in the pathological conditions with the decrease of blood flow. Angiotensin II via AT1 receptor exerts various functions to sustain blood flow to organs and maintain blood pressure. In cardiovascular system, AT1 signaling shows the potent vasocontractive ability, which results in the increase of vascular tone and decrease of the diameter of arteries, by increasing the intracellular Ca^{2+} content through Gq – phospholipase C – IP_3 signaling cascade. Angiotensin II further regulates the release of aldosterone from the cortex of adrenal gland. Aldosterone, then, functions in kidney and retain sodium to maintain intravascular water content. In the central or peripheral nervous system, angiotensin II plays the various roles to maintain the vascular tone through accumulated water in body or the activation of sympathetic nerves. Angiotensin II increases the sensation of thirst by acting on the subfornical area in brain and also increases the desire for salt. Further, angiotensin II increases the secretion of ADH from the posterior lobe and ACTH from the anterior lobe of pituitary glands. Angiotensin II also regulates and increases the release of norepinephrine from para-ganglionic sympathetic nerves.

Vasopressin is a hormone inhibiting the water excretion from kidney by acting on its receptor, V_2 receptor, on collecting ducts. Vasopressin is, therefore, important for the regulating of body water content and especially for securing the necessary water volume to maintain blood flow. In this sense, vasopressin is the important hormone for the cardiovascular function. Vasopressin can also contract arteries via acting on other receptor, V_{1A}, which is present on cytoplasmic membrane of vascular smooth muscle cells and is coupled with phospholipase C/calcium signaling cascade.

Other Factors

Several other factors can regulate the vascular tone, sometimes independent of the need of blood flow in each organ (Klabunde 2012; Mohrman and Heller 2010).

Histamine is released from mast cells present in tissues under the various pathological conditions, such as inflammation, injury and allergic response. Histamine binds to its receptors, H_1 to H_4, on target tissues and plays a role. In vascular system, histamine reduces vascular tone and relaxes vascular smooth muscle cells through H_1-Gq or H_2-Gs-cAMP-PKA signaling cascade.

Bradykinin consists of 9 amino acids and synthesized as a proteolytic cleavage of kininogen by the action of serine proteinase, kallikrein. Bradykinin acts on its receptor, B_1 and B_2. B1 expression is induced in various pathological settings, such as injury and inflammation and, therefore, B_1 signaling regulates the function of bradykinin as a modulator of inflammation, such as a recruitment of neutrophil through CXCL5 induction (Duchene et al. 2007; McLean et al. 2000). On the contrary, B_2 receptor constitutively expresses and couples with Gi and Gq. In vascular system, B_2 signaling functions in endothelial cells and stimulates the induction of vasodilators, nitric oxide and prostacyclin, resulting in the reduction of vascular tone. Notably, because the breakdown of bradykinin is regulated at least partially by the angiotensin-converting enzyme (ACE), the usage of ACE inhibitor, which is a powerful anti-hypertensive drug through the inhibition of renin angiotensin system, increases an amount of bradykinin in tissues, sometimes leading the side effects of this drug, cough and systemic edema due to the increased permeability of vascular system.

3.2.3.2 Intrinsic Factors Regulating Vascular Tone

Blood flow to each organ or tissue is also regulated *in situ*, because organ and tissue has their own regulatory mechanisms of blood flow (Klabunde 2012; Mohrman and Heller 2010). Basically, the mechanisms underlying the local control of blood flow is mediated by factors derived from vascular wall itself or surrounding cells. And many of these responsible factors for local control of blood flow are ones related with cell metabolism or biochemical signaling pathways (Fig. 3.5).

Adenosine

Adenosine is a derivative from ATP. When the oxygen consumption increases and sometimes hypoxia happens in cells, ATP is hydrolyzed resulting in the generation of ADP and AMP. AMP is, then, dephosphorylated by the action of 5′- nucleotidase and adenosine is released. Because the content of ATP in cells is much higher than that of adenosine, small amount of ATP hydrolysis greatly influences the adenosine content in cells. Adenosine released from cells acts on surrounding tissues such as arteries, sympathetic nerves and myocardium via adenosine receptor. In arteries, adenosine dilates arteries and decreases the vascular tone. Especially in coronary arteries, adenosine plays the crucial role to maintain the blood flow to myocardium when the oxygen consumption by myocardium increases. Adenosine, further, decreases the release of norepinephrine from sympathetic nerve terminals and inhibits Ca^{2+} influx in myocardium, resulting in the dilatation of arteries and decrease of oxygen consumption by myocardium.

Potassium Ion

When muscle cells, such as skeletal muscle or myocardium, are activated and contracts, the membrane potential of cells are first depolarized through the increase of Na^+ influx and finally return to the initial polarization status via the K^+ efflux after the activation. After the contraction cycle of smooth muscle cells, intracellular Na^+ and K^+ contents are restored to those before the contraction, low intracellular

Na^+ and high intracellular K^+ contents, by the action of Na^+/K^+-ATPase (Na^+/K^+ exchanger). However, Na^+/K^+ exchange by Na^+/K^+-ATPase is the relatively time-consuming step and, therefore, cannot catch up the rapid depolarization cycle by the continuously contracting muscle cells resulting in the increase of K^+ content in the extracellular space. Because the increase of K^+ content in the extracellular space leads the hyperpolarization of membrane potential, the relaxation of smooth muscle cells in surrounding vessels occurs and, as a result, blood flow to skeletal muscles and myocardium increases.

Carbon Dioxide and Hydrogen Ion

Carbon dioxide (CO_2) concentration in tissues is increased when the oxidative metabolism increases or the supply of blood flow to tissues are inadequate compared with the production of CO_2. CO_2 is a gas and, therefore, easily diffuses into surrounding cells including vascular smooth muscle cells. In vascular system, CO_2 causes vasodilatation via the formation of hydrogen ion (H^+) and resultant opening of ATP-sensitive potassium channels.

Oxygen

The partial oxygen concentration is decreased in tissues as a result of the increase of oxygen consumption and the inadequate supply of blood flow. Reduction of the partial oxygen concentration in tissues leads the vasodilatation of arteries *in situ* partially because inadequate oxygen concentration cannot maintain the contractive status of vascular smooth muscle cells (Klabunde 2012; Mohrman and Heller 2010).

Factors from Endothelial Cells

Endothelial cells are the important cell type for the regulation of vascular tone through paracrine mechanisms. Endothelial cells release various vasoactive factors in response to the change of microenvironment, such as the hypoxia or the change of hemodynamic status, and regulate vascular tone through acting on vascular smooth muscle cells. Among these factors released from endothelial cells, nitric oxide (NO) and prostacyclin (prostaglandin I_2) reduce and endothelin-I increases vascular tone (Fig. 3.5).

Nitric oxide is one of the most important factors regulating vascular tone in physiological settings. Nitric oxide is synthesized from L-arginine by the actions of nitric oxide synthases (NOS) in both physiological condition and pathological conditions. In the regulation of vascular tone, endothelial cells release nitric oxide synthesized by the action of endothelial NOS (eNOS), which is constitutively expressed in endothelial cells, and nitric oxide diffuses into surrounding vascular smooth muscle cells. In smooth muscle cells, nitric oxide binds and activates the guanylyl cyclase resulting in the increase of intracellular cGMP concentration. Increase of intracellular cGMP concentration, then, leads the dilatation of vascular smooth muscle cells and reduces vascular tone. Here, because nitric oxide synthesis from endothelial cells are induced according to the increase of blood flow and shear stress loaded on endothelial cells, the function of nitric oxide is known as 'flow-dependent vasodilatation'. In pathological conditions, a large amount of nitric oxide is released from cells by the action of inducible nitric oxide synthase (iNOS) and increases reactive oxygen species resulting in the induction of inflammation.

Prostacyclin is derived from arachidonic acid by the sequential actions of cyclooxygenase and prostacyclin synthase in both the physiological and pathological conditions and exerts its action on the specific receptor, IP, on target cells. In vascular smooth muscle cells, prostacyclin released from endothelial cells binds to IP and, then, IP signaling increases intracellular cAMP concentration by activating adenylyl cyclase resulting in the relaxation of vascular smooth muscle cells and the reduction of vascular tone.

Endothelin-I is a peptide hormone consisting of 21-amino acids with the vasocontractive activity. Endothelin-I is synthesized by the action of endothelin-converting enzyme from its precursor in endothelial cells, acts on its receptors, ET_A and ET_B, and plays its role in vascular system. ET_A receptor couples with Gq and, therefore, ET_A signaling in vascular smooth muscle cells increases the contraction of vascular smooth muscle cells and vascular tone through activating phospholipase C-IP_3-Ca^{2+} cascade. On the contrary, ET_B is present on endothelial cells and the stimulation of this type of receptor induces the syntheses of nitric oxide and prostacyclin, resulting in the vasodilatation. These ET_B signaling cascade is considered as a negative feedback one against ET_A signaling cascade. Importantly, because endothelin-I formation from endothelial cells is regulated by various cytokine and bioactive factors, such as angiotensin II, vasopressin, thrombin and reactive oxygen species, endothelin-I is a mediator of vasoconstriction not in physiological conditions but also in pathological conditions such as pulmonary hypertension.

3.2.4 Response and Compensatory Mechanisms of Vascular System for Blood Flow and Blood Pressure Change

The adequate blood supply according to the demand of organs and tissues is fundamental to keep the homeostasis of our body. Our body, therefore, equip the mechanisms to compensate the decrease of blood supply and maintain the blood flow to organs especially in the case of the decrease of blood pressure.

When the decrease of blood pressure occurs, our body tries to restore the blood pressure and to keep the blood flow to organs and tissues. Major compensatory mechanisms responsible for the maintenance of blood flow and blood pressure are sympathetic nerve system and renin-angiotensin system (Figs. 3.6 and 3.7) (Atlas 2007; Califf and Bengtson 1994; Kemp and Conte 2012; Klabunde 2012; Lymperopoulos 2013; Mohrman and Heller 2010; Zelis et al. 1991).

3.2.4.1 Sympathetic Nerve System

Sympathetic nerve system is crucial for the regulation of systemic blood pressure (Klabunde 2012; Lymperopoulos 2013; Mohrman and Heller 2010). Systemic blood pressure is sensed by arterial baroreceptors located in the carotid sinus and

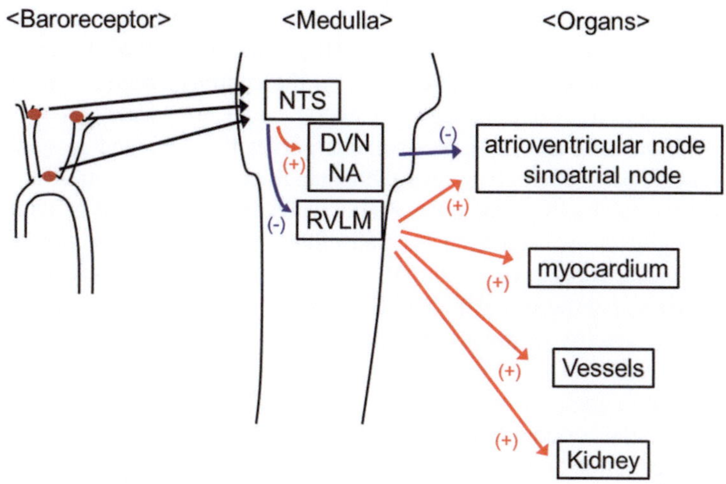

Fig. 3.6 Schema of the baroreceptor reflex. *NTS* the *nucleus tractus solitaries* of medulla, *RVLM* the *rostral ventrolaeral medulla*, *DVN* the *dorsal vagal nucleus* of medulla, *NA* the *nucleus ambiguus* of medulla

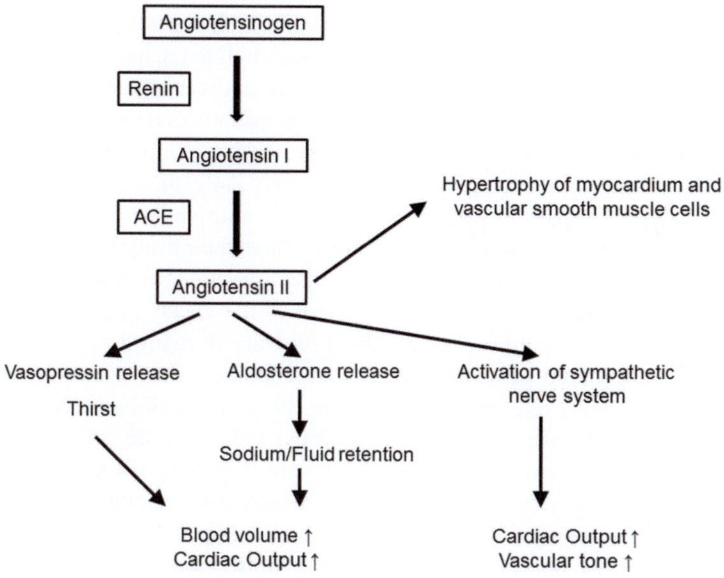

Fig. 3.7 Renin-angiotensin system. *ACE* angiotensin-converting enzyme

aorta (Fig. 3.6). Afferent fibers from the carotid sinus or aorta ascend to the brain stem via glossopharyngeal nerve (cranial nerve IX) or vagus nerve (cranial nerve X), respectively (Fig. 3.6). In brain stem, these fibers enter the medulla at the *nucleus tractus solitaries* (NTS). Then, from NTS, inhibitory interneurons project to sympathetic nerves and excitatory interneurons also project to parasympathetic nerves. As a consequence of these innervations, the activation of baroreceptors decreases sympathetic nerve activity and, on the contrary, increases parasympathetic nerve activity. Neurons of sympathetic nerve are located in the *rostral ventrolaeral medulla* (RVLM). Sympathetic nerves then descend in spinal cord, synapses at the paravertebral or prevertebral ganglions through nicotinic acetylcholine receptors and finally reach to the arterial walls. At the terminals of sympathetic nerves, norepinephrine is released as a transmitter. Norepinephrine acts on vascular system in various aspects via α1 and β1 receptors. α1 receptor is located in arterial walls and heart, and the stimulation of this receptor increases the vascular tone and the contractile activity of myocardium through the activation Gq-IP_3 pathway. In addition, the activation of β1 receptor located in pacemaker cells of heart increases heart rate, via the increase of spike frequency in atrioventricular node or sinoatrial node, and also the contractile activities of ventricles via PKA pathway. Further, in kidney, the activation of β1 receptor in juxtaglomerular cells of kidney enhances the release of renin from these cells and up-regulates renin angiotensin system as described in next section in detail.

Neurons of parasympathetic nerves are located in the *dorsal vagal nucleus* (DVN) and *nucleus ambiguus* (NA) of medulla (Fig. 3.6). Efferent fibers exit the medulla and run to heart as a cranial nerve X (Vagal nerve). In heart, parasympathetic nerve fibers function in atrioventricular node or sinoatrial node and decreases heart rate through acetylcholine-muscarinic acetylcholine receptor signaling. Parasympathetic nerves also innervate to myocardium and can reduce the contractile activity of this cell. However, the effect of parasympathetic nerve system on cardiac output is only less than that of sympathetic nerve system. Moreover, the regulation of vascular tone by parasympathetic nerves plays only minor role in systemic vascular systems though they play the crucial role in limited organs.

Baroreceptors at the carotid sinus and aorta change their firing rate in response to the changes of mean arterial pressure and pulse pressure. Because the maximum sensitivity of carotid sinus can be observed at the mean pressure of around 90 mmHg, which is the normal mean blood pressure in many people, only small changes of mean blood pressure from normal condition can be converted to the large changes of firing rate. Here, importantly, the set point of baroreceptor can be shifted according to the individual conditions. For example, in patients with hypertension, the curve is shifted to the right ward (Fig. 3.8). If the blood pressure becomes lower, firing rate of carotid sinus decreases. As a natural consequence of the reduction of firing rate in baroreceptors, sympathetic nerve system is activated and parasympathetic nerve system is, on the contrast, suppressed resulting in the increase of vascular tone and heart function (Fig. 3.6).

3.2.4.2 Renin-Angiotensin System

Renin-angiotensin system plays the crucial role for the compensation and maintenance of blood pressure (Fig. 3.7) (Atlas 2007; Klabunde 2012; Mohrman and Heller 2010). Renin is synthesized and released from juxtaglomerular cells associated with the afferent arteries of kidney under some conditions, such as the

Fig. 3.8 Correlation of the firing of baroreceptor with mean blood pressure (Note that, in hypertension, the *curve* is shifted to *right* and, as a result, the point with the maximum sensitivity is resetting to the higher point)

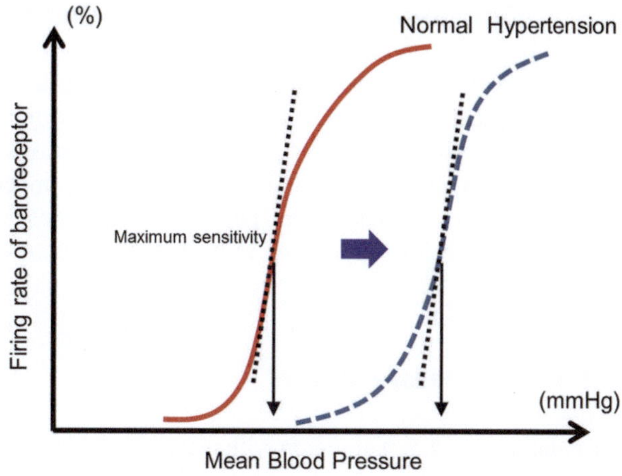

activation of sympathetic nerve system and hypo-perfusion of renal arteries. Renin released from kidney produces angiotensin I (Asp-Arg-Val-Tyr-Ile-His-Pro-Phe-His-Leu) through the proteolytic cleavage of the angiotensinogen released from liver (Fig. 3.8). Angiotensin I is, then, cleaved into angiotensin II (Asp-Arg-Val-Tyr-Ile-His-Pro-Phe) by the action of angiotensin converting enzyme (ACE) present in endothelial cells especially in lung (Fig. 3.7). Angiotensin II functions via its receptor ATI in various aspects to maintain blood pressure (Fig. 3.7). Angiotensin II contracts vascular smooth muscle cells via ATI- Gq-IP_3 cascade, activates sympathetic nerve system, increases the secretion of anti-diuretic hormone vasopressin from the posterior lobe of pituitary gland, acts upon the cortex of adrenal gland to release aldosterone, stimulates the center for thirst in hypothalamus, and induces hypertrophy in myocardium and vascular smooth muscle cells. Here, angiotensin II enhances the activities of sympathetic nerve system through multiple steps, such as the stimulation of sympathetic nerve activity in RVLM, the increase of facilitation of norepinephrine release, and the inhibition of norepinephrine reuptake, and indirectly regulates blood pressure. As a result of these multiple function of angiotensin II, renin-angiotensin system maintains the blood pressure (Atlas 2007).

Renin-angiotensin system is also important as a therapeutic target of pathological conditions (Atlas 2007). For example, in hypertension, the activation of this cascade can be detected in many cases. In such cases, the blockers of renin-angiotensin system, ACE inhibitors or AT1 blocker, are the effective drugs. Also, because the blockage of this system inhibits the remodeling of myocardium or vascular smooth muscle cells, usually the hypertrophy, at the chronic stage, the blockers of renin-angiotensin system also become the most important drugs for the treatment of patients with hypertension and heart failure (Schmieder 2005).

3.3 Development of Pathological Conditions

Breakdown of the proper regulation of vascular system results in the development of pathological conditions, circulatory shock, hypertension, and so on. The initial events leading the breakdown of regulation are various according to the situations of individuals. In each case, some compensatory mechanisms work to maintain homeostasis in vascular system. However, sometimes, a compensatory mechanism fails to keep a proper regulation in vascular system resulting in diseases.

3.3.1 Circulatory Shock

Circulatory shock is a situation with insufficient supply of blood flow to organs and, as a result, malfunctions of organs (Vincent and De Backer 2013) (Fig. 3.9). To compensate the insufficient blood supply to organs, of course, various compensatory mechanisms, such as the activation of sympathetic nerve system and the redistribution of blood to vital organs, are activated. However, these compensatory mechanisms finally fail to keep homeostasis of vascular system, leading to circulatory shock. In these conditions, body shows symptoms derived from the insufficiency of blood supply to organs, such as unconsciousness and decrease of urine (hypouresis) from acute renal failure, and excessive activation of sympathetic nerve system such as increased heart beats (tachycardia), pallor, and cold skin. Though primary causes of circulatory shock are various, mechanisms responsible for circulatory shock are either a failure of pump function of heart or a loose of vascular resistance as a corollary of *Poiseuille* equation (Fig. 3.9). The former case is derived from a trouble of heart itself (these situations are called *Cardiogenic shock*) (Califf and Bengtson 1994; Vincent and De Backer 2013) through, for example, myocardial infarction, arrhythmia (irregular rhythm of heart beat), abrupt valve malfunction, or from failure of filling of blood in heart (these situations are called *Hypovolemic shock*) (Vincent and De Backer 2013) through the sudden loss of

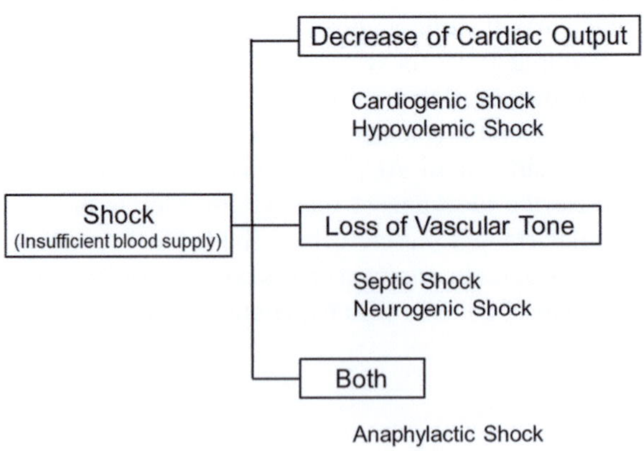

Fig. 3.9 Classification of shock by its cause (Note that shock is mainly divided in two groups (decrease of the cardiac output/loss of the vascular tone) based on its causes)

blood by the acute hemorrhage, the significant loss of body water by severe diarrhea, and the sudden arrest of pulmonary circulation by pulmonary embolism. The latter case is derived from the extrinsic vasodilators (*Septic shock*) (Angus and van der Poll 2013; Vincent and De Backer 2013), most famously endotoxin secreted by the gram-negative bacteria or from the loss of vasoconstriction by the dysfunction of sympathetic nerve system (*Neurogenic shock*). Sympathetic nerve dysfunction is a result of several events such as a traumatic injury of sympathetic nerve root, a blockage of sympathetic nerve system by anesthesia, or a vasovagal reflex by extensive outward stimuli. *Anaphylactic shock* is the other type of shock with both a decrease of cardiac output and a loss of vascular tone. In this type of shock, a certain substance occurs in the excessive allergic reaction in the individual with the sensitivity to the relevant substance. As a result of excessive allergic effect known as *immediate hypersensitivity reaction*, various reactive factors such as histamine are released from immune cells, and, through the action of these factors, vessels are extensively dilated and vascular permeability is greatly increased, resulting in the marked decrease of the blood volume and the cardiac filling.

3.3.2 Heart Failure

Heart failure is usually due to the result of inadequate blood supply from heart (Kemp and Conte 2012). The cause of heart failure is various such as the myocardial infarction, valve diseases, inflammation in myocardium (myocarditis) by infection, and cardiomyopathy. Heart failure is divided into two kinds, the right heart failure and the left heart failure, by affected ventricles and symptoms in each heart failure are different (Kemp and Conte 2012). In the left heart failure, because of the inadequate blood supply to organs, organ dysfunction occurs. Further, through the pulmonary edema as a result of inadequate cardiac output from the left ventricle, the shortness of breath during exercise or daily life activity occurs. In the right heart failure, by the insufficient output of blood from the right ventricle to lung, systemic edema and subsequent organ dysfunctions such as the disturbed function of gastrointestinal tract occur. Usually, the left heart failure is a major form of heart failure and the right heart failure is, in most cases, a consequent of the left heart failure.

In the case of heart failure, our body, of course, tries to compensate the decrease of the cardiac output to keep homeostasis (Califf and Bengtson 1994; Kemp and Conte 2012). Some compensatory mechanisms are equipped in our body. In heart *in situ*, the decrease of cardiac output is compensated by the mechanism well known as *Frank-Starling mechanism* (Fig. 3.10). When the heart failure, especially the left heart failure, occurs, Frank-Starling curve shows the rightward shift. This means that our heart is trying to maintain the amount of cardiac output by increasing the preload, shown by the left ventricular end-diastolic pressure (LVEDP) in Frank-Starling curve (Fig. 3.10). If the heart failure becomes more severe, however, the increase of LVEDP finally fails to compensate the decrease of cardiac output because, in this condition, the stretch of sarcomeres (actomyosin band) reaches

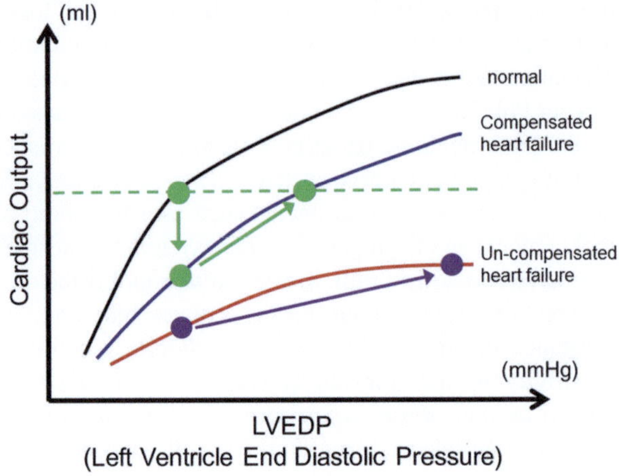

Fig. 3.10 *Frank-Starling* mechanism. *Black*, *blue* or *red line* represents the *Frank-Starling* curve from patients with normal heart function, compensated heart failure or uncompensated heart failure, respectively. *Green dot-line* represents the normal cardiac output

maximum and cannot increase the contractive activity any more. In the chronic phase of heart failure, the heart dilates as a result of new synthesis of sarcomeres to prevent the over-stretching of each sarcomere and maintain the contraction of the ventricle. However, in the excessively dilating ventricle, the compliance of the ventricle decreases and fails to effectively increase LVEDP according to the increase of left ventricle volume (Fig. 3.10). As a result, if the cause of heart failure is not resolved, in the case of severe heart failure, the heart finally cannot compensate well and blood supply to organs becomes inadequate.

Systemic compensatory mechanisms against the decrease of cardiac output under heart failure are equipped in our body (Califf and Bengtson 1994; Kemp and Conte 2012; Lymperopoulos 2013; Zelis et al. 1991). As mentioned in previous sections, major compensatory mechanisms are the activation of sympathetic nerve system and renin-angiotensin system (Lymperopoulos 2013; Zelis et al. 1991). Through these mechanisms, the body tries to increase the blood volume, cardiac output, and vascular tone. However, in some cases, all the compensatory mechanisms originally equipped in the body fail to adequately compensate the decrease of the supply of blood flow to organs leading to dysfunction of these organs.

3.3.3 Coronary Artery Disease

Coronary artery supplies the oxygen and nutrition to myocardium to maintain the function. Once the blood supply of coronary artery is disturbed by the atherosclerotic change of arterial walls or the thromboembolism, myocardium cannot perform its proper function resulting in the disturbance of the pump function of heart. If the decrease of blood flow is too significant not to support the survival of myocardium, resultant myocardium infarction occurs (Libby 2013). And if the decrease is not so significant, blood supply by coronary artery can support the myocardium at the resting state. However, once we work or exercise, the demand of blood flow in our body is significantly increased. In these situations, sometimes, the blood flow by coronary artery with atherosclerotic change or thromboembolism cannot supply the adequate oxygen and nutrition resulting in the malfunction of myocardium (Libby 2013). We feel the symptom of heart failure such as shortness of breath and an increase of heart beats and, in addition, an intense chest pain known as *angina pectoris*. Both myocardium infarction and angina pectoris fail to supply the blood flow to fulfill the demand of organs, resulting in the malfunction of various organs. Except for the disturbed myocardium function, in some cases, inadequate supply of the blood flow by coronary artery results in the malfunction of atrioventricular node or sinoatrial node. Because these two nodes regulate the rhythm of heart beats, the malfunction of nodes leads to disturbed rhythm in heart beats called *arrhythmia*. Considered the fact that proper cardiac output is achieved only when filling of blood in ventricle and the contraction of ventricle cooperatively function, arrhythmia results in inadequate cardiac output to organs.

3.4 Special Insight for the Inflammation in Pathological Condition of Vascular System

In the pathological conditions of vascular system, such as atherosclerosis and aneurysm, the triggers of disrupted homeostasis are various depending on the microenvironments, in which specific diseases develop. Nevertheless, the mechanisms contributing to the disease development, once after diseases are triggered, are somewhat common and shared among diseases. Inflammation, especially the long-lasting so-called 'chronic' inflammation, regulates the development of various diseases, not only the vascular diseases but also other socially important diseases such as cancer, diabetes mellitus, and metabolic diseases (Ben-Neriah and Karin 2011; Cruz et al. 2013; Demaria et al. 2010; Gukovsky et al. 2013; Houghton 2013; O'Shea et al. 2013; Shimizu et al. 2013; Sun and Karin 2012, 2013; Tabas and Glass 2013; Terzic et al. 2010). Therefore, the experiments of the mechanisms underlying the chronic inflammation are essential for the correct understanding of mechanisms of diseases and also the development of therapeutic strategy for diseases. Here, because the hemodynamic force is a critical and unique component of vascular system, the linkage of mechanical force (hemodynamic force) with the inflammation or other intracellular signaling pathway contributing the development of vascular disease is a proper character of vascular diseases.

In vascular diseases, the mechanisms underlying the development of atherosclerosis are well examined. In atherosclerosis, chronic inflammation in arterial walls plays the pivotal role for the development of diseases (Libby 2002, 2012; Wong et al. 2012). Cholesterol deposition in arterial walls and the engulfment of deposition by inflammatory cells are considered as a trigger of atherosclerosis. Once the inflammation in arterial walls is triggered, the active and long-lasting inflammation occurs through the cytokine production from various types of cells, including not only inflammatory cells such as macrophages but also cells consisting vascular walls such as smooth muscle cells and endothelial cells, and the cross-talk among cells in arterial walls with inflammatory cells. Further, inflammation in atherosclerotic lesion is significantly influenced by hemodynamic status in lesions, known as the 'athero-protective flow' and the 'athero-prone flow' (Gimbrone and Garcia-Cardena 2013). In other vascular disease, intracranial aneurysm, which is a main cause of subarachnoid hemorrhage, is triggered by high wall shear stress loading on the bifurcation of intracranial arteries (Meng et al. 2007; Takeuchi and Karino 2010). In intracranial aneurysm like other chronic inflammatory diseases, once the inflammation in triggered, the active and long-lasting inflammation occurs through the inflammatory cell infiltration and cytokine production basically as like in atherosclerosis (Aoki and Narumiya 2012; Aoki and Nishimura 2010). Inflammation is not only the target of therapeutic drugs but also the good target for the diagnostic marker of diseases to detect the lesions with higher probability of progression such as a vulnerable plaque. For example, in atherosclerosis, various imaging modalities and diagnostic drugs are used or now under establishment for the clinical usage for monitoring the inflammation in atherosclerotic lesion such as ^{18}F in FDG-PET and the super-paramagnetic iron oxide (SPIO) in MRI for detecting P-selectin, phosphatidylserine, or matrix metalloproteinases (McAteer and Choudhury 2013; Quillard and Libby 2012). In intracranial aneurysm, macrophage is a critical cell type for the development of disease (Aoki et al. 2009; Kanematsu et al. 2011). Based on the critical role of macrophage in aneurysm formation, the usefulness of macrophage imaging using super-paramagnetic iron nanoparticles in MRI T2* imaging to monitor the inflammatory response in aneurysm walls and the therapeutic effect of drugs against inflammation have been reported (Hasan et al. 2012a, b, 2013).

References

Angus DC, van der Poll T (2013) Severe sepsis and septic shock. N Engl J Med 369:840–851. doi:10.1056/NEJMra1208623

Aoki T, Narumiya S (2012) Prostaglandins and chronic inflammation. Trends Pharmacol Sci 33:304–311. doi:10.1016/j.tips.2012.02.004

Aoki T, Nishimura M (2010) Targeting chronic inflammation in cerebral aneurysms: focusing on NF-kappaB as a putative target of medical therapy. Expert Opin Ther Targets 14:265–273. doi:10.1517/14728221003586836

Aoki T, Kataoka H, Ishibashi R, Nozaki K, Egashira K, Hashimoto N (2009) Impact of monocyte chemoattractant protein-1 deficiency on cerebral aneurysm formation. Stroke 40:942–951. doi:10.1161/STROKEAHA.108.532556

Atlas SA (2007) The renin-angiotensin aldosterone system: pathophysiological role and pharmacologic inhibition. J Manag Care Pharm 13:9–20

Ben-Neriah Y, Karin M (2011) Inflammation meets cancer, with NF-kappaB as the matchmaker. Nat Immunol 12:715–723. doi:10.1038/ni.2060

Califf RM, Bengtson JR (1994) Cardiogenic shock. N Engl J Med 330:1724–1730. doi:10.1056/NEJM199406163302406

Cruz NG, Sousa LP, Sousa MO, Pietrani NT, Fernandes AP, Gomes KB (2013) The linkage between inflammation and Type 2 diabetes mellitus. Diabetes Res Clin Pract 99:85–92. doi:10.1016/j.diabres.2012.09.003

Demaria S, Pikarsky E, Karin M, Coussens LM, Chen YC, El-Omar EM, Trinchieri G, Dubinett SM, Mao JT, Szabo E, Krieg A, Weiner GJ, Fox BA, Coukos G, Wang E, Abraham RT, Carbone M, Lotze MT (2010) Cancer and inflammation: promise for biologic therapy. J Immunother 33:335–351. doi:10.1097/CJI.0b013e3181d32e74

Duchene J, Lecomte F, Ahmed S, Cayla C, Pesquero J, Bader M, Perretti M, Ahluwalia A (2007) A novel inflammatory pathway involved in leukocyte recruitment: role for the kinin B1 receptor and the chemokine CXCL5. J Immunol 179:4849–4856

Gimbrone MA Jr, Garcia-Cardena G (2013) Vascular endothelium, hemodynamics, and the pathobiology of atherosclerosis. Cardiovasc Pathol 22:9–15. doi:10.1016/j.carpath.2012.06.006

Gukovsky I, Li N, Todoric J, Gukovskaya A, Karin M (2013) Inflammation, autophagy, and obesity: common features in the pathogenesis of pancreatitis and pancreatic cancer. Gastroenterology 144:1199–1209.e4. doi:10.1053/j.gastro.2013.02.007

Hasan D, Chalouhi N, Jabbour P, Dumont AS, Kung DK, Magnotta VA, Young WL, Hashimoto T, Winn HR, Heistad D (2012a) Early change in ferumoxytol-enhanced magnetic resonance imaging signal suggests unstable human cerebral aneurysm: a pilot study. Stroke 43:3258–3265. doi:10.1161/STROKEAHA.112.673400

Hasan DM, Mahaney KB, Magnotta VA, Kung DK, Lawton MT, Hashimoto T, Winn HR, Saloner D, Martin A, Gahramanov S, Dosa E, Neuwelt E, Young WL (2012b) Macrophage imaging within human cerebral aneurysms wall using ferumoxytol-enhanced MRI: a pilot study. Arterioscler Thromb Vasc Biol 32:1032–1038. doi:10.1161/ATVBAHA.111.239871

Hasan DM, Chalouhi N, Jabbour P, Magnotta VA, Kung DK, Young WL (2013) Imaging aspirin effect on macrophages in the wall of human cerebral aneurysms using ferumoxytol-enhanced MRI: preliminary results. J Neuroradiol 40:187–191. doi:10.1016/j.neurad.2012.09.002

Houghton AM (2013) Mechanistic links between COPD and lung cancer. Nat Rev Cancer 13:233–245. doi:10.1038/nrc3477

Kanematsu Y, Kanematsu M, Kurihara C, Tada Y, Tsou TL, van Rooijen N, Lawton MT, Young WL, Liang EI, Nuki Y, Hashimoto T (2011) Critical roles of macrophages in the formation of intracranial aneurysm. Stroke 42:173–178. doi:10.1161/STROKEAHA.110.590976

Kemp CD, Conte JV (2012) The pathophysiology of heart failure. Cardiovasc Pathol 21:365–371. doi:10.1016/j.carpath.2011.11.007

Klabunde RE (2012) Cardiovascular physiology concepts. Lippincott Williams & Wilkins, a Wolters Kluwer business, Baltimore, MD

Libby P (2002) Inflammation in atherosclerosis. Nature 420:868–874. doi:10.1038/nature01323

Libby P (2012) Inflammation in atherosclerosis. Arterioscler Thromb Vasc Biol 32:2045–2051. doi:10.1161/ATVBAHA.108.179705

Libby P (2013) Mechanisms of acute coronary syndromes and their implications for therapy. N Engl J Med 368:2004–2013. doi:10.1056/NEJMra1216063

Lymperopoulos A (2013) Physiology and pharmacology of the cardiovascular adrenergic system. Front Physiol 4:240. doi:10.3389/fphys.2013.00240

McAteer MA, Choudhury RP (2013) Targeted molecular imaging of vascular inflammation in cardiovascular disease using nano- and micro-sized agents. Vasc Pharmacol 58:31–38. doi:10.1016/j.vph.2012.10.005

McLean PG, Ahluwalia A, Perretti M (2000) Association between kinin B(1) receptor expression and leukocyte trafficking across mouse mesenteric postcapillary venules. J Exp Med 192:367–380

Meng H, Wang Z, Hoi Y, Gao L, Metaxa E, Swartz DD, Kolega J (2007) Complex hemodynamics at the apex of an arterial bifurcation induces vascular remodeling resembling cerebral aneurysm initiation. Stroke 38:1924–1931. doi:10.1161/STROKEAHA.106.481234

Mohrman DE, Heller LJ (2010) Cardiovascular physiology. McGrawHill Companies, New York

O'Shea JJ, Holland SM, Staudt LM (2013) JAKs and STATs in immunity, immunodeficiency, and cancer. N Engl J Med 368:161–170. doi:10.1056/NEJMra1202117

Quillard T, Libby P (2012) Molecular imaging of atherosclerosis for improving diagnostic and therapeutic development. Circ Res 111:231–244. doi:10.1161/CIRCRESAHA.112.268144

Schmieder RE (2005) Mechanisms for the clinical benefits of angiotensin II receptor blockers. Am J Hypertens 18:720–730. doi:10.1016/j.amjhyper.2004.11.032

Shimizu I, Yoshida Y, Katsuno T, Minamino T (2013) Adipose tissue inflammation in diabetes and heart failure. Microbes Infect 15:11–17. doi:10.1016/j.micinf.2012.10.012

Sun B, Karin M (2012) Obesity, inflammation, and liver cancer. J Hepatol 56:704–713. doi:10.1016/j.jhep.2011.09.020

Sun B, Karin M (2013) Inflammation and liver tumorigenesis. Front Med 7:242–254. doi:10.1007/s11684-013-0256-4

Tabas I, Glass CK (2013) Anti-inflammatory therapy in chronic disease: challenges and opportunities. Science 339:166–172. doi:10.1126/science.1230720

Takeuchi S, Karino T (2010) Flow patterns and distributions of fluid velocity and wall shear stress in the human internal carotid and middle cerebral arteries. World Neurosurg 73:174–185; discussion e27. doi:10.1016/j.surneu.2009.03.030

Terzic J, Grivennikov S, Karin E, Karin M (2010) Inflammation and colon cancer. Gastroenterology 138:2101–2114.e5. doi:10.1053/j.gastro.2010.01.058

Vincent JL, De Backer D (2013) Circulatory shock. N Engl J Med 369:1726–1734. doi:10.1056/NEJMra1208943

Wong BW, Meredith A, Lin D, McManus BM (2012) The biological role of inflammation in atherosclerosis. Can J Cardiol 28:631–641. doi:10.1016/j.cjca.2012.06.023

Zelis R, Sinoway LI, Leuenberger U, Clemson BS, Davis D (1991) Time-constant adaptations in heart failure. Eur Heart J 12:2–7

Hemodynamics in Physio- and Pathological Vessels

4

Shigeru Tada and John M. Tarbell

Abstract

Hemodynamics can be defined as the part of cardiovascular physiology dealing with the forces that drive the blood circulation in mammalian cardiovascular systems. A cardiovascular system is a series of blood vessels connected to the heart. Pressure generated in the heart propels blood through the system continuously. In this chapter, basic hemodynamics essential to interpretation of arterial disease in the aspect of bio-fluid mechanics are introduced. Hemodynamics in bio-fluid mechanics plays an important role in better understanding of clinical and pathological observations and in developing new methods for diagnosis in connection with mathematical models. In particular, hemodynamic factors, such as Wall Shear Stress and Oscillatory Shear Index, correlate substantially with the generation and progression of arterial disease including intimal thickening and atherosclerosis. In the larger vessels, such as the carotid artery, interaction between the vessel wall and the blood flow affects the distribution of hemodynamic factors.

The main scope of this chapter is to introduce hemodynamic applications of mathematical modeling of fluid mechanics. Mathematical models of fluid mechanics are used to quantify the hemodynamic factors and their relationship

S. Tada (✉)
Department of Applied Physics, National Defense Academy, Yokosuka, Kanagawa, Japan
e-mail: stada@nda.ac.jp

J.M. Tarbell
Department of Biomedical Engineering, The City College of New York/CUNY, New York, NY, USA

K. Tanishita, K. Yamamoto (eds.), *Vascular Engineering*,
DOI 10.1007/978-4-431-54801-0_4

to vascular disease. The majority of all cardiovascular diseases and disorders are related to systemic hemodynamic dysfunction. Recent studies of cardiovascular diseases in relation to hemodynamic dysfunction are also briefly reviewed.

Keywords
Hemodynamics • Blood flow • Interstitial flow • Shear stress

4.1 Introduction

In hemodynamic modeling of the cardiovascular system, we can identify three types of cardiovascular subsystems in which mathematical models should be applied differently:

1. Small blood vessels, which include capillaries and arterioles.

   In small blood vessels, non-Newtonian effects are noticeable due to loss of the applicability of the continuum assumption for the dispersed multiphase system of blood. In such vessels, the effect of blood cell aggregation becomes pronounced and exhibits non-Newtonian rheological behaviors, such as the induction of elastic effects associated with the solid-mechanical properties of the red blood cells (RBCs) and their structural formation. Hence non-Newtonian rheological effects should be considered in modeling, simulating, and analyzing the flow of blood in small vessels. The typical example is blood flow in venules with diameter $> \sim 25$ μm. In this scale, the blood flow can be described by a two-phase continuum model; flow in the core is modeled as non-Newtonian, shear-thinning rheological model, or a model for thixotropic fluid that includes time-dependent effects reflecting aggregations/disaggregation kinetics.
2. Large blood vessels, which mainly apply to arteries and veins.

   In large vessels, blood essentially behaves as a Newtonian fluid. Because the blood in a large vessel is normally exposed to a relatively high shear rate and hence the non-Newtonian effects, which are basically induced at low shear rates, are eliminated. Also at this large scale, the blood appears as a homogeneous continuum medium with minor effect of blood cell aggregation. Thus the Navier-Stokes equation for incompressible viscous fluid flow is applicable. However, in some pathological situations such as arterial stenosis, non-Newtonian effects are important even in large vessels. The typical example is blood flow in artery with diameter $>\sim 4$ μm.
3. Porous tissue in the artery wall.

   With the exception of cells at the internal surface of blood vessels, blood does not come into direct contact with the cells it nourishes. For instance, as blood enters the capillaries servicing a tissue space, a large fraction of the plasma, excepting most blood cells, is filtered into the tissue space. The same situation occurs in large artery walls. This is called interstitial fluid. The interstitial fluid brings to cells all of their metabolic requirements and takes away their waste products. The well-established approach in modeling flow of interstitial fluid in biological tissue is to treat the tissue as a spongy-like porous medium and employ Darcy's law for fluid flow.

4.1.1 Overview of the Chapter

This chapter consists of three sections; Sects. 4.2, 4.3, and 4.4. In the Sect. 4.2, rheological properties of blood, which appear most relevant to slow steady flows in small vessels, are introduced. In the past, the non-Newtonian effects in blood flow originating from the shear rate dependence of the blood viscosity have been extensively studied. A clear correlation has been found between changes in blood viscosity at different shear rates and vascular flow resistance. Thus the rheological properties of steady flow of blood generally have been understood. In this section, basic principles of blood rheology and non-Newtonian effects that alter the characteristics of blood flow are briefly introduced.

In the Sect. 4.3, hemodynamics of large vessels are described. In this subsection, we outline general strategies for modeling blood flow in large arteries. The hemodynamic wall parameters, such as wall shear stress (WSS) and oscillatory shear index (OSI) intended to identify sites where intimal thickening (IT) and thrombosis formation are likely, are also introduced.

Advancing and applying the fundamentals discussed in the early part of Sect. 4.3, the following parts of Sect. 4.3 provide analysis and basic applications of exemplary hemodynamics in the carotid bifurcation. Mathematical models are based on in vivo three-dimensional geometry, pulsatile velocity profiles, and luminal pressure inputs. These data are combined with a linear and isotropic model of the mechanical properties of the carotid artery wall to create fluid-structure interaction (FSI) models of the carotid bifurcation. This model is then used to calculate hemodynamic parameters thought to promote IT and occlusions at the proximal site of the bifurcation.

In the Sect. 4.4, basics of mathematical modeling of interstitial flow in porous media and its application to analysis of the interstitial flow in biological tissues are introduced.

Interstitial fluid flow is the movement of fluid through the extracellular matrix of tissues. This flow provides a necessary mechanism for transporting large proteins through the interstitium and constitutes an important component of the microcirculation. The basic mechanism of mass transfer in artery walls is convection associated with pressure-driven interstitial flow and diffusion caused by concentration gradients. Apart from its role in mass transport, interstitial fluid flow also provides a specific mechanical environment that is important for the physiological activities of interstitial cells. For instance, several in vitro experiments showed that interstitial fluid flow was very important for cell activities and that a flow of μm/s magnitude induced physiological responses from cells.

In the following part of this section, a mathematical model of interstitial fluid flow across an artery wall is introduced. The media of an artery wall is modeled as an array of smooth muscle cells (SMCs) residing in a matrix comprised of proteoglycan and collagen fibers in the extracellular matrix. As a case study, numerical simulations of interstitial flow emphasizing how the architecture of extracellular matrix affects the WSS on cell membranes is presented to show that interstitial fluid flow imposes significant levels of fluid mechanical force on a cell imbedded in a three-dimensional matrix.

Blood flow plays an important role in guiding the physiological activities of endothelial cells (ECs) and SMCs. However, studies of the effect of interstitial fluid flow on interstitial cells are rare. The case studies are presented in an inductive fashion, i.e. building on simple examples to arrive at more general conclusions.

4.2 Rheological Properties of Blood

4.2.1 Hemodynamic Characteristics of Blood

Blood is a multi-phase mixture of cellular elements suspended in plasma, organic molecules, ions, and dissolved oxygen and carbon dioxide. Plasma is blood from which all cellular elements have been removed and is identical in composition to interstitial fluid except for the presence of some proteins.

The viscosity of blood is determined by several factors such as the hematocrit, viscosity of plasma, and the mechanical properties of blood cells. Hematocrit is the volume percentage of RBC in whole blood (Fig. 4.1). The viscosity of blood, in general, can be described as a function of shear rate and is affected by the applied deformation forces. Thus the blood as a whole behaves essentially as a non-Newtonian fluid showing rheological characteristics; deformation rate dependency of viscosity as well as viscoelasticity, shear thinning, yield stress, and thixotropy although the plasma shows a Newtonian fluid property (Fig. 4.2).

The viscoelastic properties of blood originate from microstructures formed by RBCs. The deformability of the RBCs, cell concentration, and the size of confining vessel are determining factors in the microstructure of RBCs. The typical

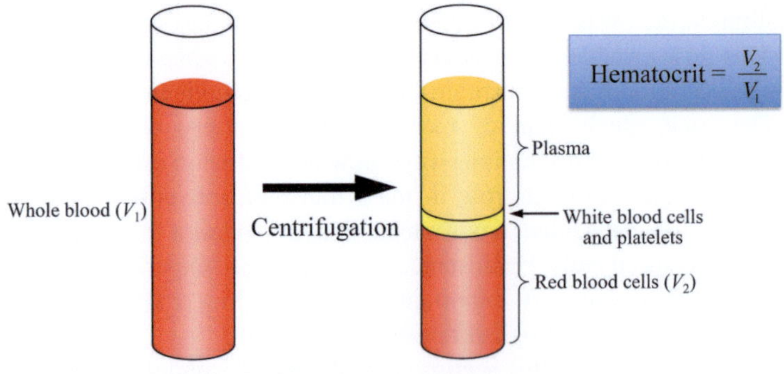

Fig. 4.1 Healthy blood contents and the hematocrit definition

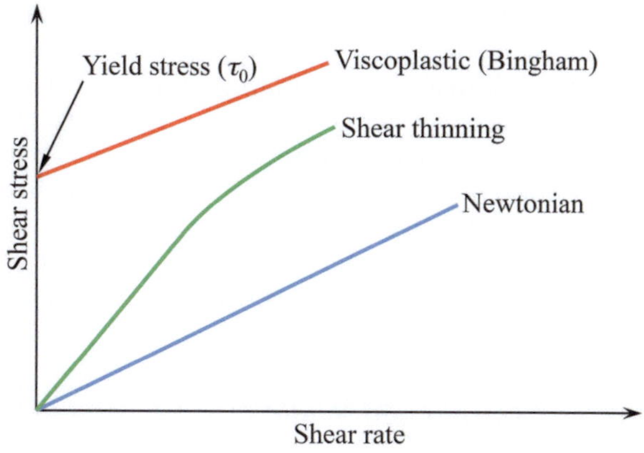

Fig. 4.2 Rheological properties of fluids

microstructure at low shear rate is rouleaux; layers of RBCs adhering face to face like piles of coins to form reversible intercellular contact. Rouleaux formation is highly dependent on the concentration of fibrinogen and globulin in plasma. Note that bovine blood does not form rouleaux because of absence of fibrinogen and globulin in plasma (Fung 1993). The shear thinning nature of blood derives from the tendency of the rouleaux aggregates to disaggregate upon the application of shear.

The source of the yield stress is also the presence of RBCs. Yield stress arises from the aggregation of RBCs at low shear rates to form the microstructures. Under physiological conditions, the RBCs in static or blood flow in a low share rate are likely to adhere to each other to form rouleaux. Therefore, in order to initiate a flow from rest, one needs to have a force large enough to break up the cohesive forces acting among the cells. Previous studies have indicated that yield stress is positively correlated to the concentration of fibrinogen protein in blood plasma and to the hematocrit level (Morris et al. 1989). Yield stress contributes to the blood clotting following injuries and subsequent healing and may also contribute to the formation of thrombi leading to occlusion in some pathological cases such as strokes. Abnormal RBC aggregation has been found to be associated with several diseases and conditions which include diabetics, cardiovascular malfunction, hypertension, and hematological disorders (Ramakrishnan et al. 1999).

The phenomenon of thixotropy results from the microstructure of blood as well. Thixotropy can be explained as a consequence of aggregation of suspended blood cells. If the suspension is at rest, the cell aggregation can form, whereas if the shear stress is added to suspension, the weak physical bonds among cells are broken, and the network among them breaks down into separate aggregates, which can disintegrate further into smaller fragments.

After some time at a given shear rate, a dynamic equilibrium is established between aggregate decomposition and growth, and at higher shear rates, the equilibrium is shifted in the direction of greater dispersion. The relatively long time required for the microstructure to stabilize following a rapid change in the rate of flow makes blood thixotropy.

When blood flows in larger vessels at high shear rates, it behaves like a Newtonian fluid as all of the microstructures are disaggregated, whereas when it flows through small vessels at low shear rates, it shows non-Newtonian behavior. The blood rheology appears most relevant to blood flows with relatively low flow velocities. In this situation, there is cell-free plasma layer adjacent to the vessel wall and core layer of suspension of RBCs. Therefore the blood can't be treated as a single non-Newtonian fluid. Hence, it is appropriate to model blood as a two-fluid model; treating the suspension of RBCs in the core region as a non-Newtonian fluid and the cell-free plasma in the peripheral layer region as Newtonian fluid.

4.2.2 Casson Fluid Model

Among numerous mathematical models of blood rheology proposed in the past, the Casson model has been widely used as the non-Newtonian fluid model to represent the suspension of RBCs in the core region of blood flow in small vessels since the Casson fluid model can predict satisfactorily the flow behavior of a non-Newtonian fluid with yield stress using only a single model parameter. The Casson fluid model could be the best description of blood when it flows through narrow arteries at low shear rates and it can be applied to human blood over a wide range of hematocrits and shear rates.

The mathematical expression of the Casson fluid model is as follows;

$$\sqrt{\tau_{ij}} = \begin{cases} 0 & \text{for } \tau \leq \tau_0 \\ \sqrt{\tau_0} + \sqrt{\mu_0}\sqrt{\dot{\gamma}_{ij}} & \text{for } \tau > \tau_0 \end{cases} \tag{4.1}$$

where τ_{ij} is the deviatoric stress tensor (stress tensor of force field per unit area as a result of the resistance to the rate of deformation of fluid element), τ_0 is the yield stress, μ_0 is the viscosity of the Casson fluid above the yield stress, and $\dot{\gamma}_{ij}$ is shear rate.

As a simple application of the Casson fluid model for the hemodynamic analysis of blood flow, let us consider steady blood flow through a small straight vessel.

As shown in Fig. 4.3, the volumetric flow rate of a Casson fluid, Q, through a small straight vessel of radius R and length L_v is given as

$$Q = \frac{\pi R^4}{8\mu_0}\left(\frac{\Delta p}{L_v}\right)\left(1 - \frac{16}{7}\sqrt{\kappa} + \frac{4}{3}\kappa - \frac{1}{21}\kappa^4\right) \tag{4.2}$$

where Δp is the pressure difference across the whole length of the vessel and κ is the non-dimensional yield stress described as

$$\kappa = \frac{2L_v\tau_0}{\Delta pR} \tag{4.3}$$

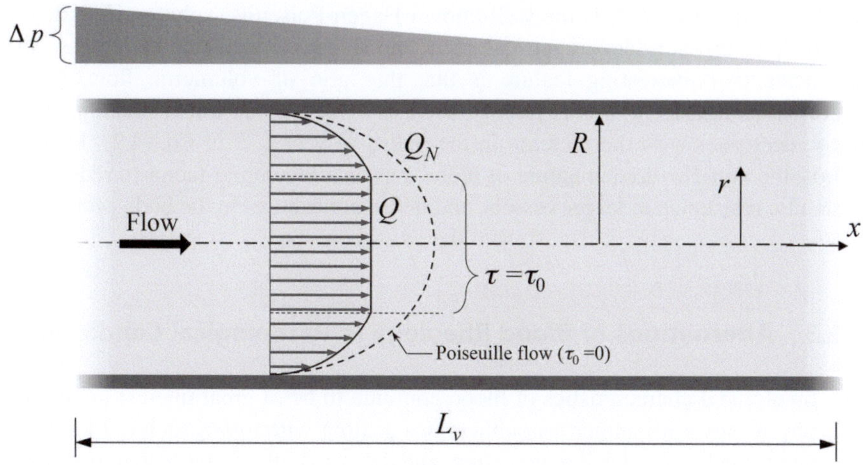

Fig. 4.3 Steady flow of the Casson fluid in a straight circular tube of radius R and length L_v

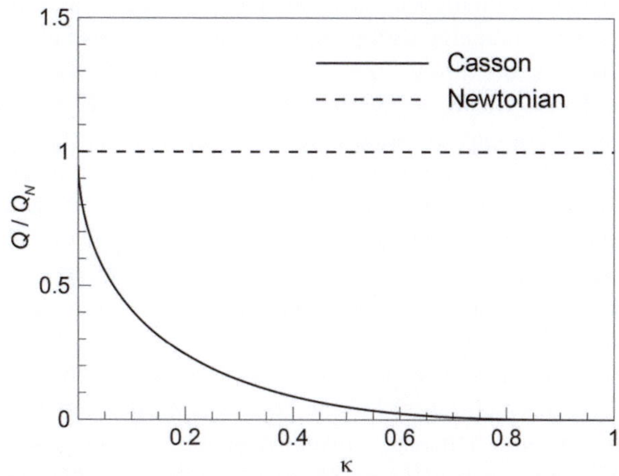

Fig. 4.4 Relationship between the volumetric flow rate of the Casson fluid, Q/Q_N, and the non-dimensional yield stress, κ. The volumetric flow of the Casson fluid is normalized to that of the Newtonian fluid, Q_N

In this model, when κ approaches 0 this model approaches the Newtonian fluid model. The relationship between the volumetric flow rate of the Casson fluid and the non-dimensional yield stress is shown in Fig. 4.4. In the figure, the volumetric flow of the Casson fluid is normalized to that of the Newtonian fluid

$$Q_N = \frac{\pi R^4}{8\mu_0}\left(\frac{\Delta p}{L_v}\right) \tag{4.4}$$

as Q/Q_N. Equation (4.4) is the well-known Hagen-Poiseuille solution. For steady blood flow, the non-Newtonian nature of blood acts as a factor to increase the resistance. An interesting feature is that the ratio of volumetric flow, Q/Q_N, decreases as the size of vessel radius increases since the non-dimensional parameter, κ, decreases with the increase in the radius of vessel, R in Eq. (4.2). In other words, the non-Newtonian nature of blood acts as a regulating factor to reduce the hydraulic resistance in larger vessels, and hence contributes to the body protection.

4.2.3 Alternations of Blood Rheology in Pathological Conditions

The rheological characteristics of blood continue to be of great interest in circulatory physiology with numerous publications dealing with topics such as blood and plasma viscosity, blood cell aggregation, and deformability. In particular, increased blood viscosity, impaired RBC deformability, and increased RBC aggregation are reported in a variety of cardiovascular diseases (Chien et al. 1987). It has been shown that in patients with myocardial ischemia and diabetes mellitus, disorders of blood rheology and endothelial dysfunction develop (Shul'man et al. 2006).

Ischemic diseases of various organs are known to be associated with the impairment of blood rheology (Kesmarky et al. 1998). Hypertension is also characterized by alterations of blood rheology. Hypertension is a complex pathphysiological process (Ajmani 1997). Especially the advanced forms of hypertension are associated with vascular damage, and this damage is claimed to be the cause of alterations of blood rheology. Recent evidence suggests that altered RBC rheological properties might be the underlying cause of some types of hypertension. RBC aggregation was found to be increased significantly in the hypertensive rats (Bor-Kucukkatay et al. 2000).

Diabetes mellitus is a disease process that is accompanied by microcirculatory disturbances. Diabetes mellitus is characterized by abnormal carbohydrate metabolism arising from insulin deficiency. Insulin is the key hormone in blood glucose homeostasis. The consequent elevation of glucose in the blood plasma affects RBCs and vascular ECs, including the walls of capillaries. Rheological changes such as augmented aggregation of RBC, decreased deformability of RBC, and increased viscosity of blood and plasma lead to disturbances of microcirculation in the terminal vascular system. A number of studies have reported increased blood and plasma viscosity, enhanced RBC aggregation, and altered RBC deformability in diabetes mellitus (McMillan 1993).

4.3 Hemodynamics of Large Vessels

4.3.1 Characteristics of Blood Flow in Large Vessels

In large vessels, the blood essentially behaves as a Newtonian fluid. The blood in large vessels is normally exposed to a relatively high shear rates, and therefore the non-Newtonian effects exhibited at low shear play a minor role. Also at this scale the blood appears as a homogeneous continuum medium with diminishing effect of blood cell aggregation because the vessel diameters are large compared with the individual cell diameters. However, in some pathological situations such as the severe stenosis where the rapid change in the shear magnitude occurs, non-Newtonian effects are important even in the large vessels and therefore they should be taken into consideration in the hemodynamic analysis.

In previous studies of hemodynamics in large vessels, the non-Newtonian phenomena in the blood have not been given sufficient attention for the reasons mentioned above. In the analysis of hemodynamic in large vessels, blood is generally assumed Newtonian, and non-Newtonian effects are ignored for simplification of the flow models and their computational implementation. When the shear rate is high, the blood viscosity is independent of shear rate. With a reduction of shear rate, the blood viscosity increases moderately until a shear rate less than ~100 1/s, where it rises extremely steeply. This behavior of blood viscosity can be well predicted by the use of the Casson fluid model. For instance, from the Casson fluid model Eq. (4.1) and general form of the constitutive equation of steady non-Newtonian fluid flow

$$\tau_{ij} = \mu(\dot{\gamma})\dot{\gamma} \tag{4.5}$$

one can obtain the following expression for the Casson viscosity as a function of the shear rate (Kleinstreuer 2006);

$$\mu(\dot{\gamma}) = \frac{1}{\dot{\gamma}}\left[C_1(Ht) + C_2(Ht)\sqrt{\dot{\gamma}}\right]^2 \tag{4.6}$$

where Ht is the hematocrit. The coefficients $C_1(Ht)$ and $C_2(Ht)$ in Eq. (4.6) were determined for $Ht = 40\%$ as $C_1(40) = 0.2$ $(\text{dyn/cm}^2)^{1/2}$ and $C_2(40) = 0.18$ $(\text{dyn/cm}^2)^{1/2}$, based on Merrill's experimental data (1969). Figure 4.5 shows the relationship between the shear rate and apparent viscosity of the Casson fluid exposed to simple shear flow. The graph implies that the Casson viscosity, Eq. (4.6), approaches the Newtonian viscosity as the shear rate increases. The shear rate range for which non-Newtonian effects are considered significant is lower than ~100 1/s. The blood is generally treated as a Newtonian fluid above this limit, which is realized at the blood flow ranges in large and medium size vessels with the diameter of about >1000 μm. In other words, Casson fluid model is adequate for the representation of the simple shear behavior of blood in vessels with the diameter of 130–1000 μm.

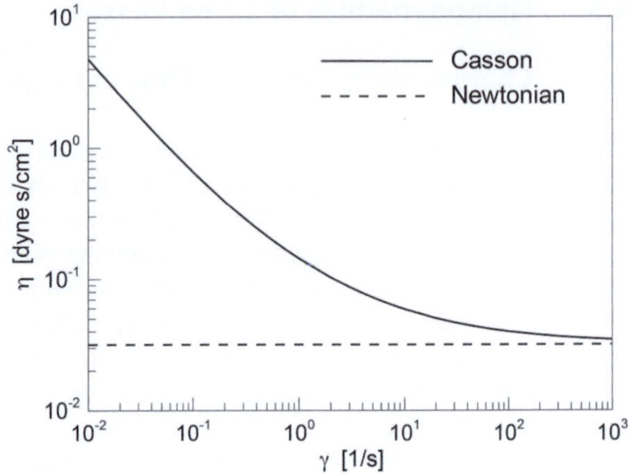

Fig. 4.5 The relationship between the shear rate, γ, and the apparent viscosity of the Casson fluid, η, exposed to a simple shear flow

In the previous sections, deformability of vessel walls has not been mentioned. However, it has been well known that vessel walls undergo mild distention with respect to the original dimensions. The rigid-wall approximation for the artery wall is justified by the observation that, under normal conditions, wall deformability does not significantly alter the velocity field. This approximation may not be valid for large and medium size vessels where wall deformations are not negligible. In such situations, large deformation theory can be applied to model artery wall deformation. It is reasonable to take the effects of wall distension on the flow field in describing hemodynamics of large and medium size vessels.

4.3.2 Mathematical Formulation of Blood Flow in Large Vessels

4.3.2.1 Blood Flow Modeling

The approximation of blood as a Newtonian fluid is an acceptable assumption and not far from reality for many circumstances such as the flow in large and medium size vessels at medium and high shear rates under non-pathological conditions. Consequently blood can be considered as a homogeneous, incompressible Newtonian fluid. Under these assumptions, the hemodynamics problem is expressed by the conservation of the two physical properties of mass and momentum, giving rise to the following set of equations:

$$\begin{cases} \dfrac{\partial u_j}{\partial x_j} = 0 \\ \dfrac{\partial u_i}{\partial t} + u_j \dfrac{\partial u_i}{\partial x_j} = -\dfrac{1}{\rho}\dfrac{\partial p}{\partial x_i} + \dfrac{\partial \tau_{ij}}{\partial x_j} \end{cases} \tag{4.7}$$

where the first equation describes the law of mass conservation and the second equation describes the law of momentum conservation. In Eq. (4.7), u_i represents velocity, while p and τ_{ij} are pressure and viscous stress tensors respectively, which are the isotropic and deviatoric parts of the stress tensor σ_{ij}:

$$\sigma_{ij} = -p\delta_{ij} + \tau_{ij} \tag{4.8}$$

For an incompressible Newtonian fluid, the constitutive relation between τ_{ij} and the rate of strain tensor e_{ij} is described as

$$\tau_{ij} = 2\mu e_{ij} \tag{4.9}$$

where μ is viscosity of a Newtonian fluid. Substituting Eq. (4.9) into the momentum conservation equation yields the Navier-Stokes equation:

$$\frac{\partial u_i}{\partial t} + u_j \frac{\partial u_i}{\partial x_j} = -\frac{1}{\rho}\frac{\partial p}{\partial x_i} + \nu \frac{\partial u_i}{\partial x_j \partial x_j} \tag{4.10}$$

4.3.2.2 Wall Motion Modeling

In the physiological range of blood pressure, the stress-strain relationship of the artery wall may be considered as linear during the pressure pulse. Large deformation theory is used to model artery wall deformation. For simple approximation, the artery can be modeled as a thick-walled linearly elastic circular cylindrical tube consisting of a homogeneous material. The equations governing the movement of the elastic wall are the momentum equations, the equilibrium conditions, and the constitutive equations, respectively:

$$\rho \frac{\partial^2}{\partial t^2} d_i = \frac{\partial}{\partial x_j}\sigma_{ij} + \rho f_i \tag{4.11}$$

$$\sigma_{ij} n_j = {}^s t_i \tag{4.12}$$

$$\sigma_{ij} = D_{ijkl}\,\varepsilon_{kl} \tag{4.13}$$

where $\partial^2 d_i/\partial t^2$ represents the acceleration of a material point (where displacement is defined as $d_i = x_i - x_{i0}$, and x_{i0} is the stress-free position) at time t, n_j is the outward pointing normal vector on the structural boundary at time t, ${}^s t_i$ is the surface traction vector at time t, D_{ijkl} is the material elasticity tensor, and ε_{ij} is the infinitesimal strain tensor.

4.3.3 Hemodynamic Wall Parameters

It has been well known that EC injury or dysfunction and the interaction of blood cells with the vascular surface play significant roles in the evolution of IT and atherosclerosis development. Hemodynamic wall parameters are intended to identify sites where IT and thrombosis formation are likely based on hemodynamic wall interactions. In other words, identifying sites of EC dysfunction has been a primary purpose of the wall parameters of WSS-based hemodynamics. In the following, some basic wall parameters are introduced.

4.3.3.1 Wall Shear Stress (WSS)

The shear stress is a tensor within the flow field and reduces to a vector on a solid surface:

$$\tau_w = n \bullet \tau \tag{4.14}$$

where n is the local surface normal vector. The magnitude of the time average WSS vector can be written as

$$\mathrm{WSS} = \overline{\tau}_w = \left| \frac{1}{T} \int_0^T \tau_w \, dt \right| \tag{4.15}$$

Alternatively, the time average of the WSS magnitude can be written

$$|\overline{\tau}_w| = \frac{1}{T} \int_0^T |\tau_w| \, dt \tag{4.16}$$

4.3.3.2 Oscillatory Shear Index (OSI)

Cyclic departure of the WSS vector from its predominant axial alignment indicates flow disruption over time and is known as the oscillatory shear index, or OSI (Ku et al. 1985). The OSI, therefore, quantifies disturbed flow interaction with the wall and is formulated as

$$\mathrm{OSI} = \frac{1}{2}\left(1 - \frac{\overline{\tau}_w}{|\overline{\tau}_w|}\right) \tag{4.17}$$

where $\overline{\tau}_w$ is the instantaneous WSS vector. The OSI can vary from 0 to 0.5, indicating the least and most severe temporal shear rate conditions, respectively.

4.3.3.3 Stress Phase Angle (SPA)

The WSS of flowing blood and the circumferential strain (CS) induced by hoop stresses that balance blood pressure are imposed on ECs. These mechanical forces are known to influence gene expression and protein and metabolite secretion of EC and are believed to play a role in the localization of atherosclerosis in regions of curvature and branching in arteries (Chien 2003). Previous studies have provided evidence that the temporal phase angle between CS and WSS, the stress phase angle (SPA), is most negative (mechanical forces are most out-of-phase) in precisely those regions where atherosclerotic plaques are localized (Thomas et al. 2002).

4.3.4 Hemodynamic Analysis of Flow in Carotid Bifurcation (Tada and Tarbell 2005)

From the physiological point of view, blood is a shear-rate dependent fluid flowing in pulsatile fashion through non-rigid bifurcating tubes with variable cross sections. While pulsatile flow of an incompressible Newtonian fluid in a bifurcation with a slightly elastic tube is a reasonable starting point when modeling the hemodynamics in actual arterial trees. A carotid bifurcation, as considered here, is the primary application site for most researchers. In modeling the hemodynamics of the carotid bifurcation, inlet and boundary conditions, in addition to the no-slip wall condition, were selected in a manner to: (a) match physiological conditions as closely as possible with available data and (b) facilitate numerical simulation.

4.3.4.1 Model Geometry, the Solution Method, and the Computational Results

The model geometry is shown in Fig. 4.6, and representative dimensions are listed in Table 4.1. For the inlet boundary conditions, a uniform pressure varying with time was applied. The waveform of the common carotid artery (CCA) wall distention was assumed to be that of the inlet pressure. Values of 110 and 76 mmHg were used for the peak systolic and minimum of diastolic pressures, respectively. For the outlet boundary conditions of the internal carotid artery (ICA) and external carotid artery (ECA), waveforms of flow obtained from phase-contrast MRI velocity measurements were used. A fully developed parabolic velocity profile was applied at the downstream end of both the ICA and ECA. For the modeling of elastic wall, an incompressible (Poisson's ratio = 0.5) linear elastic model having a Young's modulus of 5.0×10^6 dyn/cm^2 was applied. Moreover, rigid and fixed cylindrical segments of length 2.0a (a: the internal diameter of CCA) were applied to support the elastic bifurcation tube at the upstream end of the CCA and the downstream end of both the ICA and ECA. The artery wall was assumed to be a solid wall impermeable to blood plasma. The blood was assumed to be Newtonian with a density of 1.05 g/cm^3 and a viscosity of 3.5×10^{-2} dyn s/cm^2.

Fluid and solid equations Eqs. (4.10, 4.11, 4.12, and 4.13) together with coupling conditions were solved by the Arbitrary Lagrangian Eulerian (ALE) based loose coupling algorithm using FIDAP 8.72 (FLUENT) software. A gradually finer mesh toward the solid surface was taken in the grid system of the fluid phase to ensure enough spatial resolution in the boundary layer of the flow field to evaluate WSS accurately as shown in Fig. 4.6. The element numbers for the fluid and wall models were 60,000 and 36,000, respectively. The governing equations for both the fluid and solid were transformed into algebraic equations using the finite element method. The resulting algebraic equations were solved by the direct Gaussian elimination method.

Figure 4.7 shows the instantaneous streamlines of flow colored according to the flow velocity at the time of peak systole (a) and end diastole (b). The times of peak systole and end diastole are presented as dotted lines in the inset graphs showing the flow rates vs. time (t) for one cardiac time period (T). The starting points (seeds) of the streamlines in the cross-section of the CCA entrance were given as shown in the right-hand-side of panel (b). The cross-sectional velocity profiles at the CCA

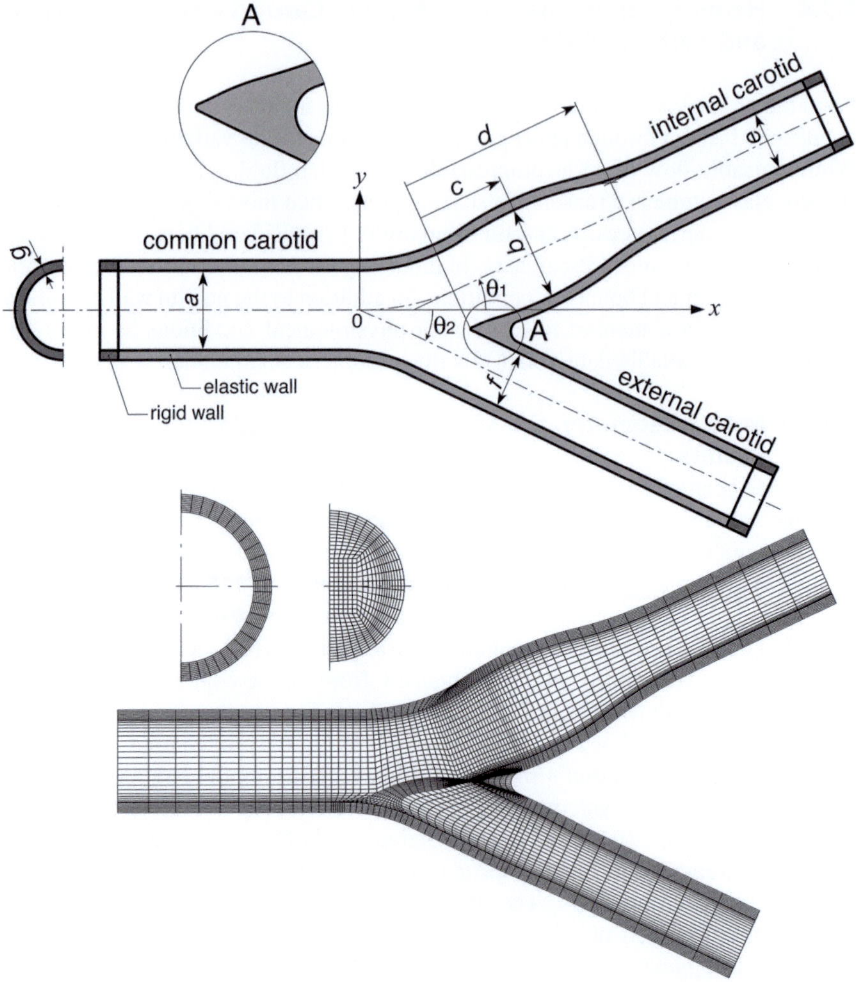

Fig. 4.6 (*Top*) Three-dimensional model geometry of the carotid bifurcation. (*Bottom*) The structured grid system consists of a grid system for the fluid phase and a grid system for the solid phase (Reproduced from Tada and Tarbell 2005 with permission from Springer)

Table 4.1 Representative values of bifurcation geometry

Location in Fig. 4.6	Labels	$\times a$ ($a = 0.6$ cm)
CCA internal diameter	a	1.000
Max. sinus internal diameter	b	1.110
	c	0.910
	d	2.300
ICA internal diameter	e	0.740
ECA internal diameter	f	0.650
Thickness of artery wall	g	0.117
ICA bifurcation angle	θ_1	25.0 °
ECA bifurcation angle	θ_2	25.0 °

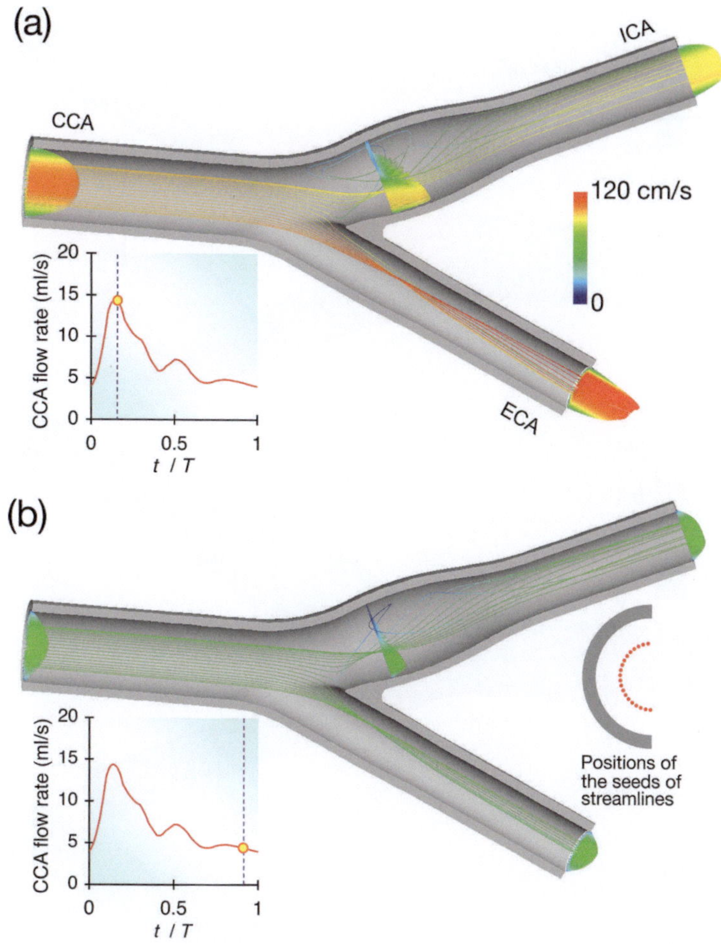

Fig. 4.7 Instantaneous streamlines colored according to the flow velocity are shown at the times of (**a**) peak systole and (**b**) end diastole. The starting points of the streamlines over the CCA entrance cross-section are distributed as presented in the right-hand side of panel (**b**). The velocity profiles at the CCA entrance, at the ICA and ECA exits, and in the middle of the ICA sinus are also depicted

entrance, at the exits of both ICA and ECA, and at the middle of the ICA sinus, are also shown. The flow pattern in the ICA sinus at the time of peak systole is complex because of flow separation, reversal, and swirling caused by the strong secondary flow. The flow divider generates vortices along the inside wall of the ICA sinus, although flow reversal is generated along the outside wall at the sinus. The size and position of the recirculation zone change only slightly during a single cardiac cycle. However, the flow velocity in the ICA sinus changes considerably with time. The flow velocity near the wall in the recirculation zone is very low for most of the time period of one cardiac cycle, except at the time of peak systole as shown in (b)

In Fig. 4.8, the distribution of the mean WSS, OSI, and SPA over the internal surface of the wall were displayed for comparison. The SPA is the phase angle

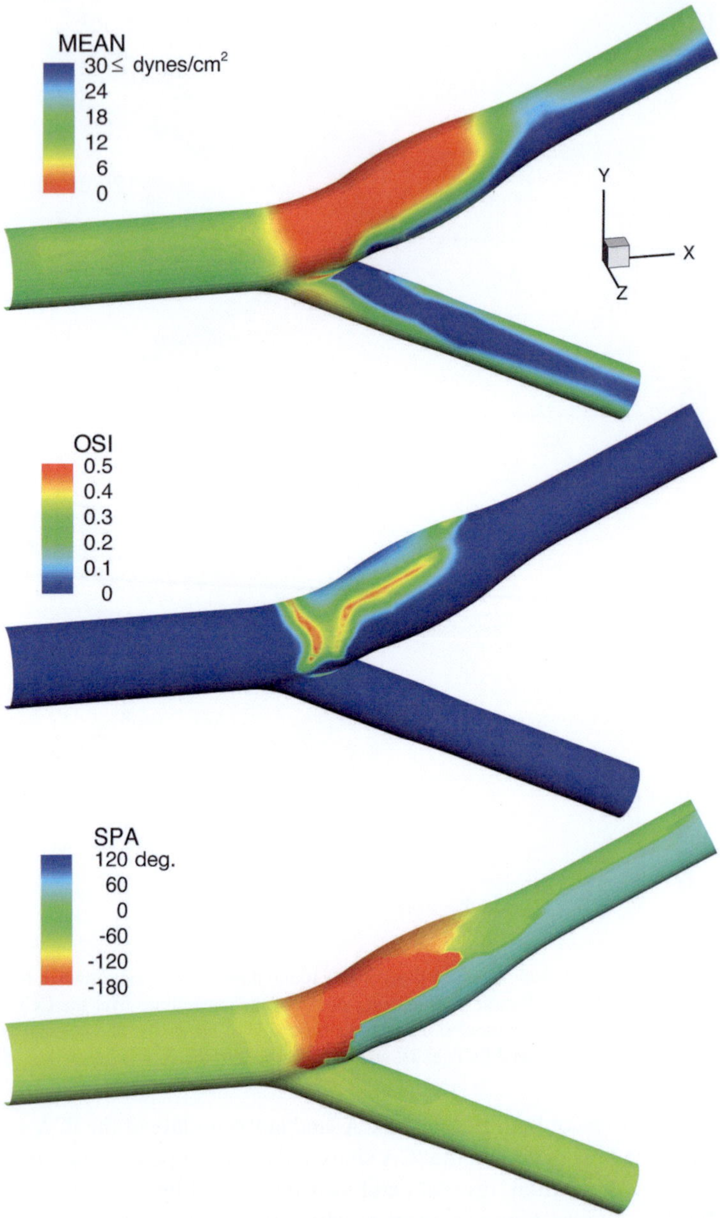

Fig. 4.8 Color contour-plots of time-average wall shear stress (mean WSS [dyn/cm^2]), oscillatory shear index (OSI, dimensionless), and stress phase angle of the first harmonic (SPA [deg.]) over the cardiac cycle (Reproduced from Tada and Tarbell 2005 with permission from Springer)

between the first harmonic components of the CS and WSS waves. The zone of high negative SPA ($< -90°$) at the sinus (bottom) is well correlated with the low mean WSS (top) and high OSI (middle) domains. The low WSS region appears at the ICA outer wall in the sinus region and is a common feature of the WSS distribution in the carotid sinus regardless of the detailed geometric features of the bifurcation and flow waveforms.

4.3.5 Arterial Disease Formation and Hemodynamic Factors

One of the major consequences of carotid artery disease is stenosis. Carotid artery stenosis occurs when the carotid arteries become narrowed. The narrowing of the carotid arteries is most commonly related to atherosclerosis. Atherosclerosis is generally characterized by the accumulation of fatty deposits right beneath the innermost endothelial linings: leading to complex cellular reactions and the formation of IT (Kohler et al. 1992). IT forms non-randomly at locations in bends and bifurcations, where the local flow can be classified as 'disturbed flow' and is associated with low WSS and oscillatory or reciprocating flow. Numerous in vitro studies have indicated that low WSS may contribute to IT and development of atherosclerosis.

If the disease process progresses, plaque formation may take place. Plaque is made up of SMCs, monocyte macrophages, foam cells, and cholesterol. The growth of plaque narrows the luminal space of arteries and can decrease blood flow or completely block the flow of blood to the brain. The interested reader is referred to Tarbell et al. (2014) for the details of the recent advances in pathological hemodynamics.

4.4 Interstitial Flow in Arterial Tissues

Interstitial flow plays important roles in the pathogenesis in arterial tissues. To investigate these roles, a comprehensive consideration of the flow in the interstitial space and how it affects such processes is critical.

As shown in Fig. 4.9, a typical mammalian artery consists of the following four structurally distinct regions: endothelium, intima, media, and adventitia. The endothelium is the innermost thin layer of arterial tissue consists of a single layer of ECs. The endothelium not only provides a structural barrier between the circulation and surrounding tissue, but ECs also secrete mediators that influence vascular hemodynamics in the physiologic state, such as nitric oxide (NO), prostacyclin (PGI_2), platelet-activating factor (PAF), and endothelin-1 (ET-1). The intima is a thin layer of basement matrix, separated from the media by the internal elastic lamina (IEL), which is a fenestrated layer of elastic tissue. The media is viewed as a porous heterogeneous medium consisting of a continuous extracellular matrix phase with embedded SMCs. The adventitia is the outermost layer of the artery wall typically

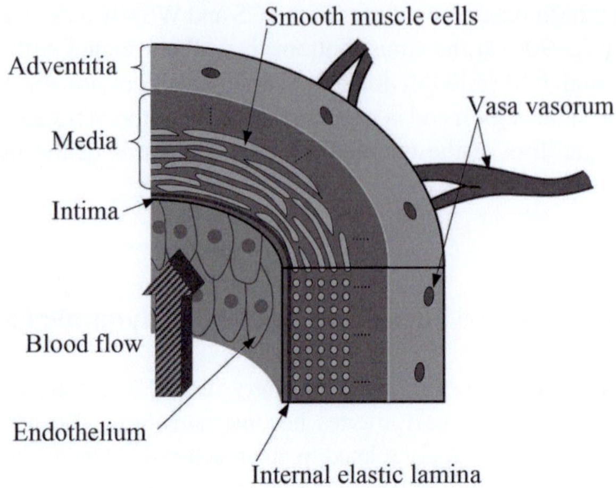

Fig. 4.9 Transverse sectional view of a blood vessel wall. The bulk of the blood vessel wall consists of the media and adventitia. There are some capillaries (vasa vasorum) in adventitia. The innermost layer of vessel wall (intima) is a single layer of endothelial cells (ECs) and a thin layer of basement matrix. The internal elastic lamina (IEL), an impermeable barrier to water flow except for fenestral pores, separates the intima from media. Between smooth muscle cells (SMCs) in the media is an interstitial phase containing proteoglycan and collagen fibers (Reproduced from Tada and Ozono 2011 with permission from Springer)

comprised of loose connective tissue and fibroblasts. There are some capillaries (e.g., vasa vasorum) in the adventitia.

Interstitial fluid that has gained entry to the intima by crossing the endothelium passes through the fenestrae in the IEL and enters the media toward the adventitia and embedded SMCs. It is well known that WSS associated with blood flow can affect EC biology and may play an important role in arterial disease. For instance, WSS on ECs can stimulate the release of Nitrogen Oxide (NO) that causes smooth muscle relaxation. This represents an indirect effect of flow on SMCs. SMCs are also exposed directly to interstitial flow.

Here we introduce a mathematical model of interstitial flow in arterial tissue that describes the hemodynamics of the interstitial flow in arterial tissues. The importance of hemodynamic factors on cell biology will be discussed in the following sections.

4.4.1 Mathematical Formulation of the Flow in Porous Media

Most of arterial tissue, e.g. intima and media, are biphasic materials, comprised mainly of water enclosed within an elastic matrix of elastin and collagen fibers. These layers can be approximated as isotropic porous media layers with different

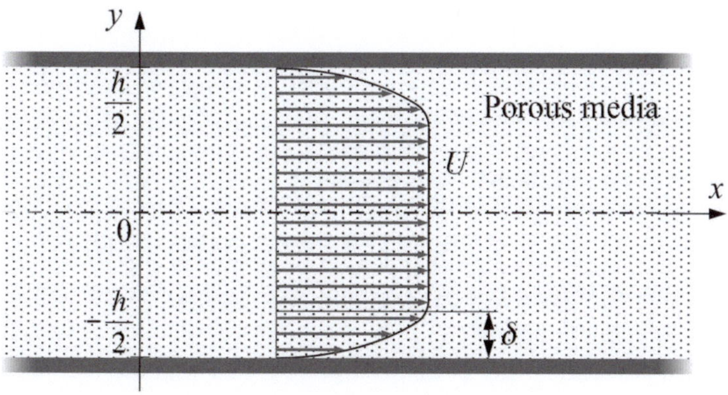

Fig. 4.10 A Newtonian fluid flow in a parallel-plate channel filled with a homogeneous porous medium. The channel with the separation distance h was aligned with the x-axis. The flow was assumed to be steady and axisymmetric about the center axis of the channel

values of permeability to fluid. The flow of interstitial fluid through a porous media is described by Darcy's Law:

$$\nabla P = -\frac{\mu}{K_p}U \tag{4.18}$$

where P is the pressure, U is the averaged (macroscopic) velocity, μ is the viscosity of interstitial fluid, K_p is the hydraulic permeability of the porous media, which has dimension of (length)^2 and depends on the geometry of the porous media. Although this model has been utilized successfully in several biomedical applications such as blood perfusion through soft tissues, it is not valid when the boundaries of the porous media should be taken into account. Because Darcy's model ignores the boundary effects on the flow, the Brinkman model is usually employed:

$$\nabla P = -\frac{\mu}{K_p}U + \mu\nabla^2 U \tag{4.19}$$

The second term on the right-hand side of Eq. (4.19) represents the viscous force which is required to satisfy a no-slip boundary condition on the solid surface.

The most general case in which the viscous force plays an important role is for flow over a plane solid surface as illustrated in Fig. 4.10. The equation governing a fully developed flow in a two-dimensional porous channel bounded by planar solid walls as shown in Fig. 4.10 is given by:

$$\frac{\partial P}{\partial x} = -\frac{\mu}{K_p}U + \mu\frac{\partial^2 U}{\partial y^2} \tag{4.20}$$

with the boundary conditions:

$$U|_{y=-h/2} = 0, \qquad \left.\frac{\partial U}{\partial y}\right|_{y=0} = 0 \tag{4.21}$$

The solution of Eqs. (4.20 and 4.21) becomes:

$$U = \frac{K_p}{\mu}\left(-\frac{\partial P}{\partial x}\right)\left[1 - \frac{cos\ h\frac{y}{\sqrt{K_p}}}{cos\ h\frac{h}{2\sqrt{K_p}}}\right] \tag{4.22}$$

The velocity distribution, Eq. (4.22), can be rewritten with the use of the mean interstitial flow velocity:

$$U_m = \frac{1}{d}\int_{-2/h}^{2/h} U dy \tag{4.23}$$

By introducing non-dimensional velocity, non-dimensional channel width, and a non-dimensional parameter, the Darcy number:

$$U^* = \frac{U}{U_m}, \qquad y^* = \frac{y}{h/2}, \qquad Da = \frac{2K_p}{h^2} \tag{4.24}$$

the non-dimensional form of the flow velocity:

$$U^* = \frac{1}{1 - \sqrt{Da}\ \tan h\frac{1}{\sqrt{Da}}}\left(\frac{1 - \cos h\frac{y^*}{\sqrt{Da}}}{\cos h\frac{1}{\sqrt{Da}}}\right) \tag{4.25}$$

can be obtained.

The velocity distribution across the channel width is shown as a function of Darcy number, Da, in Fig. 4.11. The shape of the flow distribution becomes parabolic for greater Da, namely the Hagen-Poiseuille solution:

$$\lim_{Da\to\infty} U^* = \frac{3}{2}\left(1 - y^{*^2}\right), \tag{4.26}$$

while it becomes rectangular with very thin boundary layer at the wall surface for smaller value of Da:

$$lim_{Da\to 0} U^* = 1. \tag{4.27}$$

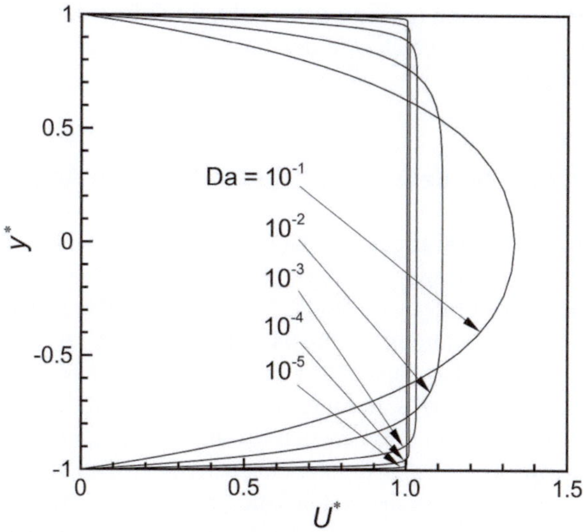

Fig. 4.11 Variations of velocity distribution of steady flow in porous channel with the Darcy number, Da. Values are normalized to the average flow velocity

The thickness of boundary layer of the interstitial flow, δ, can be roughly estimated by the dimensional analysis of the governing Eq. (4.20):

$$\frac{\partial P}{\partial x} = -\underbrace{\frac{\mu}{K_p}U}_{1/K_p} + \underbrace{\mu\frac{\partial^2 U}{\partial y^2}}_{1/\delta^2} \tag{4.28}$$

The first term in the right-hand-side of Eq. (4.28) has the order of magnitude, $\mu U/K_p$, while the second term has order of magnitude $\mu U/\delta^2$. These two terms should have comparable orders of magnitude. Hence, the order of the boundary layer thickness, δ, can be estimated as a function of the hydraulic permeability of porous media, K_p:

$$\delta \sim \sqrt{K_p} \tag{4.29}$$

Therefore, the order of magnitude of WSS at the wall surface can be estimated in a similar way:

$$\tau_w \sim \mu \frac{U}{\sqrt{K_p}} \tag{4.30}$$

The WSS of the interstitial flow can take on extremely large values even though the interstitial flow velocity is very low. For instance, the viscosity of interstitial fluid (water) is on the order of 10^{-3} Pa s, the hydraulic permeability of the tissue is on the order of 10^{-14} cm^2, the velocity of the interstitial flow is on the order of 10^{-6} cm/s.

By accounting for these values, the WSS, Eq. (4.30), is found to be on the order of 0.1 dyn/cm^2, which is in the range known to affect the biology of ECs exposed to blood flow.

4.4.2 Modeling of Interstitial Flow in Artery Wall

4.4.2.1 Analytical Solution for the Flow in the Media

In the following, a mathematical model of the interstitial flow in the media is introduced for a case study. The present model is certainly an idealization of the true architecture of the media of the artery wall (Clark and Glagov 1985).

The media, depicted in Fig. 4.12, is modeled as a periodic square array of cylindrical SMCs residing in a matrix comprised of proteoglycan and collagen fibers. The cells are treated as impermeable obstacles to flow due to the cell membrane's low hydraulic conductance, which is typically two orders of magnitude lower than the conductance of the interstitium (Levick 1987). The flow is assumed to be perpendicular to the longitudinal axis of the cylindrical SMCs. Namely the corresponding flow field can be treated as a two-dimensional flow problem.

The non-dimensional governing equations for the flow field are the Brinkman equation and the law of mass conservation:

$$\begin{aligned} \nabla p &= -\boldsymbol{u} + \frac{1}{\alpha^2}\nabla^2 \boldsymbol{u} \\ \nabla \bullet \boldsymbol{u} &= 0 \end{aligned} \tag{4.31}$$

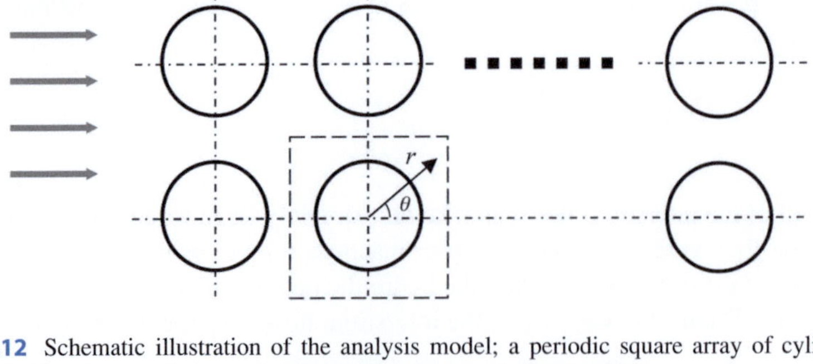

Fig. 4.12 Schematic illustration of the analysis model; a periodic square array of cylindrical SMCs residing in a matrix comprised of proteoglycan and collagen fibers

where p and u are the non-dimensional pressure and velocity, $p = K_pP/a\mu U_0$, $\boldsymbol{u} = U/U_0$, and a is the radius of SMC, U_0 is the representative flow velocity, $\alpha = a^2/K_p$. The solution of Eq. (4.31) must satisfy a no-slip boundary condition on the SMC surface. The velocity field outside the boundary layer is determined by solving the outer flow defined by the Darcy equation, and the velocity field inside the boundary layer is determined by the first equation of Eq. (4.31), the Brinkman equation. These two velocity fields must match at the edge of the boundary layer. The asymptotic solutions for $\alpha \gg 1$ can be obtained in the form of a sum of infinite power series of radial and circumferential coordinate variables (Wang and Tarbell 1995):

$$p = \sum_{n=0}^{\infty} A_n \left[r^{2n+1} + \left(\frac{1}{r}\right)^{2n+1} \right] \cos\,(2n+1)\,\theta \tag{4.32}$$

$$u_r = -\sum_{n=0}^{\infty} (2n+1)\,A_n \left[r^{2n} - \left(\frac{1}{r}\right)^{2n+2} \right] \cos\,(2n+1)\,\theta \tag{4.33}$$

$$u_\theta = \sum_{n=0}^{\infty} (2n+1)\,A_n \left[r^{2n} + \left(\frac{1}{r}\right)^{2n+2} - 2e^{-\eta} \right] \sin\,(2n+1)\,\theta \tag{4.34}$$

where u_r, u_θ are the radial and circumferential components of non-dimensional flow velocity for a single cell, respectively, and $\eta = (r-1)/(1/\alpha)$. The coefficient A_n in Eqs. (4.32, 4.33, and 4.34) is:

$$\begin{aligned} A_n &\cong A_0\left(\tfrac{a}{L}\right)^{2n+2}\left[-S_{2n+2} + (2n+3)\,(n+1)\,S_{2n+4}S_4\left(\tfrac{a}{L}\right)^6\right] \\ S_n &= 2^{-n} \sum_{k,\,m \neq 0} (k+im)^{-n} \qquad (k, m; \quad \text{integer}) \end{aligned} \tag{4.35}$$

where L is center-to-center distance of neighboring SMCs. Omitting the mathematical details, the effect of the volume fraction of a square array of SMCs, F, on the average WSS on a SMC is obtained using the analytical solutions Eqs. (4.32, 4.33, 4.34, and 4.35) as

$$\begin{aligned} \overline{\tau}^* &= \frac{\overline{\tau}}{\sqrt{U_0\mu(-\Delta P/\Delta L)}} \\ &= \frac{4}{\pi}\,\frac{1 - 0.319285F^2 - 0.043690F^4}{\sqrt{\left(1 - F - 0.305828F^4\right)\left(1 + F - 0.305828F^4\right)}} \end{aligned} \tag{4.36}$$

where * represents the value is non-dimensional, ΔP is the pressure drop across the media, and ΔL is the thickness of the media. Equation (4.36) indicates that the WSS on an SMC is of the order $[U_0\mu(-\Delta P/\Delta L)]^{1/2}$. To estimate the magnitude of the

WSS imposed by interstitial flow on SMCs, two sets of experimental data (Tedgui and Lever 1984) in rabbit thoracic aortas at 70 and 180 mmHg: (70 mmHg) $\mu = 1.0 \times 10^{-3}$ Pa s, $U_0 = 2.8 \times 10^{-6}$ cm/s, $-\Delta P = 52$ mmHg, $\Delta L = 125 \times 10^{-4}$ cm; and (180 mmHg) $\mu = 1.0 \times 10^{-3}$ Pa s, $U_0 = 4.39 \times 10^{-6}$ cm/s, $-\Delta P = 152$ mmHg, $\Delta L = 100 \times 10^{-4}$ cm were used. Notice that the pressure drop across the media is lower than the imposed pressure due to the resistance of the EC layer to the interstitial flow. The calculated average values of WSS were 0.55 dyn/cm^2 (70 mmHg) and 1.26 dyn/cm^2 (180 mmHg) when $F = 0.4$. The maximum values are about twice these values. On the other hand, mean values of WSS on ECs are on the order of 10 dyn/cm^2 in arteries and may be higher in capillaries and lower in postcapillary venules (Lipowsky 1995). The level of the WSS predicted in the present analysis was found to be similar to that on ECs, which can affect SMC biology (Shi and Tarbell 2011; Shi et al. 2009).

4.4.2.2 Numerical Computation of the Flow in Artery Wall

As seen in the previous subsection 4.4.2.1, SMCs could be subjected to significant levels of WSS associated with interstitial flow, even though the velocity of interstitial flow is very low (on the order of 10^{-6} cm/s). In this analytic model, the assumption of uniform flow velocity at the upstream end of an array of cylindrical SMCs neglected the more complex entrance conditions that exist at the intima-media boundary (Fig. 4.13). However, for the SMCs bordering the intima, the presence of IEL with leaky fenestral pores can greatly alter the flow field. Because the area fraction of fenestral pores is in the range 0.002–0.01 and the mean diameter of the fenestral pores varies from 0.4 to 2.1 μm (Huang et al. 1998), the velocity of fluid issuing from an individual pore could be 100-fold greater than the interstitial flow velocity in the media, resulting in a significant change in the distribution of

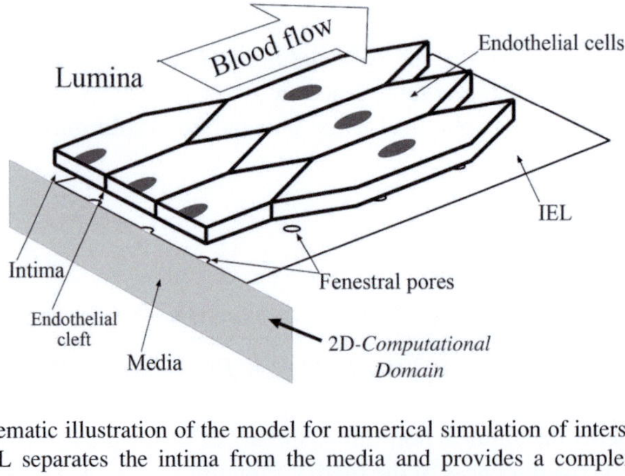

Fig. 4.13 Schematic illustration of the model for numerical simulation of interstitial flow in the media. The IEL separates the intima from the media and provides a complex entrance flow condition through the fenestral pores. The media consists of a periodic array of cylindrical SMCs embedded in a continuous, porous interstitial phase. Interstitial flow is distributed into the media from the intima through the fenestral pores in the IEL

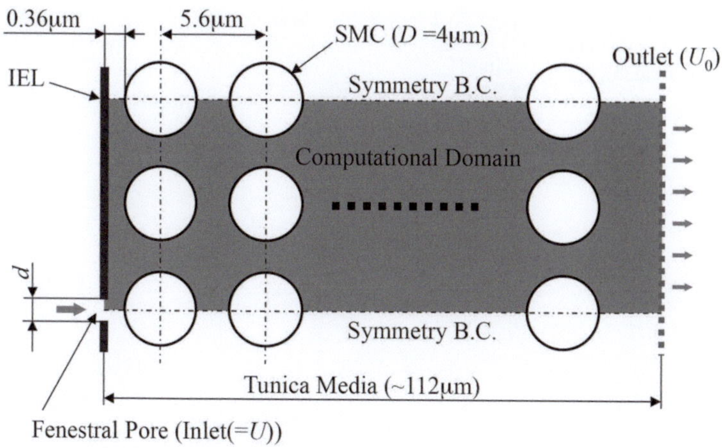

Fig. 4.14 Two-dimensional model of interstitial flow in the media (Modified from Tada and Tarbell 2000)

flow and associated WSS on the most superficial layer of SMCs that lie right beneath the IEL.

The geometry of the two-dimensional flow problem is shown in Fig. 4.14. The fenestral pore center is aligned with the SMC center to investigate the extreme case in which the WSS on SMCs attains the maximum possible value. SMCs are arranged in a square array configuration. To simplify the problem, the following assumptions are invoked: (1) The IEL is a rigid impermeable wall with zero thickness except for its fenestral openings. (2) The circular fenestral pores are distributed with a periodic square array configuration over the IEL surface. (3) The no-slip condition is applied on all surfaces of the SMCs and IEL. (4) A uniform velocity profile is specified at the entrance to the fenestral pores.

Once the SMC diameter, D ($=2a$), and volume fraction for SMC, F, are given, the distance between centers of neighboring SMCs, L, can be calculated. In the same manner, the pore spacing is also calculated when the pore diameter, d, and area fraction of pores, f, are given. The distance between the IEL and upstream end of the SMC, a, is taken as $a = 0.36\ \mu$m in accordance with experimental data. All the physical constants and parameters used are listed in Table 4.2.

For the numerical simulations, FIDAP software package (version 7.62, FLUENT) was used. The set of governing equations, Eq. (4.31), was solved by direct Gaussian elimination. The finite element mesh realizations used in the numerical analysis is shown in Fig. 4.15. A symmetry boundary condition was applied on the top and bottom end lines of the computational domain. At the downstream end, a gradient-free ($dU/dx = 0$) outlet boundary condition was applied. In the vicinity of solid boundaries, a finer mesh of grid points taken perpendicular to each SMC surface was employed to ensure sufficiently high resolution within the boundary layer. The mesh size used was ~75,000.

Figure 4.16 shows streamlines of the interstitial flow in the media. The fluid coming from the fenestral pore is distributed into the whole region of the media. Near

Table 4.2 Physical constants and parameters used in numerical simulations

Parameter	Value
a [μm]	0.36
D [μm]	4.0
d [μm]	0.4–3.2
F	0.4
f	0.004, 0.016
K_p [cm^2]	1.43×10^{-14}
U_0 [cm/s]	5.80×10^{-6}
μ [Pa s]	6.86×10^{-3}
ρ [g/cm^3]	1.0

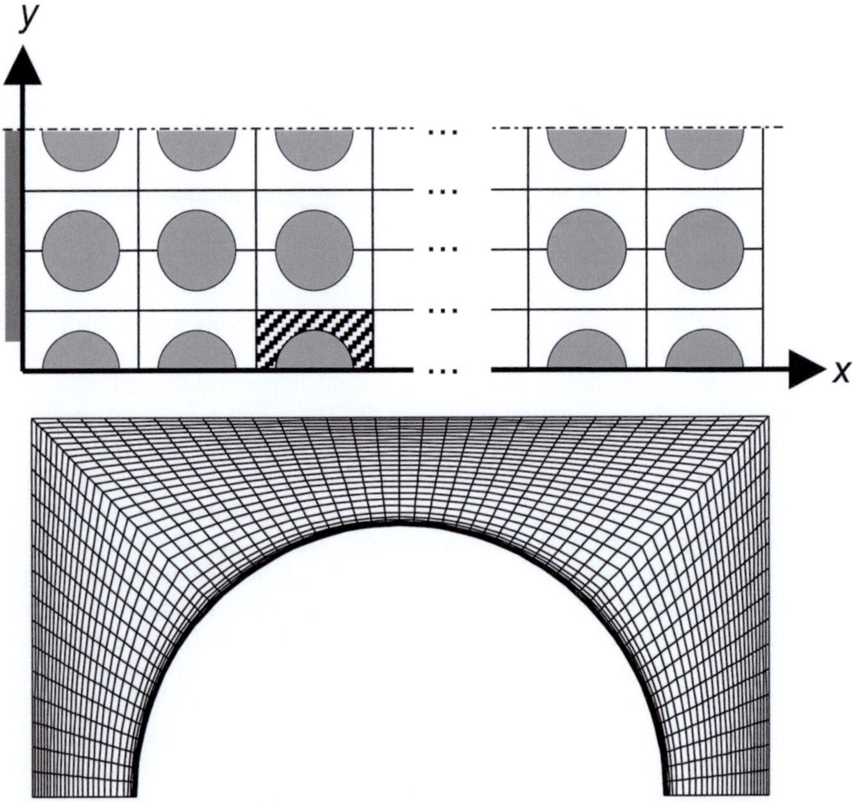

Fig. 4.15 The finite element mesh and the mesh configuration used in the numerical simulations (Reproduced from Tada and Ozono 2011 with permission from Springer)

the IEL, fluid spreads laterally into the upstream region as soon as it enters from the fenestral pore. At the narrowest point in the path between neighboring SMCs, the intervals between streamlines are almost equal, suggesting that a uniform velocity distribution is established except in a thin boundary layer near the SMC surfaces.

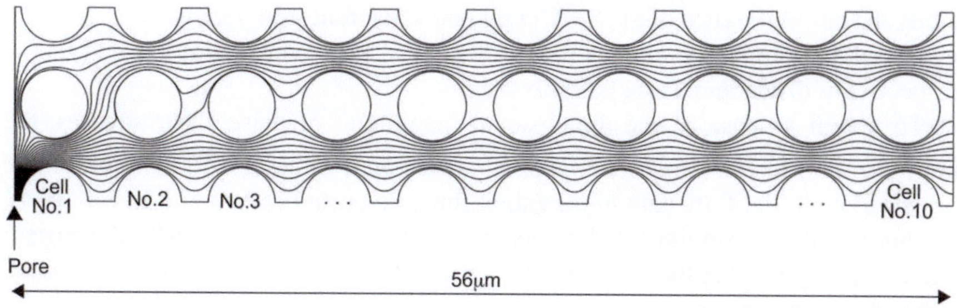

Fig. 4.16 A set of streamlines of the interstitial flow in the media. The figure is depicted only for the first 10 lines of the SMC array from the IEL (the *left end*) to show details of flow at the fenestral pore

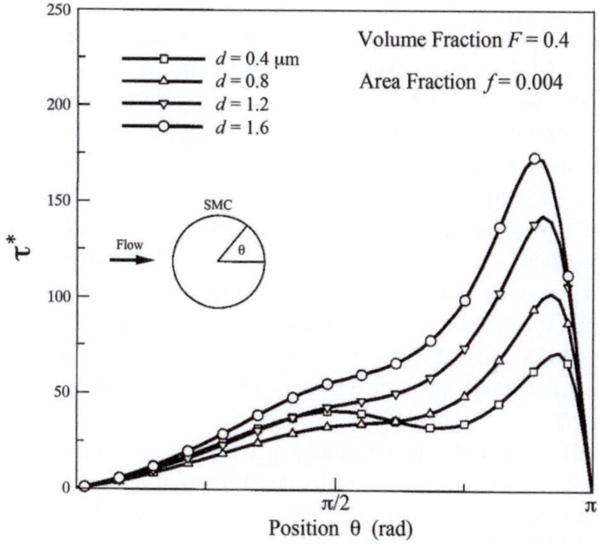

Fig. 4.17 Distributions of non-dimensional local shear stress along the circumference of the SMC right beneath a fenestral pore for four different values of fenestral pore diameter, d. The volume fraction of the SMC, F, and the area fraction of the fenestral pore, f, are kept constant at 0.4 and 0.004, respectively (Modified from Tada and Tarbell 2000)

This feature is quite different from conventional Stokes flow, in which the fluid displays a parabolic velocity distribution. Furthermore, streamlines penetrate more deeply into the wake region behind each SMC in contrast to those of Stokes flow.

Figure 4.17 shows the distribution of the non-dimensional local shear stress:

$$\tau^*\left(=\frac{\tau}{\sqrt{U_0\mu(-\Delta p/\Delta L)}}\right) = -\mu\left[r\frac{\partial}{\partial r}\left(\frac{u_\theta}{r}\right)+\frac{1}{r}\frac{\partial u_r}{\partial \theta}\right]\Bigg|_{r=1} \tag{4.36}$$

along the circumference of the SMC right beneath a fenestral pore for four different values of fenestral pore diameter, d. F and f are kept constant at 0.4 and 0.004, respectively (Tada and Tarbell 2000).

To obtain a sense of the dimensional magnitude of these shear stresses, the normalization factor, $[U_0\mu(-\Delta P/\Delta L)]^{1/2}$, takes on a value of 0.52 dyn/cm^2 on the basis of interstitial flow data in the rabbit thoracic aorta (Tedgui and Lever 1984). Predictions show a significant elevation of shear stress on the first SMC. A constant value of f implies that the incoming flow velocity at the pore entrance is constant, independent of d.

4.4.3 Hemodynamics in Diseased Site of Artery Wall

The progression of intimal hyperplasia (IH) is affected by hemodynamic conditions at the diseased site of an artery wall, where SMCs are likely to migrate toward the sub-endothelial layer. A recent study showed that arterial levels of WSS act to suppress SMC migration by up-regulating the cellular production of NO, which in turn inhibits matrix metalloproteinase-2 (MMP-2) activity. It has been also shown in balloon catheter injury animal models (Tada and Tarbell 2000) where superficial layers of SMC are exposed directly to blood flow that intimal thickening in the absence of the endothelium is accelerated in low WSS environments, suggesting that SMCs exposed to blood flow respond directly to changes in blood flow. However, other recent studies have shown that even in the case of an intact endothelium, SMCs are subjected to WSS through exposure to interstitial flow driven by the transmural pressure gradient (Wang and Tarbell 1995; Tada and Tarbell 2000, 2002).

Shi et al. (2011) showed experimentally in a 3D collagen gel model of suspended SMC exposed to interstitial flow that increased interstitial flow, such as that which occurs at sites of elevated EC hydraulic conductivity in vascular disease, leads to an enhancement SMC migration that could contribute to enhanced IT. They also demonstrated the role of the surface layer of proteoglycans bound to the SMC (the glycocalyx) in mediating the mechanotransuction of interstitial flow shear stress by the SMC. The glycocalyx represents the additional thin matrix layer (order 1 μm thick) on the cell surface that is involved in transmitting fluid shear stress to the cell surface through the solid core proteins of the bound proteoglycan molecules in the surface glycocalyx layer. In a related theoretical study, Tarbell and Shi (2013) deduced that the solid shear stress component transmitted to the cell could be 10 times the imposed fluid shear stress on the surface of the glycocalyx. This suggests WSS values of 5–10 dyn/cm^2 on SMC surface imposed by interstitial flow. These stresses are essentially the same as imposed by blood flow on EC.

Although the biological evidence is still limited at the present time, it is reasonable to speculate that interstitial flow affects the function of SMCs in addition to their migration. While in the realm of speculation, it should also be noted that alterations in blood pressure and endothelial hydraulic conductivity will affect interstitial flow and in turn the WSS on SMCs. The above issues have been extensively reviewed by Shi and Tarbell (2011).

References

Ajmani RS (1997) Hypertension and hemorheology. Clin Hemorheol Microcirc 17:397–420

Bor-Kucukkatay M, Yalcin O, Gokalp O et al (2000) Red blood cell rheological alterations in hypertension induced by chronic inhibition of nitric oxide synthesis in rats. Clin Hemorheol Microcirc 22:267–275

Chien S (2003) Molecular mechanical bases of focal lipid accumulation in arterial wall. Prog Biophys Mol Biol 83:131–151

Chien S, Dormandy J et al (eds) (1987) Clinical hemorheology. Martinus Nijhoff Publishers, Dordrecht

Clark JM, Glagov S (1985) Transmural organization of the arterial media: the lamellar unit revisited. Arterioscler Thromb Vasc Biol 5:19–34. doi:10.1161/01.ATV.5.1.19

Fung YC (1993) Biomechanics: mechanical properties of living tissues, 2nd edn. Springer, New York

Huang Y, Jan KM, Rumschitzki D et al (1998) Structural changes in rat aortic intima due to transmural pressure. Trans ASME J Biomech Eng 120:476–483

Kesmarky G, Toth K, Habon L et al (1998) Hemorheological parameters in coronary artery disease. Clin Hemorheol Microcirc 18:245–251

Kleinstreuer C (2006) Biofluid dynamics: principles and selected applications. CRC Press, Boca Raton

Kohler TR, Jawien A (1992) Flow affects development of intimal hyperplasia after arterial injury in rats. Arterioscler Thromb 12:963–971

Ku DN, Giddens DP et al (1985) Pulsatile flow and atherosclerosis in the human carotid bifurcation: positive correlation between plaque location and low oscillating shear stress. Arteriosclerosis 5:293–302

Levick JR (1987) Relation between hydraulic resistance and composition of the interstitium. Adv Microcirc 13:124–133

Lipowsky HH (1995) Shear stress in the circulation. In: Bevan JA et al (eds) Flow-dependent regulation of vascular function. Oxford University Press, New York

McMillan DE (1993) Hemorheological studies in the diabetes control and complications trial. Clin Hemorheol 13:147–154

Merrill EW (1969) Rheology of blood. Physiol Rev 49(4):863–888

Morris CL, Rucknagel DL et al (1989) Evaluation of the yield stress of normal blood as a function of fibrinogen concentration and hematocrit. Microvasc Res 37(3):323–338

Ramakrishnan S, Grebe R et al (1999) Aggregation of shape altered erythrocytes: an in vitro study. Curr Sci 77:805–808

Shi ZD, Tarbell JM (2011) Fluid flow mechanotransduction in vascular smooth muscle cells and fibroblasts. Ann Biomed Eng 39(6):1608–1619

Shi ZD, Ji XY, Qazi H et al (2009) Interstitial flow promotes vascular fibroblast, myofibroblast, and smooth muscle cell motility in 3-D collagen I via upregulation of MMP-1. Am J Physiol Heart Circ Physiol 297(4):H1225–H1234

Shi ZD, Wang H et al (2011) Heparan sulfate proteoglycans mediate interstitial flow mechanotransduction regulating MMP-13 expression and cell motility via FAK-ERK in 3D collagen. PLos One. doi:10.1371/journal.pone.0015956

Shul'man ZP, Mansurov VA et al (2006) Rheological changes in the blood and plasma of patients with myocardial ischemia and diabetes mellitus and dysfunction of their endothelium. J Eng Phys Thermophys 79(1):99–104

Tada S, Ozono H (2011) Computational study of LDL mass transport in the artery wall. J Biorheol 25(1–2):27–35

Tada S, Tarbell JM (2000) Interstitial flow through the internal elastic lamina affects shear stress on arterial smooth muscle cells. Am J Physiol Heart Circ Physiol 278:H1589–H1597

Tada S, Tarbell JM (2002) Flow through internal elastic lamina affects shear stress on smooth muscle cells (3D simulations). Am J Physiol Heart Circ Physiol 282:H576–H584

Tada S, Tarbell JM (2005) A computational study of flow in a compliant carotid bifurcation – stress phase angle correlation with shear stress. Ann Biomed Eng 33:1202–1212

Tarbell JM, Shi ZD (2013) Effect of the glycocalyx layer on transmission of interstitial flow shear stress to embedded cells. Biomech Model Mechanobiol 12(1):111–121. doi:10.1007/s10237-012-0385-8

Tarbell JM, Shi ZD et al (2014) Fluid mechanics, arterial disease, and gene expression. Annu Rev Fluid Mech 46:591–614. doi:10.1146/annurev-fluid-010313-141309

Tedgui A, Lever MJ (1984) Filtration through damaged and undamaged rabbit thoracic aorta. Am J Physiol Heart Circ Physiol 247:H784–H791

Thomas JB, Milner JS et al (2002) On the influence of vessel planarity on local hemodynamics at the human carotid bifurcation. Biorheology 39:443–448

Wang DM, Tarbell JM (1995) Modeling interstitial flow in an artery wall allows estimation of wall shear stress on smooth muscle cells. Trans ASME J Biomech Eng 117:358–363

Cyclic Stretch-Induced Reorganization of Stress Fibers in Endothelial Cells

5

Roland Kaunas and Shinji Deguchi

Abstract

Vascular endothelial cells (ECs) are subjected to cyclic circumferential stretching generated by the cardiac cycle. This force contributes to the alignment and elongation of ECs along the long axis of the vessel. Thus, ECs align perpendicular to the principal direction of cyclic stretch. At arterial bifurcations and regions of curvature, however, ECs do not exhibit an aligned morphology. The absence of aligned cell morphology at non-straight regions has been attributed to the response of cells to the local mechanical environment. Importantly, these regions show a level of chronic activation of inflammatory signals. This inflammation can become severe under certain conditions and finally contribute to the development of atherosclerotic plaques. Thus, the responses of ECs to their local mechanical environment that include changes in cell structure and intracellular signaling are important in maintaining an anti-atherogenic phenotype. In this chapter, we review studies that focused on the response and signaling of ECs subjected to cyclic stretch with particular focus on actin stress fibers.

Keywords

Endothelial cells • Cyclic stretch • Stress fibers • Mitogen-activated protein kinase • Tensional homeostasis

R. Kaunas (✉)
Department of Biomedical Engineering, Texas A&M University, Emerging Technologies Building, MS 3120, College Station 77843-3120, TX, USA
e-mail: rkaunas@bme.tamu.edu

S. Deguchi
Department of Nanopharmaceutical Sciences, Nagoya Institute of Technology, Nagoya, Japan

K. Tanishita, K. Yamamoto (eds.), *Vascular Engineering*,
DOI 10.1007/978-4-431-54801-0_5

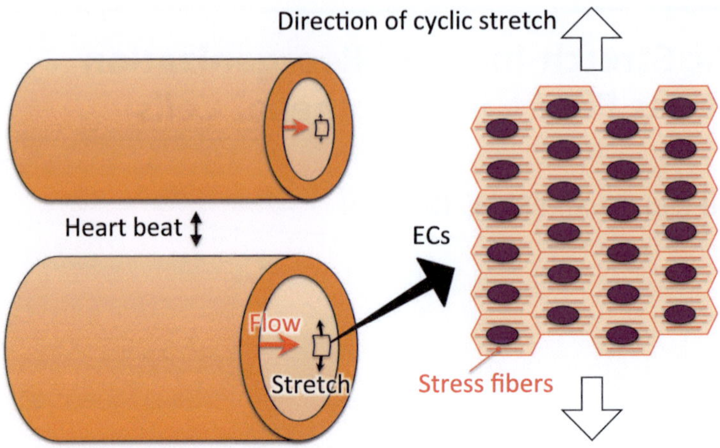

Fig. 5.1 ECs align perpendicular to the principal direction of cyclic stretch

5.1 Introduction

The effects of stretch on the morphology (Fig. 5.1) and signal transduction of ECs have been studied in cell culture models that allow for control of cyclic stretch patterns. These experiments have revealed that cell alignment requires a directional cue. Specifically, there must be a principal direction of stretch for the cells to orient away from. The realignment of cells in response to cyclic stretch is normally preceded by the reorganization of actin stress fibers and associated cell adhesions to the underlying matrix (Hayakawa et al. 2001). The alignment of cells and stress fibers perpendicular to the direction of cyclic stretching is a ubiquitous phenomenon, which is actually observed with many cell types that include vascular and airway smooth muscle cells, osteoblasts, fibroblasts, keratinocytes and mesenchymal stem cells (Takei et al. 1997; Hayakawa et al. 2001; Neidlinger-Wilke et al. 2001; Smith et al. 2003; Park et al. 2004).

5.2 Structure of Stress Fibers

Here we pay particular attention on the regulation of stress fibers in response to cyclic stretch (Kaunas and Deguchi 2011). Stress fibers in nonmuscle cells consist of bundles of actin filaments cross-linked mainly by myosin II and α-actinin of nonmuscle isoforms (Lazarides and Burridge 1975; Cramer et al. 1997). Using myosin S1 fragments to label the polarity of F-actin, Cramer et al. (1997) observed that actin filaments are organized in distinct patterns in stress fibers within motile and non-motile cells. In motile cells, actin filaments are predominantly oriented with their barbed ends toward the cell adhesions, with a gradation of polarity along the length of the stress fibers. In non-motile cells, actin filaments alternate in polarity all along the length of the stress fibers. Accompanying the periodic alterations in actin polarity, myosin II and α-actinin appear alternately along the length of the stress fibers (Sanger et al. 1983). These findings led to early models of stress fiber structure and contraction very similar to the organization of skeletal muscle sarcomeres (Lazarides and Burridge 1975). In electron micrographs, bands of α-actinin in stress fibers are observed as electron-dense striations, with myosin-rich regions appearing as interleaving electron-light regions (Langanger et al. 1986). In a sarcomeric model of the contraction of stress fibers, intercalating arrays of myosin II slide between adjacent actin filaments to result in overall shortening or lengthening of the stress fibers during concentric and eccentric contractions, respectively.

5.3 Assembly of Stress Fibers

Integrin-based cell adhesions that connect to stress fibers are categorized into two types according to the size and composition: small focal complexes and larger focal adhesions. The formation of stress fibers involves the clustering of integrins into focal complexes that can grow into focal adhesions that mechanically support the stress fibers via structural proteins such as talin and vinculin (Schoenwaelder and Burridge 1999). The assemblies of stress fibers and focal adhesions are correlated with the generation of contractile force via the motor activity of myosin II (Chrzanowska-Wodnicka and Burridge 1996; Deguchi et al. 2011). The process of stress fiber assembly is described by Hotulainen and Lappalainen (2006), in which they classified stress fibers into ventral stress fibers, dorsal stress fibers, and transverse arcs. The most commonly observed actin bundles are ventral stress fibers, which are located along the basal surface of the cells and are anchored to the underlying matrix via focal adhesions at each end. Transverse arcs are curved actin bundles that are not directly attached to focal adhesions, but are instead attached to other actin bundles. Dorsal stress fibers are anchored to a focal adhesion at one end and to a transverse arc at the other end. Ventral stress fibers form by end-to-end joining of dorsal stress fibers, either directly or including a transverse arc in between.

Rho family small GTPases are key regulators of actin organizations and are required for cyclic stretch-induced responses of ECs. Of these proteins, RhoA is closely linked to the assembly of stress fibers (Ridley and Hall 1992; Chrzanowska-Wodnicka and Burridge 1996). Various soluble signals, including growth factors, cytokines, and hormones, induce the activation of RhoA and resultant stress fibers via the activation of specific guanine nucleotide-exchange factors. Rho-kinase (or ROCK) and mDia are two of the main downstream effectors of RhoA that work concurrently for proper assembly of stress fibers (Watanabe et al. 1999). Rho-kinase regulates the phosphorylation of myosin regulatory light chain (MLC) at ser19 and thr18 (Kimura et al. 1996). The first phosphorylation at ser19 of MLC, which is a 20 kDa protein that attaches to myosin II, induces the formation of bipolar filaments of myosin II and increases the ATPase activity of myosin II. The second phosphorylation at thr18, together with the first phosphorylation, allows for further enhancement of the ATPase activity of myosin II. The myosin ATPase is required for cross-bridge cycling with actin filaments, which consequently induces the assembly of stress fibers or more specifically the bundling of actin filaments into stress fibers. At the same time, with the enhancement of cross-bridge cycling, contractile forces are generated in the actin bundles and are applied to the focal adhesions located at their ends.

Myosin light chain kinase (MLCK) mediates the phosphorylation of MLC upon activation of Ca^{2+}–calmodulin that is independent of the RhoA-associated pathways. Interestingly, MLCK and Rho-kinase appear to regulate different subpopulations of stress fibers, with MLCK primarily regulating peripheral stress fibers and Rho-kinase regulating central stress fibers (Totsukawa et al. 2000; Katoh et al. 2001) Using laser scissors, Tanner et al. (2010) demonstrated that MLCK-dependent stress fibers located along the cell periphery display much lower elastic moduli and higher retraction distances than Rho-kinase-dependent stress fibers located in the cell center. These findings suggest that stress fibers regulated by different myosin II activators may exhibit different biophysical properties and have different functions.

Thus, MLC phosphorylation-dependent ATPase activity of myosin II or acto-myosin contractility has been believed to be a requirement for the maintenance of stress fibers. While much attention has thus been focused on the regulation of stress fibers through molecular signaling pathways associated with MLC phosphorylation, applied tension clearly plays a role as well. Hirata et al. (2007) permeabilized fibroblasts that had been pretreated with a Rho-kinase inhibitor to disassemble stress fibers. Addition of ATP caused the assembly of stress fibers in the permeabilized cells through a process involving the centripetal movement of actin filaments. These authors then demonstrated that applied force could also induce the assembly of stress fibers in the same preparation even in the absence of ATPase activity. We have also shown that actin bundles assemble in response to cyclic stretch in ECs treated with inhibitors of RhoA, mDia, Rho-kinase or MLCK (Kaunas et al. 2005; Lee et al. 2010). These actin bundles only formed in the direction parallel to the stretch, indicating the necessity for tension provided by applied stretching to induce the formation of actin bundles. Further, only peripheral actin bundles formed when Rho-kinase was inhibited, while only central actin bundles formed when MLCK was inhibited. Based on time-lapse microscopy of GFP-actin in cyclically-stretched U2OS cells (a human osteogenic sarcoma cell line), the assembly of actin bundles in cells treated with a Rho-kinase inhibitor appeared to involve the bundling of individual thin fibers resembling transverse arcs (Lee et al. 2010). Together, these studies are consistent with the model that stress fibers assemble in response to tension, and the tension can be either internally or externally generated.

Meanwhile, mDia, one of the main effectors of RhoA, facilitates the nucleation of new actin filaments at the ends of stress fibers. Riveline et al. (2001) demonstrated that externally applied forces induce the growth of focal adhesions in the direction of force in a manner dependent on mDia, but not Rho-kinase. In addition, it has recently been reported that force-facilitated actin polymerization at focal adhesions (Hirata et al. 2008) and subsequent maintenance of stress fiber integrity (Smith et al. 2010) are dependent on the recruitment of zyxin to focal adhesions and stress fibers. Thus, forces serve to reinforce both stress fibers and their associated adhesions.

5.4 Regulation of Tension in Stress Fibers

The phosphorylation of MLC allows for the interaction between actin and myosin II, which generates tension in stress fibers. Rho-kinase, MLCK, and MLC phosphatase are two of the major kinases and the principal phosphatase that phosphorylate and dephosphorylates MLC, respectively. The level of MLC phosphorylation, which directly affects the level of cellular tension, is somehow maintained within stationary cells. Thus, there is a dynamic balance between kinase and phosphatase activities that regulates the tension in stress fibers.

Myosin II-generated contraction leads to the development of isometric tension or prestress (Polte et al. 2004). This pre-stress extends SFs beyond their unloaded lengths (Deguchi et al. 2005; Lu et al. 2008). The level of SF pre-extension is maintained at a set-point value with measured values ranging from 1.10 to 1.35 (Lu et al. 2008). Precise measurements of forces generated at cell attachments indicate that cells tend to maintain a constant level of stress applied to focal adhesions (Balaban et al. 2001).

Cell stretching and shortening cause transient changes in intracellular tension and strain that gradually recover over time (Mizutani et al. 2004). A similar response is observed for cells in collagen matrices, where the cells react to changes in tension in a direction toward re-establishment of a set-point value through a process that Brown and colleagues termed "tensional homeostasis" (Brown et al. 1998).

5.5 Disassembly of Stress Fibers

Elevation of cyclic AMP (cAMP) and consequent activation of PKA (cAMP-dependent protein kinase) induces the disassembly of stress fibers and focal adhesions by inhibiting actomyosin contractility in several ways (Schoenwaelder and Burridge 1999; DeMali et al. 2003). An increase in cAMP levels decreases the phosphorylation of the cell adhesion-associated proteins FAK and paxillin, thus preventing the maturation of the cell adhesions into focal adhesions. Activated PKA phosphorylates MLCK, which inhibits its activity for phosphorylating MLC and then is followed by a decrease in MLC phosphorylation. Another substrate for PKA is RhoA whose phosphorylation leads to decreased association with Rho-kinase and hence its inactivation. Vasodilater-stimulated phosphoprotein (VASP) is involved in the zyxin-mediated enhancement of actin polymerization at focal adhesions

(Hirata et al. 2008; Smith et al. 2010). VASP is inhibited by PKA (Schoenwaelder and Burridge 1999; DeMali et al. 2003), indicating that the stabilization of stress fibers is achieved without PKA. Thus, the overall effects of cAMP/PKA-mediated signal pathways are to inhibit cellular contractility, which results in loss of both stress fibers and focal adhesions (Chrzanowska-Wodnicka and Burridge 1996). Retraction of cell processes from the substrate or a release of cellular tension causes an increase in cAMP levels (He and Grinnell 1994; Schoenwaelder and Burridge 1999; DeMali et al. 2003). In addition, new integrin ligation with the extracellular matrix activates cAMP/PKA transiently, which likely allows cells to promptly adapt to their environment through disassembly and subsequent remodeling of stress fibers and focal adhesions (DeMali et al. 2003). However, it remains to be elucidated how cAMP/PKA levels change in response to cyclic stretch and whether they contribute to the direction-dependent disassembly of stress fibers and focal adhesions observed at the early stage of the cellular response to cyclic stretch.

A prevailing hypothetical mechanism for stress fiber disassembly is unbalanced actin depolymerization over polymerization. Actin filament depolymerization is regulated by ADF/cofilin (Maciver et al. 1991) and gelsolin (Yin and Stossel 1979), which are capable of severing actin filaments at their minus ends composed of ADP-actin by distinct mechanisms. Hayakawa and colleagues proposed that stress fiber tension stabilizes actin filaments against severing by ADF/cofilin (Hayakawa et al. 2011). Another possible hypothesis is that stress fiber disassembly primarily involves stress fiber unbundling (Matsui et al. 2011). Our experiments indicate that both mechanisms of stress fiber disassembly are important, depending on the circumstances. Inhibition of actin polymerization with cytochalasin D results in the gradual disassembly of stress fibers from their ends, which is consistent with the gradual depolymerization of the actin at the distal ends of the stress fibers (Matsui et al. 2010). Inhibition of Rho-kinase, on the other hand, results in the unbundling of stress fibers via detachment of individual actin filaments from the edges of existing stress fibers, which presumably depolymerize shortly thereafter (Matsui et al. 2010).

Although myosin II is conventionally associated with stress fiber contractility and assembly, a number of studies indicate that myosin II also plays a role in stress fiber disassembly. As would be expected, a decrease in myosin II activity results in the loss of stress fibers (Chrzanowska-Wodnicka and Burridge 1996; Hotulainen and Lappalainen 2006) and decreases in integrin clustering (Kawakami et al. 2001). Interestingly, active myosin II can also promote the disassembly of stress fibers. Bernheim-Groswasser and co-workers observed that myosin II contributes to the unbundling of actin bundles *in vitro*, followed by their subsequent depolymerization (Haviv et al. 2008). Wilson and colleagues demonstrated that myosin II has a direct role in the disassembly of actin networks in motile fish keratocytes (Wilson et al. 2010).

Using detergent-extracted stress fibers activated with ATP, we observed that stress fibers transitioned from conventional stress fiber shortening to rapid disassembly as ATP concentration is increased from 2 to 5 mM, which is within the physiological range of intracellular ATP concentrations (Matsui et al. 2010). These precarious conditions are predicted to facilitate the transition of stress fibers between states of stable tension-bearing and rapid remodeling.

Release of intracellular tension due to rapid cell shortening leads to stress fiber disassembly (Costa et al. 2002; Sato et al. 2005). Stress fibers that oriented in the direction of shortening are preferentially disassembled, indicating only the unloaded fibers are destabilized (Sato et al. 2005). Release of tension by localized disruption of focal adhesions with trypsin/EDTA results in the selective disassembly of the associated stress fibers without affecting nearby fibers (Sato et al. 2006). Since Rho-kinase inhibition decreases the contractile force of stress fibers, it is possible that a loss of tension leads to the fraying of stress fibers before depolymerization.

Cyclic stretch causes the selective disassembly of stress fibers oriented in the direction of stretch, followed by their preferential assembly perpendicular to the stretch direction (Lee et al. 2010). Interestingly, stretch-induced stress fiber reorganization does not require focal adhesion turnover. Kemkemer and colleagues (Goldyn et al. 2009) demonstrated that adhesions labeled with GFP-vinculin translocated centripetally via a sliding motion in response to cyclic stretch. Consistent with this result, we observed that stress fibers disassemble centrally without necessarily disassembling at their distal ends. These relatively stable distal ends, where focal adhesions are located, served as the starting material for new stress fibers oriented progressively more perpendicular to the direction of stretch. The Kemkemer group also showed that stress fibers labeled with GFP-actin rotate in response to cyclic stretch (Chen et al. 2012). Thus, stretch-induced stress fiber alignment can occur through multiple mechanisms.

5.6 Modulation of MAPK Activation by Reorganization of Stress Fibers

There is much evidence supporting the role of focal adhesions as mechanosensors of tension (Bershadsky et al. 2003). Integrin is activated by stretch to increase binding affinity with extracellular matrix ligands, which is required for the subsequent stretch-induced activation of JNK (c-Jun N-terminal kinase) (Katsumi et al. 2005). JNK is a member of the mitogen-activated protein kinase (MAPK) family of signal transduction proteins and is implicated in pro-inflammatory signaling. Evidence suggests that the pathway leading from integrin activation to JNK activation involves several proteins localized to focal adhesions, including FAK, Cas, Ras, Rac1, Pak1, and MKK4 (Almeida et al. 2000). Sheetz and colleagues proposed that the transduction of mechanical signals into intracellular biochemical signals occurs through force-induced conformational changes in focal adhesion and cytoskeletal proteins (Sawada and Sheetz 2002). In support of this concept, Sawada et al. (2006) demonstrated that the phosphorylation of Cas by Src involved mechanical extension of Cas.

The activation of JNK induced by cyclic stretch has a positive correlation with the reorganization of stress fibers (Kaunas et al. 2006). This correlation depends on

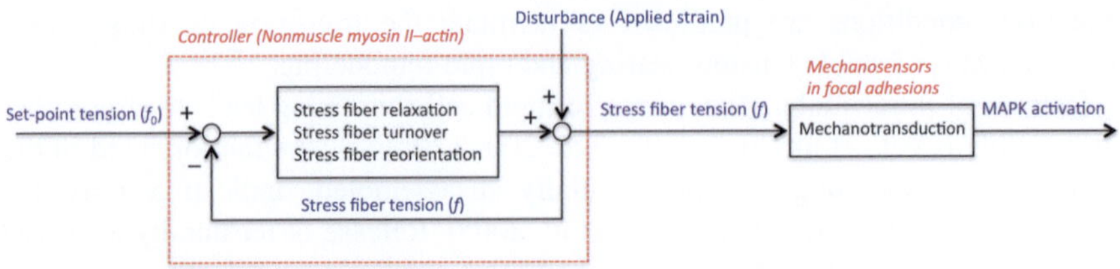

Fig. 5.2 Negative feedback model for the regulation of stretch-induced JNK activation through the activity of myosin II

the spatiotemporal pattern of the stretching. Figure 5.2 illustrates a conceptual model of stretch-induced signal transduction in terms of control theory (Kaunas and Deguchi 2011). In this scheme, myosin II plays multiple roles including the controller and sensor to regulate the process variable (f). Applied strain (or disturbance to the system) functions to perturb the process variable from the set-point level (f_0), to which myosin responds via cross-bridge cycling. Cross-bridge cycling contributes to stress fiber relaxation and/or stress fiber turnover. Each of these responses acts to bring the process variable back toward the set-point level. Stress fiber tension is directly transmitted to the associated mechanoreceptors at focal adhesions that transduce the mechanical signal into a biochemical signal including the activation of JNK.

Figure 5.3 illustrates predicted changes in average stress fiber tension and JNK activation in response to different patterns of cyclic stretch (Kaunas et al. 2010). In the case of cyclic uniaxial stretch at high frequency (e.g., 1 Hz), the tension in stress fibers cyclically changes as the stress fibers are cyclically stretched. Further, the amplitude of tension waveforms decreases gradually as the stress fibers align perpendicular to the direction of stretch. The increase in JNK activity is predicted to increase initially and then gradually decrease as the amplitude of stress fiber tension falls (Fig. 5.3a). In the case of cyclic equibiaxial stretch at high frequency, stress fibers cannot orient away from the direction of stretch. Thus, tension amplitude and JNK activity are predicted to remain elevated (Fig. 5.3b). These results are consistent with the actual time courses of JNK activation and stress fiber alignment measured in ECs subjected to 10 % cyclic uniaxial and equibiaxial stretches at 1 Hz (Kaunas et al. 2006). Stretch-induced stress fiber reorganization and JNK activation are dependent on the rate of strain for both uniaxial and equibiaxial cyclic stretch (Hsu et al. 2010). At low stretch frequencies, cyclic stretch (uniaxial and equibiaxial) is predicted to result in negligible tension amplitude, stress fiber reorganization, and JNK activation (Kaunas and Deguchi 2011). A step change in stretch is predicted to cause a transient increase in stress fiber tension and JNK activities, while the subsequent release of the stretch is predicted to cause a transient decrease in both tension and JNK activities. These results are consistent with experimental measurements on other MAPKs (Gavara et al. 2008; Hsu et al. 2010) as well and thus strongly indicate that stress fiber reorganization caused

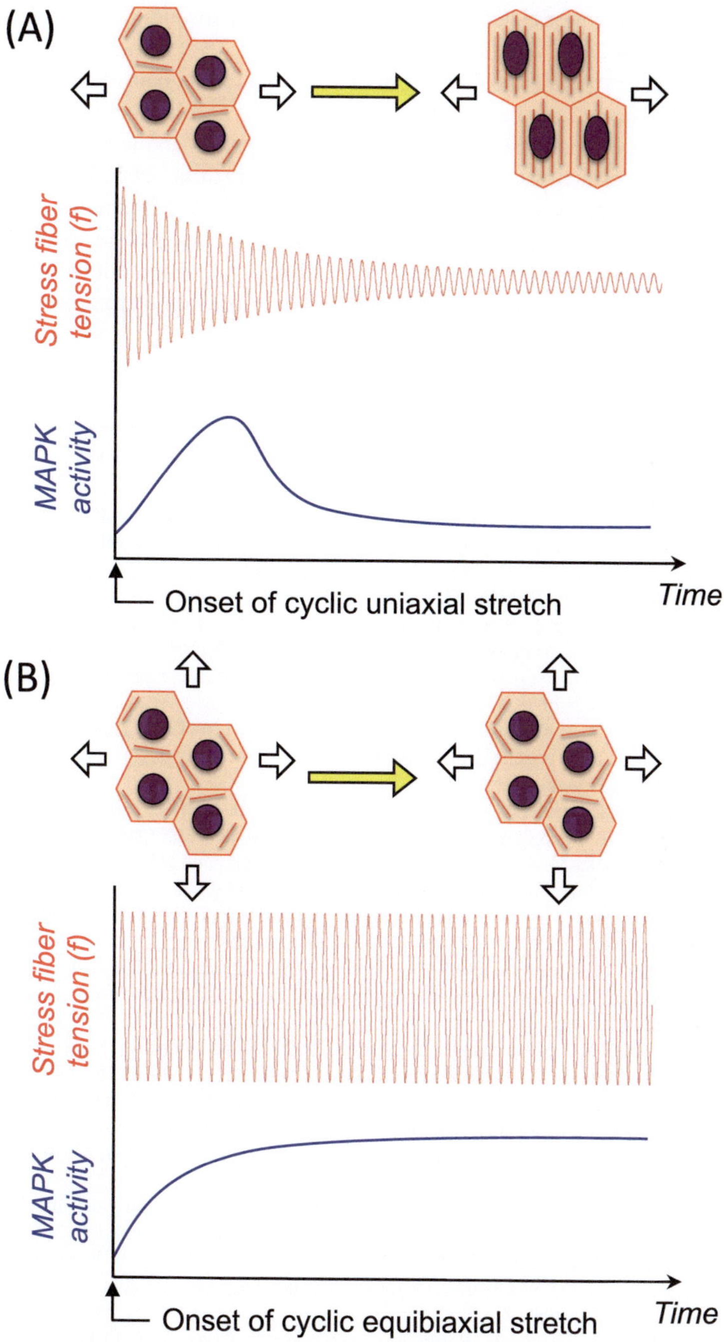

Fig. 5.3 Predicted changes in average stress fiber tension and JNK activity in response to cyclic uniaxial (**a**) and equibiaxial (**b**) stretch

by cyclic stretching modulates stretch-induced signal transduction through changes in cellular tension.

In summary, experimental data together with theoretical considerations provide important insight into the mechanisms by which cells regulate cellular tension and signal transduction via turnover of structural constituents and active relaxation of stress fibers. Myosin II, as the force generator within stress fibers, has emerged as playing a central role in each of these adaptive processes to help maintain tensional homeostasis in a dynamic mechanical environment subjected to cyclic stretch. Regulation of cytoskeletal tension in turn allows the endothelium to maintain functional homeostasis that may contribute to circumventing chronic activation of inflammatory signals.

References

Almeida EA, Ilic D, Han Q, Hauck CR, Jin F, Kawakatsu H, Schlaepfer DD, Damsky CH (2000) Matrix survival signaling: from fibronectin via focal adhesion kinase to c-Jun NH(2)-terminal kinase. J Cell Biol 149(3):741–754

Balaban NQ, Schwarz US, Riveline D, Goichberg P, Tzur G, Sabanay I, Mahalu D, Safran S, Bershadsky A, Addadi L, Geiger B (2001) Force and focal adhesion assembly: a close relationship studied using elastic micropatterned substrates. Nat Cell Biol 3(5):466–472

Bershadsky AD, Balaban NQ, Geiger B (2003) Adhesion-dependent cell mechanosensitivity. Annu Rev Cell Dev Biol 19:677–695

Brown RA, Prajapati R, McGrouther DA, Yannas IV, Eastwood M (1998) Tensional homeostasis in dermal fibroblasts: mechanical responses to mechanical loading in three-dimensional substrates. J Cell Physiol 175(3):323–332

Chen B, Kemkemer R, Deibler M, Spatz J, Gao H (2012) Cyclic stretch induces cell reorientation on substrates by destabilizing catch bonds in focal adhesions. PLoS One 7(11):e48346

Chrzanowska-Wodnicka M, Burridge K (1996) Rho-stimulated contractility drives the formation of stress fibers and focal adhesions. J Cell Biol 133(6):1403–1415

Costa KD, Hucker WJ, Yin FC (2002) Buckling of actin stress fibers: a new wrinkle in the cytoskeletal tapestry. Cell Motil Cytoskeleton 52(4):266–274

Cramer LP, Siebert M, Mitchison TJ (1997) Identification of novel graded polarity actin filament bundles in locomoting heart fibroblasts: implications for the generation of motile force. J Cell Biol 136(6):1287–1305

Deguchi S, Matsui TS, Iio K (2011) The position and size of individual focal adhesions are determined by intracellular stress-dependent positive regulation. Cytoskeleton 68(11):639–651

Deguchi S, Ohashi T, Sato M (2005) Evaluation of tension in actin bundle of endothelial cells based on preexisting strain and tensile properties measurements. Mol Cell Biomech 2(3):125–133

DeMali KA, Wennerberg K, Burridge K (2003) Integrin signaling to the actin cytoskeleton. Curr Opin Cell Biol 15(5):572–582

Gavara N, Roca-Cusachs P, Sunyer R, Farre R, Navajas D (2008) Mapping cell-matrix stresses during stretch reveals inelastic reorganization of the cytoskeleton. Biophys J 95(1):464–471

Goldyn AM, Rioja BA, Spatz JP, Ballestrem C, Kemkemer R (2009) Force-induced cell polarisation is linked to RhoA-driven microtubule-independent focal-adhesion sliding. J Cell Sci 122(Pt 20):3644–3651

Haviv L, Gillo D, Backouche F, Bernheim-Groswasser A (2008) A cytoskeletal demolition worker: myosin II acts as an actin depolymerization agent. J Mol Biol 375(2):325–330

Hayakawa K, Sato N, Obinata T (2001) Dynamic reorientation of cultured cells and stress fibers under mechanical stress from periodic stretching. Exp Cell Res 268(1):104–114

Hayakawa K, Tatsumi H, Sokabe M (2011) Actin filaments function as a tension sensor by tension-dependent binding of cofilin to the filament. J Cell Biol 195(5):721–727

He Y, Grinnell F (1994) Stress relaxation of fibroblasts activates a cyclic AMP signaling pathway. J Cell Biol 126(2):457–464

Hirata H, Tatsumi H, Sokabe M (2007) Dynamics of actin filaments during tension-dependent formation of actin bundles. Biochim Biophys Acta 1770(8):1115–1127

Hirata H, Tatsumi H, Sokabe M (2008) Mechanical forces facilitate actin polymerization at focal adhesions in a zyxin-dependent manner. J Cell Sci 121(Pt 17):2795–2804

Hotulainen P, Lappalainen P (2006) Stress fibers are generated by two distinct actin assembly mechanisms in motile cells. J Cell Biol 173(3):383–394

Hsu HJ, Lee CF, Locke A, Vanderzyl SQ, Kaunas R (2010) Stretch-induced stress fiber remodeling and the activations of JNK and ERK depend on mechanical strain rate, but not FAK. PLoS One 5(8):e12470

Katoh K, Kano Y, Amano M, Kaibuchi K, Fujiwara K (2001) Stress fiber organization regulated by MLCK and Rho-kinase in cultured human fibroblasts. Am J Physiol Cell Physiol 280(6): C1669–C1679

Katsumi A, Naoe T, Matsushita T, Kaibuchi K, Schwartz MA (2005) Integrin activation and matrix binding mediate cellular responses to mechanical stretch. J Biol Chem 280 (17):16546–16549

Kaunas R, Deguchi S (2011) Multiple roles for myosin II in tensional homeostasis under mechanical loading. Cell Mol Bioeng 4(2):182–191

Kaunas R, Nguyen P, Usami S, Chien S (2005) Cooperative effects of Rho and mechanical stretch on stress fiber organization. Proc Natl Acad Sci U S A 102(44):15895–15900

Kaunas R, Usami S, Chien S (2006) Regulation of stretch-induced JNK activation by stress fiber orientation. Cell Signal 18(11):1924–1931

Kaunas R, Huang Z, Hahn J (2010) A kinematic model coupling stress fiber dynamics with JNK activation in response to matrix stretching. J Theor Biol 264(2):593–603

Kawakami K, Tatsumi H, Sokabe M (2001) Dynamics of integrin clustering at focal contacts of endothelial cells studied by multimode imaging microscopy. J Cell Sci 114(Pt 17):3125–3135

Kimura K, Ito M, Amano M, Chihara K, Fukata Y, Nakafuku M, Yamamori B, Feng J, Nakano T, Okawa K, Iwamatsu A (1996) Regulation of myosin phosphatase by Rho and Rho-associated kinase (Rho-kinase). Science 273(5272):245–248

Langanger G, Moeremans M, Daneels G, Sobieszek A, De Brabander M, De Mey J (1986) The molecular organization of myosin in stress fibers of cultured cells. J Cell Biol 102(1):200–209

Lazarides E, Burridge K (1975) Alpha-actinin: immunofluorescent localization of a muscle structural protein in nonmuscle cells. Cell 6(3):289–298

Lee CF, Haase C, Deguchi S, Kaunas R (2010) Cyclic stretch-induced stress fiber dynamics – dependence on strain rate, Rho-kinase and MLCK. Biochem Biophys Res Commun 401 (3):344–349

Lu L, Feng Y, Hucker WJ, Oswald SJ, Longmore GD, Yin FC (2008) Actin stress fiber pre-extension in human aortic endothelial cells. Cell Motil Cytoskeleton 65(4):281–294

Maciver SK, Zot HG, Pollard TD (1991) Characterization of actin filament severing by actophorin from Acanthamoeba castellanii. J Cell Biol 115(6):1611–1620

Matsui TS, Ito K, Kaunas R, Sato M, Deguchi S (2010) Actin stress fibers are at a tipping point between conventional shortening and rapid disassembly at physiological levels of MgATP. Biochem Biophys Res Commun 395(3):301–306

Matsui TS, Kaunas R, Kanzaki M, Sato M, Deguchi S (2011) Non-muscle myosin II induces disassembly of actin stress fibres independently of myosin light chain dephosphorylation. Interface Focus 1(5):754–766

Mizutani T, Haga H, Kawabata K (2004) Cellular stiffness response to external deformation: tensional homeostasis in a single fibroblast. Cell Motil Cytoskeleton 59(4):242–248

Neidlinger-Wilke C, Grood ES, Wang J-C, Brand RA, Claes L (2001) Cell alignment is induced by cyclic changes in cell length: studies of cells grown in cyclically stretched substrates. J Orthop Res 19(2):286–293
Park JS, Chu JS, Cheng C, Chen F, Chen D, Li S (2004) Differential effects of equiaxial and uniaxial strain on mesenchymal stem cells. Biotechnol Bioeng 88(3):359–368
Polte TR, Eichler GS, Wang N, Ingber DE (2004) Extracellular matrix controls myosin light chain phosphorylation and cell contractility through modulation of cell shape and cytoskeletal prestress. Am J Physiol Cell Physiol 286(3):C518–C528
Ridley AJ, Hall A (1992) The small GTP-binding protein rho regulates the assembly of focal adhesions and actin stress fibers in response to growth factors. Cell 70(3):389–399
Riveline D, Zamir E, Balaban NQ, Schwarz US, Ishizaki T, Narumiya S, Kam Z, Geiger B, Bershadsky AD (2001) Focal contacts as mechanosensors: externally applied local mechanical force induces growth of focal contacts by an mDia1-dependent and ROCK-independent mechanism. J Cell Biol 153(6):1175–1186
Sanger JW, Sanger JM, Jockusch BM (1983) Differences in the stress fibers between fibroblasts and epithelial cells. J Cell Biol 96(4):961–969
Sato K, Adachi T, Matsuo M, Tomita Y (2005) Quantitative evaluation of threshold fiber strain that induces reorganization of cytoskeletal actin fiber structure in osteoblastic cells. J Biomech 38(9):1895–1901
Sato K, Adachi T, Shirai Y, Saito N (2006) Local disassembly of actin stress fibers induced by selected release of intracellular tension in osteoblastic cells. J Biomech Sci Eng 1(1):204–214
Sawada Y, Sheetz MP (2002) Force transduction by Triton cytoskeletons. J Cell Biol 156 (4):609–615
Sawada Y, Tamada M, Dubin-Thaler BJ, Cherniavskaya O, Sakai R, Tanaka S, Sheetz MP (2006) Force sensing by mechanical extension of the Src family kinase substrate p130Cas. Cell 127 (5):1015–1026
Schoenwaelder SM, Burridge K (1999) Bidirectional signaling between the cytoskeleton and integrins. Curr Opin Cell Biol 11(2):274–286
Smith PG, Roy C, Zhang YN, Chauduri S (2003) Mechanical stress increases RhoA activation in airway smooth muscle cells. Am J Respir Cell Mol Biol 28(4):436–442
Smith MA, Blankman E, Gardel ML, Luettjohann L, Waterman CM, Beckerle MC (2010) A zyxin-mediated mechanism for actin stress fiber maintenance and repair. Dev Cell 19 (3):365–376
Takei T, Rivas-Gotz C, Delling CA, Koo JT, Mills I, McCarthy TL, Centrella M, Sumpio BE (1997) Effect of strain on human keratinocytes in vitro. J Cell Physiol 173(1):64–72
Tanner K, Boudreau A, Bissell MJ, Kumar S (2010) Dissecting regional variations in stress fiber mechanics in living cells with laser nanosurgery. Biophys J 99(9):2775–2783
Totsukawa G, Yamakita Y, Yamashiro S, Hartshorne DJ, Sasaki Y, Matsumura F (2000) Distinct roles of ROCK (Rho-kinase) and MLCK in spatial regulation of MLC phosphorylation for assembly of stress fibers and focal adhesions in 3T3 fibroblasts. J Cell Biol 150(4):797–806
Watanabe N, Kato T, Fujita A, Ishizaki T, Narumiya S (1999) Cooperation between mDia1 and ROCK in Rho-induced actin reorganization. Nat Cell Biol 1(3):136–143
Wilson CA, Tsuchida MA, Allen GM, Barnhart EL, Applegate KT, Yam PT, Ji L, Keren K, Danuser G, Theriot JA (2010) Myosin II contributes to cell-scale actin network treadmilling through network disassembly. Nature 465(7296):373–377
Yin HL, Stossel TP (1979) Control of cytoplasmic actin gel-sol transformation by gelsolin, a calcium-dependent regulatory protein. Nature 281(5732):583–586

Mechanical Characterization of Vascular Endothelial Cells Exposed to Fluid Shear Stress

6

Toshiro Ohashi

Abstract

Vascular endothelial cells are constantly exposed to shear stress generated by blood flow and respond by exhibiting alterations in morphology and cytoskeletal structures as well as by modulating cell physiological functions. Since endothelial cell responses to fluid shear stress have been implicated in the localization of atherosclerosis, the effects of fluid shear stress on endothelial cell morphology and functions have been exclusively studied. Interestingly, atherosclerosis occurs primarily at branching and curving regions of arterial walls, where endothelial cells would experience complex blood flow patterns. So far, a lot of efforts have been made to study endothelial mechanotransduction to flow, indicating the fact that after applying fluid shear stress, endothelial cells exhibit marked elongation and orientation in the direction of flow. The need for experimental techniques for studying endothelial cell responses to flow has lead to the development of different types of flow chambers. Conventional flow chambers include a cone-and-plate flow chamber and a parallel-plate flow chamber, while more recently, microfluidic flow chambers have emerged with a great potential for a high throughput analysis. In this chapter, many types of flow chambers are first summarized. Stiffness change of sheared endothelial cells has been be of great interest in particular for mechanical engineering researchers because endothelial cells may change their morphology and cytoskeletal structures in response to fluid shear stress. Therefore, AFM (Atomic Force Microscopy)

T. Ohashi (✉)
Division of Human Mechanical Systems and Design, Faculty of Engineering, Hokkaido University, Kita 13, Nishi 8, Kita-ku, Sapporo, Hokkaido 060-8628, Japan
e-mail: ohashi@eng.hokudai.ac.jp

K. Tanishita, K. Yamamoto (eds.), *Vascular Engineering*,
DOI 10.1007/978-4-431-54801-0_6

stiffness measurement of sheared endothelial cells is then described. Lastly, stiffness change of sheared endothelial cell nuclei measured with a pipette aspiration test is also presented.

Keywords
Vascular endothelial cells • Remodeling • Fluid shear stress • Cell stiffness • Nucleus stiffness

6.1 Vascular Endothelial Cells and Fluid Shear Stress

Vascular endothelial cells form a monolayer lining the luminal surface of blood vessel walls and are always exposed to shear stress generated by blood flow. It is well known that endothelial cells respond to fluid shear stress, exhibiting alterations in morphology and cytoskeletal structures (Dewey et al. 1981; Levesque and Nerem 1985; Kataoka et al. 1998; Mott and Helmke 2007; Helmke and Davies 2002; Hsu et al. 2005) as well as physiological functions (White et al. 2001; Garin and Berk 2006; Ogunrinade et al. 2002). Since endothelial cell responses to fluid shear stress have long been implicated in the localization of atherosclerosis, the effects of fluid shear stress on endothelial cell remodeling and associated functional changes have been exclusively studied. Atherosclerosis is a disease where plaque accumulates inside of arterial walls and can potentially induce serious pathological conditions including heart attack, stroke, etc. Interestingly, atherosclerosis occurs primarily at branching and curving regions of arterial walls, where endothelial cells would experience irregular blood flow patterns. Therefore, the effects of fluid shear stress on endothelial cell responses have been studied over the last few decades.

It is widely accepted that after exposure to fluid shear stress, endothelial cells become elongated and aligned in the direction of flow. In fact, in *in vivo* experiments, Flaherty et al. (1972) studied the relationship between orientation of endothelial cell nuclei and blood flow pattern using a canine artery. In straight regions, the nuclei became elongated and oriented parallel to the longitudinal axis of the artery, while in entrance regions, less ordered nuclei orientation and elongation were observed. Nerem et al. (1981) demonstrated the quantitative study of size and shape of endothelial cells using a vascular cast of rabbit aorta and suggested that endothelial cell morphology and orientation around a branch vessel might be an indicator of the detailed features of blood flow. To date, a lot of attempts have been also made to apply well-controlled fluid shear stress on an endothelial cell monolayer *in vitro*. Dewey et al. (1981) demonstrated bovine endothelial cell alignment after applying steady flow. Levesque and Nerem (1985) further reported that the elongation and orientation of bovine endothelial cells to flow direction were dependent on flow exposure time and magnitude of shear stress.

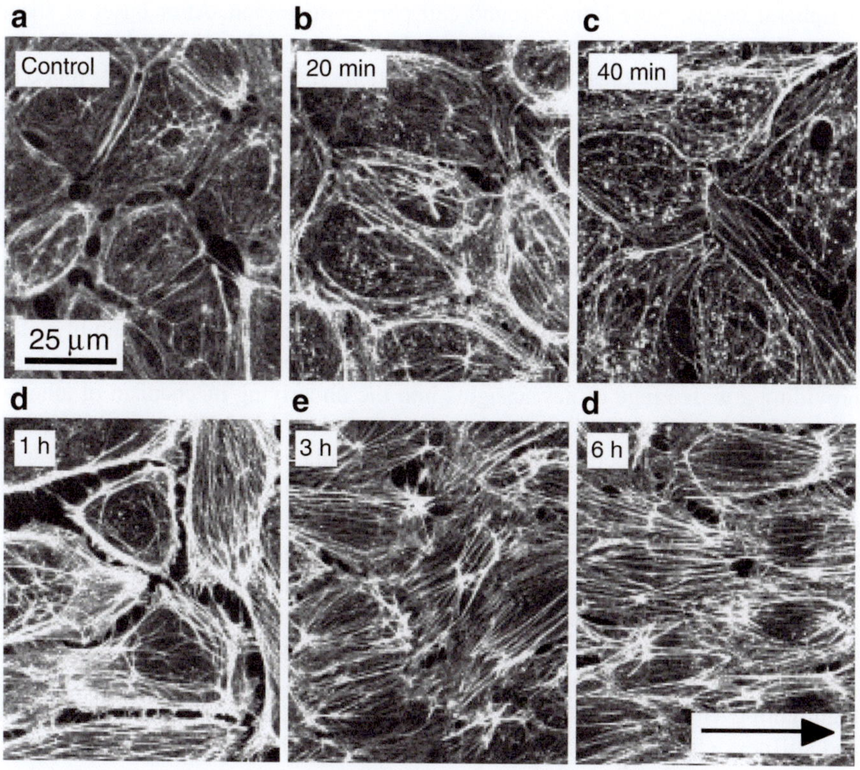

Fig. 6.1 Typical fluorescent images of actin filaments in bovine aortic endothelial cells before and after exposure to shear stress of 2 Pa. (**a**) control, (**b**) after 20 min, (**c**) 40 min, (**d**) 1 h, (**e**) 3 h, and (**f**) 6 h (From Kataoka et al. 1998 with permission)

As aforementioned, endothelial cells exposed to fluid shear stress exhibit marked elongation and orientation in the direction of flow. It is known that the morphological change is accompanied by the reorganization of cytoskeletal structure represented by thick stress fibers of actin filaments (Levesque and Nerem 1985; Kataoka et al. 1998). For instance, Kataoka et al. (1998) exposed fluid shear stress of 2 Pa for 20, 40 min, 1, 3, and 6 h on bovine endothelial cells. After the flow-imposed experiments, actin filaments were fixed with formaline, stained with rhodamine-phalloidin, and observed with a LSM (Laser Scanning Microscope). Typical fluorescent images of actin filaments are shown in Fig. 6.1 for statically cultured and sheared cells, where actin filaments were stained with rhodamine-phalloidin. Under static conditions, endothelial cells showed a cobblestone appearance with a relatively rounded shape. Thick bundles of actin fibers, called dense peripheral bands, were found around cell peripheral region. After onset of shear stress, time course of cell remodeling can be summarized as follows. At 20 min, stress fibers, large bundles of actin filaments, along the cell long axis were found in some elongated cells. At 40 min, stress fibers appeared in more cells. Between 1 and 3 h, many cells started to elongate and orient in the flow direction. Finally, most cells displayed marked elongation and orientation in the flow direction with centrally located thick stress fibers. This fact may lead to an idea that the stiffness of endothelial cells could increase under fluid shear stress due to the structural remodeling, resulting in modulation in cell physiological functions. It has been revealed that endothelial cells have many flow-induced functional changes including proliferation, migration, differentiation, secretion, adhesion, permeability, etc.

Studies on endothelial cell remodeling in response to fluid shear stress have been providing a wide range of new insights into the underlying mechanism of atherosclerosis, which has drawn a great attention of not only biomedical researchers but also mechanical engineering researchers. In particular, stiffness change of sheared endothelial cells would be of great interest because endothelial cells sense fluid shear stress and convert it into intracellular biochemical signals through a complex process termed mechanotransduction (Davies 1995; Chien 2007). External mechanical forces can be sensed by mechanosensors located at apical membrane, at focal adhesions and cell-cell junctions, and are then potentially transmitted inside of cells. In this chapter, AFM (Atomic Force Microscopy) stiffness measurement of endothelial cells before and after exposure to fluid shear stress is described, highlighting changes in morphology and cytoskeletal structures. Morphological and stiffness change in nuclei of endothelial cell nuclei measured with a pipette aspiration test is also presented.

6.2 Fluid Flow-Imposed Systems

Conventional fluid-flow imposed systems include a parallel-plate flow system (Sato et al. 2000; Ohashi et al. 2000, 2002; Ohashi and Sato 2005). A flow circuit and a flow chamber are schematically shown in Fig. 6.2. The flow circuit consists of the flow chamber, a reservoir, a roller pump and a damping chamber (Fig. 6.2a). In the system, fluid flow between two parallel flat plates separated by a narrow gap is created by an imposed pressure gradient between both ends of the rectangular chamber, causing a fully developed parabolic laminar flow profile, as shown in Fig. 6.2b. The flow chamber has a constant rectangular cross section along the length of the chamber, which can provide a constant wall shear stress over the entire chamber. The pressure gradient is generated by the roller pump. The parallel-plate flow system has many practical advantages, such as simplicity of the system, homogeneity of stress stimulus, and easy observation under microscope. In experiments, endothelial cells are cultured in a cell culture dish with a diameter

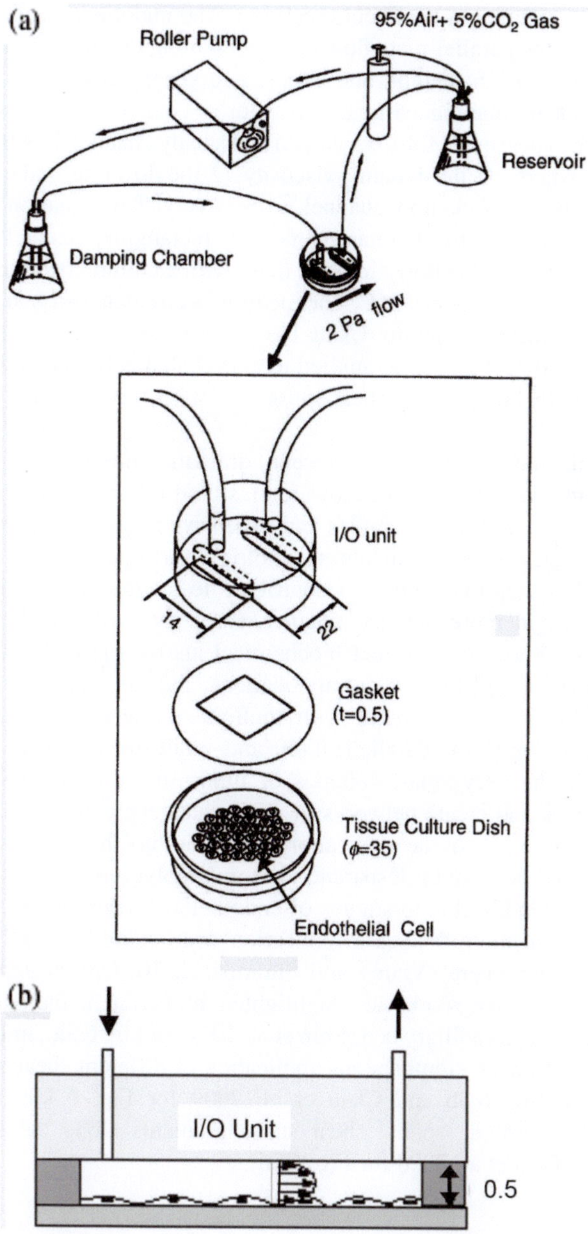

Fig. 6.2 Schematic diagram of (**a**) flow circuit and flow chamber (From Ohashi et al. 2000 with permission) and (**b**) flow section of the flow chamber

of 35 mm until reaching a confluent condition. The endothelial cell monolayer is then loaded into the parallel-plate flow chamber having the flow channel of 14 mm in width and 0.5 mm in height, and exposed to shear stress. Statically cultured endothelial cells are maintained in an incubator as controls. The shear stress, τ, to which cells are exposed in a fully developed and steady channel flow is expressed as $\tau = 6\,\mu Q/wh^2$, where μ is the dynamic viscosity, Q, the flow rate, and w and h are the height and the width of the flow channel, respectively. This equation is derived on the assumption that the chamber has a very wide rectangular channel with an ideal case of infinite width. This flow circuit is filled with a culture medium equilibrated with 95 % air and 5 % CO_2 at 37 °C. The medium is circulated around the flow loop with the help of the roller pump. Using the present system, a wide range of fluid shear stress (up to 15 Pa) can be applied to endothelial cell monolayer in laminar flow. It is generally known that a shear stress of 2 Pa is *in vivo* average value in large arteries.

It is significant to introduce recent dramatic progress in microfluidic technologies miniaturizing devices towards fast and reliable high-throughput cell analysis, low consumption of samples and expensive reagents and low fabrication cost. Microfluidics is a promising technology for realizing microscale flow chambers and is therefore hoped to contribute to studies with wider variety of cell experiments in more detailed manner. Over the past decade, microfluidic technologies such as chemical and biochemical analysis have been increasingly used for biological and biomedical applications. Indeed, some cell experiments need to be performed in parallel with multiple sequences, microfluidics is a powerful technology for parallelization and multifunctionality. Microfluidic devices handle the very small volumes of fluids including micro-, nano-, and picolitre volumes and create networks of microchannels with dimensions ranging from several mm to μm. In the application to cell biology, microfluidic devices are typically made of glass and transparent, elastomic polydimethylsiloxane (PDMS), and are therefore applicable to studies of endothelial cell responses to fluid shear stress. Reviews for microfluidics inconnection with endothelial cell studies have been published elsewhere (Young and Simmons 2010; Ohashi and Sato 2012). Some of representative works are highlighted in Fig. 6.3, including a PDMS network resembling a capillary bed (Shin et al. 2004 for Fig. 6.3a), multiple parallel microchannels allowing simultaneous application of different shear stress (Young et al. 2007 for Fig. 6.3b and Chau et al. 2009 for Fig. 6.3c) and a tapered microchannel to provide spatial shear stress gradients along the length of the microchannel (Tsou et al. 2008 for Fig. 6.3d).

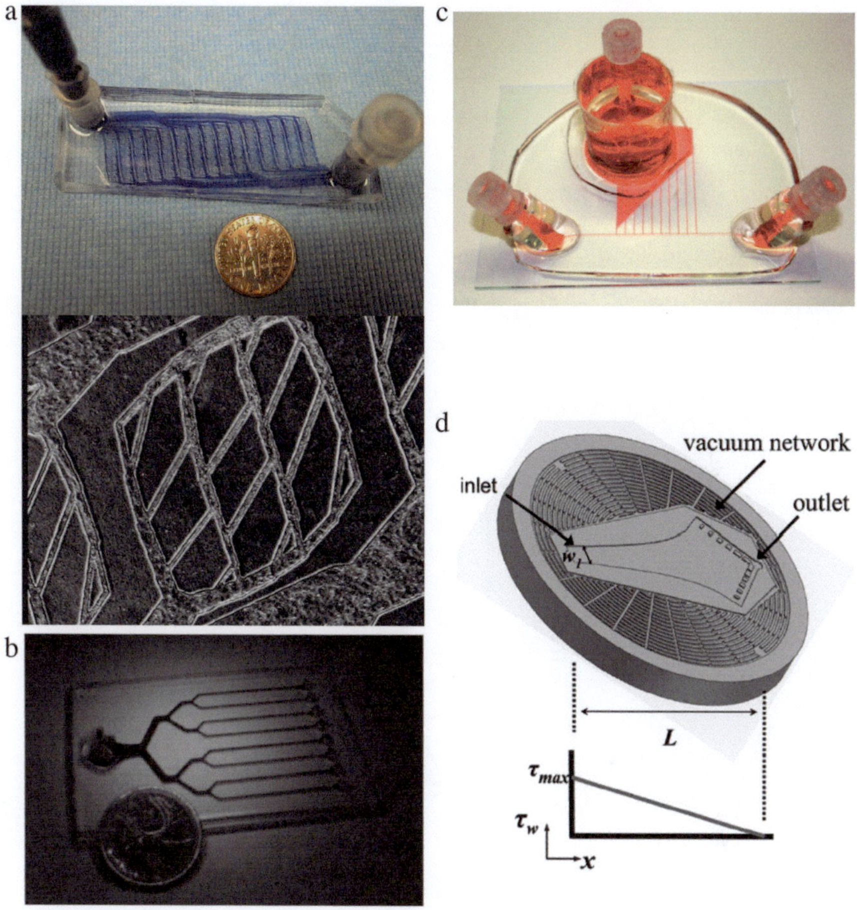

Fig. 6.3 Microfluidic channels for flow-imposed experiments. (**a**) Microchannel resembling a capillary bed (Shin et al. 2004), (**b**) eight parallel microchannels (Young et al. 2007), (**c**) microchannel with linearly increased lengths (Chau et al. 2009), and (**d**) tapered microchannel (Tsou et al. 2008) (From Ohashi and Sato 2012 with permission)

6.3 Changes in Mechanical Properties of Sheared Endothelial Cells

After the important *in vivo* finding by Flaherty et al. (1972) where they found orientation of endothelial cell nuclei parallel to the direction of blood flow, a large number of studies have been reported by many researchers to study the effects of fluid shear stress on endothelial cell morphology and cytoskeletal structures. Interesting characteristic of endothelial cell responses to fluid flow is that endothelial cells become stiffer, which was first reported by a micropipette aspiration test by

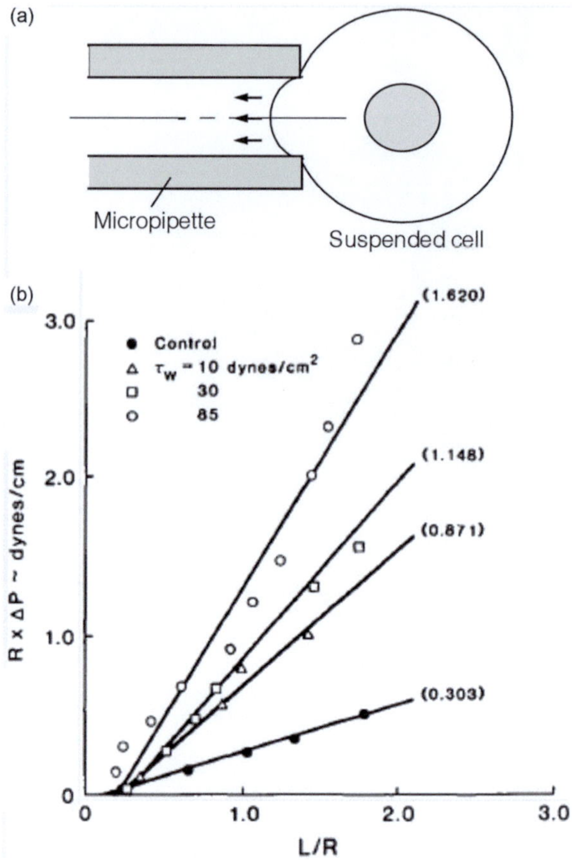

Fig. 6.4 (**a**) Schematic diagram of pipette aspiration test. (**b**) The effect of fluid shear stress on stiffness of endothelial cells. R × ΔP is the tension imposed on the cell and L/R the nondimensional aspiration length, where L is the aspirated length of the cell in the micropipette and R is the radius of the micropipette (From Sato et al. 1987 with permission)

Sato et al. (1987) and later by an AFM indentation test by Sato et al. (2000) and Ohashi et al. (2002).

In the micropipette aspiration test, a blunt ended, polished glass micropipette with a diameter of few μm is positioned on the surface of cells with a tare pressure. Negative pressure is then applied in the micropipette to aspirate the cell surface, as shown in Fig. 6.4a. The aspiration length of cells into the micropipette is measured from brightfield images. Sato et al. (1987) applied a specific steady shear stress of either 1, 3, or 8.5 Pa to cultured bovine aortic endothelial cells for a duration ranging from 0.5 to 24 h and aspirated the suspended bovine endothelial cell surface by using a glass micopipette with an inner diameter of a few μm and estimated a stiffness through the negative pressure-aspirated length curve. The stiffness of cells exposed to shear stress significantly greater than that of control cells (Fig. 6.4b), which was dependent on both the level of shear stress and the duration of exposure. The micropipette aspiration test gives a global stiffness, which is reflected by an averaged stiffness of a region corresponding to the diameter of the micropipette to be used. While in the AFM indentation test, very soft cantilever tip locally indents

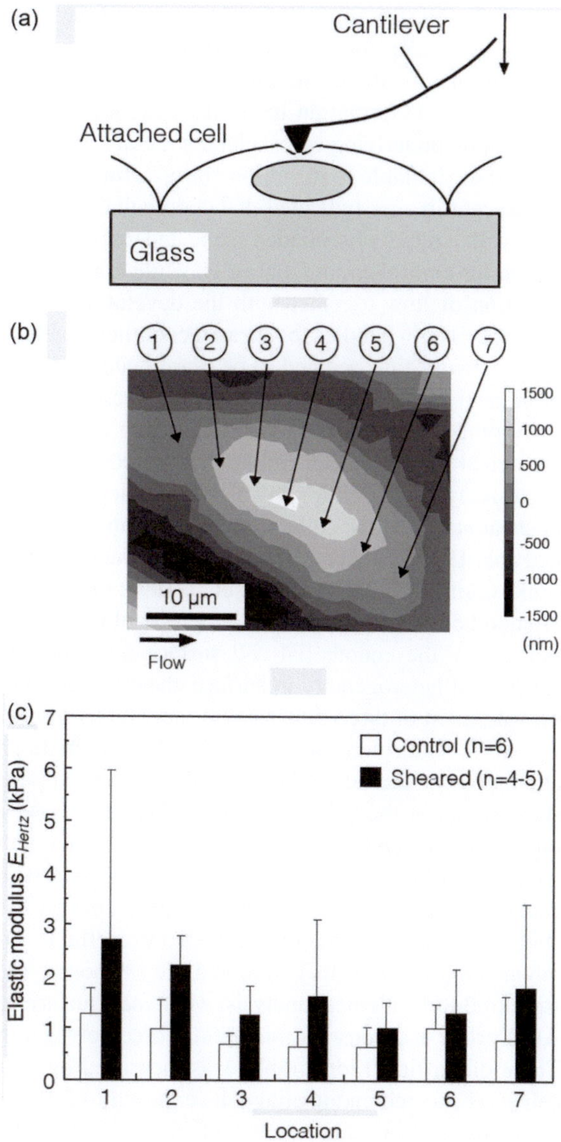

Fig. 6.5 (**a**) Schematic diagram of AFM indentation test. (**b**) Typical contour map measured by AFM for a sheared cell and measured locations (1–7) and (**c**) elastic modulus using the Hertz model for control and sheared cells (From Ohashi et al. 2002 with permission)

the surface of the attached cells (Fig. 6.5a) and the local stiffness can be obtained through the resisting force-indentation depth relationship.

Sato et al. (2000) and Ohashi et al. (2002) have utilized the AFM indentation test to measure local mechanical properties of bovine endothelial cells exposed to shear stress. In Ohashi et al. (2002), the Hertz model (Radmacher et al. 1995) was applied to the force-indentation curve to estimate the elastic modulus. Typical contour map for a sheared endothelial cell measured by the AFM and the locations (1–7) where the AFM indentation measurement was carried out are shown in Fig. 6.5b. The averaged elastic modulus obtained through all the locations indicated 0.87 ± 0.23 kPa for control and 1.75 ± 0.43 kPa for sheared cells ($p < 0.01$ vs. control), as shown in Fig. 6.5c. Since it can be considered that the major resistance to deformation induced by the indentation, stress bearing components, is given by underlying cytoskeletal structure, the increase in mechanical properties may be attributable to development of actin stress fibers for sheared cells. In fact, fluorescent images indicated that endothelial cells exhibited a cobblestone appearance with a relatively rounded shape and thick bundles actin fibers at cell peripheral region for control endothelial cells, while marked elongation and orientation in the direction of flow together with the development of centrally located thick stress fibers of F-actin bundles for sheared endothelial cells, as can be seen in Fig. 6.1. The actin network thus seemed to be rearranged from circumferential fibers to central prominent stress fibers with fluid shear stress.

Another effort to study the remodeling of sheared endothelial cells includes a numerical approach. Several studies by computational fluid dynamics have been exclusively attempted for characterizing the local flow field, the spatial distribution of shear stress, at the cellular level using an idealized cell topography (Dewey and DePaola 1989; Satcher et al. 1992; Yamaguchi et al. 2000) and real cell geometries (Fukushima et al. 2001; Barbee et al. 1995; Ohashi et al. 2000). In fact, the surfaces of endothelial cells have three-dimensional topography and wall shear stress experienced by the endothelial cell surface may thus vary locally, possibly causing substantial heterogeneity in surface shear stress. The use of AFM allows accurate measurement of three-dimensional topography of endothelial cells. On the basis of the AFM data, pioneering works with computational fluid dynamics have been attempted for characterizing local flow fields, in particular spatial distributions of shear stress, at the cellular level (Barbee et al. 1995). They measured the surface topography of endothelial cell monolayer under no-flow and flow conditions by using an AFM, and then simulated shear stress distributions, resulting in a significant decrease in the peak shear stress from no-flow to flow conditions. In a separate study, Ohashi et al. (2000) studied intracellular stress distributions in cultured bovine endothelial cells before and after exposure to fluid shear stress using fluid-structural finite element analysis, where cell surface topography was scanned by an AFM probe at a constant interaction force between the probe and the cell surface. Three-dimensional topography was measured by the AFM for control endothelial cells and sheared endothelial cells exposed to 2 Pa for 6 h. The control cells were polygonal and the cell surfaces were relatively smooth. In contrast, the sheared cells showed marked elongation and orientation in the flow direction. The peak cell height significantly decreased with fluid flow from 2.2 ± 0.4 μm (mean ± SD, $n = 4$) to 1.2 ± 0.4 μm ($n = 4$, $p < 0.05$ vs. control). On the basis of the AFM-mapped cell surface, the three-dimensional finite element model was generated, which consisted of two major domains: a fluid domain and a solid domain representing the flow field and the endothelial cells, respectively. The flow field was assumed to be a laminar, steady-state, and viscous fluid. The solid domain was assumed to be homogeneous, isotropic, linear elastic, and the zero-stress state. The results showed that although higher stress appears in the central region above the nuclei for both cells, intracellular stress gradients seem to be reduced by the alignment response. These results may indicate that endothelial cells change their cytoskeletal structures to increase their stiffness and reduce stress concentration in the cells. The spatial distributions of shear stress in cells may locally induce differential mechanotransduction events and furthermore differences in shear stress between neighboring cells may lead to heterogeneous cell responses.

6.4 Remodeling of Endothelial Cell Nucleus Under Fluid Shear Stress

Endothelial cells exposed to shear stress exhibit not only cell elongation and orientation but also orientation of nuclei parallel to the direction of blood flow, as can be seen by Flaherty et al. (1972). Although endothelial cell responses to mechanical stimuli have been well understood so far, little is still known of how endothelial cell nuclei may remodel themselves. The major function of cell nuclei is to control gene expression and mediate the replication of DNA during cell cycle, which could be potentially modulated by external forces. For example, twisting integrin receptors with RGD-containing peptide-coated beads increased endothelin-1 gene expression under the hypothesis that the molecular structural linkage of integrin-cytoskeleton is an important pathway for the mechanotransduction (Chen et al. 2001). Furthermore, when integrins were pulled by the RGD-coated microbeads, cell nuclei distorted along the direction of the applied force (Maniotis et al. 1997). In fact, it has been known that nucleus deformation can be associated with cell deformation when cells are exposed to fluid shear stress (Deguchi et al. 2005), compression (Guilak 1995), stretching (Caille et al. 1998), and flatten (Ingber 1990). These results could possively indicate direct mechanical linkages between cell surface receptors, cytoskeletons, and nuclei. For better understanding of endothelial cell responses to mechanical forces, cell nucleus responses should be thus explored.

Ohashi et al. (2008) applied fluid shear stress of 2 Pa to a monolayer of confluent bovine aortic endothelial cells for 24 h using a conventional flow circuit and observed the nuclei morphology. Fluorescence images of nucleus and cell membrane after the application of flow are shown in Fig. 6.6a, where cell membrane and nuclei are stained with CM-DiI and SYTO 13, respectively. Changes in nucleus morphology were clearly observed in response to the mechanical stimuli, showing a close association with cell shape change. The angles of orientation of cells and their nuclei before and after exposure to flow were shown in Fig. 6.6b, c respectively. The angle of orientation is defined as the deviation of the major axis of the equivalent ellipse from the direction of flow, by setting the cell orientation to a value ranging from 0° to $\pm$90°. For control cells, the angle of cell orientation was uniformly distributed between angles approaching −90° or 90°. The similar tendency was observed for the angle of nucleus orientation. The percentages of sheared cells aligning between −30° and +30° were approximately 60 % and 70 % for the cells and their nuclei, respectively, indicating a close association of the cells with the nuclei in morphological change. As Maniotis et al. (1997) pointed out that endothelial cell nuclei and cell membrane are hard-wired with actin filament network. It can

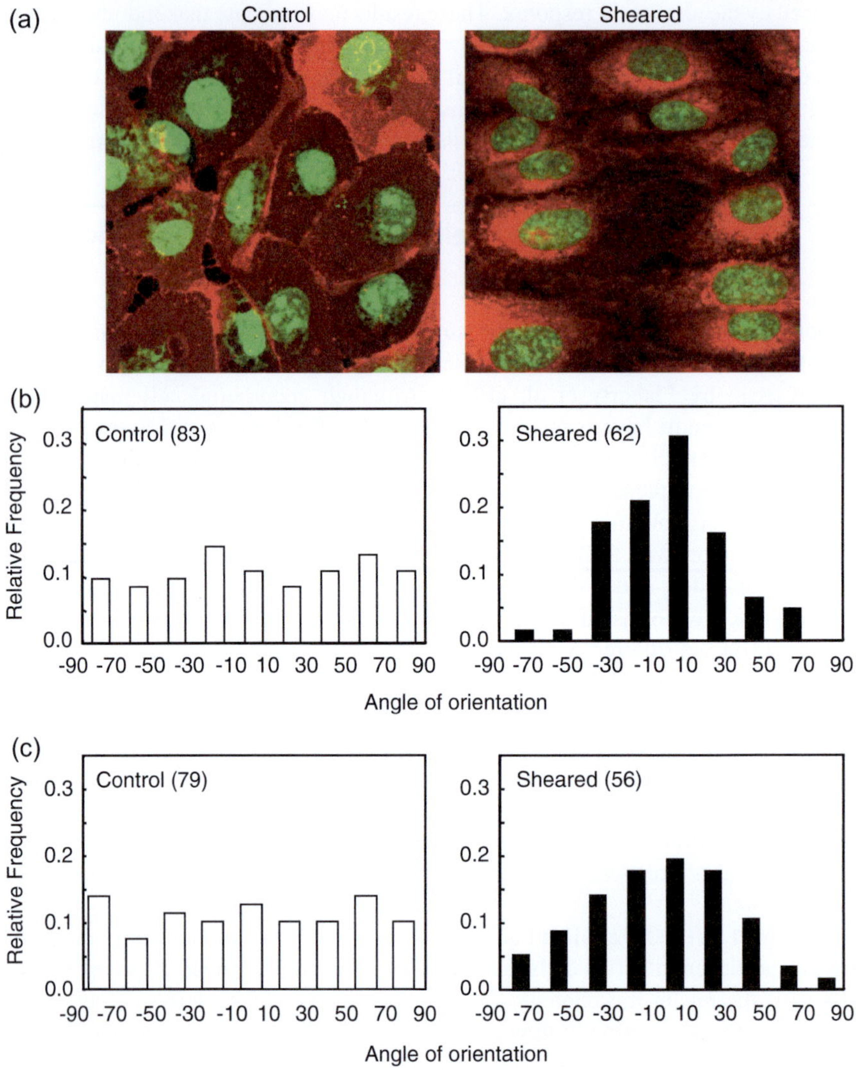

Fig. 6.6 (**a**) Fluorescence images of cell nuclei and membrane for statically cultured control (*left*) and sheared endothelial cells (*right*). Bars = 30 μm. Change in the angle of orientation for (**b**) cells and (**c**) their nuclei after application of flow (Modified from Ohashi et al. 2008 with permission)

be hypothesized from their experimental result that external mechanical forces are transmitted to nuclei via actin structure, possibly leading to remodeling of cell nuclei. To test the hypothesis, Deguchi et al. (2005) isolated nuclei from bovine aortic endothelial cells exposed to fluid shear stress with a chemical isolation technique and measured the Young's modulus of cell nuclei with the micropipette aspiration test with a micropipette having a diameter of a few μm. Typical phase-contrast and fluorescence (stained with SYTO 13) images are show in Fig. 6.7a. The

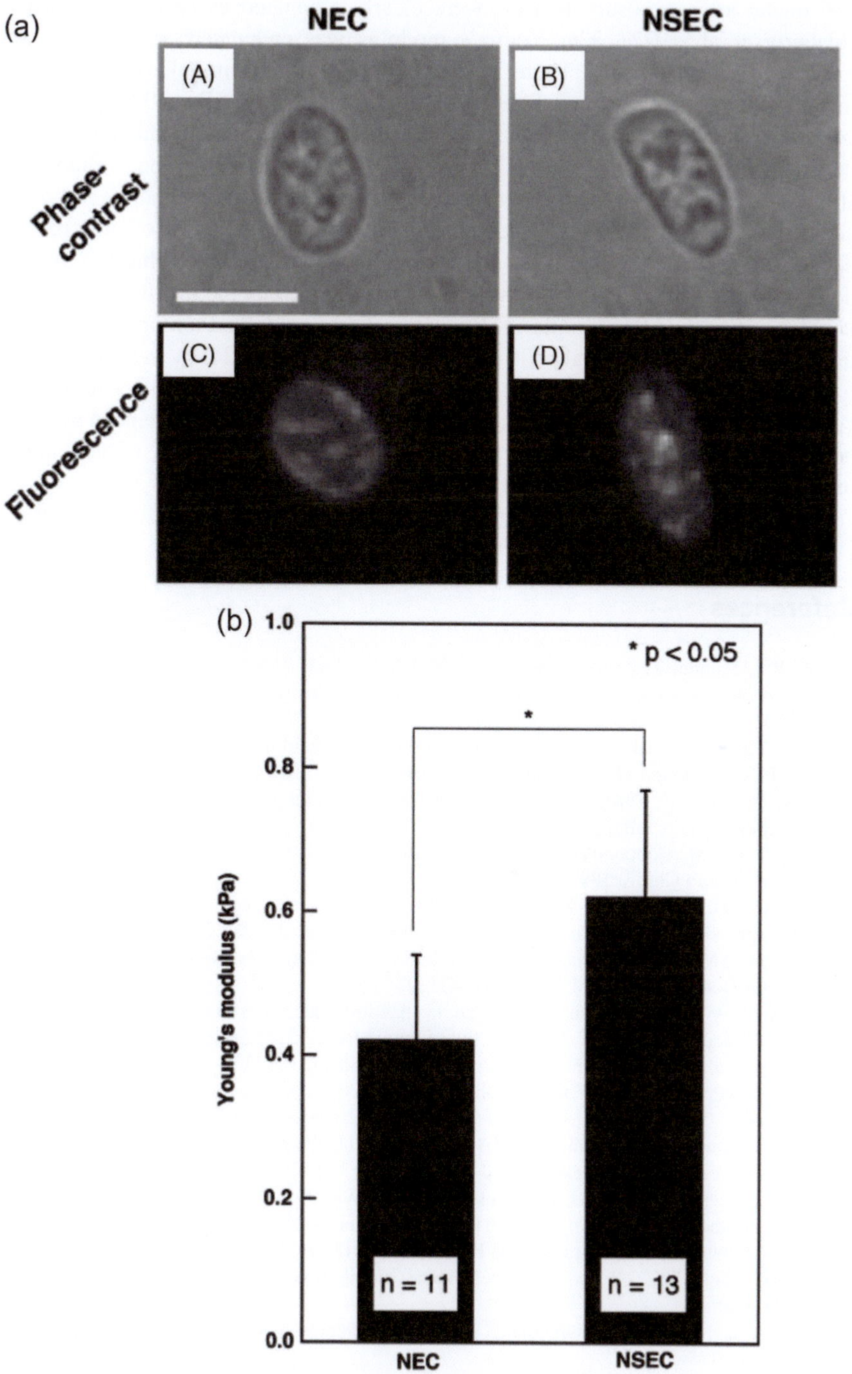

Fig. 6.7 (**a**) Phase contrast and fluorescence images of nuclei isolated from control (*A*, *C*) and sheared endothelial cells (*B*, *D*). (**b**) Change in the Young's modulus of cell nuclei before and after application of flow (Modified from Deguchi et al. 2005 with permission)

cell nuclei after exposure to flow were clearly elongated as *in vivo* experiment (Flaherty et al. 1972). The Young's modulus was calculated according to the previously reported theory (Theret et al. 1988). The Young's modulus of cell nucleus increased significantly from 0.42 ± 0.12 kPa ($n = 11$, mean $\pm$ SD) to 0.62 ± 0.15 kPa ($n = 13$) with the fluid shear stress. The findings of these studies indicate that cell nuclei change their shape and stiffness in accordance with application of fluid shear stress. Since it has been reported that the nucleus is 9 times stiffer than the rest of the cell (Maniotis et al. 1997), it seems difficult for the nuclei to exhibit passive deformation. Moreover, Caille et al. (1998, 2002) have reported that the nucleus deforms less than the cytoplasm, which can be explained by the fact that the nucleus is much stiffer than the cytoplasm. It is interesting to note that a numerical simulation showed that endothelial cells may respond to the applied flow in such a way as to minimize the total force on their nuclei (Hazel and Pedley 2000). It is possible that cell nuclei remodel themselves so as to minimize strain energy experienced by them; however, further studies are needed to elucidate the mechanism of cell nuclei hardening in response to flow.

References

Barbee KA, Mundel T, Lal R, Davies PF (1995) Subcellular distribution of shear stress at the surface of flow-aligned and nonaligned endothelial monolayers. Am J Physiol 268:H1765–H1772

Caille N, Tardy Y, Meister J-J (1998) Assessment of strain field in endothelial cells subjected to uniaxial deformation of their substrate. Ann Biomed Eng 26:409–416

Caille N, Thoumine O, Tardy Y, Meister J-J (2002) Contribution of the nucleus to the mechanical properties of endothelial cells. J Biomech 35:177–187

Chau L, Doran M, Cooper-White J (2009) A novel multishear microdevice for studying cell mechanics. Lab Chip 9:1897–1902

Chen J, Fabry B, Schiffrin EL, Wang N (2001) Twisting integrin receptors increases endothelin-1 gene expression in endothelial cells. Am J Physiol Cell Physiol 280:C1475–C1484

Chien S (2007) Mechanotransduction and endothelial cell homeostasis: the wisdom of the cell. Am J Physiol Heart Circ Physiol 292:H1209–H1224

Davies PF (1995) Flow-mediated endothelial mechanotransduction. Physiol Rev 75:519–560

Deguchi S, Maeda K, Ohashi T, Sato M (2005) Flow-induced hardening of endothelial nucleus as an intracellular stress-bearing organelle. J Biomech 38:1751–1759

Dewey CF, DePaola N (1989) Exploring flow-cell interactions using computatioal fluid dynamics. In: Woo SL-Y, Seguchi Y (eds) Tissue eng. ASME, New York, pp 31–33

Dewey CF Jr, Bussolari SR, Gimbrone MA Jr, Davies PF (1981) The dynamic response of vascular endothelial cells to fluid shear stress. J Biomech Eng 103:177–185

Flaherty JT, Pierce JE, Ferrans VJ, Patel DJ, Tucker WK, Fry DL (1972) Endothelial nuclear patterns in the canine arterial tree with p reference to hemodynamic events. Circ Res 30:23–33

Fukushima S, Ngatsu A, Kaibara M, Oka K, Tanishita K (2001) Measurement of surface topography of endothelial cell and wall shear stress distribution on the cell. JSME Int J Ser C 44:972–980

Garin G, Berk BC (2006) Flow-mediated signaling modulates endothelial cell phenotype. Endothelium 13:375–384

Guilak F (1995) Compression-induced changes in the shape and volume of the chondrocyte nucleus. J Biomech 28:1529–1541

Hazel AL, Pedley TJ (2000) Vascular endothelial cells minimize the total force on their nuclei. Biophys J 78:47–54

Helmke BP, Davies PF (2002) The cytoskeleton under external fluid mechanical forces: hemodynamic forces acting on the endothelium. Ann Biomed Eng 30:284–296

Hsu S, Thakar R, Liepmann D, Li S (2005) Effects of shear stress on endothelial cell haptotaxis on micropatterned surfaces. Biochem Biophys Res Commun 337:401–409

Ingber DE (1990) Fibronectin controls capillary endothelial cell growth by modulating cell shape. Proc Natl Acad Sci U S A 87:3579–3583

Kataoka N, Ujita S, Sato M (1998) Effect of flow direction on the morphological responses of cultured bovine aortic endothelial cells. Med Biol Eng Comput 36:122–128

Levesque MJ, Nerem RM (1985) The elongation and orientation of cultured endothelial cells in response to shear stress. Trans ASME J Biomech Eng 107:341–347

Maniotis AJ, Chen CS, Ingber DE (1997) Demonstration of mechanical connections between integrins, cytoskeletal filaments, and nucleoplasm that stabilize nuclear structure. Proc Natl Acad Sci U S A 94:849–854

Mott RE, Helmke BP (2007) Mapping the dynamics of shear stress-induced structural changes in endothelial cells. Am J Physiol Cell Physiol 293:C1616–C1626

Nerem RM, Levesque MJ, Cornhill JF (1981) Vascular endothelial morphology as an indicator of the pattern of blood flow. Trans ASME J Biomech Eng 103:172–176

Ogunrinade O, Kameya GT, Truskey GA (2002) Effect of fluid shear stress on the permeability of the arterial endothelium. Ann Biomed Eng 30:430–446

Ohashi T, Sato M (2005) Remodeling of vascular endothelial cells exposed to fluid shear stress: experimental and numerical approach. Fluid Dyn Res 37:40–59

Ohashi T, Sato M (2012) Chapter 22, Endothelial cell responses to fluid shear stress: from methodology to applications. In: Dias R, Martins AA, Lima R, Mata TM (eds) Single and two-phase flows on chemical and biomedical engineering. Bentham Science Publishers, Oak Park, pp 372–385

Ohashi T, Sugawara H, Matsumoto T, Sato M (2000) Surface topography measurement and intracellular stress analysis of cultured endothelial cells exposed to fluid shear stress. JSME Int J Ser C 43:780–786

Ohashi T, Ishii Y, Ishikawa Y, Matsumoto T, Sato M (2002) Experimental and numerical analyses of local mechanical properties measured by atomic force microscopy for sheared endothelial cells. Bio-Med Mater Eng 12:319–327

Ohashi T, Hanamura K, Azuma D, Sakamoto N, Sato M (2008) Remodeling of endothelial cell nucleus exposed to three different mechanical stimuli. J Biomech Sci Eng 3(2):63–74

Radmacher M, Fritz M, Hansma PK (1995) Imaging soft samples with the atomic force microscope: gelatin in water and propanol. Biophys J 69:264–270

Satcher RL Jr, Bussolari SR, Gimbrone MA Jr, Dewey CF Jr (1992) The distribution of fluid forces on model arterial endothelium using computational fluid dynamics. Trans ASME J Biomech Eng 114:309–316

Sato M, Levesque MJ, Nerem RM (1987) Micropipette aspiration of cultured bovine aortic endothelial cells exposed to shear stress. Arterioscler Thromb Vasc Biol 7:276–286

Sato M, Nagayama K, Kataoka N, Sasaki M, Hane K (2000) Local mechanical properties measured by atomic force microscopy for cultured endothelial cells exposed to shear stress. J Biomech 33:127–135

Shin M, Matsuda K, Ishii O et al (2004) Endothelialized networks with a vascular geometry in microfabricated poly (dimethyl siloxane). Biomed Microdevices 6:269–278

Theret DP, Levesque MJ, Sato M, Nerem RM, Wheeler LT (1988) The application of a homogeneous half-space model in the analysis of endothelial cell micropipette measurements. Trans ASME J Biomech Eng 110:190–199

Tsou JK, Gower RM, Ting HJ et al (2008) Spatial regulation of inflammation by human aortic endothelial cells in a linear gradient of shear stress. Microcirculation 15:311–323

White CR, Haidekker M, Bao X, Frangos JA (2001) Temporal gradients in shear, but not spatial gradients, stimulate endothelial cell proliferation. Circulation 103:2508–2513

Yamaguchi T, Yamamoto Y, Liu H (2000) Computational mechanical model studies on the spontaneous emergent morphogenesis of the cultured endothelial cells. J Biomech 33:115–126

Young EWK, Simmons CA (2010) Macro-and microscale fluid flow systems for endothelial cell biology. Lab Chip 10:143–160

Young EWK, Wheeler AR, Simmons CA (2007) Matrix-dependent adhesion of vascular and valvular endothelial cells in microfluidic channels. Lab Chip 7:1759–1766

Tensile Properties of Smooth Muscle Cells, Elastin, and Collagen Fibers

7

Takeo Matsumoto, Shukei Sugita, and Kazuaki Nagayama

Abstract

Artery walls change their dimensions as well as mechanical properties adaptively in response to mechanical stimulation. Because these responses are caused by the vascular smooth muscle cells (VSMCs) in the media, detailed understanding of the mechanical environment of the VSMCs is indispensable to know the mechanism of the adaptation. Artery wall has been often assumed to be homogenous in conventional mechanical analyses from macroscopic viewpoint. At a microscopic level, however, it is highly heterogeneous, and conventional mechanical analyses using homogeneous models are far from satisfactory to estimate the mechanical environment of the VSMCs. Thus, the mechanical properties of each element composing the artery wall, i.e., VSMCs, elastin, and collagen fibers, should be measured directly. In this chapter, we first introduce the experimental techniques used for the tensile testing of tissues and cell at a microscopic scale and review the tensile properties of VSMCs in detail, and then, those of elastin and collagen fibers. In contrast to elastin and collagen fibers that are simple passive materials, VSMCs are alive and their mechanical properties are highly complicated. Their mechanical properties are reviewed from the viewpoints of smooth muscle contraction, anisotropy in cytoskeletal structure, and viscoelasticity.

Keywords

Anisotropy • Contraction • Relaxation • Tensile test • Viscoelasticity

T. Matsumoto (✉) • S. Sugita • K. Nagayama
Department of Mechanical Engineering, Biomechanics Laboratory, Nagoya Institute of Technology, Gokiso-cho, Showa-ku, Nagoya 466-8555, Japan
e-mail: takeo@nitech.ac.jp

K. Tanishita, K. Yamamoto (eds.), *Vascular Engineering*,
DOI 10.1007/978-4-431-54801-0_7

7.1 Microscopic Tensile Test of the Components in the Vascular Tissues

7.1.1 Needs for Tensile Test

Artery wall has been often assumed to be homogenous in conventional mechanical analyses from macroscopic viewpoint. At a microscopic level, however, it is highly heterogeneous. For example, the medial wall of elastic arteries has a layered structure called a lamellar unit, a pair of elastic lamina mainly composed of elastin and a smooth muscle-rich layer mainly composed of VSMCs and collagen (Wolinsky and Glagov 1967), and Young's modulus of elastin, VSMCs, and collagen is about 0.6 MPa (Fung 1981), 10–100 kPa (Matsumoto and Nagayama 2012), and 1 GPa (Fung 1981), respectively. Thus, it is very important to know the mechanical behavior of these components in detail to know the intramural mechanical environment at a microscopic level. Since artery walls stretched by 30–50 % both in the circumferential and longitudinal directions in the physiological state with reference to no load condition, the mechanical properties of the elements in the artery wall under large deformation are indispensable. Conventional methodology used for the measurement of the mechanical properties of cells such as micropipette aspiration (Hochmuth 2000), magnetic particle twisting (Wang et al. 1993), and nanoindentation with atomic force microscopy (AFM) (Hoh and Schoenenberger 1994; Sato et al. 2000) can only provide the mechanical properties of local regions of cells under small deformation and it is not appropriate for the microscopic mechanical analysis of the artery wall. We, therefore, need to develop another method that can be used to measure the mechanical properties under large deformation. For this purpose, tensile test is useful (Matsumoto and Nagayama 2012).

7.1.2 Tensile Tester

Figure 7.1 shows an example of tensile tester developed in our laboratory for microscopic specimens (Nagayama and Matsumoto 2008). The specimen was observed with an inverted microscope (TE2000E, Nikon Japan), and both ends of the specimen were gripped with two glass micropipettes, an operation pipette and a deflection pipette, with tip diameters of approximately 5–8 μm. There are several methods for specimen gripping. If the specimen is long enough (several 100 mm), it can be gripped by knotting each end around a micropipette tip (Warshaw and Fay 1983). For a shorter specimen, one can grip it by aspirating it at both ends into micropipettes, or by gently pressing the micropipettes, microrods, or microplates onto its lateral surface. Since the specimen is stretched at a portion of their surface in the latter method, shear deformation might appear in the specimen. However, the stiffness obtained in the two methods has been reported to be similar (Nagayama et al. 2006). The surface of the microtools is often coated with various adhesives, such as a cellular adhesive (Cell-Tak, Becton Dickinson, USA) and a urethane adhesive (Polycel, Macklanburg-Duncan, USA), to improve the adhesion.

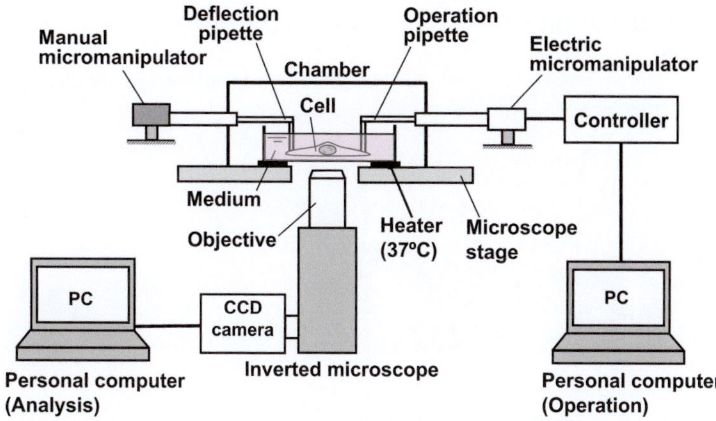

Fig. 7.1 An example of a micro tensile tester

The operation pipette was then moved with a computer-controlled electrical manipulator (MMS-77, Shimadzu, Japan) to stretch the cell horizontally. The force applied to the cell was measured by the deflection of a cantilever part of a deflection pipette that was calibrated after each test. The cell stretching process was observed with a CCD camera (ORCA-ER, Hamamatsu Photonics, Japan) connected to the microscope for an off-line analysis performed later. Spatial resolution of the images taken into the computer was 0.18 μm. Spring constant of the cantilever was 0.006–0.182 N/m, and thus the force resolution was smaller than 0.1 μN.

7.1.3 Data Analyses

After mechanical test, recorded images were fed to a computer to measure the positions of the pipette tips with image analysis software ImageJ (NIH, USA). Figure 7.2 shows an example of a smooth muscle cell during stretch. The position of the pipette tips was measured to obtain gauge length L and the displacement of the deflection pipette x. The force applied to the cell F was calculated by multiplying the displacement x with the spring constant of the deflection pipette k measured after each test. The elongation ΔL was calculated as the increment of L. The nominal stress was calculated by dividing force F with the original cross-sectional area, A_0, which is obtained from the diameter before the stretch assuming circular cross section, *i.e.*, $A_0 = \pi\ (D_0/2)^2$. The nominal strain was obtained by normalizing the elongation ΔL with the original gauge length L_0.

Mechanical properties of the cells can be evaluated with tension-elongation and nominal stress-nominal strain curves. Slope of the tension-elongation curves is used to evaluate stiffness, while that of stress-strain curves can be used for elastic modulus. Two different elastic moduli can be obtained from the nominal stress-nominal strain curves. An overall elastic modulus E_{all} is obtained by fitting a

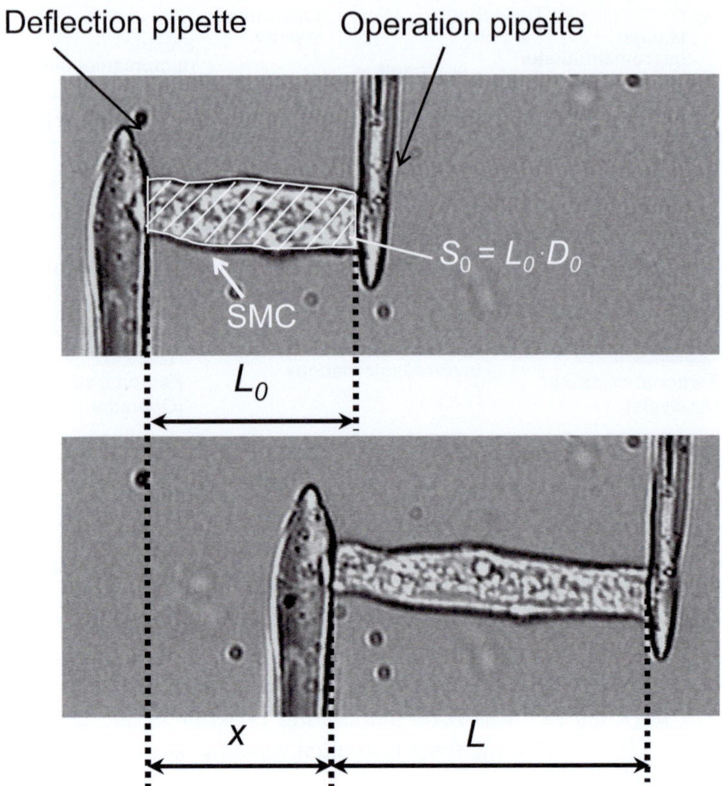

Fig. 7.2 Measurement of deformation of a freshly isolated aortic smooth muscle cell during tensile test. Mean diameter D_0 was calculated from the hatched area S_0 divided by the gauge length L_0

straight line from the origin to the entire segment of each curve. An initial elastic modulus E_{init} is obtained by fitting a straight line from the origin to the low strain region (nominal strain < 0.2) of each curve.

7.2 Tensile Properties of Smooth Muscle Cells: Relaxed vs Contractile State

7.2.1 Contraction and Relaxation of Vascular Smooth Muscle Cells

Vascular smooth muscle cells (VSMCs) contract in response to nervous, hormonal, and mechanical stimuli. Their contraction reduces vascular diameter especially at arteriolar level and thus plays vital role in controlling local blood flow. It has been pointed out that the characteristic impedance of the artery is maintained at minimum levels with the contraction of VSMCs (Cox 1975). It has also been suggested that the artery wall might control its intramural stress distribution through smooth muscle contraction and relaxation (Matsumoto et al. 1996). Since the state of contraction of VSMCs (tonus) in a physiological condition lies between relaxed and fully contracted states, we need to know their mechanical properties under the both states. In this subsection, the mechanical properties of VSMCs obtained from rat thoracic aortas are introduced.

7.2.2 Isolation of Vascular Smooth Muscle Cells and Tensile Test

Rat aortic smooth muscle cells (RASMs) were isolated from aortic tissue by enzymatic digestion (Nagayama et al. 2006). Briefly, the descending thoracic aortas excised from male Wistar rats were placed in a physiological saline solution and loose connective tissue on their surface was removed. The aortas were placed in 3 ml of a Ca^{2+}-Mg^{2+}-free Hanks' balanced salt solution (HBSS, Gibco) at 37 °C containing 300U of collagenase type III (Worthington Biochemical Corp.) and 1.8U of elastase type I (Sigma), and agitated at about 1 Hz for 120–150 min to get cell suspension. The cell suspension was immediately dropped into a Petri dish filled with a Ca^{2+}-Mg^{2+}-free phosphate buffered saline (Nissui, Japan) to reduce the effect of enzymes. The Petri dish containing specimen cells was then set on the microscope stage whose temperature was controlled at 37 °C. A cell in the dish was held with the two pipettes by pressing the micropipettes coated with the urethane adhesive gently onto the cell. To omit cells that had already contracted, RASMs whose axial length was longer than 20 μm were used for the tensile test. The cell was then stretched by moving the pipette connected to the electric manipulator stepwise every 5 s with an increment of 1 μm until fracture or until the cell began slipping off from the pipette. The rate of the pipette movement was set to keep the strain rate in each cell within the range of 0.2–4 %/s, in which the elastic modulus did not change significantly with the strain rate in our system (Matsumoto et al. 2005).

7.2.3 Tensile Test Results

Figure 7.3 shows examples of the tensile properties of relaxed and contracted VSMCs isolated from rat thoracic aortas. A maximal contraction was induced with 10^{-5} M serotonin, and the initial normalized stiffness, i.e., initial elastic modulus, increased drastically from 14.8 ± 9.6 kPa (mean $\pm$ SEM, $n = 5$) to 88.1 ± 26.6 ($n = 4$) sixfold in response to contraction (Nagayama and Matsumoto 2004). Similar results were obtained for VSMCs isolated from rabbit thoracic aortas: 15.5 ± 6.9 kPa ($n = 6$) for relaxed cells and 156.4 ± 105.9 ($n = 8$) for maximally contracted cells (Matsumoto and Nagayama 2012). These results may indicate that the tension borne by the VSMCs in the aortic wall increases several times upon maximal contraction. In contrast, the contraction decreases

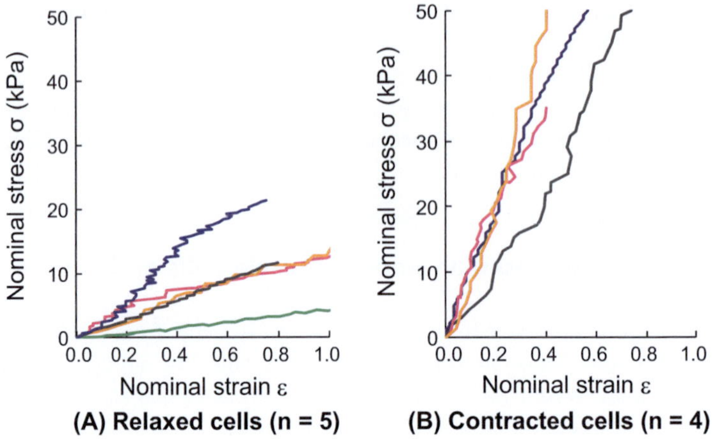

Fig. 7.3 Tensile properties of smooth muscle cells freshly isolated from rat thoracic aortas. Nominal stress-nominal strain curves of untreated cells (**a**) and of cells fully contracted with 10^{-5} M serotonin (**b**)

circumferential strain and stress applied to collagen and elastin. Taken together, the proportion of the stress borne by VSMCs may change significantly depending on the tonus of the VSMCs.

7.3 Tensile Properties of Smooth Muscle Cells: Anisotropy

7.3.1 Anisotropy of Smooth Muscle Cells

Vascular smooth muscle cells are spindle-shaped and aligned almost parallel to the circumferential direction in the arterial wall. Their intracellular contractile apparatus, i.e., stress fibers, run mostly parallel to their major axis and their contraction takes place in that direction. According to Deguchi et al. (2006), Young's modulus of the stress fibers is 1.45 MPa, *i.e.*, approximately two orders of magnitude higher than that of whole cells under relaxation. Thus, VSMCs might be much stiffer in the major axis direction than in minor. To check this hypothesis, the tensile properties of VSMCs freshly isolated from rat aortas were measured in their major and minor axes.

7.3.2 Protocol of the Experiment

RASMs were isolated from the thoracic aortas of male Wistar rats (8–14 weeks of age) as shown in Sect. 7.2.2. A cell in a dish was held with the two micropipettes coated with the urethane adhesive. When the cell was stretched along its major axis, each of the glass micropipettes was gently pressed down on each end of the cell.

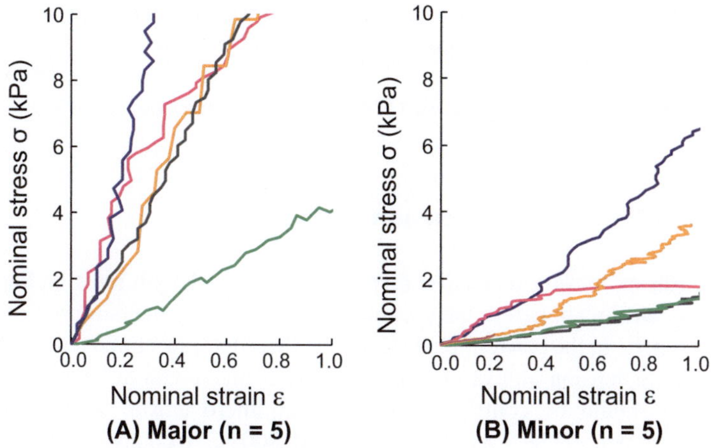

Fig. 7.4 Nonimal stress–nominal strain curves of the freshly isolated rat aortic smooth muscle cells stretched in their major (**a**) and minor (**b**) axis directions

The cell was held gently on both sides with the two pipettes when stretched in the minor axis direction. After waiting for about 5 min to make the adhesion between the cell and the pipettes firm, we stretched the cell stepwise by moving the operation pipette 1 μm every 5 s until fracture occurred, or until the cell began to slip off from the pipette.

7.3.3 Tensile Test Results

The normalized tension–nominal strain curves in major and minor axes are shown in Fig. 7.4. The relationship between the normalized tension and the nominal strain was relatively linear in the major and minor axis directions. The slopes in the major axis looked steeper than those in the minor axis. The initial elastic modulus E_{ini} of FSMCs was significantly higher in the major axis (14.8 ± 4.3 kPa, $n = 5$) than in the minor axis (2.8 ± 1.0 kPa, $n = 5$), indicating that freshly isolated SMCs are anisotropic.

7.4 Viscoelastic Properties of Smooth Muscle Cells

7.4.1 Importance of Consideration for Viscoelasticity of Smooth Muscle Cells

It is well known that intestinal smooth muscle cells are highly viscoelastic. Their viscoelastic response is not only passive but also active, *i.e.*, their response is dependent on extracellular environment such as temperature and drugs (Price et al. 1979). Vascular smooth muscle cells are no exception. Since artery wall is subjected to cyclic deformation during heartbeat, VSMCs may experience loads higher than that expected from elastic analysis. Thus, consideration of viscoelastic properties is important. In this subsection, tension relaxation test, one of the most simple viscoelastic tests, is applied to RASMs (Nagayama et al. 2007).

7.4.2 Tensile Tester for Tension Relaxation

When measuring tension relaxation response, specimen length must be kept constant. Since tension applied to the specimen is measured by the deformation of the deflection pipette in the conventional micro tensile testers, specimen length changes in response to tension relaxation. To overcome this problem, the position of micropipette should be controlled precisely. A micro tensile tester with such capability is shown in Fig. 7.5a (Nagayama et al. 2007). To control pipette position precisely, a pantagraph-type piezo actuator has been added to a conventional tensile tester. The tips of the glass micropipettes were painted black to enhance image binarization. The position of the micropipette tips was tracked using an image processor (Percept Scope, C8840, Hamamatsu Photonics, Hamamatsu, Japan) to obtain the distance between the pipettes (Fig. 7.5b). The specimen length was maintained constant by a position feedback mechanism in which the distance between the micropipettes was controlled by changing electric voltage applied to the piezo ceramics. Tension applied to the specimen was calculated from the deformation of the deflectable pipette that was derived from the position of the pipette tip and the pipette root, which was obtained from the deformation of the piezo ceramics measured by the strain gauge attached on the ceramics.

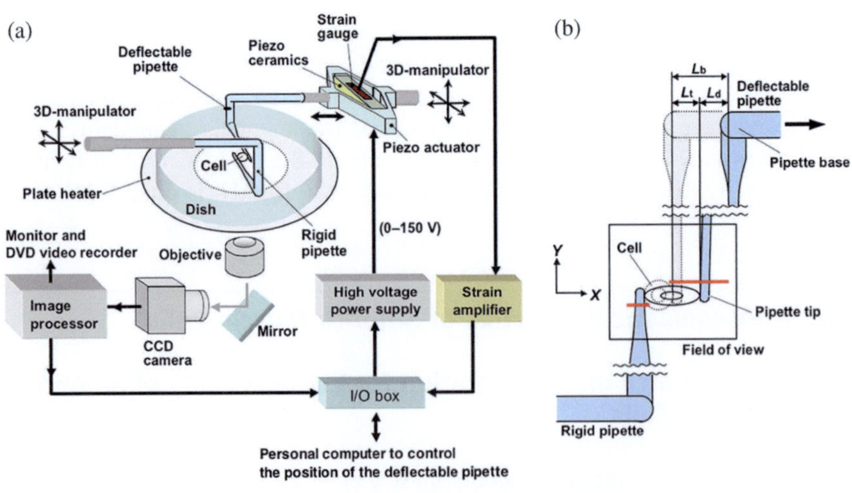

Fig. 7.5 Schematic diagram of the micro tensile tester with feed-back control (**a**), and the details of its test section (**b**). *Red lines* in the field of view indicate the binarized regions to track the outline edge of the pipettes with the image processor (Nagayama et al. 2007)

7.4.3 Tension Relaxation Test

RASMs were obtained as stated in Sect. 7.2.2 and cultured in DMEM supplemented with 10 % fetal bovine serum up to passage 3. Untreated cultured cells (SMCs) and cultured SMCs whose actin filaments (AFs) were disrupted with cytochalasin D (SMCs-CD) were used to know the effects of AFs on the viscoelastic properties of cells. Either type of cell was mounted on the tensile tester, stretched by 75 %, and its length was kept constant for 20–30 min using the position feedback control. Tension relaxation response was modeled using a 4-parameter Maxwell model (Fig. 7.6 insert) consisting of two parallel Maxwell chains, a serial combination of an elastic component with elastic parameter K^* and a dashpot with viscous parameter V^*. By assuming a step input of strain $\varepsilon(0)$, the tension normalized by the initial cross sectional area of the specimen $T^*(t)$ was expressed by the following equation:

$$T^*(t) = \left\{K_0^* \cdot \exp\left(-\frac{t}{\tau_0}\right) + K_1^* \cdot \exp\left(-\frac{t}{\tau_1}\right)\right\} \cdot \varepsilon(0), \qquad (1)$$

where the relaxation time constant τ_0 equals V_0^*/K_0^*, τ_1 equals V_1^*/K_1^*, and $\varepsilon(0)$ represents the strain at the beginning of the tension relaxation. Model parameters were determined by minimizing errors between the theoretical and experimental curves for each cell.

Figure 7.6 shows the typical examples of the relaxation responses of SMCs and SMCs-CD fitted with 4-parameter Maxwell model. Fitting of the relaxation curves obtained with the 4-parameter Maxwell model was satisfactory for both cells ($R^2 > 0.95$). Some fluctuation in tension was observed in the SMCs, while the curves for SMCs-CD were relatively smooth. The viscoelastic parameters of

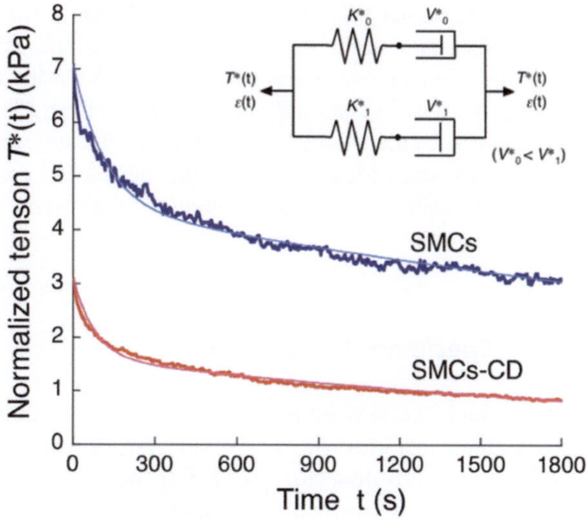

Fig. 7.6 Typical examples of the stress relaxation curves for SMCs and SMCs-CD fitted by the 4-parameter Maxwell model. *Thick lines* and *thin lines* indicate experiment data and approximate curves, respectively (Nagayama et al. 2007)

Table 7.1 Summary of the viscoelastic parameters of SMCs and SMCs-CD (mean ± SEM) (Nagayama et al. 2007)

Group	n	K_0* (kPa)	K_1* (kPa)	V_0* (kPa·s)	V_1* (kPa·s)	τ_0 (s)	τ_1 (s)
SMCs	11	6.45 ± 1.60	5.69 ± 1.01	372 ± 93	20510 ± 4330	57.1 ± 9.0	3620 ± 470
SMCs-CD	6	2.49 ± 0.56#	2.39 ± 0.43#	109 ± 24#	4740 ± 700#	46.7 ± 9.7	1820 ± 230#

K_0* and K_1*, elastic parameters; V_0* and V_1*, viscous parameters; τ_0 and τ_1, relaxation time constants (= V_n^*/K_n^*, $n = 1, 2$). $^{\#}P < 0.05$

SMCs and SMCs-CD were compared in Table 7.1. The elastic parameters of SMCs-CD were one-half of those of SMCs. The viscous parameters also decreased following AF disruption, with the decrease being larger for V_1* than V_0*. These changes in the elastic and viscous parameters caused the change in the relaxation time constant ($\tau = V^*/K^*$): The τ_0 did not change significantly with AF disruption, while τ_1 significantly decreased following the disruption.

These results indicate that AFs have crucial roles not only in elastic properties of VSMCs but also in their viscous properties, especially in the slow response of stress relaxation. AFs also have significant effects on the fluctuation in tension during stress relaxation of SMCs. Viscoelastic response of VSMCs is not passive but active response of the cells.

7.5 Isolation and Measurement of Tensile Properties of Elastic Lamina

7.5.1 Mechanical Behavior of Elastic Lamina in the Vascular Wall

Elastic lamina (EL) is a several μm-thick layer of elastic tissue mainly composed of elastin and forms boundary of intima and the media (internal EL) and of media and adventitia (external EL) of the arteries. ELs are abundant on elastic arteries, for their media has concentric layers of elastic laminas. Elastin is the most linearly elastic materials known (Fung 1981) and plays pivotal roles in maintaining elastic properties of blood vessel wall. It is well known that EL is corrugated in the section perpendicular to the axial direction in no load state, while it becomes straight in a physiological state. Measurement of the mechanical properties of the EL is important in elucidating mechanical interaction between smooth muscle cells and surrounding matrix.

7.5.2 Specimen Preparation

Ten-μm-thick sections perpendicular to the axial direction were obtained from porcine thoracic aortas with a cryotome. An EL was excised from the surrounding tissues by a micro dissector (5190, Eppendorf, Germany) and remaining tissues were removed by enzymatic digestion with purified collagenase (Worthington

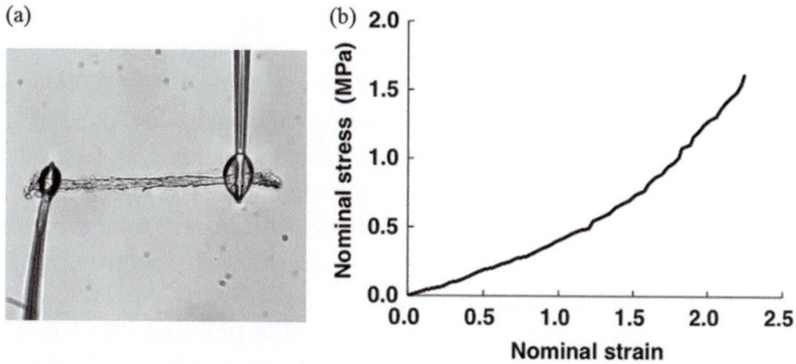

Fig. 7.7 Typical example of tensile test of elastic lamina obtained from the aorta. (**a**) Typical image of tensile test. (**b**) An example of stress-strain curve

Biochemical, USA). Shape change of the EL during isolation was observed. Tensile properties of the isolated ELs were then measured as described in Sect. 7.1. Specimens were stretched by 5 μm every 3 s to obtain tension-stretch ratio curves. Width and thickness of each EL were measured under a microscope before the tensile test to convert tension to stress.

7.5.3 Tensile Test Results

ELs became almost straightened following isolation process, indicating that the corrugation of the ELs in the unloaded aortic wall is mainly due to compressive residual stress in the ELs caused by surrounding tissues (Matsumoto et al. 2004). Nominal stress–stretch ratio relation of the ELs was almost linear up to nominal strain of 1.0 with Young's modulus of 440 kPa (Fig. 7.7). These results indicate that tension applied to the ELs in a physiological state is dependent on the degree of the corrugation. Since there is wide variation of the corrugation in the unloaded aortic walls, physiological tension borne by the ELs might also have a wide variation.

7.6 Tensile Properties of Fibrous Collagen in the Aorta

7.6.1 Mechanical Behavior of Collagen in the Vascular Wall

As described in the beginning of this section, collagen is the stiffest material in the vascular wall, and it plays important roles in the nonlinear (strain-hardening) mechanical properties of the vascular walls and in preventing overstretch of the wall. Mechanical properties of collagen fibers have been obtained mainly by uniaxial tensile test for collagen fibrils, fibers, and fascicles obtained from tendons. Stress-strain curves of the rabbit patellar tendon showed the linear relationship and tangent elastic modulus of 732 ± 200 MPa (Yamamoto et al. 1999). It should be noted that mechanical properties of collagenous tissues obtained from rabbit tendon were dependent on their size, *i.e.*, specimen diameter (*D*). Young's modulus of rabbit patellar tendon fascicles ($D \sim 300$ μm) was reported to be 216 ± 68 MPa (Yamamoto et al. 1999), while that of fibrils ($D = 1.01 \pm 0.06$ μm) was 54.3 ± 25.1 MPa (Miyazaki and Hayashi 1999). Yamamoto et al. (1999) stated that this difference might be attributable to such interactions as frictional force between collagen fascicles as well as between collagen fascicles and ground substance. Young's modulus of collagen obtained from other parts of the body ranges from several hundreds of MPa to ~1 GPa: the maximum tangent elastic modulus of rat tail tendon fascicles was 1–2 GPa (Svendsen and Thomson 1984), tangent elastic modulus of mouse tail tendon ($D \sim 100$ μm) 423.5 ± 76.6 MPa, and sea cucumber collagen fibrils ($D = 0.21$–0.45 μm) 470 ± 410 MPa (Shen et al. 2010).

7.6.2 Preparation of Collagen Fiber Specimens

To obtain the mechanical properties of collagen fibers in the aorta, other main constituents such as smooth muscle cells and elastin should be removed. Fibrous collagen in the aorta was obtained as follows: (1) thoracic aortas were excised and its adventitia and loose connective tissues on adventitial side were trimmed off; (2) the specimens were incubated in 68.5 U/ml of highly purified elastase (ES438, EPC, USA) for 2 h; (3) and they were cut into fibrous collagens of a few micrometers in diameter and ~100 μm in length by shaking the container containing the specimen. These specimens were then glued on two glass pipettes with urethane adhesive (Sista M5250, Henkel, Germany) to perform tensile test as stated in Sect. 7.4.1 (Fig. 7.8a). Stress-strain curve was obtained assuming that the specimen had a circular cross section.

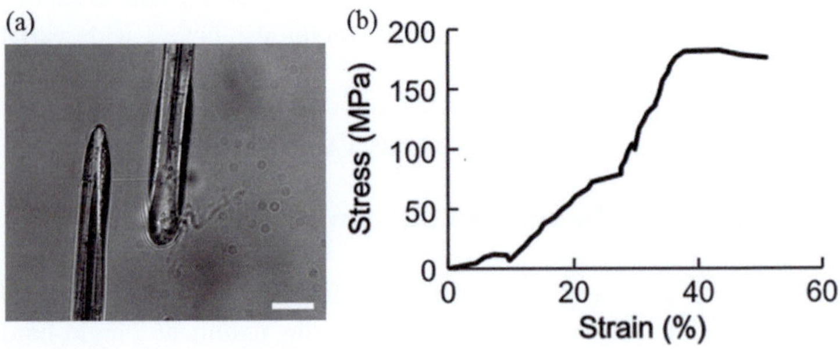

Fig. 7.8 Typical example of tensile test of fibrous collagen obtained from the aorta. (**a**) Typical image of tensile test. Scale bar = 30 μm. (**b**) An example of stress-strain curve of fibrous collagen

7.6.3 Tensile Test Results

An example of stress-strain curve of a collagen fiber obtained from a porcine thoracic aorta was shown in Fig. 7.8b. Calculated Young's modulus of collagen fibers in the aorta was ~500 MPa, which was comparable to that of collagen fibers obtained from other tissues. The curve seemed to have multiple linear regions. This may indicate that the specimen was composed of multiple thinner fibers with different slack length. Similar phenomena were observed in the collagen fibrils obtained from the sea cucumber (Shen et al. 2010). Although their elastic modulus is dependent on specimen diameter, collagen fibers are still the stiffest material in the artery wall at a microscopic level. Roughly speaking, collagen fibers are 1000 times stiffer than the ELs and 10,000 times stiffer than the VSMCs. Collagen fibers may play important roles on intramural force transmission, although they are not abundant in the media of the arteries.

References

Cox R (1975) Pressure dependence of the mechanical properties of arteries in vivo. Am J Physiol 229:1371–1375

Deguchi S, Ohashi T, Sato M (2006) Tensile properties of single stress fibers isolated from cultured vascular smooth muscle cells. J Biomech 39:2603–2610

Fung YC (1981) Bio-viscoelastic solids. In: Biomechanics. Springer, New York, pp 196–214

Hochmuth R (2000) Micropipette aspiration of living cells. J Biomech 33:15–22

Hoh J, Schoenenberger C (1994) Surface morphology and mechanical properties of MDCK monolayers by atomic force microscopy. J Cell Sci 107:1105–1114

Matsumoto T, Nagayama K (2012) Tensile properties of vascular smooth muscle cells: bridging vascular and cellular biomechanics (review). J Biomech 45:745–755

Matsumoto T, Tsuchida M, Sato M (1996) Change in intramural strain distribution in rat aorta due to smooth muscle contraction and relaxation. Am J Physiol Heart Circ Physiol 271:H1711–H1716

Matsumoto T, Goto T, Furukawa T, Sato M (2004) Residual stress and strain in the lamellar unit of the porcine aorta: experiment and analysis. J Biomech 37:807–815

Matsumoto T, Sato J, Yamamoto M, Sato M (2005) Development of a novel micro tensile tester for single isolated cells and its application to viscoelastic analysis of aortic smooth muscle cells. In: Biomechanics at micro- and nanoscale levels, vol I. World Scientific, Singapore, pp 16–25

Miyazaki H, Hayashi K (1999) Tensile tests of collagen fibers obtained from the rabbit patellar tendon. Biomed Microdevices 2:151–157

Nagayama K, Matsumoto T (2004) Mechanical Anisotropy of Rat Aortic Smooth Muscle Cells Decreases with Their Contraction: Possible effect of actin filament orientation. JSME Int J Ser C 47:985–991

Nagayama K, Matsumoto T (2008) Contribution of actin filaments and microtubules to quasi-in situ tensile properties and internal force balance of cultured smooth muscle cells on a substrate. Am J Physiol Cell Physiol 295:C1569–C1578

Nagayama K, Nagano Y, Sato M, Matsumoto T (2006) Effect of actin filament distribution on tensile properties of smooth muscle cells obtained from rat thoracic aortas. J Biomech 39:293–301

Nagayama K, Yanagihara S, Matsumoto T (2007) Actin filaments affect on not only elasticity but also late viscous response in stress relaxation of single isolated aortic smooth muscle cells (Possible effect of active reorganization of actin filaments). J Biomech Sci Eng 2:93–104
Price JM, Patitucci PJ, Fung YC (1979) Mechanical properties of resting taenia coli smooth muscle. Am J Physiol Cell Physiol 236(5):C211–C220
Sato M, Nagayama K, Kataoka N, Sasaki M, Hane K (2000) Local mechanical properties measured by atomic force microscopy for cultured bovine endothelial cells exposed to shear stress. J Biomech 33:127–135
Shen ZL, Dodge MR, Kahn H, Ballarini R, Eppell SJ (2010) In vitro fracture testing of submicron diameter collagen fibril specimens. Biophys J 99:1986–1995
Svendsen KH, Thomson G (1984) A new clamping and stretching procedure for determination of collagen fiber stiffness and strength relations upon maturation. J Biomech 17:225–229
Wang N, Butler J, Ingber D (1993) Mechanotransduction across the cell surface and through the cytoskeleton. Science 260:1124–1127
Warshaw D, Fay F (1983) Cross-bridge elasticity in single smooth muscle cells. J Gen Physiol 82:157–199
Wolinsky H, Glagov S (1967) A lamellar unit of aortic medial structure and function in mammals. Circ Res 20:99–111
Yamamoto E, Hayashi K, Yamamoto N (1999) Mechanical properties of collagen fascicles from the rabbit patellar tendon. J Biomech Eng 121:124–131

8 Mechanobiology of Endothelial Cells Related to the Formation of Arterial Disease

Noriyuki Kataoka

Abstract

Atherosclerosis is a serious disease that causes cardiovascular diseases such as cerebral infarction and myocardial infarction. Endothelial cell injury is the first step in atherogenesis by inducing increase in the production of chemoattractant proteins and adhesion molecules to leukocytes (Ross, Nature 362 (6423): 801–809, 1993). One of the key events in atherogenesis is the recruitment of blood leukocytes, especially monocytes, to proatherogenic vascular regions (Swirski et al, J Clin Invest 117:195–205, 2007; Tacke et al, J Clin Invest 117:185–194, 2007) and their subsequent transmigration across endothelial cells. Interactions between leukocytes and endothelial cells involve multi-step processes including rolling, adhesion, locomotion, and transmigration. Therefore, mechanics and dynamics of endothelial cells and monocytes are important matters to understand the whole process of atherogenesis. Here, we show and discuss the mechanobiology of endothelial cells related to the formation of arterial disease.

Keywords

Atherogenesis • Endothelial cell • Monocyte • Mechanobiology

N. Kataoka (✉)
Department of Mechanical Engineering, College of Engineering, Nihon University, 1 Nakagawara, Tokusada, Tamuramachi, Koriyama, Fukushima, Prefecture Japan, 963-8642
e-mail: kataoka@mech.ce.nihon-u.ac.jp

K. Tanishita, K. Yamamoto (eds.), *Vascular Engineering*,
DOI 10.1007/978-4-431-54801-0_8

8.1 Changes in Micromechanics of Endothelial Cells Induced by Monocyte Adhesion

8.1.1 Measurement Tools of Micromechanics of Endothelial Cells

When monocytes migrate into the intima, endothelial cells might dynamically change their shape and cytoskeletal structure. It is considered that the endothelial cell micro-mechanics that are involved in the endothelial cell micro-motion and the mechanical properties of endothelial cells are important in these processes. For the micro-motion measurement of the cultured cells, Giaever and Keese (1991) developed a morphological biosensor named "Electric Cell-substrate Impedance Sensing (ECIS) system". In this system, we can detect the change of the electrical impedance that reflects the apposition of the cells to each other and to their substrate grown on the small gold electrode. The advantage of this system is that quantitative estimation of cell-to-cell and cell-to-substrate distances can be performed separately and in real time.

Atomic force microscope (AFM) is a useful tool not only for imaging the biological tissues but also for examining mechanical properties of cells. The measurement of mechanical properties of living soft tissue is possible with a nanoindentation technique (Radmacher et al. 1996). We applied these two methods, ECIS and AFM, to evaluate the effect of the interaction between monocyte and endothelial cell on the endothelial cell micro-motion and mechanical properties (Kataoka et al. 2002).

8.1.2 Estimation of Endothelial Cell-to-Cell and Cell-to-Substrate Gaps

Human umbilical vein endothelial cells (HUVECs) were seeded on a cell culture dish (10 mm × 10 mm for each section; Applied Biophysics, Inc., NY, USA) with a micro-electrode ($\phi = 250$ μm) (Fig. 8.1). HUVEC were then stimulated with interleukin-1β (5 ng/ml) 6 h and subsequently overlayered with the 20 μl of culture medium containing THP-1 cells that is a human monocytic cell line. The electrical impedance was measured with ECIS (Applied Biophysics, Inc.). The change of impedance is attributed to the changes of the resistance in paracellular pathway (Rb) and the resistance between the ventral cell surface and the electrode (α: $\alpha = r\,(r/\mathrm{h})^{0.5}$; r, cell radius; r, the resistivity of the culture medium; h, the gap between the cells and the electrode). From the changes of Rb and a, the changes in cell-to-cell and cell-to-substrate gaps were estimated.

8.1.3 Measurement of Mechanical Property Endothelial Cells

By pressing the AFM cantilever to the cell surface, a force-curve, the relationship between the cantilever deflection and its indentation depth, was obtained. In this study, the force curve of the HUVEC was measured with AFM (NVB100;

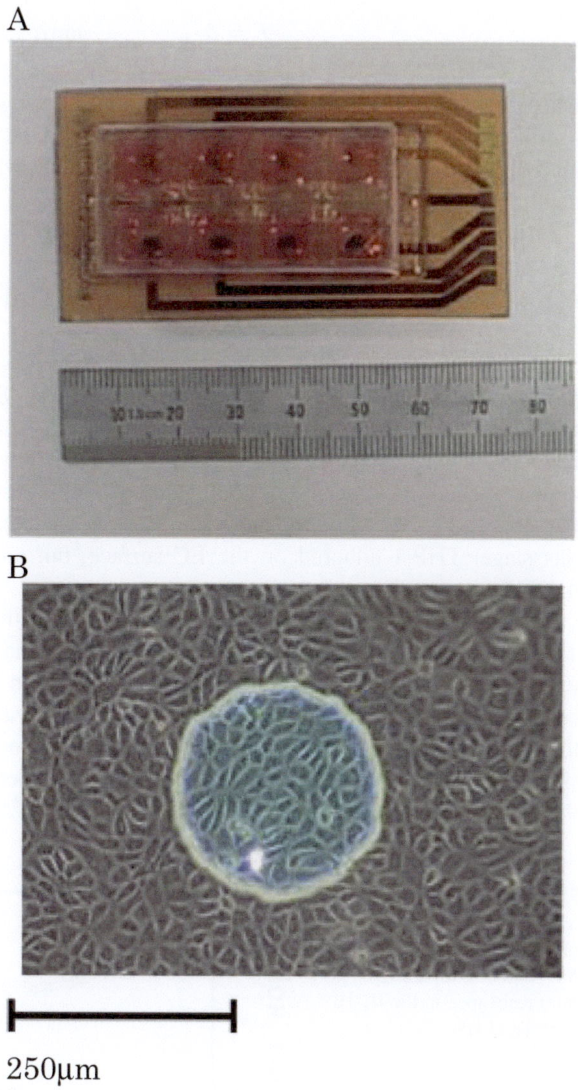

Fig. 8.1 (**a**) Cell culture dish for Electric Cell-substrate Impedance Sensing (ECIS) system, (**b**) Micro-electrode ($\phi = 250$ μm) covered with cultured endothelial cells in the center of the cell culture dish for ECIS

Olympus, Tokyo, Japan). HUVECs were seeded on a gelatin-coated cell culture dish, and THP-1 was applied. The force curves were measured after 2 h of application of THP-1 to HUVECs. Force-curve measurements were performed in the center region, just above the cell nucleus and the peripheral region of HUVEC. Elastic modulus was estimated with the Hertz model that describes the indentation of a homogeneous/semi-infinite elastic material (Radmacher et al. 1996).

$$F = \delta^2 \frac{\pi}{2} \frac{E}{(1 - \nu^2)} \tan \alpha$$

where F is the applied force (calculated from the spring constant multiplied by the cantilever deflection), E is the elastic modulus, ν is the Poisson's ratio assumed to be 0.5, α is the opening angle of the tip of the cantilever, and δ is the indentation depth.

8.1.4 Results of Changes of Endothelial Cell Micromechanics

Typical traces of the total, cell-to-cell (Rb) and cell-to-substrate (α) resistances in the IL-1β stimulated HUVECs, are shown in Fig. 8.2. Immediately after the application of THP-1 cells (time = 0 in the figure) to the IL-1β stimulated HUVECs, cell-to-substrate resistance gradually decreased, whereas cell-to-cell resistance was almost unchanged. After 1 h of application of THP-1, α was significantly decreased by -40.1 ± 20.1 %, while the change in Rb was not significant (14.9 ± 50.6 %). At this stage, THP-1 adhered on the EC surface, but did not migrate into the EC monolayer.

Typical force curves for the peripheral region of the control HUVEC and the THP-1 adhered HUVEC are shown in Fig. 8.3. The changes of the elastic modulus in both the center and peripheral regions are shown in Fig. 8.4. Under the control condition, the elastic modulus in the central region was significantly smaller than that in peripheral region. The elastic modulus of the monocytes adhered HUVECs significantly decreased in both the central and peripheral regions. Coincidentally, the density of actin filaments and the number of $p125^{FAK}$ decreased (Fig. 8.5).

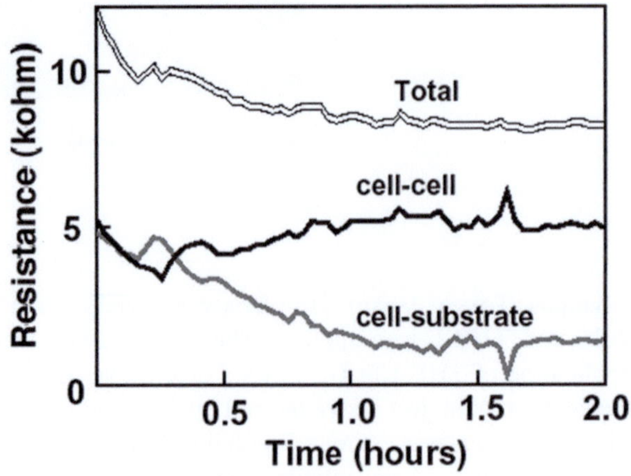

Fig. 8.2 Typical real-time traces of the total (*Double line*), cell-to-substrate (*Gray line*) and cell-to-cell (*Black line*) resistance in the IL-1β stimulated HUVECs

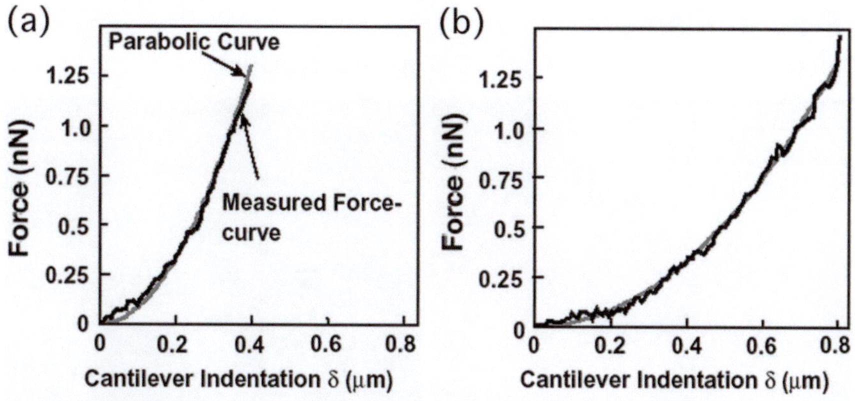

Fig. 8.3 Typical force-curves for the peripheral region of control (**a**) and monocyte adhered (**b**) HUVEC measured with AFM

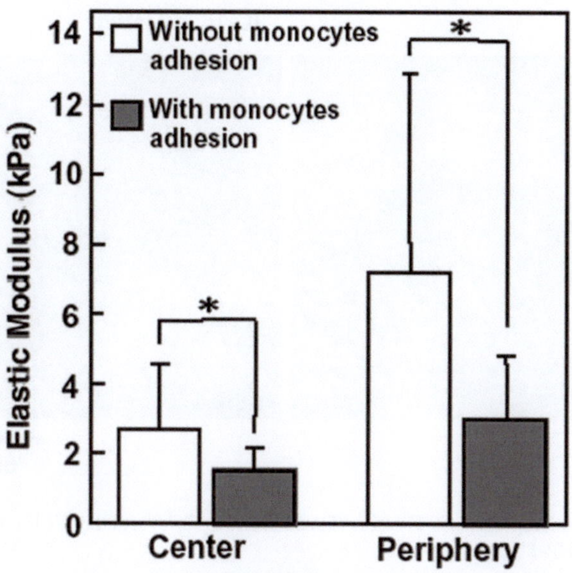

Fig. 8.4 Elastic modulus of HUVECs with and without monocytes adhesion. $*p < 0.05$

Decrease of the endothelial cell-to-substrate resistance (increase of cell-to-substrate gap) is related with the change of the adhesiveness of endothelial cells to the substrate, and the decrease of elastic modulus of endothelial cell indicates the increase of deformability. These may facilitate trans-endothelial migration of monocytes in the later stage.

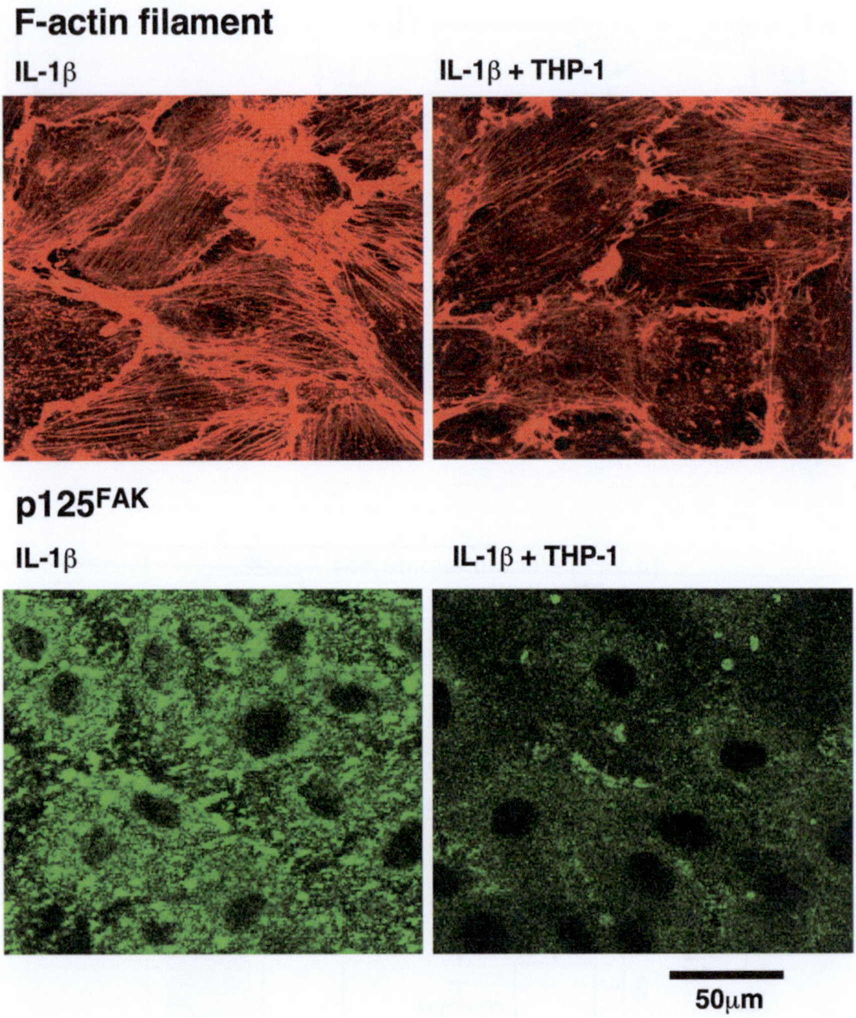

Fig. 8.5 Photomicrographs of F-actin filaments and $p125^{FAK}$ in HUVECs before and after application of THP-1

8.2 Direct Observation of Individual Living Monocytes During Transendothelial Migration

As mentioned above, the transendothelial migration (TEM) of monocytes is the crucial event at the early stage of atherosclerosis. However, considerable part of this complicated process is not well understood, and actually it is not clear how individual living monocytes behave three-dimensionally. We developed an

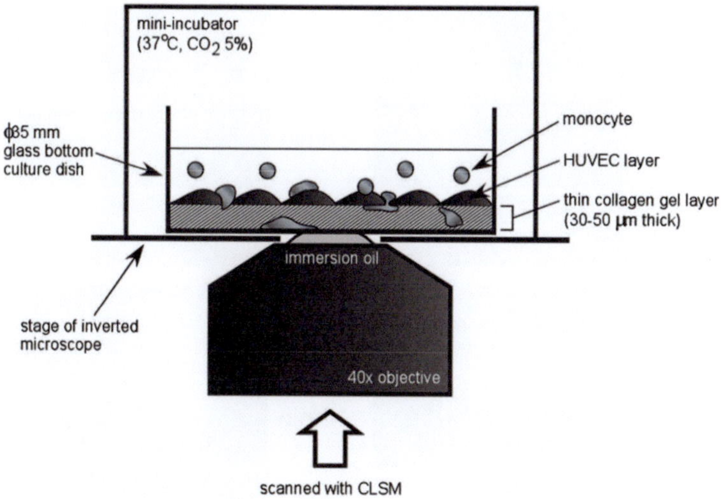

Fig. 8.6 Schematic diagram of newly-developed experimental system. HUVECs were cultured on ultra thin collagen gel layer (30–50 μm thickness) in the glass bottom culture dish. Monocyte behavior on HUVEC monolayer was observed with CLSM using 40 × oil immersion objective

experimental system using confocal laser scanning microscope (CLSM) to analyze and reveal spatiotemporal dynamics of individual living monocytes during TEM (Hashimoto et al. 2004).

Figure 8.6 shows the schematic diagram of the experimental system. Ultra thin collagen gel layer (approximately 30–50 μm thick) was constructed under HUVEC monolayer on glass bottom culture dish to provide both minimal subendothelial space for monocytes to invade, and maximal thickness for focus adjustment of the microscope with high magnification. Before transmigration assay, HUVECs were stained with red fluorescent probe 5-(and-6)chloromethyl SNARF-1, acetate (Molecular Probes, OR), and monocytes were counter-stained with green fluorescent probe CellTracker Green CMFDA (Molecular Probes).

A typical example of time course of TEM is shown in Fig. 8.7. The monocyte started invasion over 3–5 min (arrow) after application on HUVECs, but hesitated to transmigrate, and surprisingly returned onto the apical surface of HUVECs at 7 min (arrow). There were also some other monocytes that had once extended projections into subendothelium, hesitated to transmigrate, retracted them, and started to crawl on HUVECs again. A few minutes later, the monocyte deformed and extended projections into collagen gel again at 12 min ((*) at 12 min). Adhesion of the projection ((*) at 12 min) to collagen gel subsequently caused remained tail of the monocyte ((+) at 12 min) on EC to be retracted into subendothelium over 12–25 min (arrow at 18 and 25 min), and the monocyte completely transmigrated at 30 min (arrow).

We performed quantitative analysis of ratio and time course of TEM of individual monocytes (Table 8.1). Each monocyte at each time point was classified into

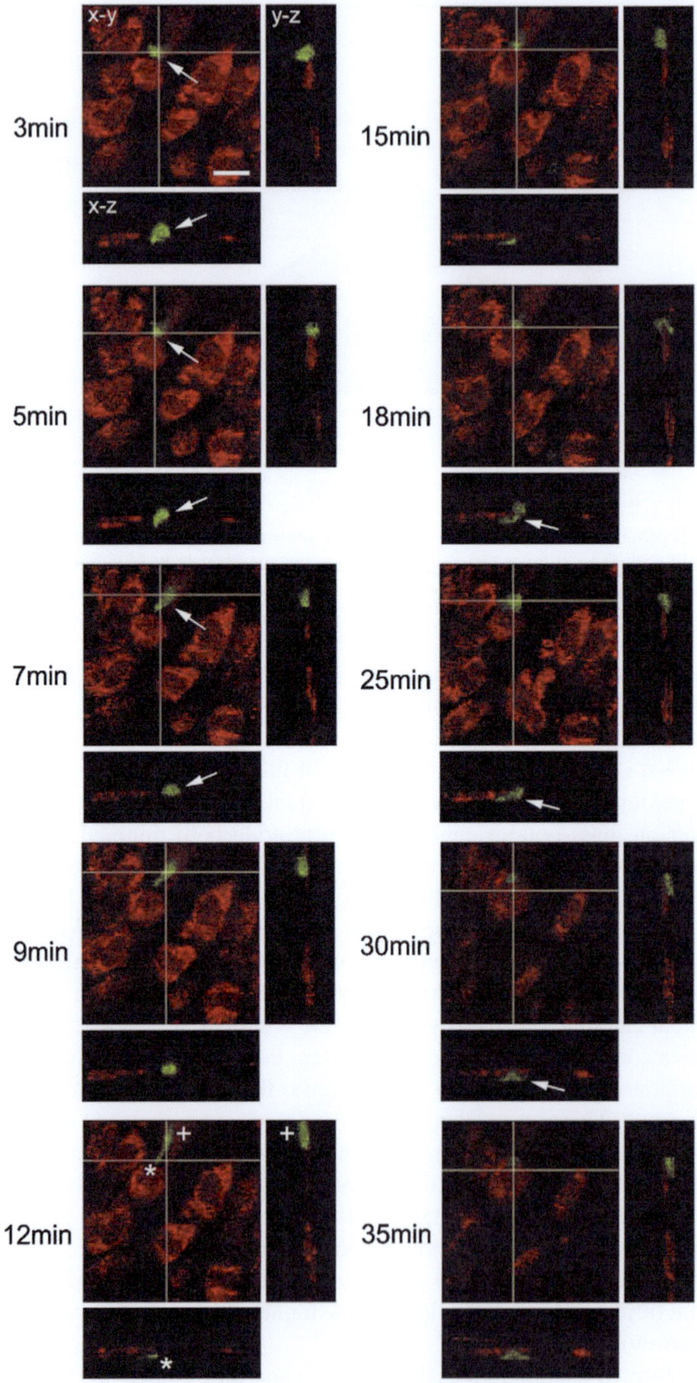

Fig. 8.7 Atypical example of TEM, HUVECs were stained with SNARF-1 (*red*) and monocytes were stained with CellTracker Green (*green*). Bar = 20 μm

Table 8.1 Results of quantitative analysis of ratio and time course of TEM of individual monocytes

	Number of monocytes	Ratio (%)
Total adhered monocytes measured (A)	96	—
Monocytes that started invasion (B)	67	69.8 (B/A)
Monocytes that completed transmigration (C)	53	55.2 (C/A)
Ratio (C/B)		79.1 (C/B)
Average time course of TEM		
	Time (min)	**(meant ± SD)**
From adhesion to start invasion (A to B)	8.6 ± 5.4	(n = 61 monocytes)
From start of invasion to finish of transmigration (B to C)	6.3 ± 3.2	(n = 53 monocytes)
From adhesion to finish transmigration (A to C)	14 ± 5.2	(n = 47 monocytes)

three categories according to the stage of TEM, i.e., adhered on endothelium (A), started invasion (B), and completed transmigration (C). Total number of 96 adhered monocytes from five separate culture experiments was measured up to 25–30 min after application on HUVECs. As to ratio of TEM, 67 out of 96 monocytes (69.8 %) started invasion, and 53 out of such 67 monocytes (79.1 %) completely finished transmigration. Times from adhesion (A) to start of invasion (B), from start (B) to finish (C) of invasion, and from adhesion (A) to finish of transmigration (C) were 8.6 ± 5.4 min (mean ± S.D., $n = 61$ monocytes), 6.3 ± 3.2 min (mean ± S.D., $n = 53$ monocytes), and 14 ± 5.2 min (mean ± S.D., $n = 47$ monocytes), respectively.

8.3 Effects of Oxidized LDL on Behavior of Monocytes and Junctional Conformation of Endothelial Cells

It is well known that the accumulation of lipid within the intima is one of principal risk factors for atherosclerosis. In particular, oxidized low density lipoprotein is a crucial factor in atherogenesis. It is also known that monocyte TEM is facilitated by oxLDL (Klouche et al. 1999). However, it is unclear how and at which stage oxLDL promotes monocyte dynamics during multi-step monocyte TEM process. We have investigated the effects of oxLDL on behavior of individual monocyte and junctional conformation of endothelial cells (Hashimoto et al. 2007).

We divided monocyte dynamics into three stages: (1) adhesion on ECs, (2) invasion, and (3) complete transmigration below ECs. Each monocyte at each time was carefully classified into the three stages from 3D-reconstructed images. Figure 8.8a–c shows the representative 3D-cross-sectional images for each stage of monocyte. Migration speed of each monocyte was calculated by tracking the x–y location of the same monocyte temporally in z-projected images (two dimensionally) during

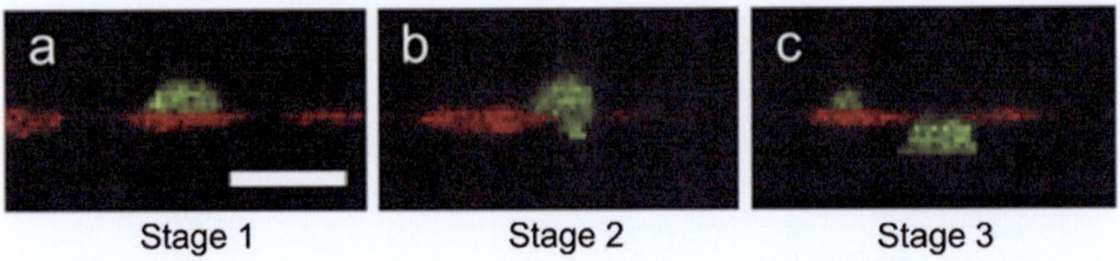

Fig. 8.8 Representative 3D-cross-sectional images at each stage of monocyte (*green*). (**a**) Stage 1: adhesion on ECs (*red*), (**b**) stage 2: invasion and (**c**) stage 3: complete transmigration below ECs. Bar = 20 μm

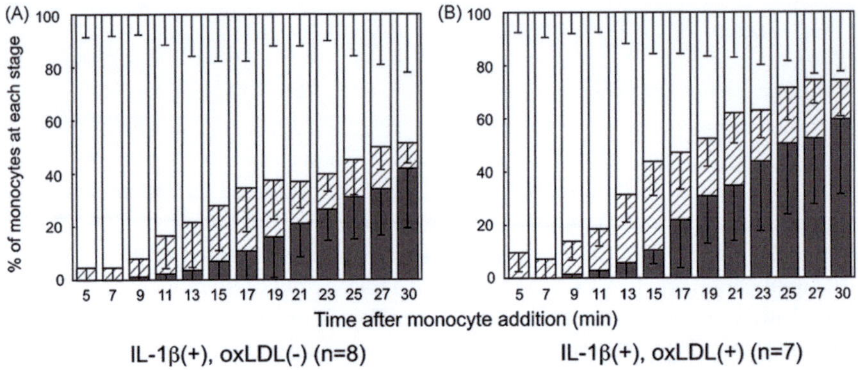

Fig. 8.9 Each monocyte at each time in our TEM assay was classified into three stages (stage 1: *open bars*, stage 2: *hatched bars* and stage 3: *gray bars*; representative images shown in Fig. 8.8) to analyze the detailed temporal kinetics of individual living monocyte. (A) without oxLDL (B) with oxLDL

stage 1 (locomotion on ECs after adhesion before invasion) and 3 (subendothelial migration after invasion).

Using our experimental system described above, detailed analysis in temporal kinetics of individual living monocyte revealed that oxLDL significantly promoted monocyte TEM process (Fig. 8.9a, b). The enhancement of transmigrated monocyte (stage 3) and the reduction of non-transmigrating monocyte (stage 1) by oxLDL were determined to be statistically significant ($p < 0.05$ and $p < 0.004$, respectively) by two-way ANOVA. The temporal change of monocyte population at stage 2 was not statistically significant. OxLDL also enhanced monocyte deformation and spreading to thin, amoeba-like shape.

Initial adhesion of moncoytes to HUVECs (stage 1) significantly increased with IL-1β stimulation as expected. But an additional exposure of subendothelial oxLDL to IL-1β-stimulated HUVECs did not promote the monocyte adhesion further.

To elucidate at which stage oxLDL affects monocyte to promote TEM, we evaluated the stage transition (stage 1 → 2, and 2 → 3) of monocytes by calculating the ratio of monocytes which have reached stage 2 out of the monocytes at stage 1, and stage 3 out of the monocytes at stage 2 in the same living samples. As shown

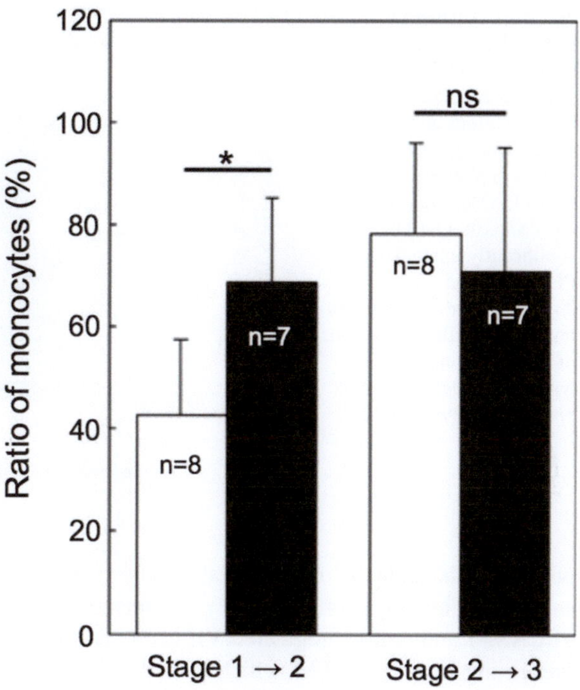

Fig. 8.10 The stage transition of monocytes (stage 1 → 2, and 2 → 3) was evaluated by calculating the ratio of monocytes which have reached stage 2 out of the monocytes at stage 1, and stage 3 out of the monocytes at stage 2 in the same living samples

in Fig. 8.10, oxLDL specifically and significantly increased the ratio of adherent monocytes that started invasion at junctions from stage 1 → 2 (68.7 ± 16.4 % with oxLDL versus 42.9 ± 14.8 % without oxLDL, $p < 0.007$). However, majority of the monocytes (70–80 %) at stage 2 completed transmigration (reached stage 3) regardless of oxLDL, indicating oxLDL has no effect on monocyte transition dynamics once invasion starts (stage2 → 3).

To explore the mechanisms of oxLDL-induced stage-specific priming of monocyte (stage 1 → 2), we evaluated the migration speed of monocytes during stage 1 (locomotion on ECs after adhesion before invasion) and 3 (subendothelial migration after invasion). The speed of monocyte was around 3–4 m/min and was not altered by oxLDL in either stage, suggesting that the oxLDL-induced stage-specific priming of monocyte was not the consequence of its enhanced migration speed.

We examined the rearrangements of two major endothelial junctional molecules, PECAM-1 and VEcadherin by oxLDL (without monocyte addition). OxLDL significantly upregulated PECAM-1 and downregulated VEcadherin (Fig. 8.11; 130 %, $p < 0.01$, and 56 %, $p < 0.0002$, respectively, versus control without oxLDL).

OxLDL upregulated PECAM-1 and downregulated VE-cadherin on the intercellular junctions of IL-1β-stimulated HUVECs before monocyte adhesion, which changed junctional conformation to more monocyte acceptable state, and led to the stage-specific promotion of monocyte TEM (stage 1 → 2; the initiation of invasion),

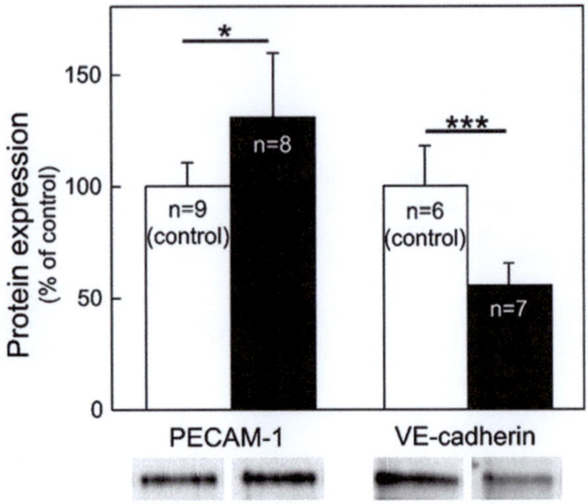

Fig. 8.11 Changes of PECAM-1 and VE-cadherin expression in EC affected with OxLDL. Protein quantifications and corresponding representative images of Western blotting for PECAM-1 and VE-cadherin. OxLDL significantly upregulated PECAM-1 and downregulated VE-cadherin on the intercellular junctions of IL-1β-stimulated ECs without monocyte addition. $*p < 0.01$, $***p < 0.0002$

with no enhancement of its initial adhesion, speeds of locomotion on ECs, or subendothelial migration. Clinically, pharmacological targeting against upregulated PECAM-1 and/or downregulated VE-cadherin could potentially be beneficial with preventing monocyte TEM, which is the key process in early atherogenesis.

8.4 Dynamic Redistribution and Recruitment of Endothelial PECAM-1 at Sites of Monocyte

Leukocytes have been shown to cross EC monolayers by both paracellular (between ECs) and transcellular (through the body of ECs) routes. For paracellular transmigration, many molecules have been reported to be involved, including platelet-endothelial cell adhesion molecule-1 (PECAM-1) (Mamdouh et al. 2003; Schenkel et al. 2002; Su et al. 2002; Muller et al. 1993), vascular endothelial cadherin (VE-cadherin) (Su et al. 2002; Shaw et al. 2001) junctional adhesion molecules (JAMs), (Keiper et al. 2005; Bradfield et al. 2007) inter-cellular adhesion molecule-2 (ICAM-2), (Huang et al. 2006) and CD99 (Schenkel et al. 2002). It is reported that part of cellular PECAM-1 exists in the lateral border recycling compartment (LBRC), which is located below the surface of the cell at the lateral (Mamdouh et al. 2003, 2008) borders in ECs, and PECAM-1 is constitutively recycled between the LBRC and the cell surface. During paracellular diapedesis of monocytes or neutrophils, this recycling PECAM-1-bearing membrane is targeted rapidly and extensively to the site of diapedesis (Mamdouh et al. 2003). This targeted recruitment is mediated by kinesin family molecular motors and normally functioning microtubules (Mamdouh et al. 2008). However, the molecular dynamics of the targeted recruitment of PECAM-1 has not been directly demonstrated in live cells. Moreover, it is currently unclear what triggers the recruitment during a series of diapedesis processes from rolling to transmigration, as has only been suggested that homophilic PECAM-1-PECAM-1 interactions between leukocytes and ECs are required for the recruitment (Mamdouh et al. 2003). It is also important to examine whether PECAM-1 plays a selective role for paracellular diapedesis or has some undefined roles in transcellular diapedesis.

We constructed a PECAM-1 fusion vector, in which the C-terminal, intracellular tail of the molecule was linked to green fluorescent protein (GFP) and transfected this PECAM-1-GFP vector into HUVECs. We analyzed the molecular dynamics of endothelial PECAM-1 during monocyte paracellular and transcellular diapedesis in three-dimensions, at a single-cell level, in these live HUVECs, which can not be achieved with a conventional end point-type assay using fixed cells (Hashimoto et al. 2012).

8.4.1 Route of Diapedesis and PECAM-1 Gaps

Of a total of 153 diapedesis events analyzed, only 5 % monocytes (7 events) transmigrated 10 by a transcellular route while approximately 95 % took a paracellular route. Of these, 68 % migrated through bicellular junctions and 27 % through multicellular junctions.

A typical example of the 3D-molecular dynamics of PECAM-1 during monocyte diapedesis through a bicellular junction is shown in Fig. 8.12 PECAM-1 staining at a bicellular junction was intact before the monocyte arrived (0 min). The monocyte adhered to the HUVEC at 6 min after the addition of monocytes started to invade the HUVEC monolayer at 14 min, forming a PECAM-1 gap de novo. Local endothelial PECAM-1 redistributed to surround the transmigrating monocyte, as clearly shown in xy, xz, and yz planes (arrows at 14 and 18 min). The migrating front of the monocyte had already reached the base of the HUVEC layer by 18 min.

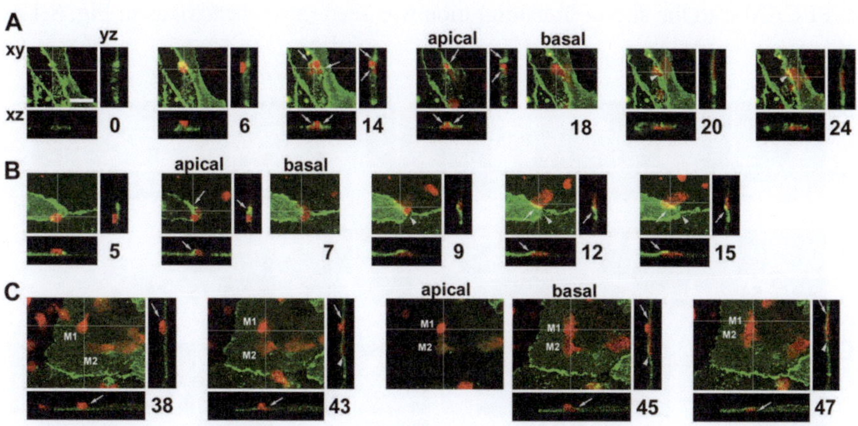

Fig. 8.12 Local redistribution of endothelial PECAM-1 during monocyte transmigration at (A) bicellular (B) multicellular and (C) transcellular

By 20 min, transmigration was virtually complete, but a PECAM-1 gap was still visible, and this had completely resealed by 24 min (arrowheads). Similar dynamics for endothelial PECAM-1 were seen around monocytes migrating through multicellular junctions (Figure 8.12), where gaps formed de novo, with PECAM-1 redistribution at 7–12 min (arrows) and the gap resealing after transmigration at 9–15 min (arrowheads). The majority of these 11 paracellular events were accompanied by PECAM-1 redistribution to surround transmigrating monocytes (77 % for bicellular and 83 % for multicellular junctions). In contrast, Fig. 8.12c shows a monocyte M1 that transmigrated transcellularly, but the redistribution of endothelial PECAM-1 was never observed. At 38 min, M1 was on the apical surface of an HUVEC (arrows), and another monocyte M2 had already transmigrated beneath the HUVEC. At 43–45 min, M2 migrated towards M1, and M1 started to invade the HUVEC transcellularly (arrows) as if attracted by M2, but no redistribution of endothelial PECAM-1 was observed. M1 had almost completed transcellular transmigration by 47 min and appeared to aggregate with M2 (arrows and arrowhead). This mode of transcellular diapedesis, in which adherent monocytes seemed to be attracted by another transmigrated monocyte, was seen in 57 % (4/7) of transcellular events observed in this system. Local recruitment of endothelial PECAM-1 to sites of transmigration after paracellular, but not transcellular diapedesis was observed. A typical example of the 3D molecular dynamics of PECAM-1 after the completion of diapedesis at a bicellular junction is shown in Fig. 8.13a. At 11 min, monocyte M1 was on the apical surface of the HUVEC (arrows). The redistribution of PECAM-1 around M1 was seen during transmigration (arrows and arrowhead at 23 min), as seen in Fig. 8.12a, b. After transmigration, PECAM-1 was dynamically recruited to the site of transmigration to form a PECAM-1 rich region which increased over time (arrowheads at 28–43 min). A similar dynamics of PECAM-1 was seen at multicellular junctions (Fig. 8.13b). At 54–62 min, monocyte M1 invaded the EC layer through a multicellular junction (arrows and arrowheads), and from 62 to 75 min, the recruitment of PECAM-1 to the site of transmigration was seen (arrowheads), as in Fig. 8.13a.

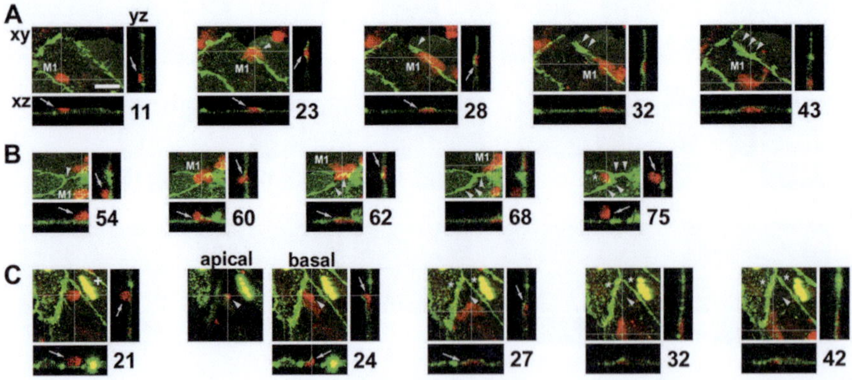

Fig. 8.13 PECAM-1 was recruited to sites of monocytes transmigration at bicellular (**a**) and multicellular (**b**), but not to the transcellular migration site (**c**)

At 75 min, M1 migrated out of 12 sight and another monocyte was observed on the apical surface of HUVEC (* and arrows), but it was not related to the diapedesis of M1. In contrast, PECAM-1 recruitment was never seen to the sites of transcellular diapedesis. Figure 8.13c shows a monocyte that invaded the EC layer transcellularly at 21–27 min (arrows and arrowheads), in which a transcellular pore was clearly visible at 24 min, but endothelial PECAM-1 did not redistribute around the monocyte or was not recruited either to the site of transmigration (arrowhead) or neighboring junctions (*), between 27 and 42 min.

Quantitative image analysis showed a significant, time-dependent increase in PECAM-1-GFP fluorescence specifically at the sites of paracellular transmigration at both bicellular and multicellular junctions. There was up to a 1.4-fold increase within 20 min of transmigration (Fig. 8.14a, b), which was accompanied by a significant reduction in fluorescence in the neighboring cytoplasmic regions (Fig. 8.14c), suggesting that PECAM-1 was recruited from the sub-junctional cytoplasm to the site of transmigration. Fluorescent intensities over the whole scanned area (250 Å ~ 250 μm) remained unchanged (Fig. 8.14c). In contrast, no change in PECAM-1-GFP fluorescence was seen at the sites of transcellular diapedesis (Fig. 8.14b). PECAM-1 recruitment was not seen at bicellular junctions where no transmigration occurred (Fig. 8.14Aa, while many added monocytes had adhered to the HUVEC layer) or in the presence of antibodies against β2-integrin

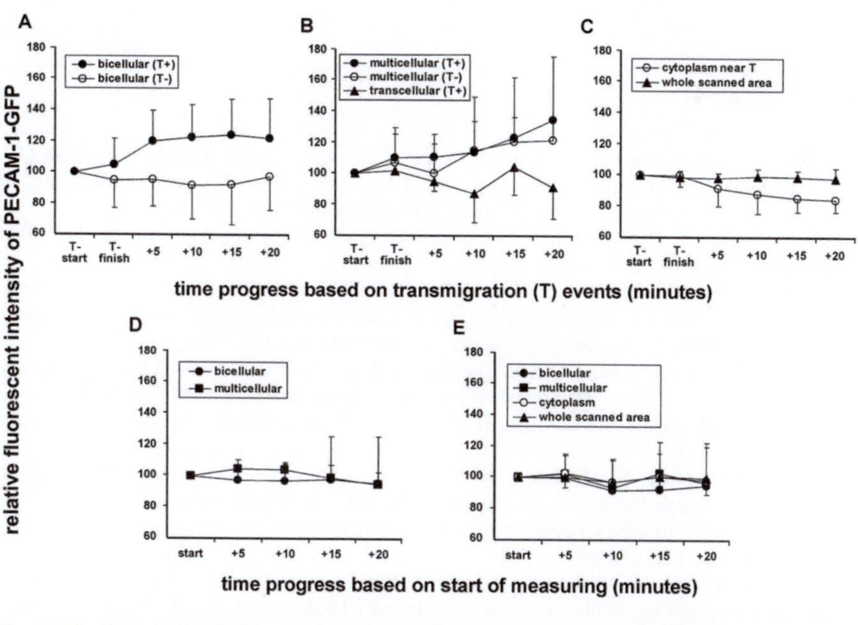

Fig. 8.14 Quantitative image analysis of PECAM-1 recruitment to transmigation sites. (**a**) At bicellular junctions, (**b**) At multicellular junctions, (**c**) PECAM-1-GFP in sub-junctional cytoplasmic regions, (**d**) PECAM-1 recruitment at bicellular ($n = 8$) or multicellular ($n = 6$) junctions when monocyte adhesion was blocked. (**e**) PECAM-1-GFP fluorescence at bicellular junctions ($n = 10$), multicellular junctions ($n = 8$), in the sub-junctional cytoplasmic regions ($n = 10$), or over the whole scanned area ($n = 3$) in control HUVECs without monocytes

and ICAM-1 which blocked monocyte adhesion (Fig. 8.14d), suggesting that the recruitment was driven by monocyte transmigration itself, rather than by adhesion or monocyte-derived soluble factors. At multicellular junctions, the recruitment appeared to occur in the regions where no transmigration occurred (Fig. 8.14b), but not when adhesion was blocked (Fig. 8.14d), suggesting that the recruitment was driven initially by adhesion, and subsequently by transmigration, but not by soluble factors. No changes were observed in any regions in control HUVECs without monocytes (Fig. 8.14e), showing that the recruitment of PECAM-1 was not caused by activities intrinsic to ECs or by artifacts such as photobleaching.

We have presented hear the importance of mechanobiology of endothelial cells and monocyte dynamics in atherogenesis. We have also shown that monocyte adhesion to endothelial cells may facilitate subsequent monocyte transendothelial migration via not only the changes of mechanical properties of EC but also the redistributution of PECAM-1 in EC. It is possible that monocyte adhesion and migration could be mechanical stimulus to endothelial cells, because endothelial cell deformation and/or strain with monocyte adhesion are much more than those by fluid shear stress. To investing the direct effects of monocyte adhesion and/or migration to endothelial cells, the real time observation of F-actin dynamics in endothelial cells and monocytes is needed. Of course, real time observation of signaling molecules in cells also needed. For further investigation, newly designed biomechanical and biophysical experimental technics should be developed.

References

Bradfield PF, Scheiermann C, Nourshargh S, Ody C, Luscinskas FW, Rainger GE, Nash GB, Miljkovic-Licina M, Aurrand-Lions M, Imhof BA (2007) JAM-C regulates unidirectional monocyte transendothelial migration in inflammation. Blood 110:2545–2555

Giaever I, Keese CR (1991) Micromotion of mammalian cells measured electrically. Proc Natl Acad Sci U S A 88:7896–7900

Hashimoto K, Kataoka N, Nakamura E, Asahara H, Ogasawara Y, Tsujioka K, Kajiya F (2004) Direct observation and quantitative analysis of spatiotemporal dynamics of individual living monocytes during transendothelial migration. Atherosclerosis 177:19–27

Hashimoto K, Kataoka N, Nakamura E, Tsujioka K, Kajiya F (2007) Oxidized LDL specifically promotes the initiation of monocyte invasion during transendothelial migration with upregulated PECAM-1 and downregulated VE-cadherin on endothelial junctions. Atherosclerosis 194:e9–e17

Hashimoto K, Kataoka N, Nakamura E, Hagihara K, Okamoto T, Kanouchi H, Mohri S, Tsujioka K, Kajiya F (2012) Live-cell visualization of the trans-cellular mode of monocyte transmigration across the vascular endothelium, and its relationship with endothelial PECAM-1. J Physiol Sci 62(1):63–69

Huang MT, Larbi KY, Scheiermann C, Woodfin A, Gerwin N, Haskard DO, Nourshargh S (2006) ICAM-2 mediates neutrophil transmigration in vivo: evidence for stimulus specificity and a role in PECAM-1-independent transmigration. Blood 107:4721–4727

Kataoka N, Iwaki K, Hashimoto K, Mochizuki S, Ogasawara Y, Sato M, Tsujioka K, Kajiya F (2002) Measurements of endothelial cell-to-cell and cell-to-substrate gaps and micromechanical properties of endothelial cells during monocyte adhesion. Proc Natl Acad Sci U S A 99:15638–15643

Keiper T, Al-Fakhri N, Chavakis E, Athanasopoulos AN, Isermann B, Herzog S, Saffrich R, Hersemeyer K, Bohle RM, Haendeler J, Preissner KT, Santoso S, Chavakis T (2005) The role of junctional adhesion molecule-C (JAM-C) in oxidized LDL-mediated leukocyte recruitment. FASEB J 19:2078–2080

Klouche M, May AE, Hemmes M, Messner M, Kanse SM, Preissner KT, Bhakdi S (1999) Enzymatically modified, nonoxidized LDL induces selective adhesion and transmigration of monocytes and T-lymphocytes through human endothelial cell monolayers. Arterioscler Thromb Vasc Biol 19:784–793

Mamdouh Z, Chen X, Pierini LM, Maxfield FR, Muller WA (2003) Targeted recycling of PECAM from endothelial surface-connected compartments during diapedesis. Nature 421:748–753

Mamdouh Z, Kreitzer GE, Muller WA (2008) Leukocyte transmigration requires kinesin-mediated microtubule-dependent membrane trafficking from the lateral border recycling compartment. J Exp Med 205:951–966

Muller WA, Weigl SA, Deng X, Phillips DM (1993) PECAM-1 is required for transendothelial migration of leukocytes. J Exp Med 178:449–460

Radmacher M, Fritz M, Kacher CM, Cleveland JP, Hansma PK (1996) Measuring the viscoelastic properties of human platelets with the atomic force microscope. Biophys J 70:556–567

Ross R (1993) The pathogenesis of atherosclerosis: a perspective for the 1990s. Nature 362 (6423):801–809

Schenkel AR, Mamdouh Z, Chen X, Liebman RM, Muller WA (2002) CD99 plays a major role in the migration of monocytes through endothelial junctions. Nat Immunol 3:143–150

Shaw SK, Bamba PS, Perkins BN, Luscinskas FW (2001) Real-time imaging of vascular endothelial-cadherin during leukocyte transmigration across endothelium. J Immunol 167:2323–2330

Su WH, Chen HI, Jen CJ (2002) Differential movements of VE-cadherin and PECAM-1 during transmigration of polymorphonuclear leukocytes through human umbilical vein endothelium. Blood 100:3597–3603

Swirski FK, Libby P, Aikawa E, Alcaide P, Luscinskas FW, Weissleder R, Pittet MJ (2007) Ly-6Chi monocytes dominate hypercholesterolemia-associated monocytosis and give rise to macrophages in atheromata. J Clin Invest 117:195–205

Tacke F, Alvarez D, Kaplan TJ, Jakubzick C, Spanbroek R, Llodra J, Garin A, Liu J, Mack M, van Rooijen N, Lira SA, Habenicht AJ, Randolph GJ (2007) Monocyte subsets differentially employ CCR2, CCR5, and CX3CR1 to accumulate within atherosclerotic plaques. J Clin Invest 117:185–194

Mechanotransduction of Shear Stress by the Endothelium

9

Peter J. Butler

Abstract

The ability of endothelial cells to convert forces from blood flow to biochemical signals underlies vascular health and disease. While the mechanisms of endothelial cell mechanotransduction are still an area of ongoing research, a picture is emerging that encompasses the temporal and spatial complexity of blood flow patterns and the highly heterogeneous and dynamic mechanical properties of endothelial cells. In this framework, cell sense blood flow-induced shear stress through specialized structures such as the glycocalyx, membrane microdomains, focal adhesions and adherens junctions, where forces are converted to biochemical signaling cascades via alterations in protein conformations and associations. The result of these processes are the production of vasodilators and the activation of genetic transcription factors that lead to changes in endothelium permeability, adhesiveness to circulating leukocytes and platelets, and changes in vascular diameter. Therefore, understanding the mechanobiology of endothelial cells is at the heart of promoting vascular health and predicting, diagnosing, treating, and preventing vascular disease.

Keywords

Endothelial cells • Mechanotransduction • Mechanobiology • Membrane • Cytoskeleton

P.J. Butler (✉)
Department of Biomedical Engineering, The Pennsylvania State University, 230 Hallowell Building, University Park, PA 16802, USA
e-mail: pbutler@psu.edu

K. Tanishita, K. Yamamoto (eds.), *Vascular Engineering*,
DOI 10.1007/978-4-431-54801-0_9

9.1 Introduction

Sensation and transduction of forces by endothelial cells are part of their inherent role as permeability barriers and regulators of blood flow in arteries, capillaries, and veins. Because of their proximity to blood flow, they are continuously exposed to fluid flow-induced mechanical forces that exhibit a wide range of spatial and temporal scales and directions. Endothelial cells (ECs) form junctions to create the endothelium, a one-cell layer thick, highly regulated, semipermeable barrier to blood-borne solutes. It is now recognized that the interactions between the endothelium and hemodynamic forces, such as shear stress, are largely responsible for regulating the endothelium's barrier function, either through modulation of the space between cells, changes in mitotic turnover rates of endothelial cells, or by affecting the endothelium's adhesiveness to circulating leukocytes, which can migrate into the sub-endothelial intima. Such mechanotransduction involves the sensation of force by several endothelial cell structures. These include a surface glycocalyx layer composed of glycosaminoglycans, glycolipids, and glycoproteins, a plasma membrane composed of a multi-phase lipid bilayer and integral and peripheral membrane proteins, a four-component cytoskeleton (fodrin, actin filaments, microtubules, intermediate filaments), focal adhesion complexes responsible for cell signaling and adhesion to the basement membrane, and cell (gap and tight) junction proteins responsible for cell-cell communication and adhesion. These structures transmit forces to proteins and lipids, which become biochemically active and transduce these mechanical signals into chemical pathways in the cell leading to formation of vasodilators and vasoconstrictors, activated transcription factors, or alterations in cell shape and endothelium permeability.

The key questions about mechanotransduction are as follows. What components of force are sensed by the cell? What are the molecular sensors that transmit force to transducers in the cell? Is transduction through physical links that transmit force from one molecular partner to another, or does force work through existing receptors, which transmit biochemical signals largely through diffusive pathways? To what extent are the force transduction partners regulated by the very force they sense? And what are the main factors produced at the terminal end of the force transduction pathways? Answers to these questions require a detailed evaluation of vascular and cellular structures and inherent engineering analysis of flow dynamics in areas experiencing vascular disease. Second, molecular biological investigations are needed that elucidate molecular force sensors and their interconnections and dynamic rearrangements in response to force. Third, it will be necessary to understand, from an engineering perspective, what components of force are sensed, and what the mechanical properties of these sensors are and how they can respond to force magnitude, direction, frequency, and duration. Finally, it is important to determine the identity of final soluble products produced as a result of mechanotransduction and how these products can lead to vascular health and disease.

The purpose of this chapter is to systematically explore these questions and provide the most plausible answers to them. As in any emerging field of molecular biology, there remains much controversy over the precise identity of molecular partners and pathways. In response to this, there have been some excellent reviews on the specific biochemical pathways that are activated by force. For example, shear stress elicits calcium increase in cells (Ando and Yamamoto 2013; Yamamoto and Ando 2010) and increases in transcription factor phosphorylation (Davies 2009; Li et al. 2005; Nigro et al. 2011). Nevertheless, there remains controversy about the precise mechanical pathways linking force and changes in biology (Huang et al. 2004). This chapter begins with an analysis of the main components of blood flow (stress and transport) that are sensed by the endothelium and then describes the mechanical linkages responsible for mechanotransduction and their dynamics. It is hoped that understanding these frameworks in the context of the role of shear stress in endothelium function will assist in the development of strategies to ameliorate vascular diseases that find their origins at the interface of engineering and biology.

9.2 Overview of Mechanotransduction and Organization of This Chapter

Mechanotransduction has been measured on temporal scales that range from seconds to hours of exposure to flow. Rapid responses include potassium channel activation (Boycott et al. 2013), intracellular calcium increase (Yamamoto et al. 2011), G-protein activation (Chachisvilis et al. 2006; Gudi et al. 1998), stimulation of mitogen-activated protein (MAP) kinases (Jalali et al. 1998), and cell junction responses (Tzima et al. 2005). A typical example of mechanotransduction occurs when forces that act on the apical surface are transmitted by the cytoskeleton to other locations where signaling can occur. For example, force can be applied to the apical surface of cells and integral membrane proteins and transmitted via the cytoskeleton to focal adhesions, intercellular junctions, cellular organelles, and the nuclear membrane. Such structures have been shown to remodel in response to force and to transduce force into biochemical signaling. Other predominant molecular players in mechanotransduction include membrane ion channels, cilia, tight and gap junction complexes, and nuclear tethered nesprins (nuclear envelope spectrin repeat proteins), each sharing the capability to sense, transduce, and adapt to force. Furthermore, studies have shown that EC function is not only affected by force but also by temporal and spatial gradients in force (Blackman et al. 2000; Butler et al. 2000, 2002; Frangos et al. 1996). Thus, the temporal qualities of force application need to be matched to the dynamics of sensing, transducing, and adaptation of the transducer elements in cells.

Most information about mechanotransduction arises from the intersection of engineering and molecular biology. Forces can be applied to cell cultures followed by assays of population of cells, single cells and single molecules, and assays that evaluate whole tissue and organ morphology. Many approaches to understanding the mechanical origins of vascular disease use flow chambers to impart

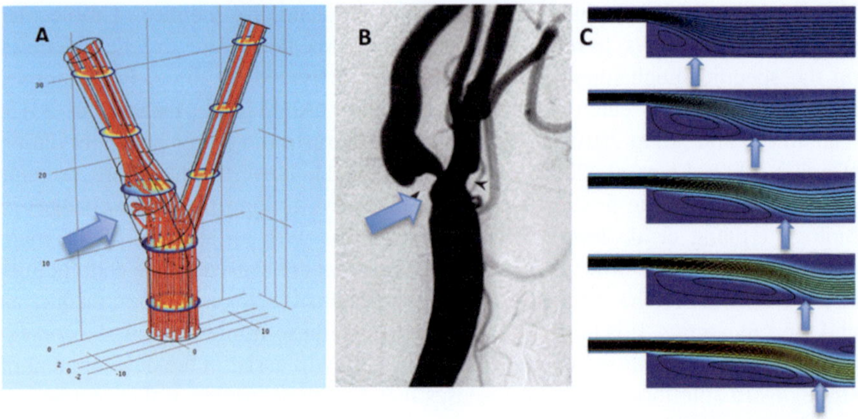

Fig. 9.1 (**a**) Streamlines and relative velocity magnitudes in the carotid bifurcation. (**b**) *In vivo* angiogram of stenosis (*arrows*) at areas predicted to experience disturbed laminar flow *in vivo* (compare with (a)) (Malek 1999). (**c**) In vitro, shear stress effects on endothelium can be studied by plating cells on the floor of flow chambers and subjecting them to fluid shear stresses that differ in spatial and temporal components (flow speeds and dimensions taken from (Haidekker et al. 2001) and solved for using COMSOL multiphysics simulation of the Navier-Stokes equations). *Arrow* indicates traveling location of the separation point as a function of time (frames are 0.1 s apart). To the *left of the arrow*, flow recirculates; to the *right* it increases to the free stream velocity

physiological levels of shear stress onto cells (e.g. see Fig. 9.1). Alternatively, some studies describe cell-generated traction forces, or consider the forces between the extracellular matrix and the endothelial cell. To begin to decipher mechanotransduction events, one must identify, characterize, and develop an engineering framework toward understanding how mechanical forces contribute to cell biology. One purpose of this chapter is to review these frameworks.

We begin this chapter with an overview of the shear effects on endothelial cells in atheroprone and atheroprotected areas of the vasculature. We then outline the frameworks for mechanotransduction of shear stress and discuss three of the main structural mechanisms by which force is transduced into biochemical signaling, namely, the glycocalyx, membrane, and cytoskeleton. We then discuss the possible mechanisms of force-induced changes in biochemical signaling and conclude with an example that focuses on force-induced nitric oxide production. When possible and helpful, we have created figures using finite element analysis of the Navier-Stokes equations, which describe the relationship between force and fluid dynamics, and equations of solid mechanics relating time-dependent force and cellular deformation. These engineering equations have been applied to realistic *in vitro* and *in vivo* geometries in order to quantitatively illustrate the relationship between cellular mechanotransduction and blood flow magnitudes and direction.

9.3 Shear Stress Is Associated with Vascular Health and Disease

9.3.1 Low Oscillatory Shear Is Atherogenic While High Unidirectional Shear Is Atheroprotective

In its most basic form, shear stress, τ, is the product of the blood velocity gradient (increase in axial velocity, u, as one moves away from a wall a distance y) and the blood viscosity, μ:

$$\tau = \mu \frac{\partial u}{\partial y}. \tag{9.1}$$

It is a frictional force in the direction of blood flow, tangent to the wall. Shear stress arises from blood flow, the temporal and spatial velocities ($\boldsymbol{u}$) of which are captured in the Navier-Stokes equations:

$$\rho \frac{\partial \boldsymbol{u}}{\partial t} + \rho \boldsymbol{u} \cdot \nabla \boldsymbol{u} = -\nabla P + \mu \nabla^2 \boldsymbol{u}. \tag{9.2}$$

In this form t is time, ρ is density, μ is viscosity, P is pressure, and $\boldsymbol{u}$ the velocity vector in three dimensions. The equation arises from equating the forces of pressure and shear with the acceleration of a differential element of fluid. In large blood vessels, inertial forces of blood flow dominate, while in small blood vessels frictional forces dominate. In general, this equation cannot be solved analytically. Rather, this equation can be implemented in complex geometries characteristic of branching blood vessels using finite element analysis, a numerical technique in which the complex flow field is divided into small elements on which the equation is linearized. The full flow field is then constructed by combining contributions from the small elements. Such simulations are indispensible in understanding the directional and temporal features of flow that are most likely responsible for the development of atherosclerosis via mechanotransduction by the endothelium. For example, in Fig. 9.1a, the flow streamlines and velocities were derived from solving the Navier-Stokes equations in a realistic geometry for the carotid bifurcation and a realistic pressure waveform.

These forces from fluid flow act on the blood vessel wall and induce deformation of the endothelial cells there. Further, flow chambers can be constructed in which the temporal and spatial gradient match the *in vivo* conditions. While it is beyond the scope of this chapter to describe the details of mechanical properties of cells, which is still an area of intense debate, evidence suggests that cells react to forces by deforming in a time-dependent manner, suggesting, in a continuum formulation, that they are viscoelastic, meaning that their deformation is proportional both to the magnitude of shear stress applied and to its rate of change. A recent study suggests that cells deform as viscoelastic liquids (Dangaria and Butler 2007), which can be described by Jeffrey's liquid model:

$$\tau + \lambda_1 \frac{\partial \tau}{\partial t} = \eta \left(\dot{\gamma} + \lambda_2 \frac{\partial \dot{\gamma}}{\partial t} \right). \quad (9.3)$$

where τ is the stress tensor, η is the shear viscosity describing the resistance to shear strain, and $\dot{\gamma}$ is the shear rate describing the time-dependent deformation of the cell (in two dimensions, $\dot{\gamma} = \frac{\partial u}{\partial y} + \frac{\partial v}{\partial x}$, with u and v being the deformations in the x and y directions, respectively). This equation allows for deformation that is proportional to the stress as well as deformation that depends on the rate of change of stress, captured by two time constants λ_1 and λ_2. In one form of this equation, the cell can be described as a dashpot in parallel to a Maxwell element (in series dashpot and spring). Deformation of a cell in response to shear stress is small (on the order of 50–100 nm) and time dependent with a time constant on the order of 1–4 s (Fig. 9.2). This result suggests that small deformations lead to mechanotransduction

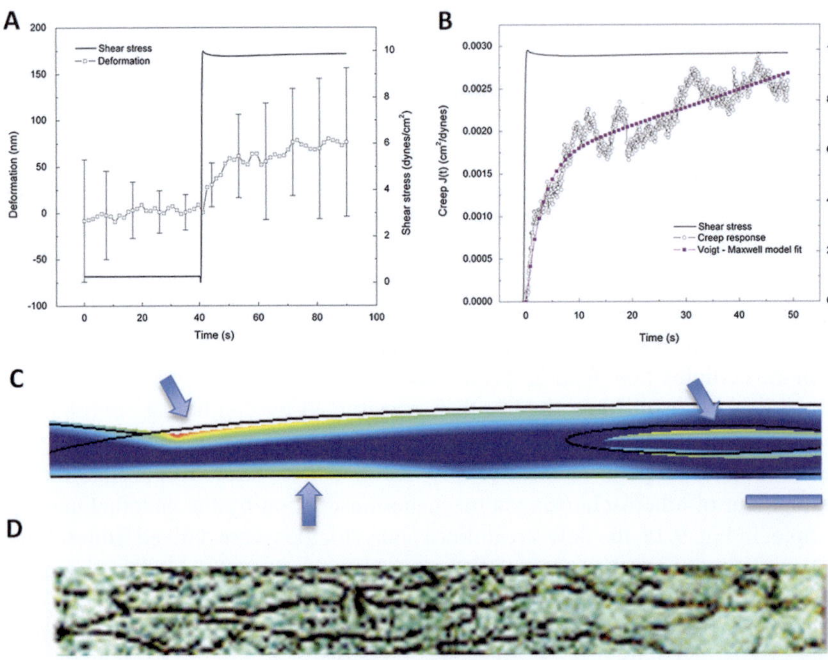

Fig. 9.2 (**a**) Shear stress results in time-dependent deformation of endothelial cells. (Dangaria and Butler 2007). (**b**) Creep in response to a step change in shear can be fit with a model for a viscoelastic liquid. (**c**) A simple viscoelastic model of a cell with a 2 s relaxation time was subjected to 1 Pa of surface shear. Deformation was on the order of 10 s of nanometers. *Arrows* denote areas of stress concentration in the membrane (*colors* represent relative von Mises stresses), at cell junctions, focal adhesions, and the nucleus. Bar = 10 μm. (**d**) Silver nitrate staining of endothelial cell borders in small arterioles (rat cremaster). Alignment of long axis of endothelial cell is in line with the prevailing flow direction. Bar = 20 μm

events and that the cell is potentially sensitive to temporal strains that are on the same time constant as flow pulsatility.

Studies that correlate flow patterns with vascular disease have determined that shear stress is the main determinant of whether the endothelium is atherogenic or atheroprotective (reviewed in (Cecchi et al. 2011)). Key features of blood flow that determine this distinction are shear magnitude and the temporal and spatial nature of shear stress. When low and oscillatory, shear stress has flow reversal that is associated with endothelial cells that are cobblestone in appearance, have elevated surface expression of leukocyte adhesion proteins such as vascular cell adhesion molecule-1 (VCAM-1) (Feaver et al. 2013), and increased production of monocyte chemotactic protein-1 (MCP-1) (Shyy et al. 1994) . The endothelium in these areas is leakier and more adhesive than similar endothelium found in areas of the vasculature that experience high unidirectional shear stress, even if it is also pulsatile. Such adhesiveness, leakiness, and cobblestone shape are strongly correlated with atherosclerotic lesion formation. In these areas, the temporal nature of the shear arises from the pumping action of the blood and elasticity of blood vessels, which behave as windkessels with phase angles between stretch and shear pulsatility that are greater than zero (Tada et al. 2007). In addition, when shear changes magnitude over short distances, spatial shear gradients exist which induce forces on cells that may be significantly different than forces on neighboring cells (DePaola et al. 1992). Regarding flow direction, areas experiencing oscillatory shear with flow reversal will see flow direction changes over the time of a heartbeat and an average shear around zero pascals.

Areas that experience atherogenic shear include large blood vessels with large curvature such as the aortic arch, coronary arteries, the carotid sinus, superficial femoral arteries near the adductor canal, femoral arteries, and aortic bifurcation, aneurisms, and areas downstream of stenoses. In addition, vascular devices intended to aid in improving vascular function and patency can induce alterations in blood flow that are atherogenic in, for example, stents (Koskinas et al. 2012) and vascular grafts (Kassab and Navia 2006). Thus, interventions often intended to ameliorate the effects of atherosclerosis can lead to relapses of the disease. Areas devoid of atherogenic shear include straight parts of the aorta, and in the microcirculation, where pulsatility and, hence, temporal flow reversal are absent.

When shear stress is unidirectional and high, pro-inflammatory mediators are downregulated (reviewed in (Davies 2009)). Such reactions to prevailing flow suggest that there are homeostatic mechanisms by which cells attempt to achieve an intracellular and intercellular tension that is optimal for health-related singling pathways. Long-term high shear downregulates the transcription factors required for MCP-1 transcription. High unidirectional pulsatile shear up-regulates KLF-2 (Kruppel-like factor 2), a molecule that is a beneficial to endothelial cell survival. Differences in KLF-2 are also seen *in vivo* with high KLF-2 on the medial flow divider where high shear stresses are predicted. Similarly, MCP-1 was found in flow divider regions where lower and oscillatory shear would be expected. These *in vivo* results corroborate *in vitro* results that suggest that areas experiencing low and oscillatory shear are atherogenic, while high shear protects the endothelium and artery wall.

Methods to determine shear dynamics *in vivo* include non-invasive computational fluid dynamics (Giddens et al. 1993), which can provide detailed shear profiles even in the face of pulsatile flow-induced distention. Ultrasound techniques include pulsed Doppler and magnetic resonance imaging phase contrast velocity mapping. Alternatively ultrasound or MRI can be used to provide detailed anatomical representations of blood vessels. These images can be reconstructed using solid modeling software followed by image segmentation, meshing, and computational reconstruction (using finite element analysis) of the fluid-structure interactions. Such computational methods provide detailed stress profiles and they can also be used to calculate mass transfer of blood borne metabolites that might be atherogenic. Approaches to reconstruction of shear profiles based on anatomical data provide the possibility of patient-specific diagnosis and treatment (Cecchi et al. 2011).

The role of shear oscillations and shear stress gradients in vascular health and disease is covered extensively in other chapters of this book. In this chapter, we focus on the mechanism by which temporal and spatial shear gradients might induce mechanosensation and transduction in endothelial cells, the main cell type reacting to these flow dynamics.

While shear stress is one of the main aspects of blood flow governing susceptibility of the vasculature to atherosclerosis, to date there has been little progress on using this observation to prevent shear-induced atherogenesis, largely because it is difficult to conceive how one might alter shear *in vivo*. However, the fact that high shear is generally seen as atheroprotective means that regular exercise, which may increase blood flow rate temporarily, may help to mitigate the progression of the disease. The connection between a healthy vasculature and exercise is well known and covered in other chapters of this book. The role of high shear in prevention of stenosis could also help in surgical procedures such as vascular graft anastomosis where angle and suture strategies could be employed that minimize the occurrence of recirculation zones.

9.3.2 Evidence for Correlation Between Low Shear and Atherogenesis

One of the early studies demonstrating the correlation of low and oscillating shear stress was by Caro and co-workers (Caro et al. 1969). While others had noticed the focal nature of atherosclerotic lesions, many had assumed that mass transfer effects (such as platelets) or high shear stress-induced endothelial cell injury were responsible for the localization of disease. Caro et al. argued that high shear has a protective effect on the endothelium, a prescient observation that has been borne out in numerous subsequent studies. In high shear regions, the boundary layer is expected to be relatively thin while low shear areas would be marked by a thicker boundary layer. Thus, Caro hypothesized that atherosclerotic lesions should occur where the boundary layer was thick, such as the outer walls (where the flow moves away) of daughter arteries. Observation of human superior mesenteric and renal arteries indeed showed that inner walls were generally spared lesions while the outer wall, immediately downstream from flow dividers, had evidence of lesions. From these observations, the authors proposed a new theory on atherosclerosis formation. They suggested that in areas of low shear, the shear rate is also low and this fact would minimize mass transfer away from these areas. The concomitant increase in local concentration of atherogenic blood borne substances (e.g. cholesterol) that would accumulate in these areas could contribute to the formation of lesions. From a clinical perspective, this suggests that exercise that increases heart rate may lead to the washing out these areas and thus explain its known health benefits (Caro et al. 1971).

To test further the role of hemodynamics in atherogenesis, Ku *et al.* developed realistic models of the carotid sinus lumen and measured flow velocities in 3D in areas of the bifurcation where cadaver specimens indicated atherosclerotic lesion formation and absence (Ku et al. 1985). In this study, the authors developed an oscillatory shear index that quantified how much of the instantaneous shear was in the direction of the prevailing shear of inner and outer walls where the midpoint was defined as the centerline of the sinus of the internal carotid. The outer wall (proximal internal carotid) was prone to atherosclerosis, while the inner wall was relatively protected. The formulation of this index recognized that not only low shear might be important but flow direction as well. Results show that the intimal thickness was positively correlated to the oscillatory shear index indicating that flow reversal at low shear stress is associated with intimal thickening.

To assess an *in vivo* link between vascular geometry, altered hemodynamics, endothelial wall shear stress and atherogenesis, Koskinas and colleagues performed an *in leporine* serial study using intravascular ultrasound to create 3D models of all major coronary arteries before and after induction of diabetes and hyperlipidemia, considered strong risk factors for atherosclerosis. They found that low wall shear stress promoted the formation of plaques (Koskinas et al. 2010). From a clinical perspective, these authors demonstrated *in vivo* correlation between low wall shear stress and expansive remodeling of the vessel wall.

However, not all studies point to a relationship between low and oscillatory shear stress and atherogenesis. Peiffer et al. conducted a systematic review of articles purporting to draw correlations between low and oscillatory shear stress and atheroma formation (Peiffer et al. 2013). Their study reviewed 27 articles out of an initial 406 that had in common the condition that lesions were formed only as a result of diet and flow patterns (not surgically induced) and that anatomical flow patterns were correlated to computational fluid dynamics simulations, so that calculations of wall shear stress came from hemodynamically accurate flows and in conjunction with true geometric properties. In this study, when looking at only the steady component of blood flow, nine of the 27 papers concluded that there was an inverse relationship between shear stress magnitude and wall thickening, confirming earlier hypotheses that low shear stress was atherogenic while high shear was atheroprotective. Even when considering shear averaged over the cardiac cycle, three additional studies concluded that low shear co-localized with atherosclerosis. Other papers include oscillation by computing an oscillatory shear index. In theses studies, both high shear oscillations and low shear coincided with plaque location. Of course, shear oscillation and averaged wall shear do not account for all properties of flow that might be involved with plaque development. Shear angle and relative residence time of blood borne solutes may also be strongly correlated with plaques. While eleven articles supported the low and oscillatory shear theory of plaque progress, these studies used qualitative mapping and correlation with calculated shear to make their conclusions. Point-by-point comparisons in some cases failed to reach a correlation between low and oscillatory shear and intimal thickening. With respect to shear magnitude, a number of studies cited suggested that a wall shear threshold of 1–1.5 Pa range was associated with atherogenesis.

One possible solution to the controversy over low/oscillatory shear and other atherogenesis predictors is by Davies and co-workers who laid out a hypothesis suggesting that low shear leads endothelial cells to operate in a low level inflammatory state (Davies et al. 2013). ECs become susceptible to atherosclerosis in areas of low and oscillating shear where eddies concentrate undesirable metabolites such as reactive oxygen species (ROS) and where lower stress fails to elicit increases in anti-inflammatory genes. These phenotypes include cells with leaky junctions, higher levels of cell division and proliferation, and higher expression of receptors for monocytes. However, these phenomena are not sufficient for atherosclerosis without the existence of additional cardiovascular risk factors, such as hypercholesterolemia, hypertension, diabetes, and smoking. Thus, hemodynamic shear stress (a mechanical viewpoint) can be seen as priming the endothelium for atherosclerosis development, while mass transfer characteristic may contribute the actual development of atheroma. Pro-inflammatory events in these areas include suppressed endothelial cell nitric oxide synthase (eNOS) activity, increase NF-κB transport to the nucleus, decreased protective transcription factors KLF-2 and KLF-4, and reduced expression of the electrically conductive connexin 43. Interestingly, the increased in protein message causes overload in the endoplasmic reticulum (ER) of protein packaging, leading to an increase in misfolded proteins. These misfolded proteins cause an increase in upregulation of ER chaperones and folding enzymes to prevent aggregation of misfolded proteins. Thus is must be recognized that the link between mechanics and biology can be circuitous and highly indirect. Mechanotransduction often includes all of these pathways simultaneously, somewhat obscuring the cause and effects between mechanical force and changes in biological function.

9.3.3 Hemodynamics Influences Mass Transfer of Blood Borne Solutes

While is now generally accepted that shear forces play a dominant role in localization of atherosclerotic lesions, there has also been considerable number of studies on mass transport properties. For example, disturbed blood flow in areas of the vasculature with abrupt geometric changes causes eddies, swirling flow, and reverse flow that can alter the degree of accumulation of blood borne lipids and blood cells, as well as the transport of substances produced by the endothelial cells themselves (e.g. ATP), leading to an autocrine activation of endogenous ATP receptors (Nollert et al. 1991; Yamamoto et al. 2003). Tarbell has reviewed the mechanism by which transport is important in atherosclerosis localization (Tarbell 2003). These mechanisms involve a reactive surface in which a species is transported to the endothelial cell surface and catalyzed by surface enzymes via Michaelis–Menten kinetics. In such a scenario, the shear rate (rather than shear stress) is responsible for modulating the local concentration of reactants, a principle modulator of the reaction rate. By comparing the mass transfer (characterized by a dimensionless Sherwood number) and the reaction rates (characterized by a Damkholer number), it is apparent that when $Da < Sh$ the process in reaction-limited (in other words, there are plenty of reactants and the rate is controlled by the reaction rate), whereas when $Da > Sh$, the reaction is transport-limited (reaction is controlled by the rate at which reactants are transported because the reaction proceeds much faster). Therefore, in the reactive surface model, flow rate, rather than force, controls the rate of reaction, particularly in areas where the flow is low and reversing.

The most important feature of transport is the fact that geometry of blood vessels creates unique patterns of blood flow. In the longer straight parts of aortae, the flow is unidirectional with an overall pulsatility governed by the heart rate. In these areas, Sh for small molecules such as O_2 and ATP is low relative to Da making the reaction fluid transport limited. Alternatively, larger molecules such as albumin and low density lipoprotein (LDL) are not transport limited. Such analysis suggests that transport may be an important factor in the localization of lesions if the mechanism of lesion formation requires small molecules. In support of a small molecule origin for atherosclerosis, hypoxia appears to be prevalent in recirculation zones and at areas of the vasculature that are predicted to have a low Sherwood number. Indeed, the areas of low shear often correlate with the areas of low concentration of small molecules such as oxygen and ATP. Thus, a vexing and outstanding problem with investigating the connection between mass transfer and focal nature of atherosclerosis is that the shear and mass transfer are linked.

Thus, the precise relationship between shear, mass transfer, and vascular disease remains uncertain (Ethier 2002). Attempts to separate the two generally focus on altering shear using viscosity-altering agents. Nollert and colleagues investigated the role of mass transfer and shear directly using cultured endothelial cells (Nollert et al. 1991). In their study, initiation of flow caused the increase in prostacylcin production, arachadonic acid incorporation, mRNA, and calcium. They then investigated the potential role of mass transfer using a computational model of ATP convection and diffusion because flow did not appear to increase calcium in the absence of ATP. Overall, the role of mass transfer in activation of cells via fluid flow may depend on the receptor pathway involved. Some pathways appear to be shear stress (mechanical) sensitive while other pathways are shear rate (convective flux) sensitive. Nevertheless, because shear is so important, at least in so far as it acts to allow or inhibit inflammation of the endothelium, depending on its spatial and temporal features, much research has focused on engineering analysis of shear-induced deformation of endothelial cells and their subcellular structures. It is hoped that by such analysis, the identity of shear stress sensors and transducers can be found.

9.4 The Main Frameworks for Mechanotransduction Include Structural Transduction and Transduction Through Diffusive Pathways

9.4.1 Sensitivity of Cells to Force Dynamics Depends on the Dynamics of the Pathway Stimulated

Hoffman and colleagues reviewed studies to determine if mechanotransduction occurs through a switch mechanism (Hoffman et al. 2011). In this framework, mechanotransduction works through mechanotransmission, mechanosensing, and mechanoresponse. Since load-bearing structures are structurally dynamic, forces may elicit mechanoresponses through alterations in intrinsic rates of subcellular responses. By understanding in the rates of cellular reactions in cells, it may be possible to determine why cells are sensitive to multiple time scales of temporally changing mechanical stimuli. In a switch model, mechanotransduction begins with mechanotransmission, by which forces are transmitted directly or remotely to mechanosensitive elements. For example, force applied on the apical side of the cell exerts effects through focal adhesions on the basal side of the cell. In this case, the entire cell acts as a mechanotransmitter.

In mechanosensing, forces applied to a structure are accommodated by alterations in either the molecular assembly of that structure, or in the conversion of a protein from one conformational state to another. Proteins sample many energy minima and can be transformed from one to another by ligand binding. It is not clear yet whether forces can directly induce a protein to convert into an active conformation that is equivalent to one know to be induced by ligand binding (Lee et al. 2007), but it is known in many instances that methods to block protein function (e.g. antibodies, siRNA knockdown) are effective if blocking certain mechanotransduction events. As discussed later in this chapter, mechanosensing can occur with changes in conformation of a protein, whole protein unfolding, or alteration in association with other proteins.

Mechanoresponses are the cellular processes that are initiated after mechanosensation. Theses processes (such as calcium signaling, phosphorylation pathways, and transcription factor binding to DNA) may not be directly force sensitive, but they participate in the most dramatic elements of mechanotransduction. As discussed later, there remains some distinction between mechanotransduction in which all forces are carried through force bearing structures to the final mechanically induced response (e.g forces transmitted to the nucleus to affect transcription of DNA) and those mechanotransduction pathways that work through diffusive elements (e.g. MAPK signaling pathways leading to activation of transcription factors). In switch-like models, there is often a direct connection of load bearing structures to the mechanosensor. A mechanosensor might be an ion channel, or a protein, whose force-induced conformational changes make it accessible to other adapter proteins (Bao et al. 2010; Lee et al. 2007). Further studies are necessary to demonstrate how such a mechanism could also depend on the temporal aspects of applied force, as have been shown to be important for many mechanoresponses.

Other models must incorporate reaction rates. In focal adhesions, where many force sensitive molecules reside, it has been found that integrins and actin are continually recycling and flow in response to force. The recycling has a time constant of seconds to minutes (Lele et al. 2006), while flow is on the order 0.1 μm/min. Therefore, observable dynamics of focal adhesions in response to force is on the order of minutes to hours. These multiple time constants make these structures differentially sensitive to the temporal features of the applied forces. Membranes have dynamic microdomains that assemble and disassemble on the order of milliseconds (Eggeling et al. 2009). Since stretch-induced ion channels can be activated by force on the order of milliseconds, one can see how there is a relationship between time course of molecular reorganization of a structure and the time courses of its mechanoactivation.

Because broken links cannot transmit force, the dynamic interaction with linking partners dictates the transmission of force to mechanosensors. Slip bonds have very short lifetimes while catch bonds have longer lifetimes and their binding strength increase with force. With respect to the whole cytoskeleton as a force transmitter, forces can cause fluidization of the cytoskeleton, therefore reducing its ability to transmit force. In other cases, reinforcement results in a nonlinear relationship between applied force and the force reaching mechanosensors. This interplay of reinforcement and fluidization can be manifested in viscoelastic behavior, which carries with it a source of sensitivity to a range of frequencies of force transmission.

Similarly, time scales of mechanosensing can dictate how temporally applied forces are interpreted by cells. When the mechanoresponse is dynamic, the feedback processes that alter mechanosensitive molecules and organization of mechanosensitive molecule complexes can result in responses that depend on the temporal nature of the applied force. For example, the formation of focal adhesions requires diffusion of integrins into the plaque, binding of integrins to extracellular matrix, reinforcement by intracellular binding of focal adhesion kinase and talin and actin. Subsequently, in mature focal adhesions, Zyxin binds to plaques that are in need of repair after force application by recruiting α-actinin leading to actin polymerization and repair. Therefore if cyclic strain was applied to these plaques, mechanosignals that occurred on the same temporal scale as the frequency of force application would experience strains that ranged from positive to negative.

As a rule of thumb, when forces are applied statically, such as the average forward component of unidirectional shear stress, the mechanoreceptor is likely to be a stronger bond with a long life and can sense force by undergoing a conformational change to a new functional shape. In contrast, with rapidly changing forces, such as the pulsatile component of blood flow, mechanosensing is likely to arise from weak linkages between proteins or weakly folded proteins. Regarding flow/

glycocalyx/caveoli/eNOS signaling pathway, research remains as to which parts are sensitive to statically applied forces and which parts are sensitive to cyclic shear applications.

9.5 Shear Stress Acts on the Apical Surface and Is Transmitted to the Cell Interior by Cellular Structures

9.5.1 Shear Stress First Acts on the Glycocalyx

The glycocalyx is a highly glycosylated structure on the surface of almost all cells (Reitsma et al. 2007). In endothelial cells, it has been visualized by electron microscopy and fluorescence confocal microscopy (Lipowsky et al. 2011) and is likely to exhibit a thickness of between 100 nm and 0.5 μm . However, it is thicker in large arteries where it grows up to as much as 4.5 μm in the carotid artery, compared to 0.5 μm in capillaries. Its net composition is a result of dynamic equilibrium between the formation of new molecules and flow-mediated washout of old molecules (Lipowsky et al. 2011). Interspersed with the glycocalyx are membrane-anchored glycoproteins. The selectins are the most studied because of their role in adhesion to circulating leukocytes. Importantly, they tend to be significantly shorter than the proteoglycans and thus are buried in the glycocalyx. Thus the glycocalyx needs to be digested away in order for leukocytes to adhere to the glycoproteins (Lipowsky et al. 2011). While there is significant evidence that glycocalyx proteins are involved in shear stress sensing, there is little evidence that the selectins or other glycoproteins are.

The glycocalyx is composed of acidic oligosaccharides anchored to the membrane via glycoproteins (Pahakis et al. 2007). In addition, there are unanchored polyanionic constituents. Overall the structure is slightly negatively charged and therefore significantly hydrated. In this structure, heparin sulfate (HS), chondroitin sulfate (CS) and hyaluronic acid (HA) link with membrane-bound glypicans and syndecans. Syndecans (1, 2, and 4) are large transmembrane proteins decorated with HS, to which they have a high affinity, with a lower binding affinity to CS. As transmembrane proteins, syndecans enable transduction of forces from outside the cell to the inside. Glypicans, in contrast, form GAG attachments close to the surface of the cells and are decorated exclusively with heparin sulfate. Interestingly they are associated with glycophoinositol (GPI) anchors, which are, in turn, associated with cholesterol-rich microdomains termed caveolae and rafts, suggesting that known involvement of the membrane with shear-induced signaling may be through glypicans and rafts. HA weaves through the glycocalyx and does not associate directly with glypicans or syndecans. Rather, HA links with membrane-bound CD44 receptor that localizes in caveolae.

The syndecans, glypicans, HS, CS, and HA make up a highly ordered periodic structure on the cell surface. To study this structure, Squire and colleagues stained the glycocalyx of frog mesentery endothelial cells using a variety of staining techniques followed by electron microscopy (Squire et al. 2001). They found that the glycocalyx had a quasi-regular meshwork with a characteristic periodic spacing of 20 nm. Further, these periodic structures were bundled and anchored to the cell surface at attachment points spaced on the order of 100 nm apart. This spacing is similar to known spectrin (fodrin) periodicity of around 100 nm suggesting that the bundles were anchored to the actin/fodrin cortical cytoskeleton through transmembrane proteins. Thus, forces transmitted to the glycocalyx may be mechanically transduced directly to the underlying cytoskeleton. Importantly, with respect to mechanotransduction, this arrangement appears to be sufficient to significantly attenuate flow within the glycocalyx suggesting that fluid shear is transduced through the solid portion of the glycocalyx rather than through fluid shear stress per se. Their data also suggested that the thickness obtained from rapid freezing and staining is on the order of 100–200 nm. Arkill and colleagues followed up on the observation of periodic spacing of 20 nm observed in frog mesentery capillaries (Arkill et al. 2011). In this study, they tested for the ubiquity of this spacing in mammalian tissue including choroid, renal tubules, glomerulus, and psoas muscle. Each of these tissues revealed a 20 nm spacing as was observed in frog, suggesting a universality of patterned spacing. Interestingly the 100 nm spacing of bundles was not always observed with some spacing being larger. This suggests the 20 nm spacing may represent the minimal spacing possible.

With respect to mechanotransduction, Ebong and colleagues investigated the roles of HS proteoglycan in shear-induced eNOS activation and cytoskeletal remodeling (Ebong et al. 2014). They specifically focused on the protein anchors used by heparin sulfate. By knocking down glypican but maintaining the syndecan anchor, they showed that shear induced cytoskeletal remodeling but failed to elicit eNOS activation. This result suggests that eNOS is immediately activated by virtue of its association (perhaps indirectly through caveolin) with GPI-anchored protein glypican. Similarly, when syndecans were knocked down and glypicans were intact, shear failed to induce remodeling, but was able to activate eNOS. This suggests that syndecans, by virtue of their direct linkage to the cytoskeleton, transmit shear to the cytoskeleton in a decentralized mechanotransduction model. In contrast, the glypican anchor is more intimately tied with eNOS activation suggesting that it plays as a role in focal mechanotransduction.

To complement observations that the glycocalyx is involved in mechanotransduction, a number of studies have attempted to quantify the forces in the glycocalyx and forces that arise as a result of its interaction with the membrane and cytoskeleton. Secomb modeled the endothelial cell surface layer as flow through vertical strands and predicted a significant attenuation of fluid flow (and hence shear) in the glycocalyx (Secomb et al. 2001). Weinbaum and colleagues idealized the glycocalyx as stalks of core proteins of varying density and length, consistent with the Squire model (Squire et al. 2001). In this model, shear stress acts on the tips of the core proteins and deflects them (Weinbaum et al. 2003). Most of the flow is attenuated before it actually reaches the cell membrane. The drag on a single core filament was calculated to be on the order of 7×10^{-4} pN for a shear of 10 dyn/cm^2. This force is likely to be insufficient to cause major rearrangement of the underlying cytoskeleton or to alter the conformation of attached core proteins. However, the drag on a collection of core glycocalyx strands anchored to a single core protein may cause the core protein to experience a force of 1.9×10^{-2} pN, sufficient to result in about 6 nm displacements. With a mechanical advantage afforded by the long strands, this displacement could be as much as 17 nm, which may be enough for mechanical activation. It is not clear, however, if cells respond to apical surface deformation or internal strain or stress of mechanosensors deep in the cell. Following Secomb's and Weinbaum's models, Ferko and colleagues estimated flow through the glycocalyx on a model of the cell that was constructed directly from fluorescence images (Ferko et al. 2006, 2007). Using finite element evaluation of a Brinkman layer whose dimensions and porosity were estimated from the Weinbaum model, they demonstrated significant transmission of stress to the cell membrane surface as well as to basal regions of the cell and nucleus. Thus the forces integrated from drag on the glycocalyx may lead to increased force production deeper in the cell, at focal adhesions, nucleus and cell junctions.

9.5.2 The Glycocalyx Is Dynamic and Spatially Non-Uniform

Complicating the role of the glycocalyx as a mechanotransducer is the fact that it is spatially non-uniform and compositionally dynamic. For example, Forbes-Dewey and colleagues have demonstrated that HS redistributes in the face of shear stress (Giantsos-Adams et al. 2013). In addition, enzymatic degradation caused the release of GAGs while repopulation of GAGs was shear stress dependent. Shear accelerated repopulation of GAGs and proteoglycans to 12 h compared to 20 h for static cells. These data further suggest that the atheroprotective effects of shear may be related to its ability to simulate cells to produce glycocalyx. Moreover, this remodeling of the glycocalyx was necessary for realignment of cells in the flow direction. The authors further developed a kinetics model to suggest that shear induced an increase in exocytosis of HS. Such studies are consistent with *in situ* imaging of the endothelial cell surface layer in which the inflammation-induced shedding of the glycocalyx could lead to regrowth that is shear stress dependent (Lipowsky et al. 2011). Using high-speed confocal intravital microscopy, Lipowsky and colleagues demonstrated that inflammation in the microvasculature is associated with glycocalyx shedding (Lipowsky et al. 2011). In this study, FMLP was used to induce inflammation. This work suggests that a link between atherosclerosis, shear and mechanotransduction may be through inflammation such that inflammation causes shedding, which causes a reduction in mechanosensation leading to further reduction in production of vascular protective molecules such as NO.

Other recent studies suggest that shear stress can induce alterations in the structure of the glycocalyx (Gouverneur et al. 2006) where exposure of endothelial cells to 24 h of atheroprotective-like shear of 10 dyn/cm^2 (1 Pa) led to increased incorporation of glycosaminoglycans in the glycocalyx. This production helps improve the integrity of the glycocalyx and may explain why high shear regions are protected from atherosclerosis. It is possible that increased GAGs make the shear-eNOS pathway more likely to be activated than in areas where shear is low and GAG synthesis is inhibited.

With respect to how the glycocalyx may transmit forces from flow to the cell, Zeng and Tarbell demonstrated that glycocalyx anchors, membrane rafts and the actin cytoskeleton may constitute a mechanosome (group of molecules that, together, form a mechanosensitive structure) (Zeng and Tarbell 2014). They showed that shear stress caused reorganization of each of these components in concert with each other. Consistent with their architecture, glypicans, which are peripheral membrane bound proteins free to move in the plane of the membrane, associate with GM-1 (lipid raft marker) and may play a role in the transport of heparin sulfate to the junctions of cells after 30 min of shear stress. Conversely, syndecans, which are transmembrane and bound to the cytoskeleton, associate with HS and CS but remain fixed during the 30 min of shear stress. After 24 h of shear stress, new caveolae were produced and glypicans-associated lipid raft redistribution contributed to a uniform distribution of HS and CS on the cell surface. This study points to the role of lateral diffusion and dynamic production of surface mechanoreceptors in the mechanotransduction of shear stress.

9.5.3 Forces from Shear Stress can Result in Force Transmission to the Membrane and the Membrane Microdomains, Rafts and Caveoli

The cellular plasma membrane is made up of a 5-nm thick lipid core that houses various carbohydrates and proteins that are in either the outer leaflet, the inner leaflet or transmembrane proteins. The core is made up of two phospholipid monolayers which are stabilized by hydrophobic and van der Waals interactions. The multiple combinations of lipid hydrophilic head groups and hydrophobic acyl chains permit the inclusion of various molecules such as glycerophospholipids, sphingolipids, glycolipids and sterols. In addition, complex lipids are not distributed randomly in the membrane but form lipid domains, where sphingolipids and cholesterol are segregated in dipalmitoyl phosphatidyl choline-rich membrane areas. These domains are involved in the trafficking and sorting of specific proteins, as well as in signal transduction processes. Therefore, there is a link between the composition of lipid membrane domains and their biological and functional properties. However, the molecular basis and implications of these links to cellular function are not yet fully understood.

Membrane rafts are small ~50 nm groups of liquid-ordered domain lipids. Small rafts are able to form larger stabilized platforms by means of protein-protein and protein-lipid interactions. Membrane rafts may also function as general signaling compartments, allowing the clustering of proteins, such as integrins, that regulate cytoskeletal organization. The composition of these domains, besides being rich in sterol- and sphingolipids, is also associated with the inclusion of gangliosides. Gangliosides are complex glycolipids that have a strong amphiphilic character because of their large saccharidic head groups and the double-tailed hydrophobic moiety. They are particularly abundant in the plasma membrane, where they are inserted into the external leaflet, with a hydrophobic ceramide moiety and with the oligosaccharide chain protruding into the extracellular space. The lipid moiety of

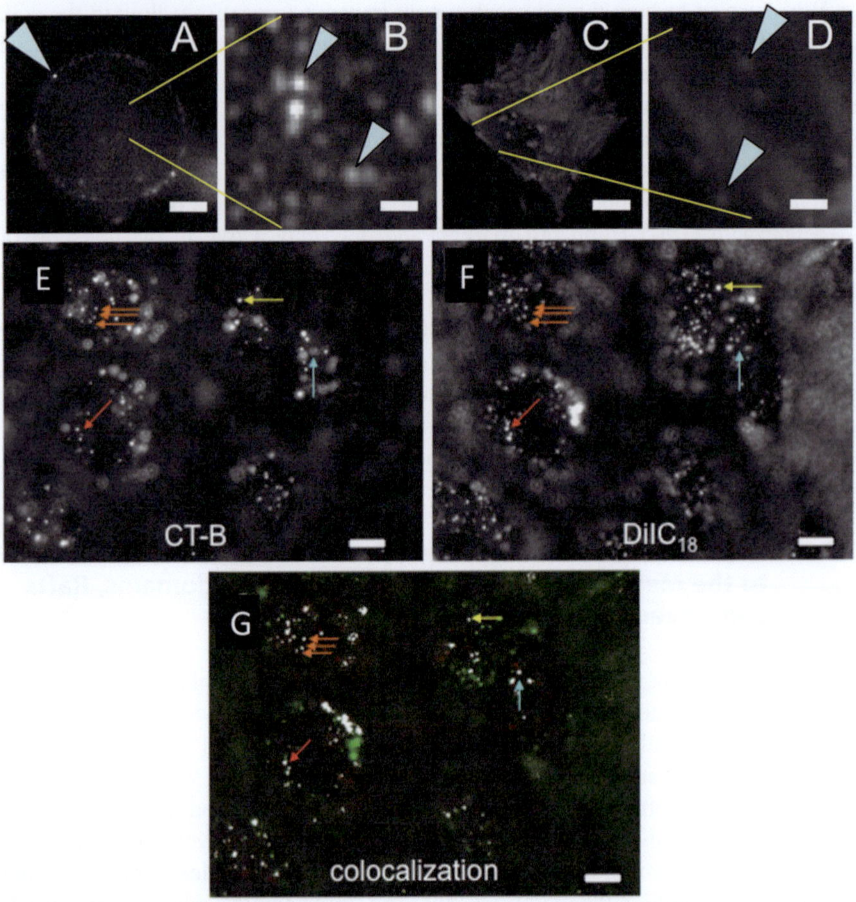

Fig. 9.3 (**a**) Phase separation in giant unilamellar vesicles (Diameter ~ 50 μm). (**b**) Close-up of domains seen in (a). Morphological similarities to domains seen in (c). (**c**) Total internal reflection fluorescence image of DiI-C_{18} stained cells. Higher intensity indicates localization within 100 nm of coverslip (focal adhesion). (**d**) Liquid ordered phase doamins. (**e–g**). Colocalization of DiI C_{18} with cholera toxin B (CT-B) indicates presence of ordered domains. Bars in (a, c, e, f, g) = 10 μm; Bar in (b, d) = 1 μm

gangliosides, shared with all sphingolipids, is called ceramide. The common name of this is sphingosine and is connected to a fatty acid by an amide linkage.

There are different oligosacharide chains associated with various gangliosides. The oligosaccharides depicted lack the membrane-anchoring ceramide, which would be attached to the terminal Glc in the GM1 ganglioside. The GM1 is the most abundant ganglioside and is commonly used to identify membrane raft using fluorescence (see Fig. 9.3) by taking advantage of the strong affinity between GM1 and cholera toxin subunit B.

Caveolae are dynamic structures that can fuse or leave the membrane through endocytosis in order to regulate their density (reviewed in (Parton and del Pozo 2013)) in response to membrane stresses. As such they are ideal candidates for mechanoreceptors of surface shear stress. Caveolae are cholesterol-rich microdomains that are curved by virtue of their association with caveolin, a major structural and scaffolding protein, as well as a newly discovered cavins, which associate with caveolin and assist in curvature generation. These 60–80 nm invaginations are composed of ~150 caveolin molecules that help form the membrane into a cup shape that is open on the exoplasmic side of the membrane. While caveolins have been implicated in mechanosensation of shear stress, there is currently very little research on cavins and even less on whether these molecules participate in mechanosensation. Caveolin molecules are located on the interfacial leaflet with a characteristic hairpin turn that integrates into one leaflet. Cavins appear to stabilize caveolin oligomers after trafficking to the plasma membrane. Importantly, the lipid make-up (e.g. cholesterol) ensures that caveoli are liquid-ordered microdomains. Potentially important for mechanosensing, when order is disrupted or the architecture of caveoli is forced into a flattened state, caveoli disassemble via caveolin-1 and cavin 1 disassembly. Cholesterol depletion causes similar dissociation. Thus, this ease of disassembly may make caveoli a metastable structure capable of sensing minor membrane forces.

Also important for mechanosensing is PIP_2 (phophatidylinositol-4,5-biphophate), which is enriched around the caveolar opening (Parton and del Pozo 2013). This lipid has long been associated with shear sensing in that conversion of PIP_2 to IP_3 leads to binding of IP_3 to receptors on the ER and opening of calcium channels (Park et al. 1998). The resultant release of calcium to the cytoplasm appears to be necessary for the production of nitric oxide via the calcium-calmodulin system. Cytoskeletal modulation of caveolin organization via actin has been demonstrated (Parton and del Pozo 2013). On the cell surface, actin appears to bind to caveolin via PKCα. In addition, caveolae are known to associate with actin and can be organized in long straight lines along stress fibers. As such, experiments that degrade stress fibers, likely alter the organization and structure of caveolae.

A main mechanism of mechanosensing may be caveolae flattening, which can arise from surface shear or through cell stretch as in capillaries under high transmural pressure (Schmid-Schönbein et al. 1995). The excess membrane afforded by caveolae results in increase in the pool of membrane needed during mechanical activation. This process can protect against membrane lysis. Whether membrane flattening is the means by which caveolin in caveolae is activated, is not known directly. However, it is known that caveolin can be phosphorylated in response to membrane stress and that this depends on β_1 integrin activation and src-kinase activity. Src-kinases are known to be involved in mechanosensing. Such phosphorylated caveolin can regulate Rho-dependent actomyosin contraction. As reviewed by Parton and del Pozo (2013), one possible mechanism of mechanosensing is when cells are rich in caveolin-1, stretch causes caveolae to flatten and cavin 1 then dissociates from caveolin 1. Without cavin 1, p190RHOGAP can be sequestered in the liquid-disordered parts of the membrane leading to loss of Rho inhibition by p190RHOGAP. When Rho is free to be active, it leads to actomyosin contraction, ECM remodeling, and integrin-dependent changes in focal adhesion assembly. Conversely, when p190RHOGAP is sequestered in the liquid-ordered parts of membrane, it is more likely to inhibit Rho leading to the preservation of actomyosin contraction.

G-proteins have long been suspected as being involved in mechanotransduction of shear stress. Oh and Schnitzer found that G-proteins reside in caveolin on the cell surface and are activated there (Oh and Schnitzer 2001). Specifically Gq proteins are associated with caveolae, and Gi and Gs proteins are associated with lipid rafts. This distribution may explain the differential activation of these proteins by shear noted by Frangos and colleagues (Bao et al. 2000). Rizzo and colleagues showed that flow activates endothelial nitric oxide synthase in caveolae (Rizzo et al. 1998). This activation was associated with dissociation of caveolin and calmodulin. Interestingly this finding is consistent with the role of the calcium-calmodulin system known to participate in shear-induced nitric oxide synthesis (Isshiki and Anderson 1999). Rizzo and colleagues exposed endothelial cells to shear stress and noted increased caveolin at the cell surface. The density of caveolae was enhanced sixfold after shear exposure. Importantly, it was only after shear that the density of caveolae resembled the *in vivo* density and distribution suggesting that chronically loaded cells are more *in vivo* like. Importantly, flow-conditioned cells exhibited higher density of caveolae at cell borders, confirming earlier reports of functional activation at cell-cell junctions (Rizzo et al. 2003). Park and colleagues showed that caveolin participates in the mechanotransduction of shear stress and extracellular signal regulated kinase (ERK) phosphorylation. In this study, the group developed a caveolin antibody that prevented shear-induced activation of ERK, suggesting that caveolin 1 participates in the pathway between shear and ERK activation (Park et al. 2000). Interesting, this antibody bound to oligomerization domains of the caveolin and prevented its oligomerization. Thus clustering of caveolin may play a role in its ability to transduce shear into ERK activation.

Yamamoto and colleagues demonstrated that shear stress can elicit ATP release and that this release coincides with the location of caveolae (Yamamoto et al. 2011). It is thought that this ATP binds to cell surface purinergic receptors to initiate increases in cytosolic calcium, a known contributor to shear-induced nitric oxide release. Importantly, ATP release occurred exactly at locations where caveolin 1 was enriched, suggesting that local autocrine mechanism may play a role in shear-induced calcium signaling. Together with research on glycocalyx, it is possible that glycocalyx connection to proteins in caveolae may lead to increased ATP production followed by calcium influx and eNOS activation. Removal of cholesterol attenuated shear-induced ATP release. Furthermore, the location of initiation of calcium waves coincided with the location of caveolin at cell junctions and ATP release was largely localized to the cell edges. These temporal and spatial correlations suggests that there may be a convergence of research on force transmission from the cell surface to the cell edges, PECAM-1, membrane order, ATP release, calcium increases, and eNOS activation.

Research on the association of mechanotransduction events with rafts is less prominent, owing to the fact that rafts have no real structural protein associated with them and they are very dynamic (Eggeling et al. 2009). The association of the glycocalyx with rafts via glipicans and GM-1 gangliosides suggest that they could be an important mechanotransducer. Recently, Fuentes and colleagues showed that rafts are the first participants in adhesion to the extracellular matrix via β_1 integrins and fibronectin (Fuentes and Butler 2012). In addition, Wang and colleagues noted that mechanically activated src activity appeared to originate from Lyn- rich microdomains, commonly thought to be rafts (Lu et al. 2008).

9.5.4 Stresses on Membrane are Non-Uniform and Increase in the Upstream Direction

Satcher et al. noted that the undulating cell shape characteristic of endothelial cells produces surface shear stresses that are non uniform (Satcher et al. 1992). Later, Barbee and colleagues measured the exact surface undulations using atomic force microscopy and showed that shear stress gradients occur over the surface of single cells (Barbee 2002). In order to understand the relationship between local and distributed models of mechanotransduction, Ferko and colleagues developed a multi-structural model of sheared endothelial cells that included focal adhesions, the nucleus and glycocalyx (Ferko et al. 2007). They found that focal adhesions caused large stress concentrations arising from having sudden changes in membrane support by the extracellular matrix (see Fig. 9.4). In addition, larger stress concentrations occurred around the cell nucleus upon shear because of the juxtaposition of low and high moduli organelles. These results suggest that the forces arising from the application of shear stress to the cell surface can be significantly amplified in part because of the unique architecture of the endothelial cell. This model was made possible because of the development of methods for accurate delineation of surface topography of the cell from fluorescence and confocal microscopy (Ferko et al. 2006).

Despite the fact that shear distribution is not uniform, there have been only a few papers that identify that this shear distribution correlates to mechanotransduction events. For example, Fung and Liu developed a model of the endothelial cells that attempted to predict the forces occurring on the membrane from shear stress (Fung and Liu 1993). They found that the stresses can accumulate appreciably as one moves upstream. This result arises because membrane stresses result from an integration of all the stress felt downstream. This accumulation of stress on the upstream side of the cell may explain why Butler and colleague found a shear-induced increase in fluidity on the upstream side of the cell and not on the downstream side (Butler et al. 2001). Recently, such subcellular stress distributions

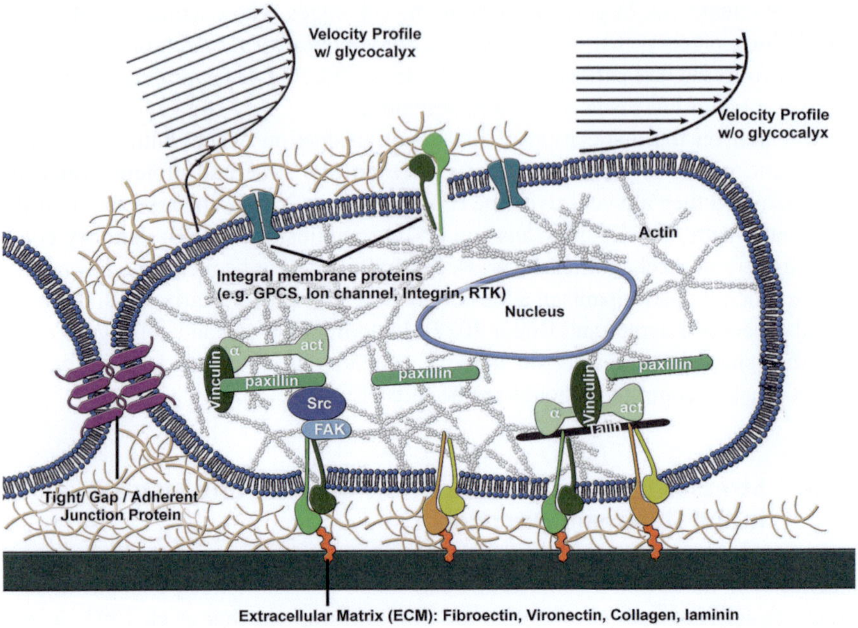

Fig. 9.4 *Top*: Shear stress impacts the glycocalyx or membrane. The forces are transmitted via the cytoskeleton to cell junctions, focal adhesions, nucleus and other organelles. Molecular complexes in these locations transduce force to soluble signals such as transcription factors or vasodilators and constrictors (Figure adapted from (Kamm and Mofrad 2009)). *Bottom*: The range of forces is in pN (result of about 1 Pa shear stress) and cause deformations on the order of 10 s of nm. This deformation results in focal strains at focal adhesions and near the nucleus (Bottom figure adapted from (Ferko et al. 2007))

have been modeled quantitatively. These results suggest that shear stress can result in high stresses at focal adhesions and cell junctions, an emerging mechanosensitive organelle (Bagi et al. 2005; dela Paz et al. 2014; Fujiwara et al. 2001; Tzima et al. 2005) in which PECAM-1 may constitute part of a mechanosome.

9.5.5 Forces in the Membrane Result in Alterations of Lipid Mobility

Butler and colleagues discovered that shear-induced stresses are likely to be non uniform as indicated by changes in the diffusion of lipids in response to applied surface shear stress (Butler et al. 2001). Accordingly, the increases in membrane fluidity on the upstream side of the cell corresponded to increased positive shear stress gradients as predicted by Barbee and colleagues (Barbee 2002). In this study, the authors used fluorescence recovery after photobleaching (FRAP) with a single scanning laser in the direction of flow to catch, within 2 ms, the bleach event and recovery of an upstream part of the cell and downstream nearly simultaneously.

Newly derived FRAP equations allowed the calculation of diffusion coefficients. These diffusion coefficients of DiI indicated changes in the membrane fluidity upon shear stress. Recently, Yamamoto and colleagues were able to validate these results and show that fluidity increased differentially in liquid-ordered domains (Yamamoto and Ando 2013).

Tabouillot and colleagues investigate shear-induced fluidity changes of membrane domains by staining the lipid domains differentially with long chain DiI (18 carbon stains rafts and caveolae) and short chain dyes (staining liquid disordered parts of the membrane using DiI C_{12}) (Tabouillot et al. 2011). In these studies, they used time correlated single photon counting (TCSPC) to determine changes due to the application of shear stress, of the fluorescence lifetime, diffusion coefficients (FCS) and number of molecules arising from a small confocal volume that intersected a small section of the membrane (system described in (Gullapalli et al. 2007)). They found that shear induced a decrease in lipid order (as assessed by a decrease in fluorescence lifetime) in the liquid-disordered part of the membrane that was rapid and transient (over a few 10s of seconds) while shear induced a later but sustained decrease order of liquid-ordered membranes. This suggests that membrane microdomains are differentially sensitive to applied stresses, which is consistent with the notion that these domains have different moduli, with liquid-ordered domains (cholesterol rich) being stiffer then liquid-disordered domains (Shamitko-Klingensmith et al. 2012). In addition to changes in lipid order, number of molecules decreased with application of shear. This decrease can be explained by a flattening of the cell membrane with shear. Such flattening may be important stimulus for the disassembly of caveolin, dissociation of cavin, sequestering of p190RHOGAP, loss of inhibition of RHO and actin-myosin contraction and mechanosensing (Parton and del Pozo 2013).

9.5.6 Forces Can Act on the Cytoskeleton, Which Has Static Structural Integrity as Well as Dynamic Reorganization

Wang and colleagues highlighted the importance of tensegrity structures, which are a combination of elastic and compressive elements that combine to give the cell its structure (Wang et al. 2009). When a tensegrity structure is deformed by an applied force it reaches a new equilibrium in which forces are transferred to other structures. Such force transfer could be a way to rapidly transmit force to distant parts of the cell where it is transduced into mechanoresponses. This framework for mechanotransduction explains some of the possible force-at-a-distance phenomenon observed in cells, but it does not include diffusive factors that were converted upstream and diffusion of activators downstream. While the focus was on structures that maintain structural integrity during force application, it must be recognized that forces can cause changes in this structural integrity (e.g. fluidization). In addition, forces applied at the membrane are indeed dissipated, but this dissipation implies energy transfer to the lipid molecules, which could manifest itself into reorganization and attendant force-dependent signaling.

In the tensegrity model, structural elements maintain their integrity because their load bearing mechanisms are stable for longer periods than the turnover time of individual monomers. In order for the tensegrity model to transmit stress it must be under pre-stress; isometric tension generated by contractile elements acting against compressive elements. Compressive elements are normally thought to be microtubules, but they can also include the extracellular matrix as well as organelles such as mitochondria and nucleoli capable of minimal deformation under compressive stress. While the tensegrity concept explains rapid transduction, many processes known to be involved in mechanotransduction are not as rapid as that. For example, MAPK phosphorylation peaks at around 5 min, about the same time scale as agonist-induced phosphorylation, suggesting that the mechanically induced pathway shares some similar molecular partners as the agonist induced one.

However, the cytoskeleton is not static but dynamically remodels. Noria and colleagues looked directly at the remodeling occurring as a result of sustained application of physiological fluid shear stress (Noria et al. 2004). These authors confirmed existing evidence that shear stress causes elongation of actin stress fibers in the direction of applied shear stress. Interestingly, by microinjecting monomeric fluorescent actin, they showed that in response to shear, actin polymerization originated at the location of an existing stress fiber end with neither the upstream or downstream part preferred. This insertion of monomeric actin into focal adhesions was independent of the classic zyxin/vasodilator-stimulated phosphoprotein (VASP) complex.

The overall cell deformability rests largely on its cytoskeleton, a crowded array of filamentous structures (Bursac et al. 2005). The cytoskeleton is comprised of actin filaments, intermediate filaments and microtubules, while other structural proteins (such as fodrin) are also present in endothelial cells. Bursac and colleagues pointed out that in glassy rheology systems, the crowding causes metastable states in which the dynamic (e.g. diffusion of an entrapped bead) is locally constrained. Periodically, the system can jump from one metastable state to another. Over time, the system samples many of these states and evolves through multiple microconfigurations with increases in their stability. This process is called aging and can be reversed by applying a highly dynamic force (strong fast oscillations) that returns an aged system back to its original state. Importantly, the group also found natural ATP-dependent oscillations in cells. This feature is radically different from passive glassy systems. ATP may provide fuel for cells so that they sample many different metastable states. Similarly applied force can drive cells to settle into different metastable states. This group focused on the role of the cytoskeleton, which has fairly accessible (slow) dynamics. It remains unexplored whether the membrane or other dynamic parts of the cell exhibit similar glassy rheology, and whether other force-dependent states arise from self assembly and reorganization.

Evidence for shear-induced transition of one structural state to another has come from using fluorescence recovery after photobleaching (FRAP). Osborn and colleagues demonstrated that net cytoskeletal depolymerization is high and that shear stress causes immediate (within minutes) remodeling after shear application (Osborn et al. 2006). Such depolymerization could arise from the loss of myosin II or other crosslinking resulting in fluidization. Danagaria and Butler found that shear stress elicits rapid softening of the cell cytoplasm as assessed by rheological investigations of endogenous vesicles (Dangaria and Butler 2007). Interestingly even though shear was maintained, the softening response was reversed by 4 min. When shear was removed the cell became stiffer. In addition, creep tests to step shear indicated that, on the time scale tested, ECs behaved like a viscoelastic fluid with a deformation time scale of about 3 s (Fig. 9.2). Thus physiological shear with pulsatility in an atheroprone area is likely to lead to continuous adaptation of cell shape.

Krishnan et al. noted that the cell can employ different strategies to modulate intrinsic mechanics in response to applied force (Krishnan et al. 2009). Whereas it has been long known that, on longer time scale, force application induces reorganization of the cytoskeleton such that the cell, in general, becomes stiffer; it is now recognized that on short time scales, the cell can fluidize and become softer. Fluidization from stretch is thought to happen because of passive breaking of crosslinks in the cytoskeleton, whereas reinforcement occurs after alteration in mechanosensitive transduction processes. Reinforcement can prepare a cell for responses to large forces, or fluidization could be employed to rapidly relieve stress in response to applied force. Which of these cases predominates is an active area of research. By combining cell stretch and traction force microscopy, Krishnan et al. used cell-mapping rheometry to investigate the time course of fluidization versus reinforcement after applied stretch. Cells were stretched with varying degrees of anisotropy for 4 s. The degree of stretch ranged from 2.5 to 10 % corresponding to stretches that might occur by large changes in pressure of blood vessels. Immediately after the stretch maneuver, the traction force of the cell was monitored. It was observed that stretch caused an immediate reduction in traction (which was thought to arise from fluidization of the cytoskeleton) that was sustained for about 100 s, after which reinforcement mechanisms took over. This fluidization did not depend on isotropy, presumably because the cells were grown with no preferred direction in the cytoskeleton.

The authors concluded that the mechanism of stretch-induced fluidization was because of many weak dynamic bonds that are responsible for overall integrity of the cytoskeleton. These bonds can be rapidly broken and reformed giving the cell the ability to respond to indiscriminate non-specific forces from the environment. Again, shear forces appear to induce similar fluidization (Dangaria and Butler 2007), but it is not clear that shear-induced fluidization is due to the breaking of weak bonds or due to signaling pathways that result in changes in crosslinking.

9.5.7 Forces Can Be Transmitted to Focal Adhesions and Cell-Cell Junctions

Focal adhesions are assembled via sequential transport of proteins via diffusion through the membrane or from the cytosol (Broday 2000) and are converted to a mechanotransducing organelle through reinforcement of simple adhesion plaques to stable adhesions. The lipid around focal adhesions is highly ordered (Gaus et al. 2006) suggesting that, with respect to the membrane, the focal adhesion is a lipid-ordered membrane domain similar to caveolae and lipid rafts. Kanchanawong and colleagues recently studied the ultrastructure of focal adhesions (Kanchanawong et al. 2010). They found that focal adhesions have a complex yet ordered interaction with the cytoskeleton, particularly in the reinforcement phase after adhesion. Focal adhesions are comprised of an integrin extracellular domain that selectively binds extracellular matrix proteins (laminin, fibronectin, collagen, etc.) depending on the type of integrin. The integrins span the plasma membrane to bind to an internal integrin-signaling domain composed of talin and paxillin. The force transduction domain is thought to attach to these proteins and be comprised of vinculin and vasodilator-stimulated phosphoprotein (VASP). Next, the actin regulatory domain governs the insertion and remodeling of actin into the focal adhesion via α-actinin. Finally, actin and myosin form stress fibers that enable the application of force to the focal adhesions. Therefore, focal adhesions can exert forces on the ECM surrounding the cell and feel force through reactions to ECM deformation.

Forces can be transmitted to the endothelial cell junctions via actin filaments (Dejana 2004; Tzima et al. 2005). Cell junctions consist of tight junctions, which are largely responsible for transcellular permeability and are composed of claudins (e.g. occuldins), gap junctions (connexins), responsible for cell-to-cell communication via ion transport, and adherens junctions (e.g E-cadherins), which participate with claudins to regulate endothelium permeability. An additional important mechanotransducing molecule is PECAM-1, which participates in eNOS signaling upon force application (Bagi et al. 2005; Fujiwara et al. 2001)

An important question remains whether forces from shear stresses are sufficient to alter molecular conformation or alter changes in assembly of protein complexes. Recently, Dabagh et al. (2014) developed a computational model that incorporated each of the main structural elements responsible for mechano-transmission. These included the glycocalyx, membrane, actin cytoskeleton, adherens junctions, focal adhesions, the nuclei, and neighboring cells. The group estimated forces on focal adhesions and adherens junctions and found them to be on the order of 8 pN, sufficient for mechanoactivation. While the precise binding partners that might feel these forces are still unknown, this model provided some of the first experimentally validated stress profiles of cells undergoing physiological shear stress. The results argue for both centralized models of mechanotransduction, in which forces act directly on specific organelles, such as the glycocalyx and cell membrane, and a decentralized model in which forces are transmitted to remote sites and transduced into biochemical signaling.

9.6 Answering the Questions

This chapter began with key questions about mechanotransduction. We answer these questions using a concrete example of shear-induced nitric oxide production. It should be recognized that the answers to these questions would be different for different mechanotransduction pathways. In addition, the answers presented here are by no means final and require significant amount of additional research. Nevertheless, these answers can help the reader form a new framework for thinking about mechanotransduction in such a way that, perhaps in the future, knowledge of mechanotransduction can be used to develop novel diagnostic tools and clinical strategies to treat cardiovascular disease.

9.6.1 Integrin Ligation May Play a Role in Assembly of Mechanosomes

A widely accepted paradigm is that shear stress imparts forces on the EC surface glycocalyx, which transmits these forces to cellular structures whose molecular activity or association is altered by force. This perturbation leads to Ca^{2+} influx and release from cellular stores. Ca^{2+} is one ingredient needed to activate calcium-calmodulin, which is required for the conversion of L-arginine to nitric oxide (NO) by EC nitric oxide synthase (eNOS). This release of NO dilates blood vessels in the microvasculature and inhibits platelet aggregation in blood. Recent studies have demonstrated that the ubiquitous mechanosensitive Ca^{2+} channel, transient receptor potential vanilloid-4 (TRPV4), is involved in the mechanosensation of shear stress in ECs and responsible for shear-induced Ca^{2+} influx (Hartmannsgruber et al. 2007). In addition, it is part of a mechanosome consisting of spectrin, protein 4.1, Lyn (a member of the Src family of protein tyrosine kinases), and β_1 integrins, suggesting that mechanically gated TRPV4 channel conductance may be related to cytoskeleton-membrane adhesion (spectrin-protein 4.1), membrane rafts (Lyn), and focal adhesion (FA) assembly (β_1 integrins) (Fig. 9.5).

There is evidence that assembly of such mechanosomes may depend on lipid membrane properties in that the very first constituents of this mechanosome, namely rafts and β_1 integrins, must diffuse to these sites of adhesion to extracellular matrix proteins in order to coordinate mechanotransduction (e.g. talin activation). We hypothesize that membrane lipid raft recruitment to locations where β_1 integrins have ligated ECM ligands, as indicated by recruitment of the ganglioside GM-1, enable further (β_3) integrin clustering. Evidence supporting these ideas comes from the measurement of raft transport to nascent adhesion plaques and assays for integrin activation by imaging talin (Fuentes et al. 2011; Fuentes and Butler 2012).

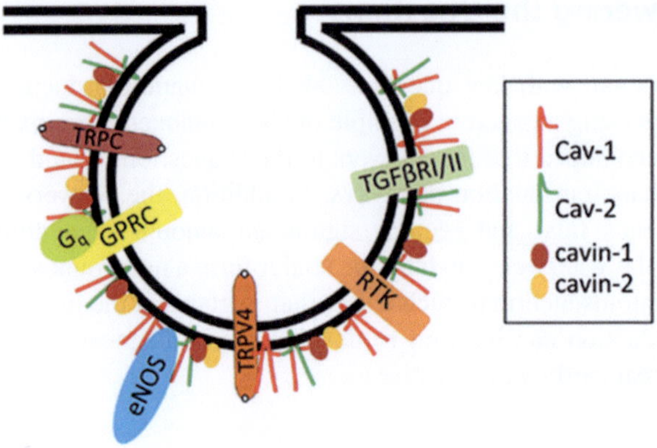

Fig. 9.5 Caveolae structure. Association of TRPV4 with eNOS and other caveolar associated proteins suggests a possible mechanotransduction mechanism in which forces applied to the cell surface and the reaction forces from actin association with caveolae cause the disruption of association of caveolin with TRPV4 and eNOS. This release allows these proteins to be active thus increasing Ca^{2+} influx into the cell as well as increased eNOS activity. TRPC, transient receptor potential cation channel; GPRC, G-protein couple receptor; eNOS, endothelial cell nitric oxide synthase; TRPV4, transient receptor potential cation channel, subfamily V, member 4; RTK, receptor tyrosine kinase; TGF, transforming (or tumor) growth factor (Figure adapted from Sowa (2012))

9.6.2 What Components of Force Are Sensed by the Cell?

Cells respond to shear stress temporal aspects. Feaver and colleagues recently described the role of other harmonics of blood flow that may impact endothelial cell mechanotransduction (Feaver et al. 2013). Although blood flow dynamics are often simplified as averages stresses (e.g. 0 ~ 2 Pa) with a superimposed oscillatory stress (with amplitude of about 0.5–1 Pa), wall shear stress actually has a number of other harmonics associated with it. These frequencies come from the rhythmic pumping off blood described by the PQRST wave of an ECG, longitudinal pulsatile waves arising from windkessel mechanics of the vascular wall, sharp closing of valves, and the impact of stochastic forces arising from blood cell interaction with the blood vessel wall. While it is difficult to capture all of these temporal variations, Feaver succeeded in capturing some of the complex stress frequency harmonics by analyzing the effects of 0th (average shear) and 1st (temporal) harmonic. Atheroprone hemodynamics were considered as low pulsatile multidirectional shear such as those that occur in the carotid sinus. These stresses up-regulated genes involved in inflammation that depended on NF-κB activity and genes which this transcription factor regulates (e.g. VCAM-1, fibronectin, E-Selectin, MCP-1, IL-8). Such activation of NF-κB was inhibited when PECAM-1 was knocked down using siRNA suggesting that PECAM-1 may be a mechanosensor. Conversely, high shear up-regulated anti-inflammatory genes such as eNOS, KLF-2, KLF-4, and BCL2-like protein A1. Analysis of all higher order frequency modes indicated that the first harmonic was largely responsible for atheroprotection via down regulation of NF-κB.

Cells may also respond to spatial shear stress gradients (DePaola et al. 1992). In these experiments, there is a large change in shear stress from one point on a cell to another or from one cell to a neighboring cell. In areas of flow recirculation, these gradients can be very large, with as much as a few dynes per cm^2 over a few microns. Such gradients could induce differential forces between one part of the cell and another part thus causing stretching between these points. Similarly, if shear on one cell was different than a neighboring cell, then resultant forces could span junctions between cells. Consistent with this idea, connexins were specifically activated when shear stress gradients were present (Chadjichristos and Kwak 2007; Johnson and Nerem 2009). In addition, consistent with activation in junctions, PECAM-1 has emerged as an important sensor of force at the cell surface (Bagi et al. 2005; Fujiwara et al. 2001; Tzima et al. 2005)

9.6.3 What Are the Molecular Sensors That Transmit Force to Transducers in the Cell?

Ueki showed that shear stress elicits shear strain in endothelial cells with a measured shear modulus of 231 Pa (Ueki et al. 2010). Such a low modulus is consistent with other measurements of shear-induced deformation by Dangaria (Dangaria and Butler 2007). In this experiment, the group used high speed laser line scanning confocal microscopy to take a rapid axial section through a cell in the direction of shearing. By focusing only on one slice, they were able to capture cellular deformation at 3–4 frames per second. Using image correlation methods, they found that shear stress deformed the cells in the direction of shear. This strain was found to be dependent on actin as cytochalasin D increased the shear strain (and decreased the modulus) dramatically. Such studies indicate that the cytoskeleton and internal organelles are mechanically linked and that shear stress can be transmitted to remote locations almost everywhere in the cell.

As a discrete example, direct mechanical linkages exist between the cell surface and organelles where mechanosensitive processes are known to happen (Maniotis 1997). Maniotis developed an assay based on micropipette manipulation of an ECM coated bead and showed that there exists strong cytoskeletal linkages between the cell surface and the nucleus (Maniotis et al. 1997). While they used deformations that were much larger and faster than *in vivo* deformations by shear, the work showed that force can act at a distance in cells. Later work showed that such a phenomenon might be relevant in shearing of endothelial cells (Helmke et al. 2001). While there has not been definitive proof that forces can directly cause changes in transcription in the nucleus, there is growing evidence for such linkages (N. Wang et al. 2009).

Alternatively, force could act directly on the cell membrane and cause clustering of mechanosensitive proteins. Rafts coalesce with force due to enhanced hydrophobic mismatch between liquid-ordered (Lo) and liquid-disordered (Ld) membrane domains. Mismatch of the hydrophobic thickness of various lipids in the membrane bilayer drives aggregation of lipid domains (Baumgart et al. 2003) which, in turn, facilitates segregation or aggregation of membrane proteins (Ayuyan and Cohen 2008; Botelho et al. 2006; Periole et al. 2007). Membrane tension induces raft clustering (Ayuyan and Cohen 2008; Heinrich et al. 2010) with a time course on the order of seconds. These studies demonstrate that rafts are poised to coalesce at physiological temperatures (Lingwood et al. 2008) or with minor alterations in the force landscape (Garcia-Saez et al. 2007).

9.6.4 Is Transduction Through Physical Links That Transmit Force from One Molecular Partner to Another, or Does Force Work Through Existing Receptors That Transmit Biochemical Signals Largely Through Diffusive Pathways?

Mechanotransduction appears to occur in specialized molecular complexes and organelles. While diffusion may carry some mechanotransduction players into active phosphorylation pathways (Dimova et al. 2000; Jalali et al. 1998; Lu et al. 2008; Wang et al. 2005), it appears that many mechanical systems are hard-wired for mechanotransduction (Na et al. 2008). One example of a protein complex involved in shear stress induced nitric oxide production is the mechanosome composed of spectrin and protein 4.1. This membrane-associated complex regulates the activity of transient receptor potential vanilloid-4 (TRPV4), a ubiquitous mechanosensitive Ca^{2+} channel. This activation is specifically regulated by spectrin and actin, and the deletion of protein 4.1 binding domain to TRP channels prevents TRP activation, consistent with early observations linking spectrin, actin, and protein 4.1 (Leto et al. 1986). TRPV4 and the spectrin skeleton are associated with TRPV4-mediated Ca^{2+} conductance in ECs as disruption of the spectrin-protein 4.1 interaction abolishes TRPV4 activity (Wu et al. 2001). In addition, the activation of TRPV4 channels by osmotic stretch depends on interaction with β_1 integrins and Lyn (a src-family kinase and a lipid raft protein) (Alessandri-Haber et al. 2008; Xu et al. 2003). Thus there appears to be a correlative, and possibly mechanistic, link between spectrin, protein 4.1, membrane rafts, integrins, TRPV4 activation, and Ca^{2+} entry through the plasma membrane. Thus, it is likely that TRPV4 is part of a mechanosome (Bidwell and Pavalko 2010) consisting of spectrin, protein 4.1, Lyn, and β_1 integrins. In support of this concept, the Ingber group showed that when TRPV4 was associated with β_1 integrins, forces applied to β_1 integrins result in ultra-rapid (within 4 msec) activation of Ca^{2+} influx through TRPV4 channels (Matthews et al. 2010), suggesting that TRPV4 may be directly mechanosensitive (Christensen and Corey 2007). The TRPV4 channels were specifically activated by mechanical strain applied to β_1 integrins and not by deformation of the lipid bilayer or submembranous cortical cytoskeleton alone. Thus, at least the β_1 integrin component of the proposed mechanosome is necessary for mechanical activation of TRPV4. It is also known that TRPV4 channels are involved in EC mechanosensing of shear stress (Hartmannsgruber et al. 2007;

O'Neil and Heller 2005; Wu et al. 2007), Ca^{2+} influx (Nilius et al. 2003), and nitric oxide (NO) generation (Kohler et al. 2006). Thus, we propose that EC mechanosensing of shear stress derives from coordination between spectrin, protein 4.1, β_1 integrin, Lyn, and TRPV4.

9.6.5 To What Extent Are the Force Transduction Partners Regulated by the Very Force They Sense?

A fruitful direction of research would be measurements of EC surface dynamics and force-dependent kinetics of lipids, integrin activation, and reinforcement (Fuentes and Butler 2012) that precede the formation of the TRPV4 mechanosome, in order to delineate the mechanical origins of EC dysfunction that underly hypertension and atherosclerosis. Previously, this chapter described how the glycocalyx and cytoskeleton remodel in the face of shear stress. Similarly, focal adhesion proteins are in constant state of flux (Lele et al. 2006). For example, new integrin ligation and clustering are major events in vascular tone regulation and shear-induced gene expression. Jalali et al. showed that shear stress caused an increase in new ligand binding of β_1 integrins in and around focal adhesions (FAs) of endothelial cells (ECs) plated on fibronectin and an increase in ligand binding of β_3 integrins in ECs plated on vitronectin (Jalali et al. 2001). In *ex vivo* arteriolar preparations, activation of the vitronectin receptor, $\alpha_v\beta_3$-integrin, and fibronectin receptor, $\alpha_5\beta_1$-integrin, induced coronary arteriolar dilation by stimulating endothelial production of cyclooxygenase-derived prostaglandins (Hein et al. 2001), which dilate blood vessels (Butler et al. 2000; Frame et al. 2007). Thus dynamic integrin-matrix interactions at FAs are required to initiate the signaling pathway leading to shear stress-induced vasodilation and blood pressure regulation.

Furthermore, integrin association with focal adhesions is modulated by force via diffusion of lipid rafts (Fuentes and Butler 2012). Lipid rafts are 10–200 nm cholesterol- and sphingomyelin-enriched liquid-ordered (Lo) membrane domains that are involved in signaling and nucleate actin polymerization [reviewed in (Levitan and Gooch 2007)] by concentrating phophatidylinositol 4,5 biphosphate (PIP_2) (Kwik et al. 2003). FAs are cholesterol rich microdomains, as are caveolae and rafts (del Pozo et al. 2004) and β_1 integrins are required for raft formation (Singh et al. 2010) and signaling through Rac-1 (del Pozo et al. 2004). Wang and colleagues found that Src-activation colocalized with Lyn, a raft marker (Lu et al. 2008) supporting an emerging picture of rafts as dynamic nanodomains that cluster the necessary critical mass of receptors (van Zanten et al. 2009) for downstream signaling of important pathways such as mitogen-activated protein kinases (MAPK) (Rotblat et al. 2010) with time scales of formation of 20 ms and length scales of 10 s of nanometers (Eggeling et al. 2009). The dynamic formation and dissolution of rafts may be related to membrane bending fluctuations which facilitate spectrin-protein 4.1 association and protein sorting (Heinrich et al. 2010).

9.6.6 What Are the Main Factors Produced at the Terminal End of Force Transduction Pathways?

The mechanism of eNOS activation is connected to its subcellular location. eNOS converts l-arginine to L-citruline and nitric oxide. NO, in turn, acts as a free radical scavenger, relaxes blood vessels, and inhibits platelet aggregation, apoptosis, and can inhibit the binding of monocytes to endothelium. Therefore, it is considered as an important atheroprotective molecule. eNOS activity is connected with the transcription factor NF-κB in that NFKB binds to the eNOS promoter and initiates the generation of eNOS. Since NO in turn can downregulate NF-κB activity, this may comprise an important negative feedback system for controlling NO bioavailability. Regarding localization in the cell, eNOS is thought to reside in caveolin and is inhibited by its association with caveolin-1 (Fleming and Busse 1999; Rizzo et al. 1998; Yamamoto et al. 2011). It is also been found to associate in an inhibitory fashion to PECAM-1. When shear is initiated, these associations are temporarily lost and eNOS is then available for the generation of NO. So an intriguing mechanism of mechanotransduction becomes simply that force pulls on the eNOS-caveolin-1-PECAM-1 complex and changes the affinity of molecules for each other thus making eNOS available for NO production. Once separated from PECAM-1 or caveolin 1, eNOS is free to catalyze the conversion of l-arginine to l-citruline and NO through Akt-mediated phosphorylation of eNOS.

9.7 Conclusion

Shear stress impacts cells via transport of macromolecules to the cell surface and through transport of blood momentum in the form of normal and shear stresses. These mechanical features of blood flow are translated into biochemical signals by mechanotransduction. All in all, conversion of force to biological changes occurs through mechanotransmission of forces to mechanosensors, which deliver forces to mechanotransducers, where force is converted to biochemical signaling. The nature of this process depends on a coordination of temporal and spatial aspects of the fluid momentum transfer, mechanical properties of transducing partners, the dynamics of cellular remodeling of these transmitting elements, the biochemical rate constants of the transducers, and the transport (active or diffusive) of the downstream factors that ultimately manifest themselves as changes in vascular biology. A thorough understanding of this process will require increased research from experts in mathematics, computational analysis, statistics, and molecular biology in order to develop predictive models of the endothelium capable of assessing the predilection of the vasculature to disease. Such a platform would provide a valuable tool to clinicians in their decisions on what types of interventions to employ when confronted with vascular disease. Importantly, it may also provide insight into strategies to prevent endothelial dysfunction and resulting atherosclerosis.

References

Alessandri-Haber N, Dina OA, Joseph EK, Reichling DB, Levine JD (2008) Interaction of transient receptor potential vanilloid 4, integrin, and SRC tyrosine kinase in mechanical hyperalgesia. J Neurosci 28(5):1046–1057. doi:10.1523/JNEUROSCI.4497-07.2008

Ando J, Yamamoto K (2013) Flow detection and calcium signalling in vascular endothelial cells. Cardiovasc Res 99(2):260–268. doi:10.1093/cvr/cvt084

Arkill KP, Knupp C, Michel CC, Neal CR, Qvortrup K, Rostgaard J, Squire JM (2011) Similar endothelial glycocalyx structures in microvessels from a range of mammalian tissues: evidence for a common filtering mechanism? Biophys J 101(5):1046–1056. doi:10.1016/j.bpj.2011.07.036

Ayuyan AG, Cohen FS (2008) Raft composition at physiological temperature and pH in the absence of detergents. Biophys J 94(7):2654–2666. doi:10.1529/biophysj.107.118596

Bagi Z, Frangos JA, Yeh J, White CR, Kaley G, Koller A (2005) PECAM-1 mediates NO-dependent dilation of arterioles to high temporal gradients of shear stress. Arterioscler Thromb Vasc Biol 25:1590–1595. doi:10.1161/01.ATV.0000170136.71970.5f

Bao X, Clark CB, Frangos JA (2000) Temporal gradient in shear-induced signaling pathway: involvement of MAP kinase, c-fos, and connexin43. Am J Physiol Heart Circ Physiol 278(5): H1598–H1605. Retrieved from http://www.ncbi.nlm.nih.gov/pubmed/10775139

Bao G, Kamm RD, Thomas W, Hwang W, Fletcher DA, Grodzinsky AJ, Zhu C, Mofrad MRK (2010) Molecular biomechanics: the molecular basis of how forces regulate cellular function. Mol Cell Biomech 3(2):91–105. doi:10.1007/s12195-010-0109-z

Barbee K (2002) Role of subcellular shear–stress distributions in endothelial cell mechanotransduction. Ann Biomed Eng 30(4):472–482. doi:10.1114/1.1467678

Baumgart T, Hess ST, Webb WW (2003) Imaging coexisting fluid domains in biomembrane models coupling curvature and line tension. Nature 425(1476–4687; 0028–0836; 6960):821–824

Bidwell JP, Pavalko FM (2010) Mechanosomes carry a loaded message. Sci Signal 3(153):pe51. doi:10.1126/scisignal.3153pe51

Blackman BR, Barbee K, Thibault LE (2000) In vitro cell shearing device to investigate the dynamic response of cells in a controlled hydrodynamic environment. Ann Biomed Eng 28 (4):363–372. Retrieved from http://www.ncbi.nlm.nih.gov/pubmed/10870893

Botelho AV, Huber T, Sakmar TP, Brown MF (2006) Curvature and hydrophobic forces drive oligomerization and modulate activity of rhodopsin in membranes. Biophys J 91 (12):4464–4477. doi:10.1529/biophysj.106.082776

Boycott HE, Barbier CSM, Eichel CA, Costa KD, Martins RP, Louault F, Dilanian G, Coulombe A, Hatem SN, Balse E (2013) Shear stress triggers insertion of voltage-gated potassium channels from intracellular compartments in atrial myocytes. Proc Natl Acad Sci U S A 110(41):E3955–E3964. doi:10.1073/pnas.1309896110

Broday DM (2000) Diffusion of clusters of transmembrane proteins as a model of focal adhesion remodeling. Bull Math Biol 62(5):891–924. doi:10.1006/bulm.2000.0183

Bursac P, Lenormand G, Fabry B, Oliver M, Weitz DA, Viasnoff V, Butler JP, Fredberg JJ (2005) Cytoskeletal remodelling and slow dynamics in the living cell. Nat Mater 4(7):557–561. doi:10.1038/nmat1404

Butler PJ, Weinbaum S, Chien S, Lemons DE (2000) Endothelium-dependent, shear-induced vasodilation is rate-sensitive. Microcirculation 7(1):53–65. Retrieved from http://www.ncbi.nlm.nih.gov/pubmed/10708337

Butler PJ, Norwich G, Weinbaum S, Chien S (2001) Shear stress induces a time- and position-dependent increase in endothelial cell membrane fluidity. Am J Physiol Cell Physiol 280 (0363–6143; 0363–6143; 4):C962–C969. Retrieved from http://www.ncbi.nlm.nih.gov/pubmed/11245613

Butler PJ, Tsou T-CC, Li JY-S, Usami S, Chien S (2002) Rate sensitivity of shear-induced changes in the lateral diffusion of endothelial cell membrane lipids: a role for membrane perturbation in shear-induced MAPK activation. FASEB J 16(2):216–218. doi:10.1096/fj.01-0434fje

Caro CG, Fitz-Gerald JM, Schroter RC (1969) Arterial wall shear and distribution of early atheroma in man. Nature 223(5211):1159–1160. Retrieved from http://www.ncbi.nlm.nih.gov/pubmed/5810692

Caro CG, Fitz-Gerald JM, Schroter RC (1971) Atheroma and arterial wall shear observation, correlation and proposal of a shear dependent mass transfer mechanism for atherogenesis. Proc R Soc B Biol Sci 177(1046):109–133. doi:10.1098/rspb.1971.0019

Cecchi E, Giglioli C, Valente S, Lazzeri C, Gensini GF, Abbate R, Mannini L (2011) Role of hemodynamic shear stress in cardiovascular disease. Atherosclerosis 214(2):249–256. doi:10.1016/j.atherosclerosis.2010.09.008

Chachisvilis M, Zhang Y-L, Frangos JA (2006) G protein-coupled receptors sense fluid shear stress in endothelial cells. Proc Natl Acad Sci U S A 103(42):15463–15468. doi:10.1073/pnas.0607224103

Chadjichristos CE, Kwak BR (2007) Connexins: new genes in atherosclerosis. Ann Med 39 (6):402–411. doi:10.1080/07853890701436757

Christensen AP, Corey DP (2007) TRP channels in mechanosensation: direct or indirect activation? Nat Rev 8(7):510–521. doi:10.1038/nrn2149

Dabagh M, Jalali P, Butler PJ, Tarbell JM (2014) Shear-induced force transmission in a multicomponent, multicell model of the endothelium. J R Soc Interface 11(98):20140431. doi:10.1098/rsif.2014.0431

Dangaria JH, Butler PJ (2007) Macrorheology and adaptive microrheology of endothelial cells subjected to fluid shear stress. Am J Physiol Cell Physiol 293(5):C1568–C1575. doi:10.1152/ajpcell.00193.2007

Davies PF (2009) Hemodynamic shear stress and the endothelium in cardiovascular pathophysiology. Nat Clin Pract Cardiovasc Med 6(1):16–26. doi:10.1038/ncpcardio1397

Davies PF, Civelek M, Fang Y, Fleming I (2013) The atherosusceptible endothelium: endothelial phenotypes in complex haemodynamic shear stress regions in vivo. Cardiovasc Res 99 (2):315–327. doi:10.1093/cvr/cvt101

Dejana E (2004) Endothelial cell-cell junctions: happy together. Nat Rev Mol Cell Biol 5 (4):261–270. doi:10.1038/nrm1357

Del Pozo MA, Alderson NB, Kiosses WB, Chiang HH, Anderson RG, Schwartz MA (2004) Integrins regulate Rac targeting by internalization of membrane domains. Science (80-.)., 303 (1095–9203; 0036–8075; 5659):839–842

Dela Paz NG, Melchior B, Shayo FY, Frangos JA (2014) Heparan sulfates mediate the interaction between platelet endothelial cell adhesion molecule-1 (PECAM-1) and the Gαq/11 subunits of heterotrimeric G proteins. J Biol Chem 289(11):7413–7424. doi:10.1074/jbc.M113.542514

DePaola N, Gimbrone MA, Davies PF, Dewey CF (1992) Vascular endothelium responds to fluid shear stress gradients [published erratum appears in Arterioscler Thromb 1993 Mar;13(3):465]. Arterioscler Thromb Vasc Biol 12(11):1254–1257. doi:10.1161/01.ATV.12.11.1254

Dimova R, Pouligny B, Dietrich C (2000) Pretransitional effects in dimyristoylphosphatidylcholine vesicle membranes: optical dynamometry study. Biophys J 79(1):340–356. doi:10.1016/S0006-3495(00)76296-5

Ebong EE, Lopez-Quintero SV, Rizzo V, Spray DC, Tarbell JM (2014) Shear-induced endothelial NOS activation and remodeling via heparan sulfate, glypican-1, and syndecan-1. Integr Biol (Camb). doi:10.1039/c3ib40199e

Eggeling C, Ringemann C, Medda R, Schwarzmann G, Sandhoff K, Polyakova S, Belov VN, Hein B, von Middendorff C, Schonle A (2009) Direct observation of the nanoscale dynamics of membrane lipids in a living cell. Nature 457(7233):1159–1162. doi:10.1038/nature07596

Ethier CR (2002) Computational modeling of mass transfer and links to atherosclerosis. Ann Biomed Eng 30(4):461–471. doi:10.1114/1.1468890

Feaver RE, Gelfand BD, Blackman BR (2013) Human haemodynamic frequency harmonics regulate the inflammatory phenotype of vascular endothelial cells. Nat Commun 4:1525. doi:10.1038/ncomms2530

Ferko MC, Patterson BW, Butler PJ (2006) High-resolution solid modeling of biological samples imaged with 3D fluorescence microscopy. Microsc Res Tech 69(8):648–655

Ferko MC, Bhatnagar A, Garcia MB, Butler PJ (2007) Finite-element stress analysis of a multicomponent model of sheared and focally-adhered endothelial cells. Ann Biomed Eng 35(2):208–223

Fleming I, Busse R (1999) Signal transduction of eNOS activation. Cardiovasc Res 43(3):532–541

Frame MD, Rivers RJ, Altland O, Cameron S (2007) Mechanisms initiating integrin-stimulated flow recruitment in arteriolar networks. J Appl Physiol 102(8750–7587; 0161–7567; 6):2279–2287

Frangos JA, Huang TY, Clark CB (1996) Steady shear and step changes in shear stimulate endothelium via independent mechanisms — superposition of transient and sustained nitric oxide production. Biochem Biophys Res Commun. 1996 Jul 25;224(3):660-5.

Fuentes DE, Butler PJ (2012) Coordinated mechanosensitivity of membrane rafts and focal adhesions. Cell Mol Bioeng 5(2):143–154. doi:10.1007/s12195-012-0225-z

Fuentes DE, Bae C, Butler PJ (2011) Focal adhesion induction at the tip of a functionalized nanoelectrode. Cell Mol Bioeng 4(4):616–626. doi:10.1007/s12195-011-0214-7

Fujiwara K, Masuda M, Osawa M, Kano Y, Katoh K (2001) Is PECAM-1 a mechanoresponsive molecule ? Possible Sites Mechanosensing 17:11–17

Fung YC, Liu SQ (1993) Elementary mechanics of the endothelium of blood vessels. J Biomech Eng 115(1):1–12. Retrieved from http://www.ncbi.nlm.nih.gov/pubmed/8445886

Garcia-Saez AJ, Chiantia S, Schwille P, García-Sáez AJ (2007) Effect of line tension on the lateral organization of lipid membranes. J Biol Chem 282(46):33537–33544. doi:10.1074/jbc.M706162200

Gaus K, Le LS, Balasubramanian N, Schwartz MA, Le Lay S (2006) Integrin-mediated adhesion regulates membrane order. J Cell Biol 174(0021–9525; 0021–9525; 5):725–734. doi:10.1083/jcb.200603034

Giantsos-Adams KM, Koo AJ-A, Song S, Sakai J, Sankaran J, Shin JH, Garcia-Cardena G, Dewey CF (2013) Heparan sulfate regrowth profiles under laminar shear flow following enzymatic degradation. Cell Mol Bioeng 6(2):160–174. doi:10.1007/s12195-013-0273-z

Giddens DP, Zarins CK, Glagov S (1993) The role of fluid mechanics in the localization and detection of atherosclerosis. J Biomech Eng 115(4B):588. doi:10.1115/1.2895545

Gouverneur M, Spaan JAE, Pannekoek H, Fontijn RD, Vink H (2006) Fluid shear stress stimulates incorporation of hyaluronan into endothelial cell glycocalyx. Am J Physiol Heart Circ Physiol 290(1):H458–H462. doi:10.1152/ajpheart.00592.2005

Gudi S, Nolan JP, Frangos JA (1998) Modulation of GTPase activity of G proteins by fluid shear stress and phospholipid composition. Proc Natl Acad Sci U S A 95(5):2515–2519. Retrieved from http://www.pnas.org/content/95/5/2515.long

Gullapalli RR, Tabouillot T, Mathura R, Dangaria JH, Butler PJ (2007) Integrated multimodal microscopy, time-resolved fluorescence, and optical-trap rheometry: toward single molecule mechanobiology. J Biomed Opt 12(1083–3668; 1083–3668; 1):14012

Haidekker MA, White CR, Frangos JA (2001) Analysis of temporal shear stress gradients during the onset phase of flow over a backward-facing step. J Biomech Eng 123(5):455–463

Hartmannsgruber V, Heyken WT, Kacik M, Kaistha A, Grgic I, Harteneck C, Liedtke W, Hoyer J, Kohler R (2007) Arterial response to shear stress critically depends on endothelial TRPV4 expression. PLoS One 2(9):e827. doi:10.1371/journal.pone.0000827

Hein TW, Platts SH, Waitkus-Edwards KR, Kuo L, Mousa SA, Meininger GA (2001) Integrin-binding peptides containing RGD produce coronary arteriolar dilation via cyclooxygenase activation. Am J Physiol Heart Circ Physiol 281(0363–6135; 0363–6135; 6):H2378–H2384

Heinrich M, Tian A, Esposito C, Baumgart T (2010) Dynamic sorting of lipids and proteins in membrane tubes with a moving phase boundary. Proc Natl Acad Sci U S A 107 (16):7208–7213. doi:10.1073/pnas.0913997107

Helmke BP, Thakker DB, Goldman RD, Davies PF (2001) Spatiotemporal analysis of flow-induced intermediate filament displacement in living endothelial cells. Biophys J 80 (1):184–194. doi:10.1016/S0006-3495(01)76006-7

Hoffman BD, Grashoff C, Schwartz MA (2011) Dynamic molecular processes mediate cellular mechanotransduction. Nature 475(7356):316–323. doi:10.1038/nature10316

Huang H, Kamm RD, Lee RT (2004) Cell mechanics and mechanotransduction: pathways, probes, and physiology. Am J Physiol Cell Physiol 287(1):C1–C11. doi:10.1152/ajpcell.00559.2003

Isshiki M, Anderson RG (1999) Calcium signal transduction from caveolae. Cell Calcium 26 (5):201–208. doi:10.1054/ceca.1999.0073

Jalali S, Li YS, Sotoudeh M, Yuan S, Li S, Chien S, Shyy JY (1998) Shear stress activates p60src-Ras-MAPK signaling pathways in vascular endothelial cells. Arterioscler Thromb Vasc Biol 18(2):227–234

Jalali S, del Pozo MA, Chen K, Miao H, Li Y, Schwartz MA, JY Shyy, Chien S (2001) Integrin-mediated mechanotransduction requires its dynamic interaction with specific extracellular matrix (ECM) ligands. Proc Natl Acad Sci U S A 98(0027–8424; 0027–8424; 3):1042–1046

Johnson TL, Nerem RM (2009) Endothelial connexin 37, connexin 40, and connexin 43 respond uniquely to substrate and shear stress. Retrieved from http://informahealthcare.com/doi/abs/10.1080/10623320701617233

Kamm R, Mofrad M (2009) Cellular mechanotransduction diverse perspectives molecules to tissues. Cambridge University Press, Cambridge

Kanchanawong P, Shtengel G, Pasapera AM, Ramko EB, Davidson MW, Hess HF, Waterman CM (2010) Nanoscale architecture of integrin-based cell adhesions. Nature 468(7323):580–584. doi:10.1038/nature09621

Kassab GS, Navia JA (2006) Biomechanical considerations in the design of graft: the homeostasis hypothesis. Annu Rev Biomed Eng 8:499–535. doi:10.1146/annurev.bioeng.8.010506.105023

Kohler R, Heyken WT, Heinau P, Schubert R, Si H, Kacik M, Busch C, Grgic I, Maier T, Hoyer J (2006) Evidence for a functional role of endothelial transient receptor potential V4 in shear stress-induced vasodilatation. Arterioscler Thromb Vasc Biol 26(7):1495–1502. doi:10.1161/01.ATV.0000225698.36212.6a

Koskinas KC, Feldman CL, Chatzizisis YS, Coskun AU, Jonas M, Maynard C, Baker A, Papafaklis M, Edelman ER, Stone PH (2010) Natural history of experimental coronary atherosclerosis and vascular remodeling in relation to endothelial shear stress: a serial, in vivo intravascular ultrasound study. Circulation 121(19):2092–2101. doi:10.1161/CIRCULATIONAHA.109.901678

Koskinas KC, Chatzizisis YS, Antoniadis AP, Giannoglou GD (2012) Role of endothelial shear stress in stent restenosis and thrombosis: pathophysiologic mechanisms and implications for clinical translation. J Am Coll Cardiol 59(15):1337–1349. doi:10.1016/j.jacc.2011.10.903

Krishnan R, Park CY, Lin Y-C, Mead J, Jaspers RT, Trepat X, Lenormand G, Tambe D, Smolensky AV, Knoll AH, Butler JP, Fredberg JJ (2009) Reinforcement versus fluidization in cytoskeletal mechanoresponsiveness. PLoS One 4(5):e5486. doi:10.1371/journal.pone.0005486

Ku DN, Giddens DP, Zarins CK, Glagov S (1985) Pulsatile flow and atherosclerosis in the human carotid bifurcation. Positive correlation between plaque location and low oscillating shear stress. Arterioscler Thromb Vasc Biol 5(3):293–302. doi:10.1161/01.ATV.5.3.293

Kwik J, Boyle S, Fooksman D, Margolis L, Sheetz MP, Edidin M (2003) Membrane cholesterol, lateral mobility, and the phosphatidylinositol 4,5-bisphosphate-dependent organization of cell actin. Proc Natl Acad Sci U S A 100(0027–8424; 0027–8424; 24):13964–13969

Lee SE, Kamm RD, Mofrad MR (2007) Force-induced activation of talin and its possible role in focal adhesion mechanotransduction. J Biomech 40(0021–9290; 0021–9290; 9):2096–2106

Lele TP, Pendse J, Kumar S, Salanga M, Karavitis J, Ingber DE (2006) Mechanical forces alter zyxin unbinding kinetics within focal adhesions of living cells. J Cell Physiol 207(1):187–194. doi:10.1002/jcp.20550

Leto TL, Pratt BM, Madri JA (1986) Mechanisms of cytoskeletal regulation: modulation of aortic endothelial cell protein band 4.1 by the extracellular matrix. J Cell Physiol 127(3):423–431. doi:10.1002/jcp.1041270311

Levitan I, Gooch KJ (2007) Lipid rafts in membrane-cytoskeleton interactions and control of cellular biomechanics: actions of oxLDL. Antioxid Redox Signal 9(1523–0864; 1523–0864; 9):1519–1534

Li Y-SJ, Haga JH, Chien S (2005) Molecular basis of the effects of shear stress on vascular endothelial cells. J Biomech 38(10):1949–1971. doi:10.1016/j.jbiomech.2004.09.030

Lingwood D, Ries J, Schwille P, Simons K (2008) Plasma membranes are poised for activation of raft phase coalescence at physiological temperature. Proc Natl Acad Sci U S A 105 (1091–6490; 0027–8424; 29):10005–10010

Lipowsky HH, Gao L, Lescanic A (2011) Shedding of the endothelial glycocalyx in arterioles, capillaries, and venules and its effect on capillary hemodynamics during inflammation. Am J Physiol Heart Circ Physiol 301(6):H2235–H2245. doi:10.1152/ajpheart.00803.2011

Lu S, Ouyang M, Seong J, Zhang J, Chien S, Wang Y (2008) The spatiotemporal pattern of Src activation at lipid rafts revealed by diffusion-corrected FRET imaging. PLoS Comput Biol 4 (1553–7358; 7):e1000127

Malek AM (1999) Hemodynamic shear stress and its role in atherosclerosis. JAMA 282(21):2035. doi:10.1001/jama.282.21.2035

Maniotis AJ, Chen CS, Ingber DE (1997) Demonstration of mechanical connections between integrins, cytoskeletal filaments, and nucleoplasm that stabilize nuclear structure. Proc Natl Acad Sci 94(3):849–854. doi:10.1073/pnas.94.3.849

Matthews BD, Thodeti CK, Tytell JD, Mammoto A, Overby DR, Ingber DE (2010) Ultra-rapid activation of TRPV4 ion channels by mechanical forces applied to cell surface beta1 integrins. Integr Biol (Camb) 2(9):435–442. doi:10.1039/c0ib00034e

Na S, Collin O, Chowdhury F, Tay B, Ouyang M, Wang Y, Wang N (2008) Rapid signal transduction in living cells is a unique feature of mechanotransduction. Proc Natl Acad Sci U S A 105(18):6626–6631. doi:10.1073/pnas.0711704105

Nigro P, Abe J, Berk BC (2011) Flow shear stress and atherosclerosis: a matter of site specificity. Antioxid Redox Signal 15:1405–1414

Nilius B, Droogmans G, Wondergem R (2003) Transient receptor potential channels in endothelium: solving the calcium entry puzzle? Endothelium 10(1):5–15

Nollert MU, Diamond SL, McIntire LV (1991) Hydrodynamic shear stress and mass transport modulation of endothelial cell metabolism. Biotechnol Bioeng 38(6):588–602. doi:10.1002/bit.260380605

Noria S, Xu F, McCue S, Jones M, Gotlieb AI, Langille BL (2004) Assembly and reorientation of stress fibers drives morphological changes to endothelial cells exposed to shear stress. Am J Pathol 164(4):1211–1223. doi:10.1016/S0002-9440(10)63209-9

O'Neil RG, Heller S (2005) The mechanosensitive nature of TRPV channels. Pflugers Arch 451 (1):193–203. doi:10.1007/s00424-005-1424-4

Oh P, Schnitzer JE (2001) Segregation of heterotrimeric G proteins in cell surface microdomains. G(q) binds caveolin to concentrate in caveolae, whereas G(i) and G(s) target lipid rafts by default. Mol Biol Cell 12(3):685–698

Osborn EA, Rabodzey A, Dewey CF, Hartwig JH (2006) Endothelial actin cytoskeleton remodeling during mechanostimulation with fluid shear stress. Am J Physiol Cell Physiol 290(2):C444–C452. doi:10.1152/ajpcell.00218.2005

Pahakis MY, Kosky JR, Dull RO, Tarbell JM (2007) The role of endothelial glycocalyx components in mechanotransduction of fluid shear stress. Biochem Biophys Res Commun 355(1):228–233. doi:10.1016/j.bbrc.2007.01.137

Park H, Go YM, St John PL, Maland MC, Lisanti MP, Abrahamson DR, Jo H (1998) Plasma membrane cholesterol is a key molecule in shear stress-dependent activation of extracellular signal-regulated kinase. J Biol Chem 273(48):32304–32311

Park H, Go YM, Darji R, Choi JW, Lisanti MP, Maland MC, Jo H (2000) Caveolin-1 regulates shear stress-dependent activation of extracellular signal-regulated kinase. Am J Physiol Heart Circ Physiol 278(4):H1285–H1293

Parton RG, del Pozo M (2013) Caveolae as plasma membrane sensors, protectors and organizers. Nat Rev Mol Cell Biol 14(2):98–112. doi:10.1038/nrm3512

Peiffer V, Sherwin SJ, Weinberg PD (2013) Does low and oscillatory wall shear stress correlate spatially with early atherosclerosis? A systematic review. Cardiovasc Res 99(2):242–250. doi:10.1093/cvr/cvt044

Periole X, Huber T, Marrink S-JJ, Sakmar TP (2007) G protein-coupled receptors self-assemble in dynamics simulations of model bilayers. Biophys J 129(33):10126–10132. doi:10.1021/ja0706246

Reitsma S, Slaaf DW, Vink H, van Zandvoort MAMJ, oude Egbrink MGA (2007) The endothelial glycocalyx: composition, functions, and visualization. Pflugers Arch 454(3):345–359. doi:10.1007/s00424-007-0212-8

Rizzo V, McIntosh DP, Oh P, Schnitzer JE (1998) In situ flow activates endothelial nitric oxide synthase in luminal caveolae of endothelium with rapid caveolin dissociation and calmodulin association. J Biol Chem 273(52):34724–34729

Rizzo V, Morton C, DePaola N, Schnitzer JE, Davies PF (2003) Recruitment of endothelial caveolae into mechanotransduction pathways by flow conditioning in vitro. Am J Physiol Heart Circ Physiol 285(4):H1720–H1729. doi:10.1152/ajpheart.00344.2002

Rotblat B, Belanis L, Liang H, Haklai R, Elad-Zefadia G, Hancock JF, Kloog Y, Plowman SJ (2010) H-Ras nanocluster stability regulates the magnitude of MAPK signal output. PLoS One 5(1932–6203; 1932–6203; 8)

Satcher RL, Bussolari SR, Gimbrone MA, Dewey CF (1992) The distribution of fluid forces on model arterial endothelium using computational fluid dynamics. J Biomech Eng 114 (3):309–316

Schmid-Schönbein GW, Kosawada T, Skalak R, Chien S (1995) Membrane model of endothelial cells and leukocytes. A proposal for the origin of a cortical stress. J Biomech Eng 117 (2):171–178

Secomb TW, Hsu R, Pries AR (2001) Effect of the endothelial surface layer on transmission of fluid shear stress to endothelial cells. Biorheology 38(2-3):143–150

Shamitko-Klingensmith N, Molchanoff KM, Burke KA, Magnone GJ, Legleiter J (2012) Mapping the mechanical properties of cholesterol-containing supported lipid bilayers with nanoscale spatial resolution. Langmuir 28(37):13411–13422. doi:10.1021/la302705f

Shyy YJ, Hsieh HJ, Usami S, Chien S (1994) Fluid shear stress induces a biphasic response of human monocyte chemotactic protein 1 gene expression in vascular endothelium. Proc Natl Acad Sci U S A 91(11):4678–4682

Singh RD, Marks DL, Holicky EL, Wheatley CL, Kaptzan T, Sato SB, Kobayashi T, Ling K, Pagano RE (2010) Gangliosides and beta1-integrin are required for caveolae and membrane domains. Traffic 11(1600–0854; 1398–9219; 3):348–360

Sowa G (2012) Caveolae, caveolins, cavins, and endothelial cell function: new insights. Front Physiol 2:120. doi:10.3389/fphys.2011.00120

Squire JM, Chew M, Nneji G, Neal C, Barry J, Michel C (2001) Quasi-periodic substructure in the microvessel endothelial glycocalyx: a possible explanation for molecular filtering? J Struct Biol 136(3):239–255

Tabouillot T, Muddana HS, Butler PJ (2011) Endothelial cell membrane sensitivity to shear stress is lipid domain dependent. Cell Mol Bioeng 4(2):169–181. doi:10.1007/s12195-010-0136-9

Tada S, Dong C, Tarbell JM (2007) Effect of the stress phase angle on the strain energy density of the endothelial plasma membrane. Biophys J 93(9):3026–3033. doi:10.1529/biophysj.106.100685

Tarbell JM (2003) Mass transport in arteries and the localization of atherosclerosis. Annu Rev Biomed Eng 5:79–118. doi:10.1146/annurev.bioeng.5.040202.121529

Tzima E, Irani-Tehrani M, Kiosses WB, Dejana E, Schultz DA, Engelhardt B, Cao G, DeLisser H, Schwartz MA (2005) A mechanosensory complex that mediates the endothelial cell response to fluid shear stress. Nature 437(1476–4687; 0028–0836; 7057):426–431

Ueki Y, Sakamoto N, Sato M (2010) Direct measurement of shear strain in adherent vascular endothelial cells exposed to fluid shear stress. Biochem Biophys Res Commun 394(1):94–99. doi:10.1016/j.bbrc.2010.02.115

Van Zanten TS, Cambi A, Koopman M, Joosten B, Figdor CG, Garcia-Parajo MF (2009) Hotspots of GPI-anchored proteins and integrin nanoclusters function as nucleation sites for cell adhesion. Proc Natl Acad Sci U S A 106(1091–6490; 0027–8424; 44):18557–18562

Wang Y, Botvinick EL, Zhao Y, Berns MW, Usami S, Tsien RY, Chien S (2005) Visualizing the mechanical activation of Src. Nature 434(7036):1040–1045. doi:10.1038/nature03469

Wang N, Tytell JD, Ingber DE (2009) Mechanotransduction at a distance: mechanically coupling the extracellular matrix with the nucleus. Nat Rev Mol Cell Biol 10(1):75–82. doi:10.1038/nrm2594

Weinbaum S, Zhang X, Han Y, Vink H, Cowin SC (2003) Mechanotransduction and flow across the endothelial glycocalyx. Proc Natl Acad Sci U S A 100(13):7988–7995

Wu S, Sangerman J, Li M, Brough GH, Goodman SR, Stevens T (2001) Essential control of an endothelial cell ISOC by the spectrin membrane skeleton. J Cell Biol 154(6):1225–1233. doi:10.1083/jcb.200106156

Wu L, Gao X, Brown RC, Heller S, O'Neil RG (2007) Dual role of the TRPV4 channel as a sensor of flow and osmolality in renal epithelial cells. Am J Physiol 293(5):F1699–F1713. doi:10.1152/ajprenal.00462.2006

Xu H, Zhao H, Tian W, Yoshida K, Roullet JB, Cohen DM (2003) Regulation of a transient receptor potential (TRP) channel by tyrosine phosphorylation. SRC family kinase-dependent tyrosine phosphorylation of TRPV4 on TYR-253 mediates its response to hypotonic stress. J Biol Chem 278(13):11520–11527. doi:10.1074/jbc.M211061200

Yamamoto K, Ando J (2010) Blood flow sensing mechanism via calcium signaling in vascular endothelium. Yakugaku Zasshi 130(11):1407–1411

Yamamoto K, Ando J (2013) Endothelial cell and model membranes respond to shear stress by rapidly decreasing the order of their lipid phases. J Cell Sci 126(Pt 5):1227–1234. doi:10.1242/jcs.119628

Yamamoto K, Sokabe T, Ohura N, Nakatsuka H, Kamiya A, Ando J (2003) Endogenously released ATP mediates shear stress-induced Ca2+ influx into pulmonary artery endothelial cells. Am J Physiol Heart Circ Physiol 285(2):H793–H803. doi:10.1152/ajpheart.01155.2002

Yamamoto K, Furuya K, Nakamura M, Kobatake E, Sokabe M, Ando J (2011) Visualization of flow-induced ATP release and triggering of Ca2+ waves at caveolae in vascular endothelial cells. J Cell Sci 124(Pt 20):3477–3483. doi:10.1242/jcs.087221

Zeng Y, Tarbell JM (2014) The adaptive remodeling of endothelial glycocalyx in response to fluid shear stress. PLoS One 9(1):e86249. doi:10.1371/journal.pone.0086249

Mechanobiology of Endothelial Cells Related to the Pathogenesis of Arterial Disease

10

Susumu Kudo

Abstract
Arteriosclerosis occurs preferentially at the inner curvature of blood vessels where shear stress induced by blood flow is low. Endothelial cells, which line the inner surface of blood vessels, transport all substances from vessels to tissues. It is well known that many functions of endothelial cells are affected by fluid shear stress, and it has been hypothesized that the dysfunction of endothelial transport is related to arteriogenesis. Endothelial transport is divided into two primary types: paracellular transport, which is passive, and transcellular transport, which is active. In this chapter, we introduce these endothelial transport processes and discuss the ways they are influenced by various shear stress conditions.

Keywords
Endothelial cell • Shear stress • Paracellular transport • Transcellular transport

10.1 Introduction

Endothelial cells (ECs) line the inner surface of all types of blood vessels, from capillaries to arteries. Oxygen and nutrients, which are delivered by blood vessels throughout the body, are transported from vessels to tissues through ECs. It is well known that arteriosclerosis occurs preferentially at the inner curvature of blood vessels, where mean fluid shear stress is low (Caro et al. 1969) (Ku et al. 1985) (Yoshida et al. 1990), and that low-density lipoprotein (LDL) and albumin accumulate in these regions (Yoshida et al. 1990; Somer and Schwartz 1972) (Packham

S. Kudo (✉)
Department of Mechanical Engineering, Kyushu University, 744 Motooka Nishi-ku, Fukuoka, Fukuoka 819-0376, Japan
e-mail: kudos@mech.kyushu-u.ac.jp

K. Tanishita, K. Yamamoto (eds.), *Vascular Engineering*,
DOI 10.1007/978-4-431-54801-0_10

et al. 1967). Thus, the permeability of these macromolecules through ECs is considered to be closely related to atherosclerosis. ECs are affected by fluid shear stress, and their function changes under shear stress loading (Li et al. 2005; Chien 2007). The impact of shear stress on EC permeability is particularly important with regard to atherogenesis. In this chapter, we introduce the ways in which EC permeability is affected by fluid shear stress.

10.2 Transport Pathways Through ECs

Mass transport through ECs is considered to occur through several pathways (Mehta and Malik 2006). As shown in Fig. 10.1, these pathways are divided into the transcellular vesicular and paracellular junctional pathways. The transcellular vesicular pathway involves active transport, meaning that metabolic energy is needed. Paracellular transport involves passive transport, indicating that pressure or concentration differentials, and not metabolic energy, are the driving forces for transport. In transcellular vesicular transport, vesicles pinch off of the cell membrane and enclose extracellular material, then move to the other side of the membrane where they fuse and release their contents. Some reviews have drawn attention to the still incompletely understood phenomena of endocytosis and transcytosis, especially forms that are mediated by caveolae or clathrin (Mehta and Malik 2006; Tuma and Hubbard 2003).

10.3 The Quantification of Endothelial Transport

Endothelial transport is quantified by the Kedem-Katchalsky equation (Kedem and Katchalsky 1958).

$$J_\text{v} = L_\text{p}(\Delta p - \sigma \Delta \pi) \tag{10.1}$$

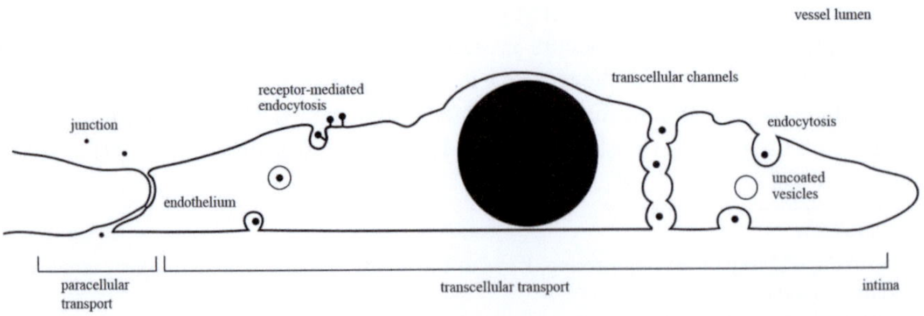

Fig. 10.1 Schematic of transport pathways across endothelial cells (Ogunrinade et al. 2002)

$$J_s = P_s \Delta C + J_v(1-\sigma)\overline{C} \qquad (10.2)$$

J_v: volume flux across endothelial cell monolayer
J_s: solute flux
Δp: pressure difference across endothelial cell monolayer
$\Delta \pi$: corresponding osmotic pressure differential
ΔC: solute concentration differential
σ: reflection coefficient
L_p: hydraulic conductivity
P_s: diffusive permeability
$\overline{C}$: mean intramembrane solute concentration

Besides quantitatively describing the vessel wall permeability to liquid and solutes, these equations have been useful to experimentally measure endothelial permeability. Δp and $\Delta \pi$ are driving forces for volume transport that convects solute. ΔC is the driving force for diffusive solute transport. P_s, L_p, and σ are transport properties. In this chapter, overall transport across the endothelial monolayer is evaluated by L_p or P_s.

10.4 The Effect of Shear Stress on Endothelial Transport

10.4.1 The Effect of Shear Stress on Endothelial Uptake Properties

It is well known that endothelial cell function is affected by fluid shear stress, and that atherosclerosis occurs at the inner curvature of blood vessels where mean shear stress is low and the oscillation frequency is high. Thus, the shear stresses generated by blood flow are considered to affect arteriogenesis. Endothelial transport is divided into transcellular and paracellular transport. However, it is difficult to evaluate both forms of transport at the same time. Thus we first introduce pinocytosis (fluid-phase endocytosis), the uptake process that is the first step in transcellular transport and is not involved in paracellular transport.

Davies et al. (1984) studied the relationship between fluid shear stress and pinocytosis in bovine aortic ECs (BAECs) by monitoring horse radish peroxidase (HRP) uptake into cells. They showed that continuous exposure to steady shear stresses (0.1–1.5 Pa) in laminar flow stimulated time- and amplitude-dependent increases in the pinocytic rate, which returned to control levels after several hours. Sprague et al. (1987) studied the effect of shear stress on LDL uptake into ECs. They showed that the amount of LDL taken up into ECs increased after 3 Pa of shear stress loading for 24 h. We studied the effect of shear stress on albumin uptake into cultured porcine aortic EC (PAEC) monolayers (Kudo et al. 1997, 1998a, 2000). The ECs were cultured on coverslips that were assembled in a flow chamber, and then exposed to fluid shear stress for 48 h. The coverslips were then dismounted from the chamber and dipped in PBS with tetramethylrhodamine-5-(and 6)-isothiocyanate (TRITC) albumin for 1 h. The cells were washed with PBS and observed by

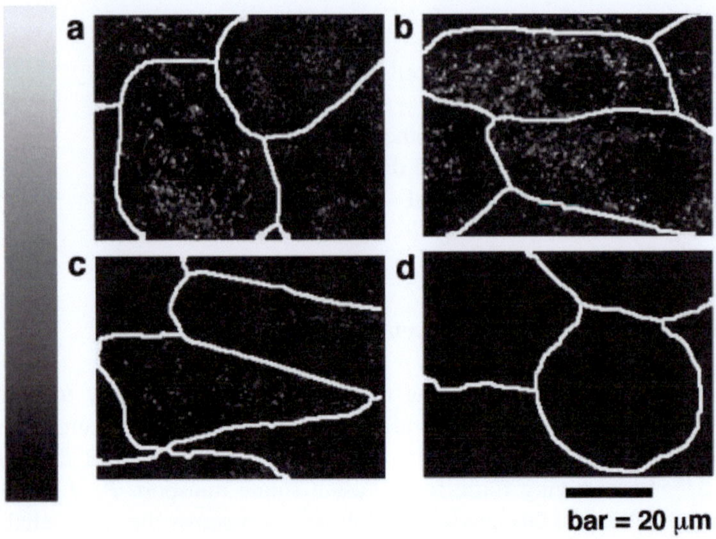

Fig. 10.2 Effect of shear stress and metabolic inhibitors on albumin uptake into endothelial cells. Endothelial cells take up albumin under conditions characterized as (**a**) no-stress (no flow); (**b**) low-stress, 10 dyn/cm^2; (**c**) high-stress, 60 dyn/cm^2; and (**d**) with the addition of 1 mM FCCP and 2 mM rotenone. *White lines* are cell contours. *Bright spots* are TRITC-albumin. Fluorescent intensity is represented by a 256-level *gray scale* (*black* is 0, *white* is 255) (Kudo et al. 2000)

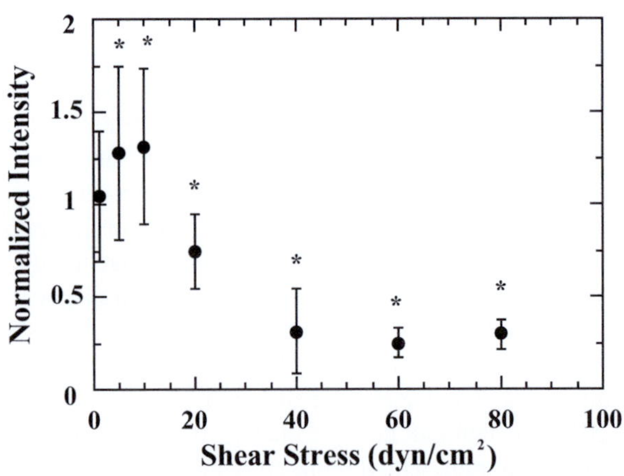

Fig. 10.3 Effect of shear stress on albumin uptake into endothelial cells (*$P < 0.05$ vs. no-flow condition) (Kudo et al. 1998a)

confocal laser microscopy. Figure 10.2 shows that the ECs took up TRITC-albumin after 48 h of shear stress loading, and the amount of uptake could be estimated based on the images (Fig. 10.3). Compared with no-flow samples, albumin uptake into ECs increased under 1-Pa conditions, but then decreased with greater shear

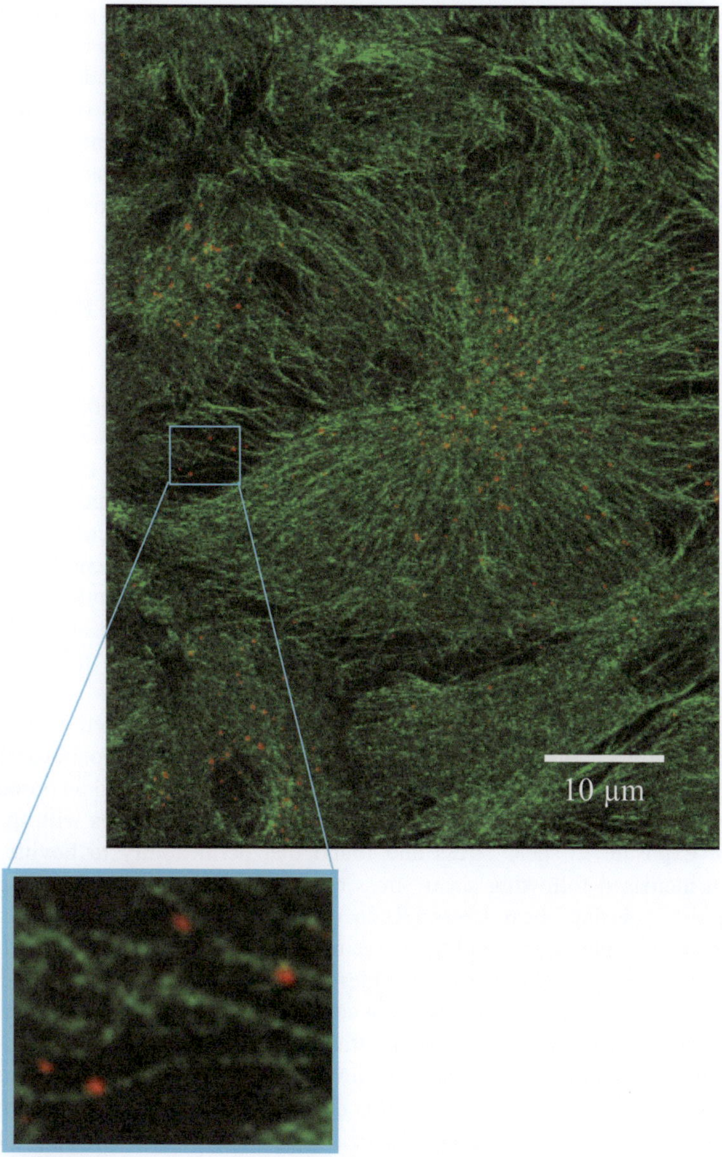

Fig. 10.4 Fluorescent images of the microtubule network and TRITC-albumin in the same cell. TRITC-albumin (*red spots*) is localized on the microtubule network structure (*green* structure) (Kudo et al. 2004)

stress. Similarly, Ueda et al. (2004) showed that an increase in shear stress from 0 to 1.0 Pa caused albumin uptake to increase by 16 %, whereas uptake decreased by 73 % when shear stress was raised from 0 to 3.0 Pa; both differences were statistically significant. Furthermore, we previously used confocal laser microscopy to show that vesicles containing TRITC-albumin were attached to microtubules (Kudo et al. 2004) (Fig. 10.4). Also, albumin uptake was found to be decreased at 4 °C and by the addition of FCCP, an inhibitor of ATP synthesis (Kudo et al. 1998b); these findings indicate that macromolecule uptake, the first step of transcellular transport, is an active transport mechanism that uses metabolic energy. Although uptake studies cannot provide complete information on transport across the entire endothelial monolayer, the first transport step appears to be affected by shear stress.

10.4.2 The Effect of Steady Shear Stress on Endothelial Permeability *In Vitro*

Several groups have measured the permeability of endothelial monolayers under shear stress. Jo et al. (1991) reported the effect of shear stress on endothelial transport using BAEC monolayers cultured on a porous substrate mounted on the wall of a parallel plate flow chamber. They found that albumin permeability (P_s Eq. 10.2) increased by fourfold at 0.1 Pa of shear stress and by tenfold at 1 Pa of shear stress after 60 min. Sill et al. (1995) examined hydraulic conductivity (L_p Eq. 10.1) under shear stress loading. They cultured ECs on a membrane that was then mounted in a rotating disc-type flow chamber. The pressure difference between the upper and lower sides of the membrane could be adjusted. They showed that L_p increased with less than 0.05 Pa of shear stress loading, while 2 Pa of shear stress loading for 2 h increased L_p by fourfold compared with a control sample. Ohshima and Ookawa (1994) evaluated transendothelial permeability (P_s Eq. 10.2) after 2 Pa of shear stress using PAEC monolayers cultured on porous filter membranes. They compared the P_s under static culture conditions with that after 24 h of exposure to shear stress and found that permeability to bovine serum albumin increased following shear stress exposure. We studied EC permeability (P_s Eq. 10.2) during shear stress (Kudo et al. 2005). We first cultured an EC monolayer on a membrane, and then assembled the membrane in a flow chamber (Fig. 10.5). The pressures on the two sides of the membrane were adjusted to the same value since J_v in Eq. (10.2) was not considered, and thus P_s was measured. In this experimental system, the permeabilities of TRITC-albumin and FITC-inulin were measured simultaneously, as albumin can be delivered both by transcellular and paracellular transport, while inulin is delivered only by paracellular transport (Schnitzer and Oh 1994). During 1 Pa of shear stress loading, the P_s of albumin increased with longer loading durations: the Ps at 36 h was 2.5 times higher than at 0 h, a statistically significant difference. On the other hand, the P_s of albumin did not change during 4 Pa of shear stress loading by 48 h (Fig. 10.6). Figure 10.7 shows the average permeability of albumin and inulin at selected time periods under lower shear stress conditions (1 Pa). The average albumin P_s increased with increasing time. The average values were 0.67×10^{-5} cm/s, 1.6×10^{-5} cm/s, and 2.9×10^{-5} cm/s for periods 1 (0–6 h), 2 (12–30 h), and 3 (36–48 h), respectively. In contrast,

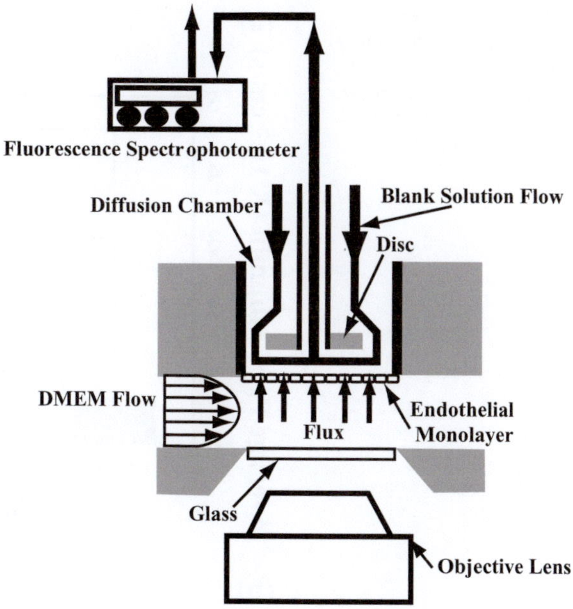

Fig. 10.5 Scheme of the experimental system used to measure macromolecule permeability across endothelial cells subjected to flow (Kudo et al. 2005)

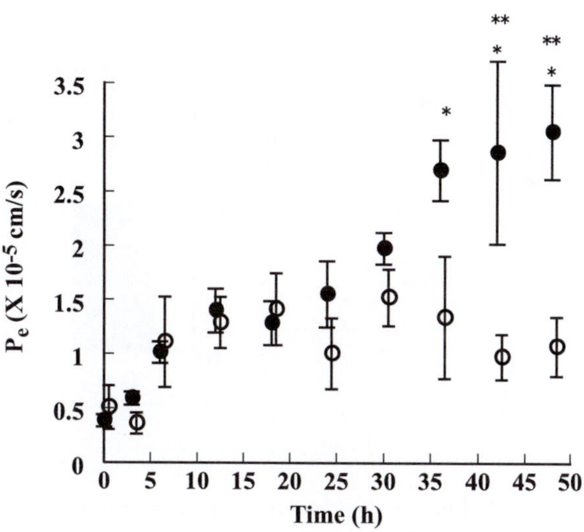

Fig. 10.6 Effect of shear stress on endothelial permeability to albumin. A *black dot* (●) indicates a shear stress of 1 Pa and a *white dot* (○) indicates a shear stress of 4 Pa. Values are mean ± SE (*$P < 0.05$ compared with the value at 0 h. **$P < 0.05$, 1 Pa vs. 4 Pa at the same time point) (Kudo et al. 2005)

the average inulin P_e showed a different pattern, with average values of 10×10^{-5} cm/s, 16×10^{-5} cm/s, and 16×10^{-5} cm/s for periods 1, 2, and 3, respectively.

Phelps and DePaola (2000) used an electrical cell-substrate impedance sensing system, developed by Giaever et al. (Giaever and Keese 1991), to measure endothelial permeability to dextran and the binding between ECs under shear stress loading. They showed that 1 Pa of shear stress increased dextran permeability by

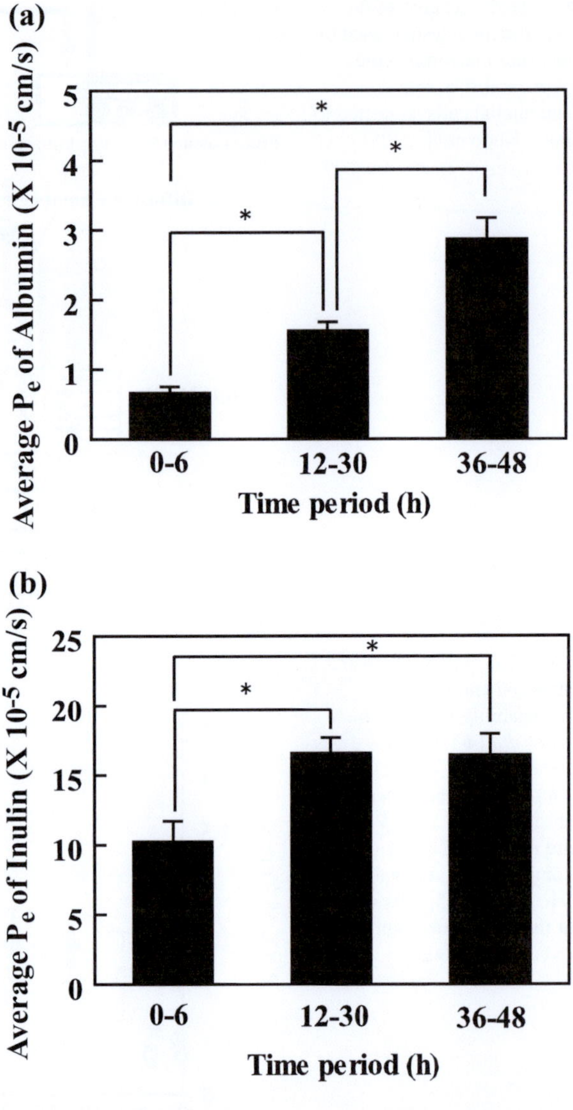

Fig. 10.7 Average permeability of albumin (**a**) and inulin (**b**) for selected time periods at a low shear stress of 1 Pa ($*P < 0.05$) (Kudo et al. 2005)

threefold after 5 h, and that the gap connection between ECs increased after shear stress loading. They therefore concluded that the increase in dextran permeability was due to the increased gap between cells. Warboys et al. (2010) investigated the permeability of albumin through ECs under 0.18 Pa of shear stress for 1 week, and found that while albumin permeability was increased by 1 h of shear stress, it was decreased by shear stress exposure lasting 1 week. Based on the above results, the amplitude and duration of shear stress appear to affect the permeability of ECs.

10.4.3 The Effect of Unsteady Shear Stress on Endothelial Permeability *In Vitro*

The functioning of ECs is well known to be affected by disturbed shear stress (Chien 2008). We investigated the permeability (P_s Eq. 10.2) of albumin through an endothelial monolayer under a pulsatile shear stress of 1 ± 0.14 Pa (Fig. 10.8) (Kudo et al. 2013). The permeability under steady shear stress increased with time and attained a maximum value of 2.8×10^{-5} cm/s. On the other hand, the permeability increased significantly after 6 h of pulsatile shear stress, reaching a maximum of 4.2×10^{-5} cm/s at 18 h. With additional time, the permeability then decreased, to 2.2×10^{-5} cm/s at 48 h, which was close to the value observed under steady shear stress. Although the average magnitude of the pulsatile shear stress was the same as that of the steady shear stress, the permeability was different. Phelps and DePaola (2000) found that under disturbed flow, the permeability of dextran through an EC monolayer was highest and the intercellular gap was widest at the reattachment flow area. We studied the albumin uptake into ECs under disturbed flow and showed that the uptake was highest at the reattachment flow area (Fig. 10.9) (Kudo et al. 1999). Davies et al. (1984) investigated the effect of oscillating shear stress on pinocytosis using HRP. While the endothelial pinocytotic rate did not change during oscillating shear stress at 1 Hz, it increased with a longer cycle duration (15 min). Based on these results, paracellular and transcellular transports are increased in the reattachment flow area.

10.5 Concluding Summary

Endothelial transport is affected by shear stress, but the mechanism whereby this occurs is still not clear. It is also not fully known how shear stress affects transcellular transport, specifically the role of caveolae and coated pits in vesicular transport, or paracellular transport, particularly the molecular basis of intercellular

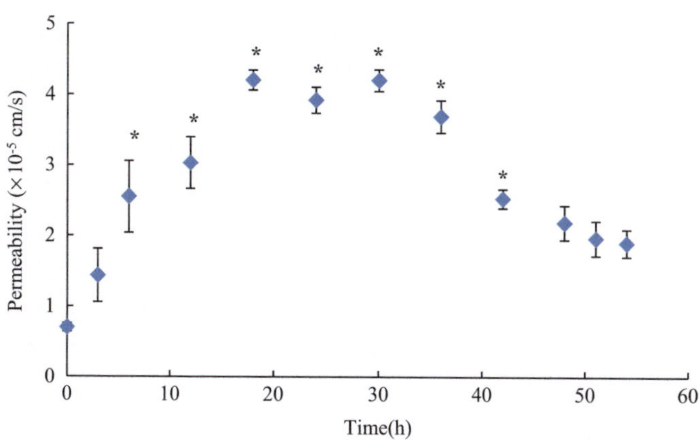

Fig. 10.8 Effects of pulsatile shear stress on albumin permeability across endothelial cell monolayers. The data are the average of four experiments. * $P < 0.05$ vs. 0 h (Kudo et al. 2013)

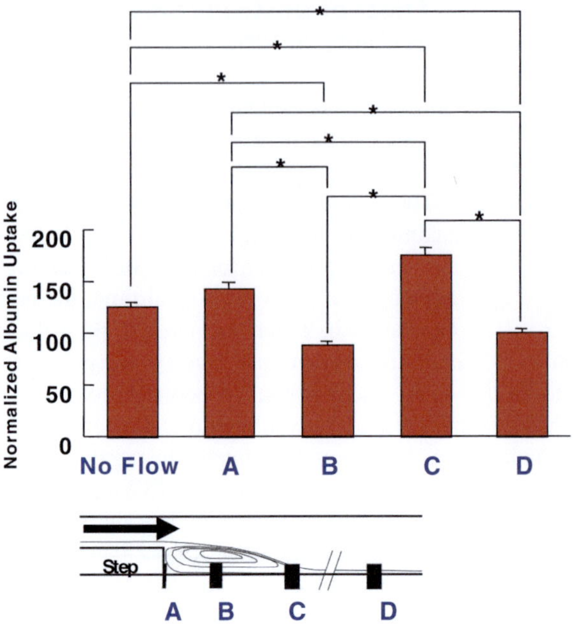

Fig. 10.9 Total albumin uptake into endothelial cells in each area. (**a**), stagnant flow area; (**b**), reversed flow area; (**c**), reattachment flow area; (**d**), fully developed flow area (Kudo et al. 1999)

junctions. Further study of the endothelial transport system under shear stress conditions will lead to a new understanding of the atherosclerosis process.

References

Caro C, Fitz-Gerald J, Schroter R (1969) Arterial wall shear and distribution of early atheroma in man. Nature 223:1159–1161

Chien S (2007) Mechanotransduction and endothelial cell homeostasis: the wisdom of the cell. Am J Physiol Heart Circ Physiol 292(3):H1209

Chien S (2008) Effects of disturbed flow on endothelial cells. Ann Biomed Eng 36(4):554–562

Davies PF, Dewey CF Jr, Bussolari SR, Gordon EJ, Gimbrone MA Jr (1984) Influence of hemodynamic forces on vascular endothelial function. In vitro studies of shear stress and pinocytosis in bovine aortic cells. J Clin Investig 73(4):1121

Giaever I, Keese CR (1991) Micromotion of mammalian cells measured electrically. Proc Natl Acad Sci 88(17):7896–7900

Jo H, Dull R, Hollis T, Tarbell J (1991) Endothelial albumin permeability is shear dependent, time dependent, and reversible. Am J Physiol Heart Circ Physiol 260(6):H1992–H1996

Kedem O, Katchalsky A (1958) Thermodynamic analysis of the permeability of biological membranes to non-electrolytes. Biochim Biophys Acta 27:229–246

Ku DN, Giddens DP, Zarins CK, Glagov S (1985) Pulsatile flow and atherosclerosis in the human carotid bifurcation. Positive correlation between plaque location and low oscillating shear stress. Arterioscler Thromb Vasc Biol 5(3):293–302

Kudo S, Ikezawa K, Ikeda M, Oka K, Tanishita K (1997) Albumin concentration profile inside cultured endothelial cells exposed to shear stress. ASME Publ Bed 35:547–548

Kudo S, Ikezawa K, Matsumura S, Ikeda M, Oka K, Tanishita K (1998a) Effect of wall shear stress on macromolecule uptake into cultured endothelial cells. Trans Jpn Soc Mech Eng B 64 (618):367–374

Kudo S, Ikezawa K, Matsumura S, Ikeda M, Oka K, Tanishita K (1998b) Relationship between energy-dependent macromolecule uptake and transport granules in the endothelial cells affected by wall shear stress. Trans Jpn Soc Mech Eng B 64(623):2123–2131

Kudo S, Sato M, Machida K, Yamaguchi R, Kurokawa M, Matsuzawa T, Ikeda M, Oka K, Tanishita K (1999) Macromolecule uptake into the cultured endothelial cells and the cell morphology in separated flow. Trans Jpn Soc Mech Eng B 65(639):3705–3712

Kudo S, Morigaki R, Saito J, Ikeda M, Oka K, Tanishita K (2000) Shear-stress effect on mitochondrial membrane potential and albumin uptake in cultured endothelial cells. Biochem Biophys Res Commun 270(2):616–621

Kudo S, Ikezawa K, Ikeda M, Tanishita K (2004) Relationship between microtubule network structure and intracellular transport in cultured endothelial cells affected by shear stress. JSME Int J Ser C 47(4):977–984

Kudo S, Tsuzaka M, Ikeda M, Tanishita K (2005) Albumin permeability across endothelial monolayers under long-term shear stress. JSME Int J Ser C 48(4):419–424

Kudo S, Kawarabayashi M, Ikeda M, Tanishita K (2013) Albumin permeability across endothelial monolayers under pulsatile shear stress. J Biorheol 26(1):38–43

Li Y-SJ, Haga JH, Chien S (2005) Molecular basis of the effects of shear stress on vascular endothelial cells. J Biomech 38(10):1949–1971

Mehta D, Malik AB (2006) Signaling mechanisms regulating endothelial permeability. Physiol Rev 86(1):279–367

Ogunrinade O, Kameya GT, Truskey GA (2002) Effect of fluid shear stress on the permeability of the arterial endothelium. Ann Biomed Eng 30(4):430–446

Ohshima N, Ookawa K (1994) Effect of shear stress on cytoskeletal structure and physiological functions of cultured endothelial cells. In: Mow VC, Guilak F, Tran-Son-Tay R, Hochmuth RM (eds) Cell mechanics and cellular engineering. Springer-Verlag New York, Inc., New York, pp 107–117

Packham M, Rowsell H, Jorgensen L, Mustard J (1967) Localized protein accumulation in the wall of the aorta. Exp Mol Pathol 7(2):214–232

Phelps JE, DePaola N (2000) Spatial variations in endothelial barrier function in disturbed flows in vitro. Am J Physiol Heart Circ Physiol 278(2):H469–H476

Schnitzer JE, Oh P (1994) Albondin-mediated capillary permeability to albumin. Differential role of receptors in endothelial transcytosis and endocytosis of native and modified albumins. J Biol Chem 269(8):6072–6082

Sill HW, Chang YS, Artman JR, Frangos J, Hollis T, Tarbell J (1995) Shear stress increases hydraulic conductivity of cultured endothelial monolayers. Am J Physiol Heart Circ Physiol 268(2):H535–H543

Somer J, Schwartz C (1972) Focal (3 H) cholesterol uptake in the pig aorta. 2. Distribution of (3 H) cholesterol across the aortic wall in areas of high and low uptake in vivo. Atherosclerosis 16 (3):377–388

Sprague EA, Steinbach BL, Nerem RM, Schwartz CJ (1987) Influence of a laminar steady-state fluid-imposed wall shear stress on the binding, internalization, and degradation of low-density lipoproteins by cultured arterial endothelium. Circulation 76(3):648–656

Tuma PL, Hubbard AL (2003) Transcytosis: crossing cellular barriers. Physiol Rev 83(3):871–932

Ueda A, Shimomura M, Ikeda M, Yamaguchi R, Tanishita K (2004) Effect of glycocalyx on shear-dependent albumin uptake in endothelial cells. Am J Physiol Heart Circ Physiol 287(5): H2287–H2294

Warboys CM, Berson RE, Mann GE, Pearson JD, Weinberg PD (2010) Acute and chronic exposure to shear stress have opposite effects on endothelial permeability to macromolecules. Am J Physiol Heart Circ Physiol 298(6):H1850–H1856

Yoshida Y, Sue W, Okano M, Oyama T, Yamane T, Mitsumata M (1990) The effects of augmented hemodynamic forces on the progression and topography of atherosclerotic plaques. Ann N Y Acad Sci 598(1):256–273

Vascular Engineering of Blood Coagulation 11

Shinya Goto

Abstract

Blood coagulation and platelet play crucial role in the maintenance of homeostasis in vascular system. Majority of major players and their roles in hemostasis and thrombus formation are known in details, but mostly qualitatively. For stabling the new concept of "Vascular Engineering", quantitative biological experiments and new technology of "Computer Simulation" will be helpful.

Keywords

Coagulation • Platelet • Endothelium • Anticoagulant • Antithrombin • Anti-Xa

11.1 Basic Rule of Blood Coagulation

Human blood plays a crucial role in keeping homeostasis of human body. Each organ of human body, including brain, heart, and kidney, require huge amount of oxygen to generate necessary energy as adenosine 5′-triphosphase by mitochondria. Erythrocytes in human blood should reach to peripheral organ to supply enough amount of oxygen for them. For this purpose, human blood should be extremely smooth, just like pure water, in the vessel system. Indeed, the viscosity of human blood is not so high in condition with peripheral circulation. Low viscosity of human blood is an important characteristic as long as the vessel system is in physiological conditions. From the viewpoint of blood, vessel wall maintains integrity with antithrombotic endothelial cell in the physiological setting (Fig. 11.1) (Mateen and Shuaib 2007; Kawamura et al. 2009).

S. Goto (✉)
Department of Medicine (Cardiology), Tokai University School of Medicine, 143 Shimokasuya, Isehara, Kanagawa 259-1193, Japan
e-mail: shinichi@is.icc.u-tokai.ac.jp

K. Tanishita, K. Yamamoto (eds.), *Vascular Engineering*,
DOI 10.1007/978-4-431-54801-0_11

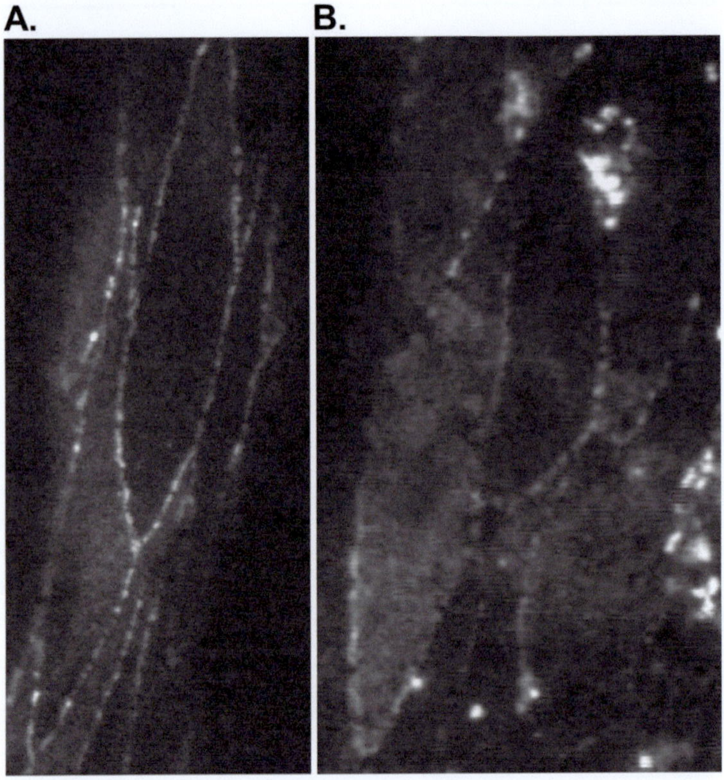

Fig. 11.1 *Vascular endothelial cell. Panel A* demonstrated the shape of vascular endothelial cell in mice cremasteric artery (Kawamura et al. 2009). *Panel B* demonstrated prompt platelet adhesion on the endothelial cells stimulated by oxidative stress

An every content of human blood is important to keep the life of human being stable. But, the low viscosity of human blood sometimes cause problem. Since the viscosity of human blood is so low, lot amount of blood might be lost upon damaging blood vessel. It is known that human being cannot keep their life when they lose one third of their blood. It is extremely important for human being to keep their life by immediate stoppage of blood loss from the site of vessel damage. For this purpose, blood coagulation system plays an important role to increase the viscosity of human blood only at the site of vessel damage.

The primary role of blood coagulation is to prevent blood loss at the site of vessel damage. However, the recent prolongation of human life turns their primary role into unnecessary additional principal. Human being became old as their blood vessels. Blood vessels in aged population are not the same as the ones in healthy young adults. Antithrombotic potentials of endothelial cells were lost or weakened when they become old (Goto 2008). Blood coagulation sometimes occurs in arteries playing crucial role for tissue perfusion such as brain and heart. Those thrombi formed in arteries to perfuse heart results in the onset of myocardial infarction (Hoshiba et al. 2006; Goto and Tomita 2014). The same is true for the onset of ischemic stroke at the time of cerebral thrombus formation (Goto 2004). Thus, we have to understand a conflicting role of blood coagulation; one is crucial role for keeping the maintenance of homeostasis in human body by immediate stoppage of blood loss and the other is the role to cause fatal disease of thrombosis.

The crucial role of blood coagulation system can easily be understood by the presence of inherited hemorrhagic disorders such as "hemophilia" (Hoyer 1994). We now understand that the phenotype of "hemophilia" occurs in patients deficient with coagulant factor VIII or IX (Hoyer 1994). Bleeding disorders also occur in patients deficient with von Willebrand factor that play crucial role for the transportation of coagulation factor VIII (Ruggeri 2007) and platelet adhesion at the site of vascular damage (Goto 2001). In addition, there are several direct inhibitors of coagulation factors II (Connolly et al. 2009; Schulman et al. 2009; Eikelboom et al. 2013) and Xa (Hokusai et al. 2013; Giugliano et al. 2013; Alexander et al. 2011; Connolly et al. 2011; Granger et al. 2011; Agnelli et al. 2013a, b; Mega et al. 2012; Patel et al. 2011; Eriksson et al. 2008) recently developed clinically. Those agents that block specific function of coagulation factors are aimed to develop for the prevention of "thrombotic diseases". However, as shown by their name, all currently available anti II and Xa induce serious bleeding events even when their main role is preventing thrombotic diseases. With the use of those agents, we understand more clearly that the function of coagulation factor which is deficient results with the risk of serious bleeding sometime results even in death.

Blood coagulation system is an extremely highly engineered system, which was not fully understood yet. Plasma concentrations of each "coagulation factor" are very low (plasma concentration of coagulation factor IX, V, and VII is around 5 μg/ml and of factor VIII is just 0.1 μg/ml). Blood coagulation occurs easily once human blood was taken out from vessel system, but it is hard to imagine the same thing happening in human blood with extremely low concentration of activated coagulation factor in flowing blood. As shown by us and others, blood cells such as platelet should play a crucial role for local activation in an efficient manner (Goto and Tomita 2014; Goto et al. 2003; Tamura et al. 2009). Blood coagulation in the test tube was prevented by the addition of calcium chelating agents. The addition of overdose calcium in combination with tissue factor or agents activating factor XII restarts blood coagulation in the test tube (Goto et al. 2003). This observation supports the idea that "blood stasis", "endothelial cell function", "contact activation", "calcium", and "tissue factor" play a crucial role for the activation of coagulation cascade in addition to their original contributing "coagulation factors". We are proposing the model of cell-based blood coagulation under blood flow condition as shown in Fig. 11.2. The essential points including our model is to consider the role of platelet cell activation, blood flow, platelet adhesion on the damaged endothelium, and blood flow for human blood coagulation.

Our knowledge regarding the component of blood coagulation occurring in human body increased over time. For further engineering "blood coagulation" in human body, all the potential players and their detailed roles should be clarified. Recently developed specific inhibitor of thrombin (factor II) and activated factor X (Xa) (Connolly et al. 2009; Schulman et al. 2009; Eikelboom et al. 2013) and Xa (Hokusai et al. 2013; Giugliano et al. 2013; Alexander et al. 2011; Connolly

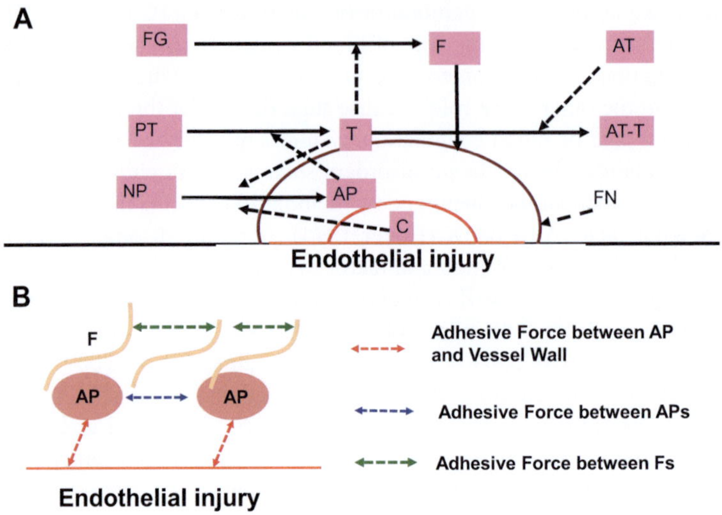

Fig. 11.2 *Model of blood coagulation.* In this model, we have considered the contribution role of platelet, coagulation cascade, and blood flow in a quantitative manner. *Panel A* shows the chemical reaction to form clot. *Panel B* demonstrated platelet adhesion, activation, and induction of fibrin, under blood flow conditions. *FG* fibrinogen, *PT* prothrombin, *NP* non activated platelet, *F* fibrin, *T* thrombin, *AP* activated platelet, *C* clot, *AT* antithrombin, *AT-T* antithrombin thrombin complex

et al. 2011; Granger et al. 2011; Agnelli et al. 2013a, b; Mega et al. 2012; Patel et al. 2011; Eriksson et al. 2008) should be a nice tool for further investigating the mechanism of blood coagulation.

11.2 Platelet Cells and Blood Coagulation

Blood coagulation occurring in human body is complicated than that occurring in test tubes outside of the body. Blood coagulation cascade is an old-established concept to explain the blood clotting occurring in test tube. Even these simple models are still helpful to understand the pathophysiological condition, such as hemophilia or strength of anticoagulantion, with the use of specific anticoagulant agents (Goto 2012).

For understanding the mechanism of blood coagulation occurring in human body, we have to realize the importance of blood cells in controlling local coagulation within human body. Indeed, platelet cells are known to play an important role not only for initial primary hemostasis (Ikeda et al. 2000) but also for controlling local activation of coagulation (Hoshiba et al. 2006; Goto et al. 2003; Tamura et al. 2009; Yamashita et al. 2006) and inflammation (Hagihara et al. 2004). Platelet cells are circulating throughout the body, interacting with vessel wall and influenced by antithrombotic potentials of endothelial cells (Goto 2008). In healthy young adults, platelet cells are not activated even though they interact with endothelial cells in whole body circulation. Local release of nitric oxide (NO) or prostaglandin (PG)-I_2 keeps platelets in quiescent states (Goto 2005). On the surface of quiescent platelet, coagulant cascade will not be activated. Upon activation, platelets express negatively charged phospholipid that is necessary for the activation of coagulation (Goto et al. 2002). Thrombin generated on the surface of activated platelet converts fibrinogen to fibrin (Tamura et al. 2009) and also further activates platelet through thrombin receptor such as protease activated receptor (PAR)-1 (Angiolillo et al. 2010). Platelets also activate coagulant system by releasing microparticles, which also contain negatively charged phospholipids (Goto et al. 2002, 2003). Our coagulation model shown in Fig. 11.2 includes activated platelet-induced procoagulant activity, but still not includes the effects of microparticles though inflammatory cells also have a crucial role in coagulation. We have shown previously that platelet has an influence on the function of inflammatory cell (Hagihara et al. 2004). Our current platelet-based coagulation model does not include the effect of inflammatory cells. Further experimental research is necessary to justify our platelet-based coagulation model.

Majority of *in vitro* coagulation experiment were conducted in the absence of blood cells. Even now we are using coagulation occurring in test tube in medical practice to measure the quality of anticoagulation therapy; e.g. activated partial thrombin time (a-PTT) for the adjustment of unfractionated heparin and prothrombin time (PT) for adjusting warfarin. These coagulation tests were conducted in the absence of blood cells. Thus, the relationship between the measured values and clinical outcomes is based only upon our previous experience. In another words, these measures might only be interesting artifacts that never reproduced in human body. Our current understanding of coagulation in human body is too primitive to be theoretical. New models should include the role of endothelial cell, platelets, and inflammatory cells. May be we have to consider the quantitative role of erythrocytes also.

11.3 Reconstruction of Blood Coagulation Within Super-Computer

The major focus in vascular engineering is to understand the exact mechanism to keep the homeostasis in circulation system. Vascular system is relatively fragile. Arterial endothelium expressed to huge change in pressure from 60 to 120 mmHg in every second. Biochemical connection among endothelial cell (shown in Fig. 11.2) is not so strong to tolerate against fluctuated blood pressure. Approximate estimation of the detaching force for endothelial cells reach to 0.6–0.7 μN under blood flow conditions. It is easy to imagine that small gaps are generated among endothelial cells in each second.

Blood cells, especially platelet cells, play a crucial role to maintain the integrity of endothelial cells against this fragility. Indeed, platelet cells accumulate immediately when the integrity of endothelial cells are lost (Kawamura et al. 2009). As shown in Fig. 11.3, we are speculating that the accumulation of platelet at the site of

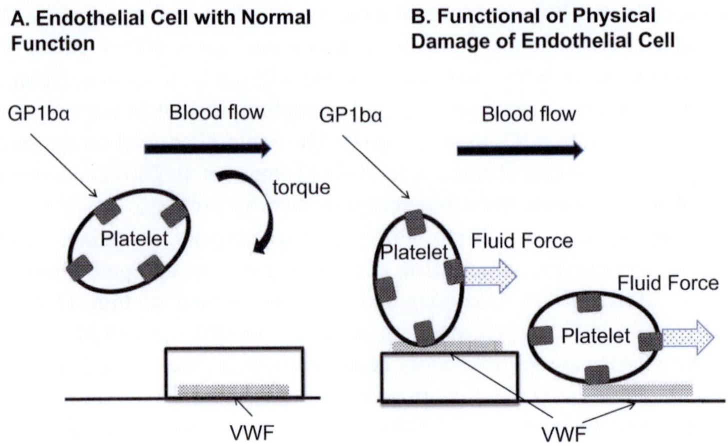

Fig. 11.3 *Mechanism of platelet adhesion on the functionally damaged endothelial cell or at sites of endothelial injury. Panel A*: the relationship between flowing platelets and vascular endothelial cells. A healthy endothelial cell covers thrombogenic subendothelial matrix such as collagen or von Willebrand factor (VWF). *Panel B*: once antithrombotic function of endothelial cells were lost, the location of VWF changed to express on the vascular lumen of endothelial cell. When endothelial cells are detached, thrombogenic subendothelial matrix such as VWF was expressed directly to the vessel. In both case, flowing platelets capture VWF through GPIbα to adhere

endothelial damage is a kind of physical phenomena. As shown in Panel A of Fig. 11.3, circulating platelets express glycoprotein (GP) Ibα with the number estimated around 15,000 molecule/platelet (Goto et al. 1995). Unlike GPIIb/IIIa, GP Ibα has a potential to bind with A1 domain of von Willebrand factor (VWF) without the activation of platelet cells. Though the blood flow is predominantly parallel to vessel wall, there are components of blood flow directing to the wall mostly due to the presence of erythrocytes. By this flow components directing to the vessel wall, some of the platelets reach to endothelial cell or at site of endothelial damage as shown in Panel B of Fig. 11.3. Once GPIbα capture VWF express on the surface of damaged endothelial cells or on the surface of damage endothelium, platelets were adhered on the vessel wall through molecular interaction between GPIbα and VWF. The interaction between GPIbα and VWF generates force strong enough to capture platelet against fluid force (Goto et al. 1998). These interaction can partly be models to reproduce in computer simulation (Tomita et al. 2014).

As modeled in Fig. 11.2 and shown by histological examination (Hoshiba et al. 2006; Yamashita et al. 2006), activation of coagulation cascade in human body occur predominantly on the surface of activated platelet (Tamura et al. 2009; Yamashita et al. 2006). Thus, we believe it reasonable to generate a computer model of blood coagulation that will be regulated by mostly by platelet activation. Further quantitative and biologically detailed data are necessary for generating a theory of "Vascular Engineering" from coagulation view point.

References

Agnelli G, Buller HR, Cohen A, Curto M, Gallus AS, Johnson M, Masiukiewicz U, Pak R, Thompson J, Raskob GE, Weitz JI, Investigators A (2013a) Oral apixaban for the treatment of acute venous thromboembolism. N Engl J Med 369:799–808

Agnelli G, Buller HR, Cohen A, Curto M, Gallus AS, Johnson M, Porcari A, Raskob GE, Weitz JI, Investigators P-E. (2013b) Apixaban for extended treatment of venous thromboembolism. N Engl J Med 368:699–708

Alexander JH, Lopes RD, James S, Kilaru R, He Y, Mohan P, Bhatt DL, Goodman S, Verheugt FW, Flather M, Huber K, Liaw D, Husted SE, Lopez-Sendon J, De Caterina R, Jansky P, Darius H, Vinereanu D, Cornel JH, Cools F, Atar D, Leiva-Pons JL, Keltai M, Ogawa H, Pais P, Parkhomenko A, Ruzyllo W, Diaz R, White H, Ruda M, Geraldes M, Lawrence J, Harrington RA, Wallentin L, Investigators A- (2011) Apixaban with antiplatelet therapy after acute coronary syndrome. N Engl J Med 365:699–708

Angiolillo DJ, Capodanno D, Goto S (2010) Platelet thrombin receptor antagonism and atherothrombosis. Eur Heart J 31:17–28

Connolly SJ, Ezekowitz MD, Yusuf S, Eikelboom J, Oldgren J, Parekh A, Pogue J, Reilly PA, Themeles E, Varrone J, Wang S, Alings M, Xavier D, Zhu J, Diaz R, Lewis BS, Darius H, Diener HC, Joyner CD, Wallentin L, Committee R-LS, Investigators (2009) Dabigatran versus warfarin in patients with atrial fibrillation. N Engl J Med 361:1139–1151

Connolly SJ, Eikelboom J, Joyner C, Diener HC, Hart R, Golitsyn S, Flaker G, Avezum A, Hohnloser SH, Diaz R, Talajic M, Zhu J, Pais P, Budaj A, Parkhomenko A, Jansky P, Commerford P, Tan RS, Sim KH, Lewis BS, Van Mieghem W, Lip GY, Kim JH, Lanas-Zanetti F, Gonzalez-Hermosillo A, Dans AL, Munawar M, O'Donnell M, Lawrence J, Lewis G, Afzal R, Yusuf S, Committee AS, Investigators (2011) Apixaban in patients with atrial fibrillation. N Engl J Med 364:806–817

Eikelboom JW, Connolly SJ, Brueckmann M, Granger CB, Kappetein AP, Mack MJ, Blatchford J, Devenny K, Friedman J, Guiver K, Harper R, Khder Y, Lobmeyer MT, Maas H, Voigt JU, Simoons ML, Van de Werf F, Investigators R-A (2013) Dabigatran versus warfarin in patients with mechanical heart valves. N Engl J Med 369:1206–1214

Eriksson BI, Borris LC, Friedman RJ, Haas S, Huisman MV, Kakkar AK, Bandel TJ, Beckmann H, Muehlhofer E, Misselwitz F, Geerts W, Group RS (2008) Rivaroxaban versus enoxaparin for thromboprophylaxis after hip arthroplasty. N Engl J Med 358:2765–2775

Giugliano RP, Ruff CT, Braunwald E, Murphy SA, Wiviott SD, Halperin JL, Waldo AL, Ezekowitz MD, Weitz JI, Spinar J, Ruzyllo W, Ruda M, Koretsune Y, Betcher J, Shi M, Grip LT, Patel SP, Patel I, Hanyok JJ, Mercuri M, Antman EM, Investigators EA-T (2013) Edoxaban versus warfarin in patients with atrial fibrillation. N Engl J Med 369:2093–2104

Goto S (2001) Role of von willebrand factor for the onset of arterial thrombosis. Clin Lab 47:327–334

Goto S (2004) Understanding the mechanism and prevention of arterial occlusive thrombus formation by anti-platelet agents. Curr Med Chem Cardiovasc Hematol Agents 2:149–156

Goto S (2005) Cilostazol: potential mechanism of action for antithrombotic effects accompanied by a low rate of bleeding. Atheroscler Suppl 6:3–11

Goto S (2008) Blood constitution: platelet aggregation, bleeding, and involvement of leukocytes. Rev Neurol Dis 5(Suppl 1):S22–S27

Goto S (2012) Monitoring of the effects of new-generation oral anticoagulants – what does it mean? Circ J Off J Jpn Circ Soc 76:596–597

Goto S, Tomita A (2014) New antithrombotics for secondary prevention of acute coronary syndrome. Clin Cardiol 37:178–187

Goto S, Salomon DR, Ikeda Y, Ruggeri ZM (1995) Characterization of the unique mechanism mediating the shear-dependent binding of soluble von willebrand factor to platelets. J Biol Chem 270:23352–23361

Goto S, Ikeda Y, Saldivar E, Ruggeri ZM (1998) Distinct mechanisms of platelet aggregation as a consequence of different shearing flow conditions. J Clin Invest 101:479–486

Goto S, Tamura N, Eto K, Ikeda Y, Handa S (2002) Functional significance of adenosine 5′-diphosphate receptor (p2y(12)) in platelet activation initiated by binding of von willebrand factor to platelet gp ibalpha induced by conditions of high shear rate. Circulation 105:2531–2536

Goto S, Tamura N, Li M, Handa M, Ikeda Y, Handa S, Ruggeri ZM (2003) Different effects of various anti-gpiib-iiia agents on shear-induced platelet activation and expression of procoagulant activity. J Thromb Haemost JTH 1:2022–2030

Granger CB, Alexander JH, McMurray JJ, Lopes RD, Hylek EM, Hanna M, Al-Khalidi HR, Ansell J, Atar D, Avezum A, Bahit MC, Diaz R, Easton JD, Ezekowitz JA, Flaker G, Garcia D, Geraldes M, Gersh BJ, Golitsyn S, Goto S, Hermosillo AG, Hohnloser SH, Horowitz J, Mohan P, Jansky P, Lewis BS, Lopez-Sendon JL, Pais P, Parkhomenko A, Verheugt FW, Zhu J, Wallentin L, Committees A, Investigators (2011) Apixaban versus warfarin in patients with atrial fibrillation. N Engl J Med 365:981–992

Hagihara M, Higuchi A, Tamura N, Ueda Y, Hirabayashi K, Ikeda Y, Kato S, Sakamoto S, Hotta T, Handa S, Goto S (2004) Platelets, after exposure to a high shear stress, induce il-10-producing, mature dendritic cells in vitro. J Immunol 172:5297–5303

Hokusai VTEI, Buller HR, Decousus H, Grosso MA, Mercuri M, Middeldorp S, Prins MH, Raskob GE, Schellong SM, Schwocho L, Segers A, Shi M, Verhamme P, Wells P (2013) Edoxaban versus warfarin for the treatment of symptomatic venous thromboembolism. N Engl J Med 369:1406–1415

Hoshiba Y, Hatakeyama K, Tanabe T, Asada Y, Goto S (2006) Co-localization of von willebrand factor with platelet thrombi, tissue factor and platelets with fibrin, and consistent presence of inflammatory cells in coronary thrombi obtained by an aspiration device from patients with acute myocardial infarction. J Thromb Haemost JTH 4:114–120

Hoyer LW (1994) Hemophilia a. N Engl J Med 330:38–47

Ikeda Y, Handa M, Murata M, Goto S (2000) A new approach to antiplatelet therapy: Inhibitor of gpib/v/ix-vwf interaction. Haemostasis 30(Suppl 3):44–52

Kawamura Y, Takahari Y, Tamura N, Eguchi Y, Urano T, Ishida H, Goto S (2009) Imaging of structural changes in endothelial cells and thrombus formation at the site of fecl(3)-induced injuries in mice cremasteric arteries. J Atheroscler Thromb 16:807–814

Mateen FJ, Shuaib A (2007) Progress in clinical neurosciences: the ‘antiplatelet’ agents and the role of the endothelium. Can J Neurol Sci Le journal canadien des sciences neurologiques 34:270–279

Mega JL, Braunwald E, Wiviott SD, Bassand JP, Bhatt DL, Bode C, Burton P, Cohen M, Cook-Bruns N, Fox KA, Goto S, Murphy SA, Plotnikov AN, Schneider D, Sun X, Verheugt FW, Gibson CM, Investigators AAT (2012) Rivaroxaban in patients with a recent acute coronary syndrome. N Engl J Med 366:9–19

Patel MR, Mahaffey KW, Garg J, Pan G, Singer DE, Hacke W, Breithardt G, Halperin JL, Hankey GJ, Piccini JP, Becker RC, Nessel CC, Paolini JF, Berkowitz SD, Fox KA, Califf RM, Investigators RA (2011) Rivaroxaban versus warfarin in nonvalvular atrial fibrillation. N Engl J Med 365:883–891

Ruggeri ZM (2007) Von willebrand factor: looking back and looking forward. Thromb Haemost 98:55–62

Schulman S, Kearon C, Kakkar AK, Mismetti P, Schellong S, Eriksson H, Baanstra D, Schnee J, Goldhaber SZ, Group R-CS (2009) Dabigatran versus warfarin in the treatment of acute venous thromboembolism. N Engl J Med 361:2342–2352

Tamura N, Kitajima I, Kawamura Y, Toda E, Eguchi Y, Ishida H, Goto S (2009) Important regulatory role of activated platelet-derived procoagulant activity in the propagation of thrombi formed under arterial blood flow conditions. Circ J Off J Jpn Circ Soc 73:540–548

Tomita A, Tamura N, Nanazawa Y, Shiozaki S, Goto S (2014) Development of virtual platelets implementing the functions of three platelet membrane proteins with different adhesive characteristics. J Atheroscler Thromb 22:201–210

Yamashita A, Sumi T, Goto S, Hoshiba Y, Nishihira K, Kawamoto R, Hatakeyama K, Date H, Imamura T, Ogawa H, Asada Y (2006) Detection of von willebrand factor and tissue factor in platelets-fibrin rich coronary thrombi in acute myocardial infarction. Am J Cardiol 97:26–28

Vascular Engineering to Make Blood-Compatible Surface

12

Hiroshi Ujiie, Yoshiaki Suzuki, and Dieter Liepsch

Abstract

In spite of tremendous efforts in the last several decades, no artificial grafts have provided a satisfactory patency rate as small-caliber vascular alternatives (less than 5 mm diameter) in clinical applications. Owing to thrombus formation at acute stage and intimal thickening caused by compliance mismatch, the long-term patency of these small-caliber vascular grafts is still disappointing. Endothelial cell seeding has been proposed to improve the blood compatibility of small-diameter vascular grafts by creating an inner lining with similar non-thrombogenic surface characteristics as native blood vessels. To obtain consistent and firm endothelial cell linings to make hybrid vascular grafts, we review special surface modification technique ion beam implantation.

Keywords

Hybrid vascular graft • ePTFE • Surface modification technique • Ion beam implantation

H. Ujiie (✉)
Department of Neurosurgery, Tokyo Rosai Hospital, 4-13-21, Ohmori-Minami, Ohta-ku, Tokyo 143-0013, Japan
e-mail: huji@tokyoh.rofuku.go.jp

Y. Suzuki
Tissue Response Control Materials Laboratory, RIKEN (The Institute of Physical and Chemical Research), Tokyo, Japan

D. Liepsch
Institut f. Biotechnik, Munich, Germany

K. Tanishita, K. Yamamoto (eds.), *Vascular Engineering*,
DOI 10.1007/978-4-431-54801-0_12

12.1 Introduction

Rough estimation for the death of human beings was that almost one third was dying due to vascular disease such as heart attack and apoplexy. With aging, vascular disease become more common, especially heart attack and apoplexy are said to be a disease for old man. Even if they do not die, these people become disabled or are limited. Their arterial lesions are usually treated medically or surgically. For example, coronary and peripheral vascular bypass procedures are considered to be performed in approximately 1,000,000 patients in each year in the world, most commonly using the saphenous vein or the internal mammary artery. Autologous vessels are the best choice for small-diameter conduit; however, in one third of these patients satisfactory vessels are not found. Therefore, the small-caliber substitute such as synthetic graft must be necessary.

In 1954, Blakemore and Voorhess first reported the treatment of 10 patients with a synthetic arterial substitute (Blakemore and Voorhees 1954). Textile conduit was fabricated from Vinyon-N fiber. Although early clinical results were satisfactory, Vinyon-N did not result in sufficient long-term biostability. Then, in spite of tremendous efforts in many years, no artificial grafts have provided a satisfactory patency rate as small-caliber vascular alternatives (less than 5 mm diameter) in clinical applications. Owing to thrombus formation at early stage and intimal thickening caused by compliance mismatch, none of these materials have proved suitable for vascular graft smaller than 5 mm in diamcter. Practically, only 45 % of standard ePTFE grafts are patent as femoropopliteal bypass grafts at 5 years, whereas autologous vein grafts have 60–80 % patency (Johnson and Lee 2000; Xue and Greisler 2003). Early clinical trial to obtain thromboresistant barrier between polymer surface and blood flow in small caliber prosthesis was failed. For larger-diameter applications, synthetic vascular grafts brought relatively satisfactory results.

Currently, only three kinds of polymers such as expanded polytetrafluoroethylene (ePTFE) and polyethylene terephthalate (PET; Dacron) and polyurethane (PU) are used to fabricate synthetic vascular grafts. ePTFE synthesized with polymer CF2 (Fig. 12.1) is chemically stable and durable biomaterials because of its fluorine side chain it does not cause immune reaction among the living bodies and stay inert life long period. The process of expanding Teflon (PTFE) creates the typical node and fiber structure that allows various sizes of porosity between fine fibrils and stretching nodes. Then, the internodal spaces of ePTFE materials were expected to function as porosity through which transmural tissue in growth occurs.

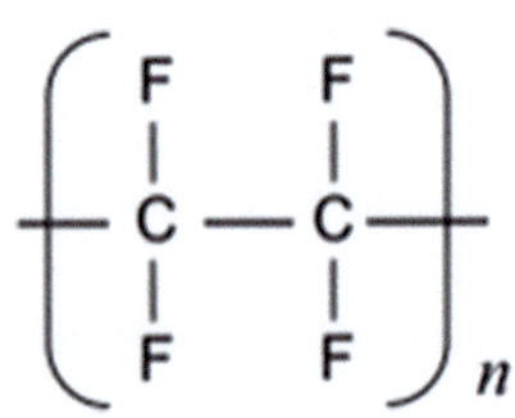

Fig. 12.1 PTFE is a polymer consists of carbon and fluorine. PTFE has a long, straight carbon back bone to which the fluorine atoms are bonded

On the other hand, Dacron consists of one basic element, namely polyester fibers. Fibers are mostly bundled into multifilament yarns that are either woven or knitted into a fabric. Weaving results in narrow spaces for tissue in growth whereas knitting leads to looser patterns containing larger spaces. Although woven material was often preferred for its high bursting strength, almost complete impermeability was also pointed out. Dacron is most commonly used for aortic replacement and to lesser extent as a conduit for small size diameter. In clinical setting, PU is used only for A-V shunt conduits of dialysis patients. The shortcoming of PU is degraded in biological environment, due to oxidation, hydrolysis, calcification, lipid sorption, and fatigue (Takahara et al. 1985). Thus, only ePTFE is used to fabricate synthetic vascular grafts less than 5 mm. ePTFE is a porous and non-degradable polymer with an electronegative luminal surface and created by thermo-mechanically expanding PTFE, which is generally known by DuPont's brand name "Teflon". Teflon possesses outstanding properties, such as excellent chemical and biological inertness, wide temperature range stability, and excellent dielectric properties. Therefore, ePTFE is widely used in both industrial and medical fields. In medical area, ePTFE is now widely used for its characteristic features of high flexibility, strength, durability, and excellent biological inertness. At present, ePTFE is really available in clinical setting as vascular graft (Soyer et al. 1972; Matsumoto et al. 1973; Kannan et al. 2005), dural substitute (Yamagata et al. 1993), pericardial substitute (Bhatnagar et al. 1998), peritoneal patch (Monaghan and Meban 1991), and suture material (Setzen et al. 1997).

Standard ePTFE graft has the fibril length or intermodal distance is about 30–90 μm (Fig. 12.2) and is highly hydrophobic, then resulting foreign surfaces re-endothelialize poorly and are chronically inflamed. A luminal surface of ePTFE

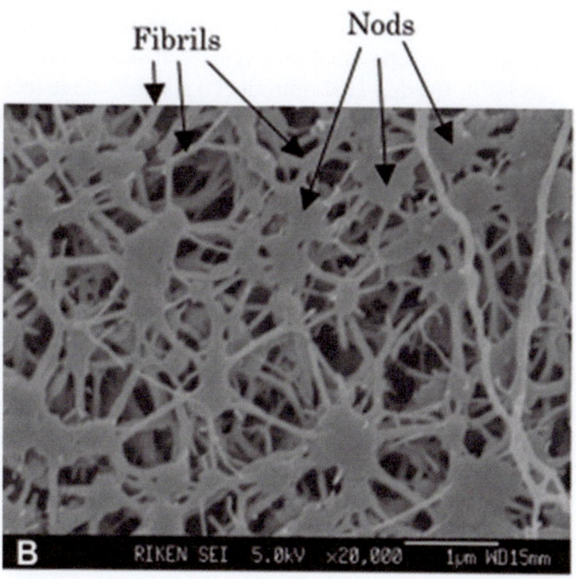

Fig. 12.2 SEM images of standard ePTFE shows a micro-porous structure consisting of submicron sized nodes and fibrils of PTFE (× 20,000)

chronically activates the components of the coagulation system. Even after long-time implantation, a persistent foreign body response dominates internodal space. Therefore, transmural migration of ECs never happens in human beings. The vast majority of synthetic grafts are not so impervious that transmural ingrowth is impossible and none of clinically used vascular graft develops a neo-intima, except for sporadically observed small islands of endothelium (Greisler 1990). Furthermore, micro-thrombus continuously builds up on the luminal surface due to constant platelet activation. In larger grafts such micro-thrombus may be washed out; however in small caliber grafts both flow velocity and volume are not enough to remove away these micro-thrombus, then often results in graft occlusions. Neoarterial wall regeneration based on tissue ingrowth from the outer surface and transanastomotic ingrowth from parent artery were only observed in animal experimental studies.

Blood compatible small caliber synthetic grafts are urgently required.

12.2 Ideal Biocompatible Material

Ideal vascular prosthesis must mimic the native arterial wall to serve as antithrombotic barrier as well as viscoelastic strength for long-term durability. To improve the biocompatibility of small diameter ePTFE vascular graft is to facilitate the formation of endothelial lining on the inner surface. It must be concluded that ideal biocompatible material is hybrid vascular graft covered with a luminal monolayer of autologous endothelial cells that is a vital component of normal vasculature, have important antiplatelet, anticoagulant, and pro-fibrinolytic properties. Hybrid artificial grafts, based on *in vitro* endothelialization technology enabling complete endothelial coverage prior to implantation, have provided a much higher patency rate than none-cell-seeded ones.

In 1978, Herring et al. introduced a single-stage technique whereby venous ECs were seeded onto the grafts with enhanced patency in a canine model (Herring et al. 1978). Although the hybrid artificial graft was a promising concept, a subsequent clinical studies revealed that insufficient initial cell density, poor adhesion under flow, and failure to achieve confluence after implant attributed a disappointing outcomes (Zilla et al. 1987; Herring et al. 1994).

Firm adhesion and proliferation of ECs on the ePTFE surface in life-long time is mandatory for success of hybrid artificial grafts. In developing hybrid vascular graft, the strength of endothelial cell attachment to the surface is important. Many investigations carried out on surface modification relative to the development of cell adhesion whose techniques include ion implantation (Buchanan et al. 1990; Suzuki et al. 1991) and chemical modifications (Grinnel and Feld 1981; Massia and Hubbell 1991).

12.3 Surface Modification Technique

The biological responses to biomaterials are largely controlled by their surface properties while bulk structure of the materials is unchanged. Therefore, it is attractive to modify only thin surface layer of materials. Surface modification should induce sufficient uniformity, stability, durability, and functionality, but minimal degradation of bulk properties.

Plasma is the fourth state of matter, formed when energy ionizes gas molecules, causing the electrons to separate from the atom. Ionization or excitation can be achieved by applying energy in the form of radiofrequency or microwave energy, or electrons from a hot filament discharge. When generated in a vacuum chamber, the highly reactive plasma species collide with the substrate and generate new reactive chemical groups on the surface of previously inert material. In this way, plasma can be used to modify the surface of polymers to create new surfaces with desired properties.

Surface can either be modified with deposited layer of nanometer-range thickness or nanostructures can be etched onto surface. The thickness of the deposited layer depends on deposition time and can vary widely, but it is usually optimal between 20 and 100 nm. ePTFE and Dacron were frequently selected for modification. Initially, improvements in cell binding and blood contact were seen as a result of dramatic reduction in hydrophobicity associated with plasma treatment.

In addition to control wettability, roughness, and surface chemistry, plasma treatment also allows the bioactivation of previously inert materials through the covalent attachment of biomolecules (Bilek and McKenzie 2010). Interactions of adhesive proteins such as fibrin, collagen, and fibronectin with an inert surface determine biocompatibility. This approach aims to achieve controlled biointegration through a covalently attached bioactive protein layer. However the chemical intermediates are changeable, the initial activation of the surface and subsequent linking steps are limited. Complete and life-long consistent surface modification must be necessary.

Ion beam irradiation, which is also called "ion irradiation", "ion (beam) implantation" (Fig. 12.3), "ion (beam) bombardment", or "ion (beam) etching" from its physical events or results, is a surface modification method using high kinetic energy (from 10 keV to MeV) ion beam. Ion implantation, a technique used in the manufacture of semiconductors and for improving the surface properties of metals, has been applied to the surface modification of polymers (Bilek and McKenzie 2010; Marletta 1990).

While the bulk structure properties are unchanged. This is of particular benefit to implantable devices requiring specific mechanical strength as biocompatible material. Ion implantation causes radiation and etching effect. The radiation effects break original chemical bonds and along the ion tracks and produce amorphous carbon structures on the ePTFE surfaces. Etching effect of ion beam implantation is the strongest in surface modification techniques. Therefore, nodes and fibrils were destroyed in 5–10 μm depth. Ion beam implantation never improves water contact angle. In general, cell attachment and spreading are significantly greater on

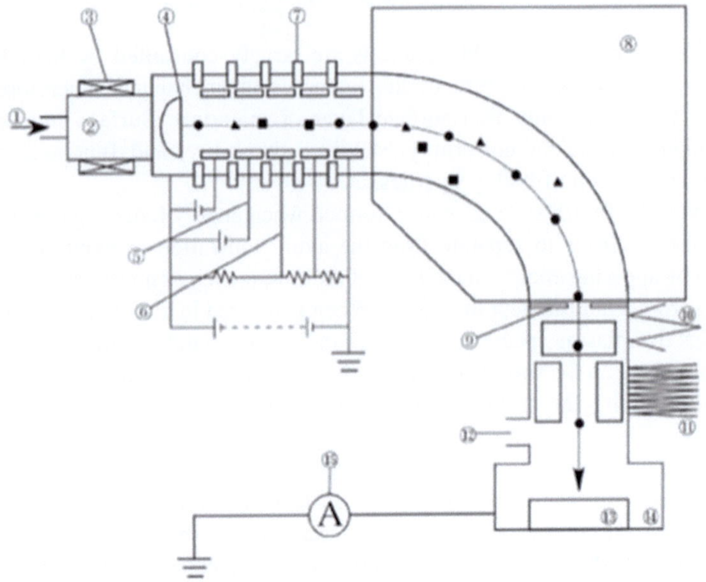

Fig. 12.3 Schematics of ion beam implanter. (*1*) Gas head (*2*) Plasma chamber (*3*) RF coil (*4*) Ion extraction electrode (*5*, *6*) Ion acceleration (*7*) Insulator (*8*) Electric magnet (*9*) Mass analyzing slit (*10*) Y-scanner (*11*) X-scanner (*12*) exhaust port (*13*) Target (*14*) Sample chamber (*15*) current integrator. Ions are extracted from the plasma chamber and electrostatically accelerated to a high energy. Ions of specific mass and charge (●) are selected from light or highly-charged ions (■) and heavy or low-charged ions (▲) in the magnetic field. The mass-separated ion beam is continuously scanned to uniformly impinge the surface of target material

hydrophilic surfaces, with moderately hydrophilic surfaces promoting the highest level of cell attachment. However, it was found that without any improvement of the water contact angle, cell attachment remarkably enhanced and cell proliferations were also observed (Fig. 12.4). One possible explanation for the attachment of ECs to the ion implanted surface could be the preferential adhesion of cell adhesive proteins from the serum containing medium to ion implanted area. (Marletta 1990; Suzuki et al. 1992) The other explanation is carbon deposition on polymer surfaces (carbonized surface). It has been reported that the adsorption of human serum albumin and fibrinogen on carbon-deposited plate is the same as that on hydrophobic surfaces such as polyethylene (Marletta 1990; Suzuki et al. 1992; Lee et al. 1993; Kaibara et al. 1996). Carbon materials such as vitreous carbon and pyrolytic carbon have been reported to be superior for biocompatibility. Thus, ECs can anchor the ion beam implanted ePTFE surface through both adhesive protein and amorphous carbon. The adhesion, growth, and proliferation of endothelial cells are anchorage dependent. Thus, ideal polymer surface where ECs anchoring and surviving long time can be created by ion beam implantation technique. The strategy that promotes a functional endothelia lining yet met with some difficulties.

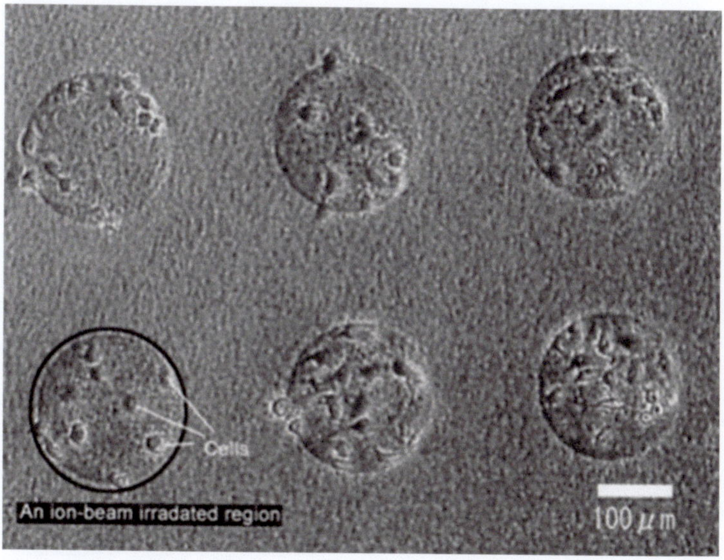

Fig. 12.4 Phase contrast micrographs of cultured cells after 24 h incubation. Cells adhered on the ion beam implanted ePTFE surfaces shown as circular areas. The cells did not anchor non-implanted areas

Problem is how to seed ECs on the ion beam implanted ePTFE tubes. Another problem is activated ECs may induce smooth muscle cell proliferation and finally result in intimal hyperplasia.

Clinical trial using endothelialized vascular graft using ion beam implanted ePTFE must be waited.

12.4 The Cell Attachment Peptide

Without surface modification, ePTFE grafts show attachment of only 10 ± 7 % of applied ECs, with EC retention of only 4 ± 3 % (Kent et al. 1992; Tseng and Edelman 1998). One method of promoting EC attachment is to coat the surface with a thin protein layer such as fibronectin. However, the adsorption and adhesion of the protein on unmodified ePTFE is limited due to interfacial characteristics. Another approach focuses on the modification of graft surfaces with domains in naturally occurring extracellular matrix (ECM) proteins, especially the Arginine-Glycine-Aspartic Acid (RGD) sequence, for specific interactions with integrin cell surface receptors (Walluscheck et al. 1996; Elloumi et al. 2006; Larson et al. 2006). RGD peptides immobilized directly onto many materials have demonstrated enhanced EC attachment. Peptides are immobilized covalently to the hydrophobic head group; a fluorocarbon tail acts as a hydrophilic anchor to the underlying fluorocarbon surface.

Endothelial attachment on ePTFE surface can be improved by specific bonding between adhesion promoting RGD-peptides and the synthetic material. Enhanced ECs attachment to the endothelialized vascular graft is necessary to prevent ECs ablation due to high shear stress caused by blood flow. Poor attachment leads to anoikis, where cells detach and undergo apoptosis. Further studies are necessary whether ECs growth kinetics on RGD-peptides pretreated ePTFE vascular grafts are suitable in clinical setting.

12.5 Ultimate Surface Modification for Small Caliber Artificial Vessel

Ion beam implanted ePTFE is considered as one of the best biocompatible surfaces where ECs adhere, anchor, and proliferate. Although hybrid vascular graft is one of the solutions for small caliber vascular graft, there is several problems; sufficient initial cell density after ECs seeding will take several weeks and also there must be some risk of infection at real clinical setting. The second best is to create the vascular graft that must be anti-thrombogenic initially and then can be covered with ECs after implanting into human bodies. Suzuki et al. (1994, 2005) have found the characteristic condition of the ion beam implantation to collagen surface that inhibits platelet adhesion and simultaneously promotes endothelialization based on tremendous experimental studies. Ion beam implantation with He^+(150 keV, 1×10^{14} ions/cm^2) to collagen surface succeeded to make the blood compatible small caliber ePTFE grafts (Fig. 12.5). A subsequent animal study (Kurotobi 2003) revealed that He^+ ion beam implanted collagen-coated ePTFE vascular grafts were patent up to 240 days that replaced mongrel dog carotid artery.

Collagens are the most abundant protein in the body. Collagens are triple-helical structural proteins. Collagens are one of major structures of basement membrane

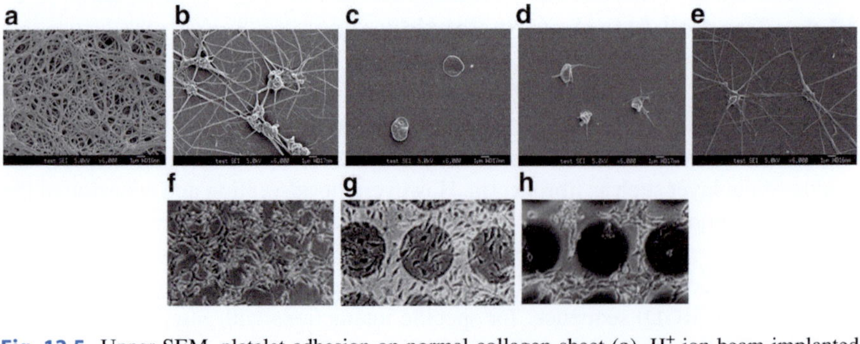

Fig. 12.5 Upper SEM, platelet adhesion on normal collagen sheet (**a**), H^+ ion beam implanted collagen surfaces with fluences of 1×10^{13} (**b**), 1×10^{14} (**c**), 1×10^{15} (**d**) and 1×10^{16} (**e**) H^+ ion/cm^2. Lower SEM, endothelial cell attachment to H^+ ion beam implanted collagen surfaces with fluences of 1×10^{13} (**f**), 1×10^{14} (**g**) and 1×10^{15} (**h**) ions/cm^2 H^+ ion beam with fluences of 1×10^{14} H^+ ion beam implanted collagen surfaces showing inhibition of platelet adhesion but still endothelial attachment retained

and play an important role to give strength to the vessel wall and function as scaffold for endothelial cells. When endothelial cells are injured, exposure collagen interacts with platelets that play a critical role to lead to thrombus formation and cessation of bleeding at sites of damage to the vasculature and to finally induce healing process. Collagen is a key extracellular matrix to start hemostasis and cell adhesion. GFOGER is an important amino sequence corresponds to platelet adhesion and RGD amino sequences that is scattered in all collagen bands are responsible for cell adhesion. RGD of only three amino acids, Arg-Gly-Asp (RGD), that show the minimal cell-recognition function through integrin on the cell surface. It is speculated that the GFOGER amino acid sequence which occupy the surface of collagen layer and corresponds to platelet adhesion was broken, while the RGD sequence which corresponds to endothelial cells attachment (Cutler and Garcia 2003; Chen et al. 2003; Jang et al. 2004) was not destroyed by He^+ ion implantation with a fluence of 1×10^{14} ions/cm^2. However, further study is necessary to verify this hypothesis.

References

Bhatnagar G, Fremes SE, Christakis GT, Goldman BS (1998) Early results using an ePTFE membrane for pericardial closure following coronary bypass grafting. J Card Surg 13:190–193

Bilek MM, McKenzie DR (2010) Plasma modified surfaces for covalent immobilization of functional biomolecules in the absence of chemical linkers: towards better biosensors and a new generation of medical implants. Biophys Rev 2:55–65

Blakemore AH, Voorhees AB Jr (1954) The use of tubes constructed from Vinyon N cloth in bridging arterial defects; experimental and clinical. Ann Surg 140:324–334

Buchanan RA, Lee IS, Williams JM (1990) Surface modification of biomaterials through noble metal ion implantation. J Biomed Mater Res 24:309–318

Chen CS, Alonso JL, Ostuni E et al (2003) Cell shape provides global control of focal adhesion assembly. Biochem Biophys Res Commun 307:355–361

Cutler SM, Garcia AJ (2003) Engineering cell adhesive surfaces that direct integrin α2β1 binding using a recombinant fragment of fibronectin. Biomaterials 24:1759–1770

Elloumi I, Kobayashi R, Funabashi H et al (2006) Construction of epidermal growth factor fusion protein with cell adhesive activity. Biomaterials 27:3451–3458

Greisler HP (1990) Interactions at the blood/material interface. Ann Vasc Surg 4:98–103

Grinnel F, Feld MK (1981) Adsorption characteristics of plasma fibronectin in relationship to biological activity. J Biomed Mater Res 15:363–381

Herring M, Gardner A, Glover J (1978) A single-staged technique for seeding vascular grafts with autogeneous endothelium. Surgery 84:498–504

Herring M, Smith J, Dalsing M et al (1994) Endothelial seeding polytetrafluoroethylene femoral popliteal bypasses: the failure of low-density seeding to improve patency. J Vasc Surg 20:650–655

Jang JH, Hwang JH, Chung CP (2004) Production of recombinant human tenascin-C module containing a cell adhesion recognition motif of RGD. Biotechnol Lett 25:1831–1835

Johnson WC, Lee KK (2000) A comparative evaluation of polytetrafluoroethylene, umbilical vein, and saphenous vein bypass grafts for femoral-popliteal above-knee revascularization: a prospective randomized Department of Veterans Affairs cooperative study. J Vasc Surg 32:268–277

Kaibara M, Iwata H, Wada H et al (1996) Promotion and control of selective adhesion and proliferation of endothelial cells on polymer surface by carbon deposition. J Biomed Mater Res 31:429–435

Kannan RY, Salacinski HJ, Butler PE et al (2005) Current status of prosthetic bypass grafts: a review. J Biomed Mater Res B Appl Biomater 74B:570–581

Kent KC, Oshima A, Whittemore AD (1992) Optimal seeding conditions for human endothelial cells. Ann Vasc Surg 6:258–264

Kurotobi K (2003) Cell induction technique by ion irradiated collagen for development of a small diameter artificial graft. Trans Mater Res Soc Jpn 28:489–494

Larson CC, Kligman F, Kottke-Marchant K, Marchant RE (2006) The effect of RGD fluorosurfctant polymer modification of ePTFE on endothelial cell adhesion, growth, and function. Biomaterials 27:4846–4855

Lee JS, Kaibara M, Iwaki M et al (1993) Selective adhesion and proliferation of cells on ion-implanted polymer domains. Biomaterials 14:958–960

Marletta G (1990) Chemical reactions and physical property modifications induced by KeV ion beams in polymers. Nucl Instrum Methods B46:295–305

Massia SP, Hubbell JA (1991) Human endothelial cell interactions with surface-coupled adhesion peptides on a nonadhesive glass substrate and two polymeric biomaterials. J Biomed Mater Res 25:223–242

Matsumoto H, Hasegawa T, Fuse K et al (1973) A new vascular prosthesis for a small caliber artery. Surgery 74:519–523

Monaghan RA, Meban S (1991) Expanded polytetrafluoroethylene patch in hernia repair: a review of clinical experience. Can J Surg 34:502–505

Setzen G, Gavin E, Williams F (1997) Tissue response to suture materials implanted subcutaneously in a rabbit model. Plast Reconstr Surg 100:1788–1795

Soyer T, Lempinen M, Cooper P, Norton L, Eiseman B (1972) A new venous prosthesis. Surgery 72:864–872

Suzuki Y, Kusakabe M, Akiba H et al (1991) In vivo evaluation of antithrombogenicity for ion implanted silicone rubber using indium-111-tropolone-platelets. Nucl Instrum Methods B59 (60):698–704

Suzuki Y, Kusakabe M, Akiba H et al (1992) In vivo evaluation of antithrombogenicity and surface analysis of ion implanted silicone rubber. Radiat Phys Chem 39:553–560

Suzuki Y, Kusakabe M, Kaibara M et al (1994) Cell adhesion control by ion implantation into extracellular matrix. Nucl Instrum Methods B91:588–592

Suzuki Y, Iwaki M, Takahashi N et al (2005) In vitro and in-vivo study of He + ion irradiated collagen for development of small diameter stent graft material. Nucl Instrum Methods B 206:538–542

Takahara A, Tashita J, Kajiyama T et al (1985) Effect of aggregation state of hard segment in segmented poly(urethaneureas) on their fatigue behavior after interaction with blood components. J Biomed Mater Res 19:13

Tseng DY, Edelman ER (1998) Effects of amide and amine plasma-treated ePTFE vascular grafts on endothelial cell lining in an artificial circulatory system. J Biomed Mater Res 42:188–198

Walluscheck KP, Steinhoff G, Kelm G et al (1996) Improved endothelial cell attachment on ePTFE vascular grafts pretreated with synthetic RGD-containing peptides. Eur J Vasc Endovasc Surg 12:321–330

Xue L, Greisler HP (2003) Biomaterials in the development and future of vascular grafts. J Vasc Surg 37:472–480

Yamagata S, Goto K, Oda Y et al (1993) Clinical experience with expanded polytetrafluoroethylene sheet used as an artificial dura mater. Neurol Med Chir 33:582–585

Zilla P, Fasol R, Deutsch M et al (1987) Endothelial cell seeding of polutetrafluoroethylene vascular grafts in humans: a preliminary report. J Vasc Surg 6:535–541

Vascular Engineering of Circulatory Assist Devices

13

Masahiro Nishida

Abstract

Heart failure is a condition in which the heart cannot pump sufficiently because its contractile force has deteriorated. Patients with heart failure can sometimes undergo heart transplantation, but in number these patients comprise less than 10 % of patients actually requiring heart transplantation. Thus, mechanical circulatory assistance plays an important role in substituting for heart transplantation wherein the pump function of the heart is assisted by an artificial blood pump called a ventricular assist device. On the other hand, cardiopulmonary bypass is a form of extracorporeal circulation that temporarily takes over the function of the heart and lungs to maintain the circulation of blood and the oxygen content of the body during surgery for heart failure and aneurysms of the thoracic aorta. This chapter describes the vascular engineering of circulatory assist devices. Particular emphasis is placed on recent progress in ventricular assist devices and cardiopulmonary bypass pumps. These important medical devices assist with human circulation at either the chronic or the acute phase. Because these devices are derived from an industrial pump, a great many studies have been conducted not only from the medical perspective but also from industrial and engineering perspectives. In this chapter, current ventricular assist devices and cardiopulmonary bypass pumps, their specifications, their classifications, and methods for their design and evaluating are presented, including flow analysis inside the pump to optimize the geometry.

Keywords

Ventricular assist device • Cardiopulmonary bypass pump • Pump design • Pump evaluation

M. Nishida (✉)
National Institute of Advanced Industrial Science and Technology, Tsukuba, Ibaraki, Japan
e-mail: masahiro.nishida@aist.go.jp

K. Tanishita, K. Yamamoto (eds.), *Vascular Engineering*,
DOI 10.1007/978-4-431-54801-0_13

13.1 Introduction

This chapter describes the vascular engineering of circulatory assist devices. Particular emphasis is placed on recent progress in ventricular assist devices and cardiopulmonary bypass pumps. These important medical devices assist with human circulation at either the chronic or the acute phase. Because these devices are derived from an industrial pump, many studies have been conducted not only from the medical perspective but also from industrial and engineering perspectives. Some achievements have been commercialized; other designs are still being studied to confirm better compatibility or lifesaving capabilities for humans (Reul and Akdis 2000; Yamane 2002; Takatani et al. 2005; Joyce et al. 2012; Kyo 2014).

13.2 Ventricular Assist Device (VAD)

Heart failure, a condition in which the heart cannot pump sufficiently because of deterioration of its contractile force, is one of the most frequent causes of death in advanced countries. Patients with heart failure can sometimes undergo heart transplantation, but there are only about 4000 heart transplantations per year in Western countries, and only about 20 per year, including those as a passage, in Japan. Those numbers are less than 10 % of patients actually requiring heart transplantation. Thus, mechanical circulatory assistance plays an important role in substituting for heart transplantation in which the pump function of the heart is assisted by an artificial blood pump called a ventricular assist device (VAD). The main purpose of a VAD is as a bridge to heart transplantation. The waiting period for patients until heart transplantation is approximately 3 months, on average, in Western countries, and more than 1 year in Japan. At present, durability for longer than 5 years is required for all VADs. Further longer-term durability is desired for these devices, as the cases of destination therapy increase assuming long-term use of VADs for patients who cannot be adapted for heart transplantation. It is also necessary to enable long-term home medical care that is more adaptable to patients that are undergoing outpatient rehabilitation while maintaining the in-hospital function for patient management. Several things are essential for this purpose, such as an implantable pump, a portable driver, and a long-life battery.

13.2.1 Specifications of VADs

The flow rate of blood circulating in an adult at rest is 80 ml/min per 1 kg body weight, which means for a person weighing 60 kg, the blood flow rate is approximately 5 l/min and the pressure head of the heart is 100 mmHg. The target design parameters of a VAD, especially a left VAD, are usually defined as a flow rate of 5 l/min and a pressure head of 100 mmHg as average values. However, the natural pulsatile flow generated by a heart should be reflected in the pump flow because the VAD bypasses the left ventricle and aorta. Although the effect of the absence of

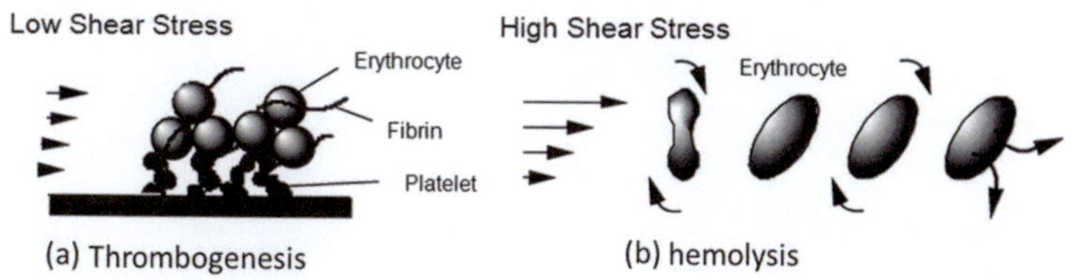

Fig. 13.1 Effect of flow on thrombogenesis and hemolysis

pulsatile flow on patients wearing continuous-flow VADs has been studied, long-term circulatory assistance was realized with continuous-flow VADs, and their effectiveness as well as that of wearing pulsatile-flow VADs, was verified by Golding et al. (1980).

The most important concern when designing VADs for long-term use is to prevent thrombus formation (blood coagulation). As thrombi usually form in areas where the shear rate is low (Fig. 13.1a), fluid dynamic design of the pump should prevent flow stagnation. More specifically, it is necessary to prevent areas of flow stagnation that cause thrombus formation, and the areas of flow stagnation should be washed out via shear stress on the surface of blood pumps. Another important aspect is prevention of hemolysis (erythrocyte fracture) caused by blood flowing through the pump. Under excessive shear stress, the erythrocyte membrane ruptures, and hemoglobin is released from the cells (Fig. 13.1b). Therefore, it is necessary to estimate the dynamic stimuli affecting erythrocytes and to design VADs that do not cause excessive shear stress.

13.2.2 Classification of VADs

13.2.2.1 Pulsatile-Flow VADs

Figure 13.2 shows various pulsatile-flow VADs. Two extracorporeal VADs [the Nipro VAD (Nipro, Japan (previously made by Toyobo, Japan)) and the Zeon VAD (Nippon Zeon and Aisin-seiki, Japan)] were covered under the health insurance system in Japan. These devices were pneumatic pulsatile-flow VADs, in which the blood is pumped via a diaphragm or sac motion caused by pneumatic pressure. At present, the Nipro VAD and the AB5000 (Abiomed, USA) are covered under the health insurance system as pulsatile-flow VADs in Japan. Pulsatile-flow VADs may be preferred from a physiological perspective, because they produce good clinical results, and the longest use has exceeded 3 years. However, these VADs require hospital management because they are extracorporeal VADs, which limits the activities of the patient, although the Movart NCVC (Senko Medical Instrument, Japan) is used with a portable compact driver that affords patients greater temporal mobility. Currently, pneumatic VADs are most often adapted as VADs for neonates and infants for reasons of their good hemocompatibility. Excor Pediatric (Berlin Heart, Germany) manufactures a number of different-sized pneumatic VADs.

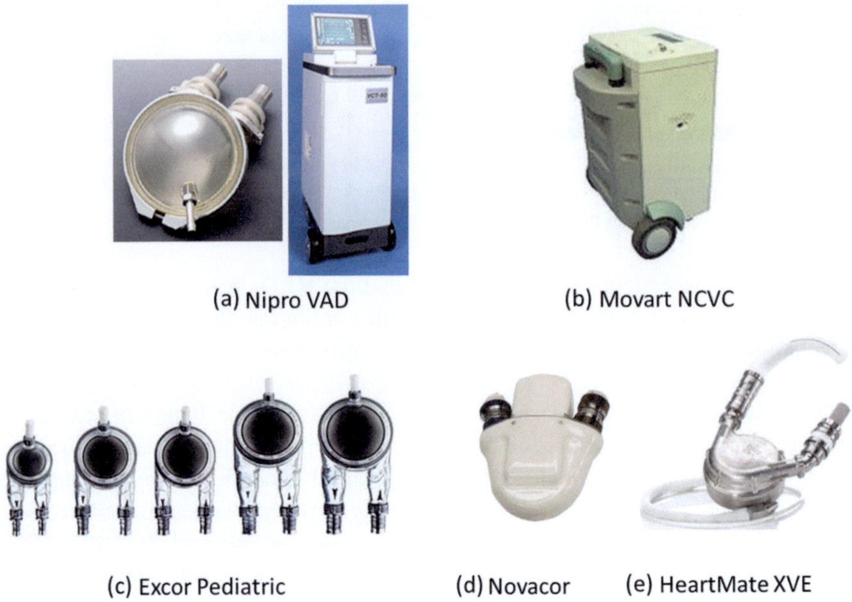

Fig. 13.2 Pulsatile-flow ventricular assist devices (VADs)

Two implantable VADs were used worldwide mainly in 1990s: the Novacor (World Heart, USA) and the HeartMate XVE (Thoratec, USA). Both are pulsatile-flow VADs that pump blood using a polyurethane blood sac moved by a magnetically driven pusher-plate and two artificial valves. These VADs have been used as the bridge to transplantation for more than 5000 patients worldwide because they dramatically improve the patients' quality of life. However, continuous-flow VADs, which are rotary pumps, have been developed because they are small, lightweight, and silent, and are expected to have longer-term durability. Pulsatile VADs are large and heavy (>1 kg), and the opening and closing of their valves is excessively noisy.

13.2.2.2 Continuous-Flow VADs

Figure 13.3 shows various continuous-flow VADs. These devices are presently the more widely implanted VADs because they have several advantages: a small size, no need for artificial valves, and increased reliability originating from their simple structure. The continuous-flow VADs originate from cardiopulmonary bypass (CPB) pumps used for heart operations. VADs are managed with the minimum use of anticoagulants because of the chronic use of the device, whereas CPB pumps are managed with the normal use of anticoagulants because these devices are used temporarily for a number of hours. Thus, anti-thrombogenicity is quite important for VADs. In light of this concern, development of continuous-flow VADs focuses on how the shaft and seal can be omitted or how the anti-thrombogenicity can be maintained around the bearing area at the center of rotation where the velocity is

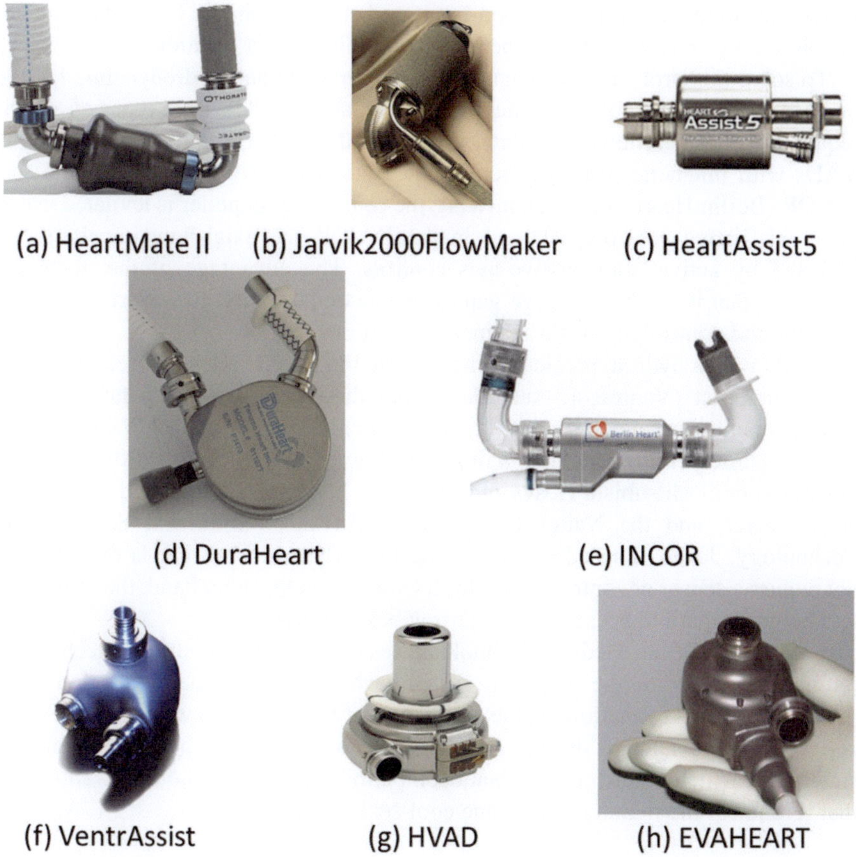

Fig. 13.3 Continuous-flow VADs

very low. Determining how a large gap can be maintained between all material surfaces including the bearing gap in the pump is also important, not only because a large gap deceases hemolysis as a result of the decrease in high shear stress caused by a narrow gap, but also because it increases anti-thrombogenicity to secure the washout leakage flow around the center of rotation, which prevents the penetration of foreign substances from outside the pump.

Development of the HeartMate II (Thoratec) and the Jarvik2000FlowMaker (Jarvik Heart, USA) began at the end of the 1990s; these were the earliest among the various VADs available today. These early VADs were adapted for practical use and were heavily marketed as implantable VADs. Their axial-flow impellers are supported by two ball-cup bearings or pivot bearings. The advantage is the small size of the axial-flow pumps, which are ϕ1.2 in. × 2.8 in. for the HeartAssist5 (MicroMed Cardiovascular, USA), ϕ1 in. × 2 in. for the Jarvik2000FlowMaker, and ϕ43 mm × 81 mm for the HeartMate II. The HeartAssist5 is used as a pediatric VAD. The disadvantage is that these axial-flow pumps have mechanical contacts, in which the contact point in the blood-immersed structure should be washed out by the blood; this may decrease the hemocompatibility at that location.

To solve this problem, the magnetic levitation VAD and hydrodynamic levitation VAD were developed to achieve a non-contact impeller. Both types of VADs require the consideration of unbalanced forces and impeller weight for their design. VADs with magnetic bearings include the DuraHeart (Terumo, Japan), and the INCOR (Berlin Heart). In the DuraHeart, the centrifugal impeller is levitated using three-axis active controls, whereas in the INCOR the axial-flow impellers are levitated by active- and passive-axis controls. The advantage of the magnetic bearing is that it can have a large gap for hemocompatibility; however, it requires a sensor and control circuit for magnetic levitation.

VADs with a hydrodynamic bearing include the HVAD (HeartWare, USA) and the VentrAssist (Ventracor, Australia). Both these VADs adopt a tapered-land hydrodynamic bearing and centrifugal impeller. There is also a VAD with a hydrodynamic bearing that adopts the axial-flow impeller developed through collaboration with Mitsubishi Heavy Industries, the National Cerebral and Cardiovascular Center, and the National Institute of Advanced Industrial Science and Technology, Japan. The advantage of the hydrodynamic bearing is that it does not require a sensor or control circuit for levitation. On the other hand, the bearing is less hemocompatible because of its narrow bearing gap.

EVAHEART (Sun Medical Technology Research, Japan) adopts the mechanical journal bearing with the cool-seal system, which was originally developed to prevent thrombus formation via the circulation of pure water to wash the mechanical seal. The advantages of this VAD include no limit on the gap design and its small size, which result in good hemocompatibility, whereas the disadvantage is that it requires the maintenance-prone cool-seal system.

13.3 Cardiopulmonary Bypass (CPB) Pump

Cardiopulmonary bypass is a form of extracorporeal circulation that temporarily takes over the function of the heart and lungs to maintain the circulation of blood and the oxygen content of the body during surgery for heart failure and aneurysms of the thoracic aorta. Recently, the westernization of the everyday lifestyle in Japan has led to an increase in ischemic heart failure requiring coronary artery bypass surgery. Off-pump coronary artery bypass grafting tends to increase during coronary artery bypass surgery of the beating heart without extracorporeal circulation. Furthermore, there has been a recent focus on percutaneous surgery to expand the narrow segment of the coronary artery using a stent without opening the heart chamber. However, coronary artery bypass surgery with extracorporeal circulation and valvular disease of the heart require opening the chambers of the heart, and this necessitates extracorporeal circulation via cardiopulmonary bypass.

Extracorporeal circulation and blood oxygenation have been experimentally investigated since the nineteenth century. Gibbon was the first to successfully utilize cardiopulmonary bypass for an animal during general perfusion in 1937

(Gibbon 1937), and also the first to successfully utilize cardiopulmonary bypass for direct visual repair of an arterial septal defect in 1953 (Gibbon et al. 1953). The latter half of the twentieth century saw remarkable developments in cardiopulmonary bypass support for heart surgery. The types of adaptable patients expanded from only infants and adults to neonates and elderly people, and the target diseases changed from simple to complicated and severe. The survival rate for patients undergoing cardiopulmonary bypass is increasing dramatically.

The main setup for cardiopulmonary bypass consists of an oxygenator, a heat exchanger, a cooling water supply system, pump, circuit, reservoir, and an artery filter. Here, focus is placed on the pump.

13.3.1 Specifications of CPB Pump

There are roller pumps and centrifugal pumps for the CPB pump. The roller pump is commonly used, and the centrifugal pump is used when the flow rate is higher than 2 l/min because it is difficult to control a flow rate lower than 0.5 l/min with these pumps.

13.3.1.1 Roller Pump

A roller pump was developed for transfusion by DeBakey in 1934 (DeBakey 1934). In principle, the roller rotates to massage tubing peristaltically and gently propels the blood through the tubing. This pump is commonly used for cardiopulmonary bypass because of its simple structure, robustness, simple operation, and reliability. Current pumps use two rollers, and the tubing is made of soft polyvinyl chloride resin. Initially, there was concern regarding peripheral circulatory failure as tissue perfusion is performed by nonphysiological blood flow. Later, however, 4–5 h of tissue perfusion by extracorporeal circulation with a roller pump was found to be without problems. The pump is now used safely for open-heart surgery.

13.3.1.2 Centrifugal Pump

A centrifugal blood pump was developed by Rafferty in 1968 (Rafferty et al. 1968). This pump ultimately became a commercialized pump named the "BioPump" in 1976. The impeller rotates at a rotational speed of several thousand revolutions per minute. Negative pressure is generated at the pump inlet (impeller rotational center), and positive pressure is generated at the pump outlet (impeller circumference) according to the spiral principle. Although a roller pump has been mainly used for open-heart surgery, the use of a centrifugal pump is gaining impetus. One of the reasons is because centrifugal pumps idle even if there is no blood in the pump, thus eliminating the possibility of air bubbles.

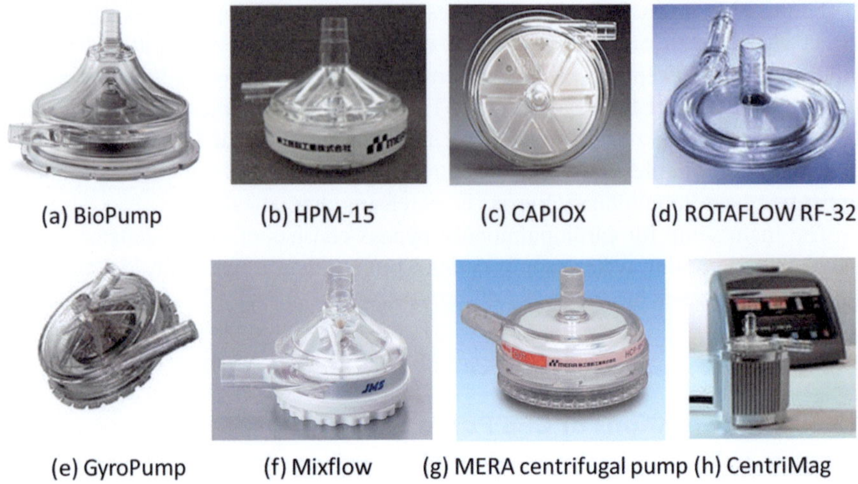

Fig. 13.4 Cardiopulmonary bypass (CPB) pumps

13.3.2 Classification of CPB Pumps

The CPB pumps usually use magnetic coupling by which the rotating force of the motor is transferred to the impeller in the pump head; this facilitates easy pump head change without any direct connection.

Figure 13.4 shows various CPB pumps. A journal bearing with a seal and a pivot bearing are usually used for CPB pumps. The BioPump (Medtronic) uses a journal bearing with a seal and pumps blood by rotation of a cone-shaped impeller. The HPM-15 (Senko Medical Instrument Manufacturing, Japan), and the CAPIOX (Terumo) use a ball bearing with a seal and pump blood by rotation of an impeller with blades. The RotaFlow RF-32 (Maquet, Germany), the MERA centrifugal pump (Senko Medical Instrument Manufacturing), the GyroPump (Medtronic), and the Mixflow (JMS, Japan) use pivot bearings. The RotaFlow RF-32 and the MERA centrifugal pump use a single pivot bearing and a centrifugal impeller, whereas the GyroPump and the Mixflow use double pivot bearings. The GyroPump uses a centrifugal impeller and the Mixflow uses a mixed-flow impeller. Currently, the magnetic bearing is used in the CentriMag (Thoratec) as a CPB pump.

13.4 Design of Continuous-Flow Pump

The pump performance required for a continuous blood pump is classified into several types by the purpose. The main uses are generally for the VAD, the CPB pump, the extracorporeal membrane oxygenation (ECMO) pump, and the percutaneous cardiopulmonary support (PCPS) pump. The VAD is an implantable or extracorporeal independent pump, whereas the CPB, ECMO, and PCPS pumps

Table 13.1 Typical clinical driving conditions of blood pumps for an adult

Condition	VAD	CPB	ECMO	PCPS
Pressure (mmHg)	100	350	325	500
Flow rate (l/min)	5	5	0.5	3

are extracorporeal multipurpose pumps. The essential driving condition of blood pumps is achieved by satisfying the pump performance, as shown in Table 13.1 (Kawahito and Nosé 1997).

13.4.1 Impeller

Centrifugal and axial-flow impellers are used in rotary blood pumps to increase the pressure by rotation. For an impeller with diameter D, a rotational speed n, and an impeller tip speed u, the flow rate Q is

$$Q \sim D^2 u \sim nD^3, \tag{13.1}$$

and the head H is

$$H \sim u^2 \sim n^2 D^2. \tag{13.2}$$

When the left side is a constant multiple of the right side in both equations, both velocity and pressure distributions are similar. Omitting D from these equations,

$$\frac{n\sqrt{Q}}{H^{3/4}} \equiv n_S \tag{13.3}$$

becomes constant. Here, n_s is the specific speed, and is a parameter used to determine impeller types that exhibit similarities in their pumping even if the pump sizes are different. In this article, the dimensioned specific speed based on the units of “rpm, m^3/min, m” is used, although the dimensionless specific speed $(= n\sqrt{Q}/(gH)^{3/4})$ is used worldwide, where g is the gravitational force. Figure 13.5 shows the relationship between the impeller geometry and the specific speed. The practical range of the centrifugal impeller is $80 < n_s < 1400$, which shows good efficiency in the low flow rate region. On the other hand, the practical range of the axial-flow impeller is $n_s > 600$, which shows good efficiency in the high flow rate region (Stepanoff 1957).

To set a pressure of 100 mmHg and a flow rate of 5 l/min for a VAD, the practical impeller rotational range of the centrifugal impeller is $1{,}400 < n < 25{,}000$ rpm, and that of the axial-flow impeller is more than 10,000 rpm in terms of efficiency. Here, the leakage flow should be secured to achieve adequate shear stress in the blood pump, which is sometimes realized by an adequate leakage flow with a large gap to wash out around the rotational center

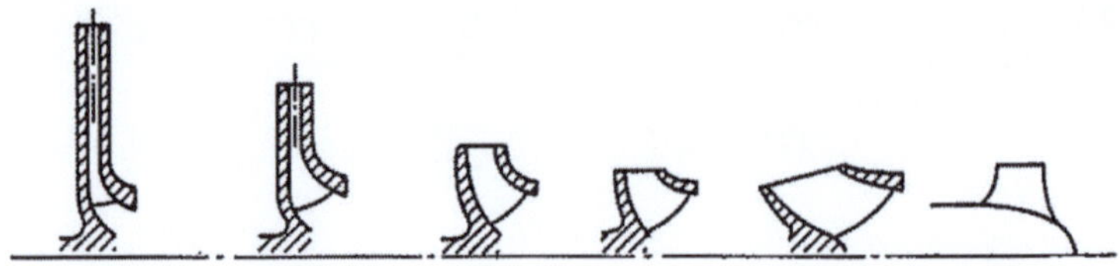

Fig. 13.5 Impeller geometry and specific speed

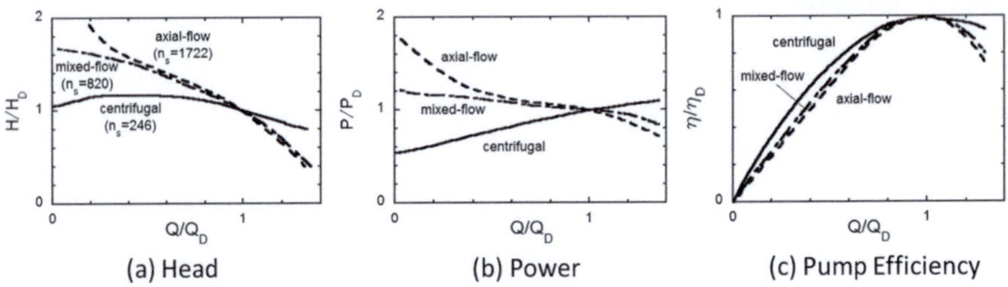

Fig. 13.6 Typical pump performance classified by impeller geometry

including bearings where stagnation tends to occur. The leakage flow is large, and usually becomes more than 1 l/min when the pump flow rate is 5 l/min.

The pump performance is closely related to the impeller geometry, which is characterized by the specific speed. Figure 13.6 shows typical pump performance classified by impeller geometry. The pressure head and power dramatically increase around the no-discharge input where the flow rate is 0 for the axial-flow impeller whose specific speed is large, because backward flow is generated when the flow rate decreases. Conversion of the mechanism to that of the centrifugal impeller increases the pressure head, and work against the backward flow increases the power. On the other hand, the head and power smoothly vary for the centrifugal impeller with a low specific speed, because the small backward flow only increases to a large backward flow when the flow rate decreases without any mechanism change.

Another important aspect about impeller geometry is that it governs the anatomical fit of the blood pump, especially for implantable VADs. The inflow is in the direction of the rotational axis, and the outflow is in the direction vertical to the rotational axis and is apart for the impeller radius from the rotational axis. On the other hand, both the inflow and the outflow are in the direction of the rotational axis in the axial-flow impeller.

13.4.2 Bearing

The bearing supports the rotating axis of the impeller. The various kinds of bearings for rotary blood pumps include journal bearings, pivot bearings, hydrodynamic bearings, and magnetic bearings (Fig. 13.7). Although a ball bearing is popular for industrial use, it is not used for a rotary blood pump because it has a short life at a high rotational speed. Tables 13.2 and 13.3 show mechanical elements of CPB pumps and VADs, respectively, which include not only the bearing but also the relationship of rotor and impeller, and the coupling.

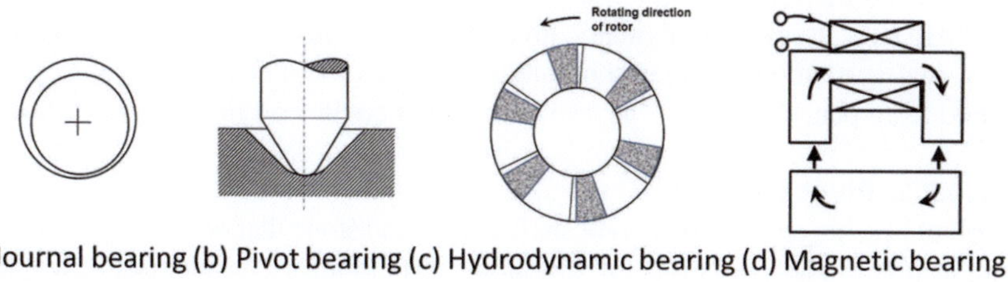

Fig. 13.7 Bearings

Table 13.2 Mechanical elements of cardiopulmonary bypass (CPB) pumps

Product	Rotor and impeller	Coupling	Bearing
BioPump	Separated	Magnetic coupling	Journal bearing and seal
HPM-15, CAPIOX	Separated	Magnetic coupling	Ball bearing and seal
RotaFlow, MERA centrifugal pump	Separated	Magnetic coupling	Pivot bearing (single)
GyroPump, Mixflow	Separated	Magnetic coupling	Pivot bearing (double)
CentriMag	Unified	No coupling	Magnetic bearing

Table 13.3 Mechanical elements of VADs

Product	Rotor and impeller	Coupling	Bearing
HeartMate II, Jarvik2000FlowMaker	Unified	No coupling	Pivot bearing (double)
HVAD, VentrAssist	Unified	No coupling	Hydrodynamic bearing
DuraHeart	Separated	Magnetic coupling	Magnetic bearing
EVAHEART	Separated	Connected by shaft	Hydrodynamic journal bearing (with cool-seal system)

13.4.2.1 Journal Bearing

A journal bearing has a simple structure, a shaft inserted into a hole. It achieves stable rotation with the support of a heavy weight. It also allows for a considerably free design of the impeller, as other bearings must have a narrow gap between an impeller and a casing. However, a seal is required to prevent blood leakage. Thus, considering hemocompatibility, a journal bearing is usually used for short-term use blood pumps, such as a CPB pump, because thrombus formation might occur around the gap if blood penetrates the gap between the shaft and the seal. For long-term use, a special anti-thrombogenicity design is required for the seal around the bearing. The BioPump, the HPM-15, and the CAPIOX use this bearing for CPB pumps, and the EVAHEART uses this bearing for a VAD. For long-term use, the EVAHEART uses a hydrodynamic journal bearing with the cool-seal system. In the cool-seal system, pure water continuously washes the mechanical seal to prevent blood from penetrating into the seal gap, which could lead to thrombus formation.

13.4.2.2 Pivot Bearing

A pivot bearing consists of a pivot shaft and an end stone that only make contact at a single point. The possibility of thrombus formation is considerably low, as the contact area in the bearing is minimized. However, the wash around a pivot bearing is important to increase its hemocompatibility, as it is not a non-contact bearing. Various improvements are made to the wash around the pivot bearing, and the wear of the bearing is also investigated. There is a single pivot type that supports only behind the impeller and a double pivot type that supports both in front of and behind the impeller. The RotaFlow and the MERA centrifugal pump are single pivot-type bearings for CPB pumps. The single pivot type is driven under the condition that the positive thrust force is loaded on the end stone. Thus, the force is adjusted with the magnetic attraction force of the magnetic coupling. Here, the fluid force should be considered, which is usually a negative thrust force, because the pressure is high in the outlet and low in the inlet. As for pumps that use double pivot-type bearings, there are the GyroPump and the Mixflow as CPB pumps and the HeartMate II and the Jarvik2000FlowMaker as VADs. The double pivot-type stably rotates with high bearing stiffness, but it has two bearings, which might increase the possibility of thrombus formation.

13.4.2.3 Hydrodynamic Bearing

A hydrodynamic bearing considerably increases hydrodynamic force by special grooves in the blood pump, although, overall, it is the same classification as that of a sliding bearing including a journal bearing, etc. The HVAD uses this bearing. The mechanism of the hydrodynamic bearing involves generating pressure by thrusting the fluid into the gap by viscous force when two planes move, and the geometry gradually narrows in the moving direction.

13.4.2.4 Magnetic Bearing

A magnetic bearing uses a feedback control system where coiled magnetic iron cores face the center rotor, and the rotor is supported by the magnetic attractive force. There is no wear on the bearing, as it achieves non-contact support that ensures high hemocompatibility. The CentriMag uses this bearing for a CPB pump and a short-term VAD, and the DuraHeart uses this bearing for a VAD.

13.5 Evaluations of Continuous-Flow Pump

The evaluations of VAD are overall established by the International Organization for Standardization (ISO) Standard for circulatory support devices. The ISO 14708-5 specifies the requirements for safety and performance of active implantable circulatory support devices. To satisfy the ISO 14708-5, the Japanese R&D guidance for VAD and total artificial heart were published by the Ministry of Economy, Trade and Industry, that describes the pump performance test including fluid dynamics analysis of blood pump, the hemolysis test, the animal test, the durability test, and many other tests of developing VADs (ISO 14708–5 2010; Imachi and Mussivand 2010; Yamane et al. 2010).

13.5.1 Pump Performance Measurement

The pump performance measurement of rotary pumps follows ISO 5198, "Centrifugal, mixed flow and axial flow pumps — Code for hydraulic performance tests — Precision grade" (ISO 5198 1987).

13.5.2 Hemolysis Test

The hemolysis evaluation follows ASTM F1841-97, "Standard practice for assessment of hemolysis in continuous flow blood pumps" (ASTM F1841-97 1997).

13.5.3 Animal Test (Anti-thrombogenesis Test)

The protocol with adequate consideration for the purpose of system use and patient safety, raw data, the observed record, and explanation and consideration of the results should be described. There must not be severe thromboembolism that is considered to have originated in the apparatus. Here, "severe" is defined as events that threaten animal life or cause conditions to deteriorate. The suspected cases of severe thromboembolism should be monitored by medical practitioners for deterioration in conditions such as the clinically unacceptable dysfunction of the kidneys or liver, as is previously defined, uncontrolled pain with any analgesic administration or analgesia treatment, and paralysis requiring assistance resulting from condition deterioration. Regular observations should be based on medical practice and diagnosis. There is currently no consensus on the appropriate number and period of animal experiments. However, it is preferable that sufficient evidence is provided that enables progression to a clinical study, with the results from animal experiments that are conducted with six animals for more than 60 days, or eight animals for more than 90 days, corresponding to the purpose of use.

13.5.4 Durability Test

The purpose of the durability test is to prove that there is no problem in the system, including reliability in daily use based on risk analysis, etc. The target patients depend on the condition that an applicant describes as the purpose of use. The fundamental goal is to record and report all events. Whether the test is stopped or continued when events occur needs to be determined. The system reliability is expressed by the number of tests and the number of failures to validate the function as intended within the specifications (period and environment) that an applicant determined. Namely, the number of test apparatuses that are necessary to attain the reliability and the confidence level is set. At least 6 months of testing are required with 80 % reliability and a 60 % confidence level for the experimental condition and the period of the durability test. Table 13.4 shows the number of tests with 80 % reliability and an 80 % confidence level, and Table 13.5 shows the number of tests with different confidence levels. A 6-month test is recommended with 80 % reliability and an 80 % confidence level when considering international harmonization. It is preferable to continue the test for more than 2 years. Test conditions are set by considering the instrument properties. It is recommended that the environment of the durability test be determined by considering physiological condition and life pattern such as pressure, flow rate, pulsatility, pH, temperature, and electrolytes.

Table 13.4 Number of tests with 80 % reliability and 80 % confidence level

Assumed number of failures	Reliability and confidence level	Number of tests
No failure allowed	80 %, 80 %	8
One failure allowed	80 %, 80 %	14
Two failures allowed	80 %, 80 %	21

Table 13.5 Number of tests with different confidence levels (Pantalos et al. 1998, Yamane et al. 2010)

Recommender	Reliability and confidence level	Number of tests (1 failure allowed)
ASAIO-STS	80 %, 60 %	9
Japanese R&D Guidance	80 %, 70 %	11
	80 %, 80 %	14
	80 %, 90 %	18

13.6 Fluid Dynamics Analysis of Blood Pump

Pump performance and hemocompatibility are important design indices in the development of blood pumps. Therefore, blood pumps were conventionally designed based on basic pump design theory, and focus was placed on avoiding flow stagnation to prevent thrombus formation and excessive high shear to prevent hemolysis. The performance was evaluated through repeated performance tests, hemolysis tests, and anti-thrombogenesis tests with animals to confirm design validation. However, in addition to the conventional methods, the inner flow state is grasped in more recent methods for developing blood pumps, even if only roughly, and is used for fluid dynamic design of the inner geometry of blood pumps because the pump performance and hemocompatibility are strongly related to the flow. The pump design is required to achieve an inner flow that avoids both excessive high shear regions and stagnation regions because hemolysis occurs in excessive high shear regions and thrombogenesis occurs in regions of stagnant flow in the blood pump. Therefore, the load of various tests related to fluid dynamics has been lightened by fluid dynamics analysis. The analyses described concern pump performance, hemolysis, anti-thrombogenesis and impeller levitation, which are considered important aspects in developing blood pumps. Discussions of various results from tests that have been conducted blindly, especially the discussions about blood test results, are supported by engineering, and the speed at which blood pumps are developed has consequently increased. Table 13.6 shows various flow analyses and their properties of methods, which can be used as follows.

(a) Computational fluid dynamics analysis

Computational fluid dynamics (CFD) analyses are used to analyze flow in blood pumps. There are now many CFD studies because of the recent development and popularity of commercial software, which makes the analysis easy for researchers who do not specialize in this type of analysis. The advantage of CFD analysis is to

Table 13.6 Properties of flow analysis methods

	Advantage	Disadvantage
Computational fluid dynamics	Easy output of various physical property (e.g., path line), simultaneous pressure analysis	Validation requirement of accuracy, difficulty for too complicated flow (e.g., flow in pulsatile-flow pump)
Qualitative flow visualization	Easy, low cost	No quantification
LDV measurement, hot-wire velocimetry	Precise, good traceability on flow	One-point measurement, error near the surface wall, high cost
PIV measurement	Multipoint measurement, high space resolution	High cost

be able to grasp easily the overall flow. However, the accuracy of CFD analysis should be validated by experimental fluid dynamics analysis, as it is not guaranteed, because geometric approximation is necessary when generating meshes, and because the most appropriate analysis method (solution method, turbulent model, etc.) is not determined at present.

(b) Qualitative flow visualization

Qualitative flow investigations that have been conducted on the inner flow of blood pumps and that are not accompanied with velocity quantification are called flow visualization (Affeld et al. 1976). Typical methods are the tracer injection method and the surface trace method. The streak line method (Schima et al. 1992) and the suspension method (Araki et al. 1993) are examples of the tracer injection method, and the oil dot method (Burgreen et al. 2001) and the oil film method (Tsukiya et al. 2002) are examples of the surface trace method. Qualitative flow visualization is sometimes conducted on the inner flow because overall goodness can be judged for the geometry design of blood pumps. For example, one can judge whether there is vortex or separation in the flow between blades of an impeller or in the flow between the impeller and the outlet by the suspension method. Moreover, one can judge whether there is stagnation in the flow in the pump by the oil dot and oil film methods. Thus, qualitative flow visualization is a meaningful experimental analysis. It has a reliability that the CFD analysis cannot achieve, and it not only reduces the work required for quantification and error estimation, but also reduces the cost for the experimental apparatus.

(c) Laser Doppler velocimetry and hot-wire velocimetry measurements

Laser Doppler velocimetry (Pinotti and Paone 1996) and hot-wire velocimetry (Chua et al. 2002) are methods for measuring the inner flow of the blood pump. These methods are well known to be highly precise in spite of the disadvantage of being labor intensive from the one-point measurements, because the tracer invasiveness is small and time resolution is high. Therefore, these methods are used for measurements that require high precision.

(d) Particle image velocimetry measurement

Particle image velocimetry (PIV) quantifies particle images obtained by the suspension method because qualitative methods cannot calculate quantitative values such as velocity, shear rate, shear stress, etc. In the 1990s, flow was analyzed in axial-flow blood pumps to improve the pump performance, which depends on the blade geometry (Kerrigan et al. 1996). However, the PIV measurement method was not adapted to analyze the flow in the axial-flow blood pump because it was difficult to assess three-dimensional flow by measuring the particle motion on a laser-light sheet. On the other hand, there are many adaptations of PIV measurement to analyze flow in the centrifugal blood pump (Ikeda et al. 1996; Nishida et al.

1999) because the main flow and secondary flow, which is a sub-flow perpendicular to the main flow, are usually two-dimensional flows.

13.6.1 CFD Analysis

In this section, current CFD analyses are described, which are established as an important development tool to optimize the geometry of blood pumps.

13.6.1.1 Methodology of CFD

The equations governing flow are the conservation equations of mass, moment, and energy that correspond to the continuity equation, the momentum equation, and the energy equation, respectively. The CFD analysis calculates the flow field by which these equations are discretized and algebraically solved (Japan Society of Mechanical Engineers 1988). A flow field and translated fluid dynamic values from the flow field in the blood pump, which are required for development of the blood pump, are calculated through the same process as CFD analysis of the flow in a general pump. Figure 13.8 shows the typical process of blood pump development using CFD

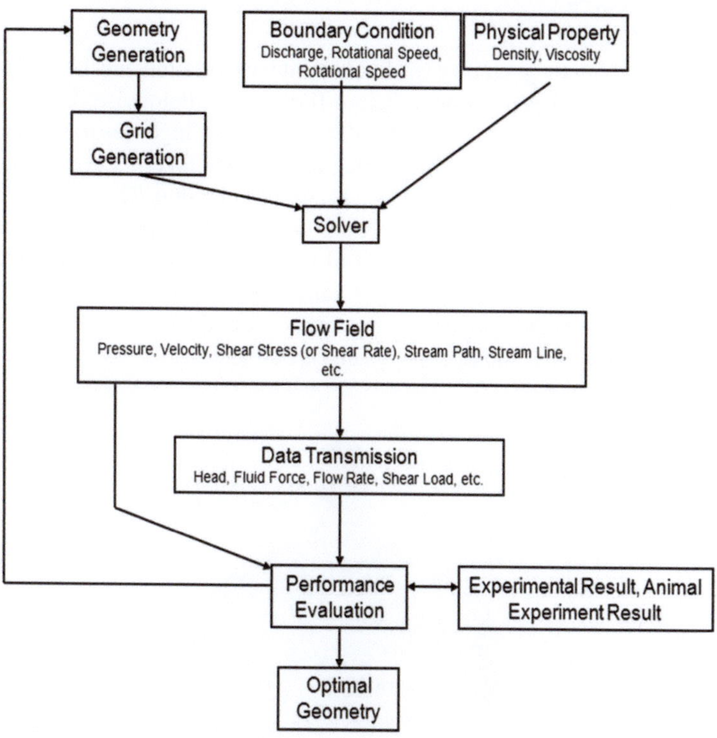

Fig. 13.8 Typical process of blood pump development using computational fluid dynamics (CFD) analysis

analysis (Antaki et al. 1995). The pump geometry is drawn using 3-D CAD software, and the meshes are generated as a result of discretization from the CAD geometry using grid-generation software. The flow is solved algebraically using CFD analysis software for the generated meshes by giving the physical properties of the fluid and the boundary conditions. Subsequently, the flow field and the fluid dynamic values translated from the flow field are calculated. Finally, the results are compared with the in vitro or in vivo experimental results, and the pump geometry is evaluated.

1. Geometry generation

The inner geometry is usually drawn using 3-D CAD software. The purpose of CFD analysis is to determine how precisely the geometry should be drawn, although it is well known that hemolytic and thrombogenic properties dramatically vary by the mere roundness of the edge (Umezu et al. 1992). There are different ways to solve the flows in the whole pump and those in a part of the pump. The way to solve the flow in a part of the pump requires a little effort and is effective when the purpose is clear, but the setting of the boundary conditions is usually difficult.

2. Grid generation

The produced inner geometry is divided into calculation meshes (Fig. 13.9) using commercial grid-generation software [Pointwise (Pointwise, USA), ICEM-CFD (ANSYS, USA), Turbogrid (ANSYS), etc.]. The high shear region in the vicinity of the wall surface, which is called a boundary layer, is divided into thin meshes comprising a prism mesh, etc. in a rotary blood pump. The comparably low shear region far from the wall surface is divided into thick meshes constituting a tetra mesh, etc. (Fig. 13.10). The number of meshes varies from 200,000 to more than 1,000,000 when solving the whole pump region. The number of meshes generated in the boundary layer is more important than that generated in the whole pump. Thus, the solution is usually conducted after the validation, for

Fig. 13.9 Meshes of a centrifugal blood pump for CFD analysis

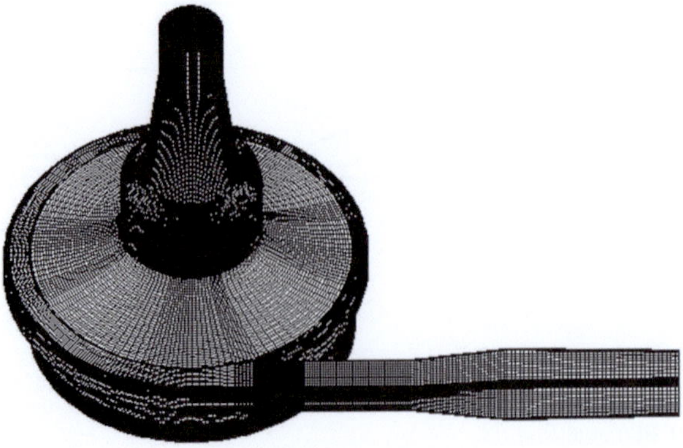

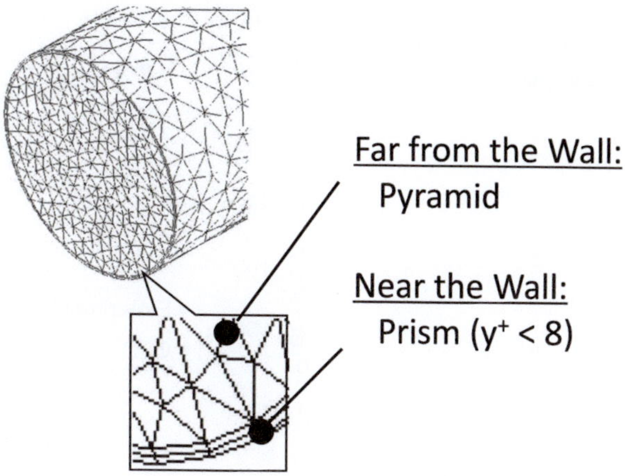

Fig. 13.10 Connection of the *thin* and *thick* meshes

example, when checking the relationship between the number of meshes and the pump head (Qian 2004).

3. Solution

The equations governing flow are algebraically solved using commercial analysis software [STAR-CCM+ (CD-Adapco, USA), Fluent (ANSYS), etc.] for generated meshes with physical properties and boundary conditions. Many studies use a finite volume method, which is the method most commonly used in commercial software, among some CFD methods, such as the differential method, finite volume method, finite element method, etc. Several studies have also used custom-written software, which sometimes achieves more precise analyses (Nakamura and Yano 1999). Commercial software is used because flow analyses in blood pumps need to be fast to be conducted in parallel with other experimental tests, such as tests of the pump performance, hemolytic properties, anti-thrombogenic properties, and fluidic force. The criteria for evaluating hemolytic and anti-thrombogenic properties are unclear, and there are many unstable factors in analyzing the flow when considering blood rheology, especially non-Newtonian and multiphase fluid dynamics that are actually almost impossible to solve. On the other hand, it is important to select the adequate turbulent model for accuracy. However, although there are many turbulent models, such as the laminar model, the k-ε model, and the large eddy simulation model, it is unclear which turbulent model is the most valid for the analysis of blood pumps. Generated meshes and turbulent models are simply validated by how much the performance curves calculated by the CFD analysis and those actually measured coincide (Fig. 13.11). These models have been recently validated by how much the velocity distributions calculated by CFD analysis and those actually measured coincide (Triep et al. 2006), although other problems with the accuracy of the velocity measurement arise. Furthermore, validation of the pressure distribution (Burgreen et al. 2001), stream path (Apel et al. 2001a), and turbulence component is required.

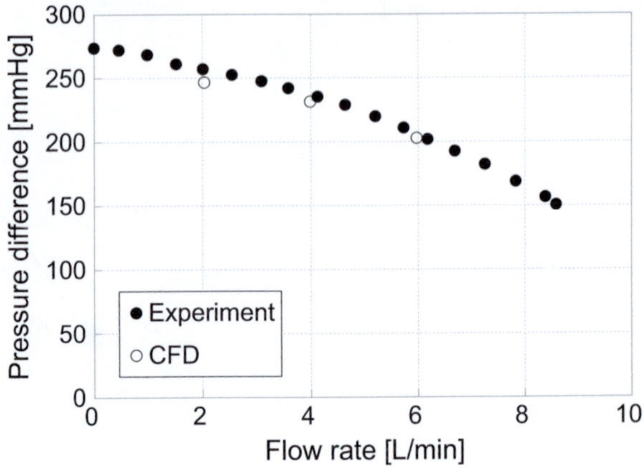

Fig. 13.11 Example of CFD validation using pump performance curve

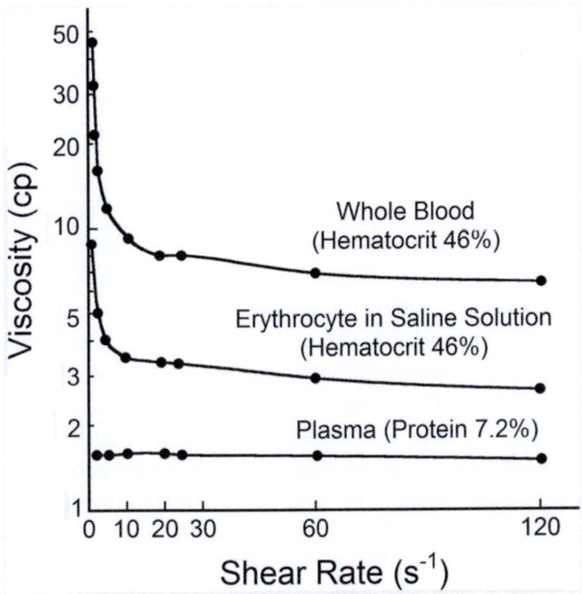

Fig. 13.12 Relationship between effective viscosity and shear rate (Wells et al. 1962)

13.6.1.2 Blood Properties

Blood has a density of 1.056 kg/m^3 and a viscosity of 0.003 Pa · s (=3 cP, at 37 °C and in high shear flow) (Cooney 1976). The viscosity from 2.4 cP to 3.6 cP is adopted, and that of 3.5 cP is generally used for CFD analysis. One reason for various viscosities is differences between individuals, and another is that it depends on the high viscosity value in the low shear region because blood is a non-Newtonian fluid approximated to be a Casson fluid (Fig. 13.12) (Wells et al. 1962; Oka 1974). The fluid in the blood pump is approximated to be a Newtonian fluid in current flow analyses because there are high shear regions in almost all regions. Blood is also a multiphase fluid with hematocytes suspended in plasma, but it is approximated to be continuous fluid for current flow analysis because the analysis considering multiphase properties is almost impossible in blood pumps. These effects should, however, be considered.

13.6.1.3 CFD of Pump Performance

The inner flow is analyzed to improve the pump performance of the blood pump, which is the same process as that in the fluid design of general pumps. The pump performance is evaluated by the pump head P, which is the pressure difference between the inlet and the outlet, as a function of the flow rate Q at the constant impeller rotational speed n.

In the 1990s, CFD was actively adapted for the design of axial-flow pumps, for which appropriate impeller design is difficult to determine (Butler et al. 1997). The analysis was then adapted for the design of centrifugal pumps (Chan et al. 2002), since the pump performance varies according to geometric factors such as the impeller geometry and the outlet position. The pump performance is the first property of blood pumps to be investigated because it also affects indirectly the hemolytic properties.

13.6.1.4 CFD of Hemolysis

Hemolysis is the rupture of erythrocytes. It is caused by fatigue failure resulting from repeated loading of high shear stress on the erythrocyte membrane in the blood pumps that sometimes exhibit excessive high shear regions. As well as the mechanism of the material fatigue failure, the amount of hemolysis is related to the shear stress and the number of loadings, that is, the exposure time (Fig. 13.13) (Helmus and Brown 1977). Thus, CFD of hemolysis is the analysis concerning the shear stress and the exposure time on erythrocytes. When the amount of hemolysis

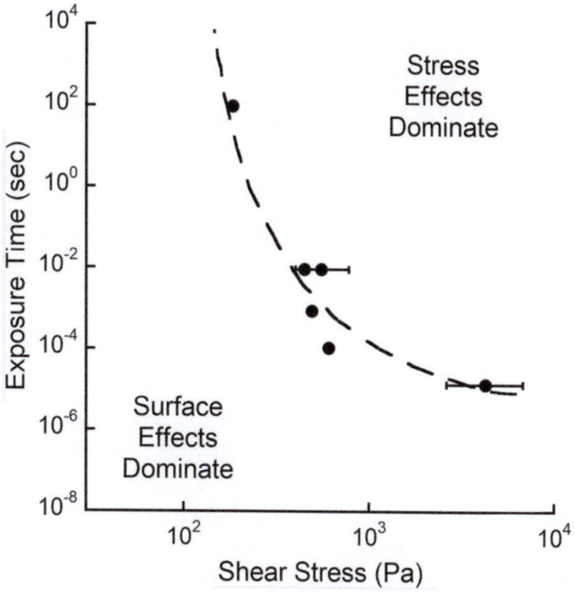

Fig. 13.13 Relationship among hemolysis, shear stress, and exposure time (Helmus and Brown 1977)

Table 13.7 Coefficients for hemolysis and shear stress (Blackshear et al. 1965; Heuser 1980; Wurzinger et al. 1985; Giersiepen et al. 1990; Song et al. 2004)

	Blackshear	Heuser	Giersiepen–Wurzinger
Blood	Canine	Porcine	Human
C	–	1.9×10^{-6}	3.62×10^{-5}
α	2	1.991	2.416
β	1	0.765	0.785
Range of τ and t	–	$40 < \tau < 700$	$57 < \tau < 255$
		$0.0034 < t < 0.6$	$0.007 < t < 0.7$

is dHb/Hb, the shear stress is τ, and the exposure time is t; their relationship is expressed by

$$\frac{dHb}{Hb} = C\tau^{\alpha}t^{\beta}, \tag{13.4}$$

where the coefficient C, and α and β, given in Table 13.7, were obtained from the results of shear loading experiments on blood using a rotational viscometer (Blackshear et al. 1965; Heuser 1980; Wurzinger et al. 1985; Giersiepen et al. 1990; Song et al. 2004). The shear stress history of an erythrocyte passing through artificial organs was first analyzed to understand the hemolysis of an artificial valve (Giersiepen et al. 1990). This technique was then adapted to understand the hemolysis of centrifugal blood pumps (Fig. 13.14) (Bludszweit 1995). Subsequently, this technique became a method for estimating hemolysis based on the shear stress for many particles (Mitamura et al. 2001).

The damage $d_{n,\,k}$ to an erythrocyte during the time interval $\Delta t_k(= t_k - t_{k-1})$ is expressed by

$$d_{n,k} = C\tau(t_{k-1})^{\alpha}\Delta t^{\beta}. \tag{13.5}$$

The damage $D_{n,k}$ to the erythrocyte between time 0 and t_k is expressed by

$$D_{n,k} = D_{n,k-1} + (1 - D_{n,k-1})d_{n,k}. \tag{13.6}$$

If the erythrocyte flows along the streamline, a shear stress history can be established. Here, the scalar shear stress τ

$$\tau = \sqrt{\frac{1}{6}\sum\left(\tau_{ii} - \tau_{jj}\right)^2 + \sum\tau_{ij}} \tag{13.7}$$

is usually used as shear stress, where τ_{ij} represents the shear stresses that are the sum of the viscous stress and the Reynolds stress. After the damage D_n of each erythrocyte is analyzed for each streamline from the inlet to the outlet of the pump, the average damage D passing through the pump is regarded as the estimated hemolysis. Many studies have examined the effect of shear stress and the exposure time for the hemolysis in the blood pump (Pinotti and Rosa 1995), whereas other

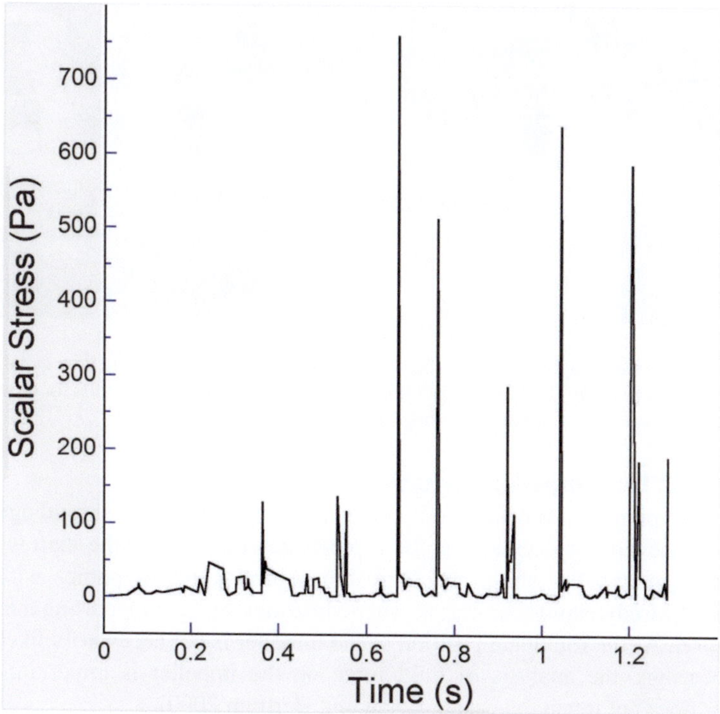

Fig. 13.14 Stress history of an erythrocyte in a centrifugal pump (Bludszweit 1995)

studies have examined additionally the effect of the pressure (Mizuguchi et al. 1995). Some earlier studies have analyzed the shear stress (Wood et al. 1999), and some studies have analyzed the relationship between the experimental results and hemolysis test results (Miyazoe et al. 1998). Various investigations have been conducted on hemolysis, including those on the effect of turbulence stress (Apel et al. 2001b), the effect of stretch (Arora et al. 2004), a method to estimate the volume integration of hemolysis (Garon and Farinas 2004), and the effect of the dissipation of the thrombocyte (Goubergrits and Affeld 2004).

13.6.1.5 CFD of Thrombus Formation

The mechanism of thrombus formation is more complicated than that of hemolysis. A thrombus is generally considered to be formed in a low shear region, that is, flow stagnation. It is sometimes formed in a narrow gap and by hemolysis. At present, CFD of thrombus formation is only analyzed by calculating the velocity distribution in the gap (Gobel et al. 2001; Tsukamoto et al. 2001; Burgreen et al. 2004) and shear stress (Nakamura et al. 1999; Nishida et al. 2006a). Figure 13.15 shows the example of geometric optimization of the blood pump using CFD analysis and antithrombogenicity. However, the estimation of thrombus formation by CFD is very important because the problem with blood pump development from a clinical perspective is thrombus formation.

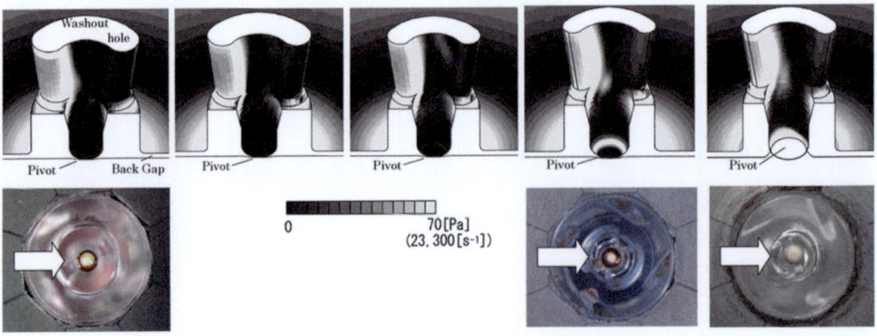

Fig. 13.15 Geometric optimization of the monopivot centrifugal pump using flow quantification and anti-thrombogenicity (Nishida et al. 2006a). *Upper* figures show CFD results for shear stress; *lower* figures show results of anti-thrombogenesis tests

13.6.1.6 CFD of Impeller Levitation

Some blood pumps do not have an impeller shaft to improve anti-thrombogenicity. Anti-thrombogenicity decreases at the gap between the seal and the shaft when the pump has an impeller shaft, and there are several kinds of pumps with pivot bearings, hydrodynamic bearings, magnetic bearings, etc. for supporting the impeller rotation. As the rotational position of the impeller is not necessarily fixed using these bearings, the analysis of fluid force on the impeller is important so the impeller does not touch the casing (Qian and Bertram 2000).

13.6.2 PIV Measurement

Although the PIV measurement, one of the experimental fluid dynamics analysis methods (Adrian and Westerweel 2011), is an effective tool to improve pump performance, and also to prevent hemolysis and thrombus formation, the PIV measurement or other experimental fluid dynamics analysis is usually regarded as a method to validate CFD analysis (Apel et al. 2001a). Namely, the PIV measurement is used to compensate for the accuracy of the CFD analysis. The development of a blood pump is now proceeding with a combination of the CFD analysis and experimental fluid dynamics analysis. In this section, the PIV measurement method of blood pumps in our laboratory is described.

13.6.2.1 Methodology of PIV

Figure 13.16 shows the experimental apparatus for flow visualization. Blood pumps are difficult to visualize because of their small size and the high rotational speed of the impeller. Thus, when developing blood pumps, selection of the scale model, working fluid, and image analysis requires several precautions. Namely, it is necessary to consider factors that affect the accuracy of the velocity measurement, for example, insufficient photographing speed of the video camera when visualizing

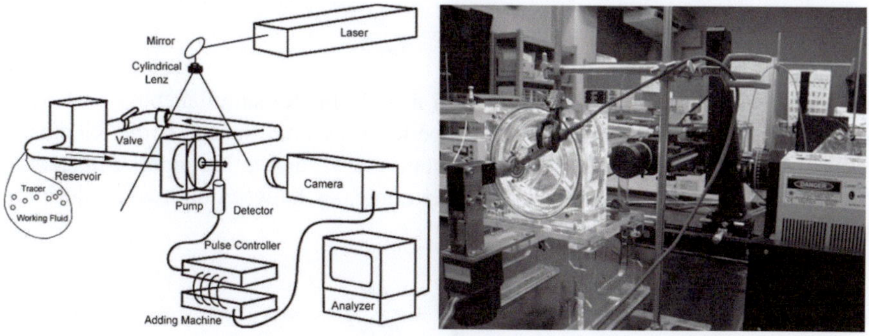

Fig. 13.16 Experimental apparatus for flow visualization

Table 13.8 Properties of the model scale

Pump model	Advantage	Disadvantage
Full-scale model	Low cost, small tasks	Low space resolution, low time resolution
Scale-up model	High space resolution, high time resolution	High cost, several tasks

full-scale models, the scaling error when visualizing scale models, and the total error when visualizing localized areas with high resolution.

1. Scale model

For a normal centrifugal blood pump, the impeller, whose overall diameter is approximately 50 mm, rotates at a speed of 2000 rpm. Thus, observation of overall flow is possible along with the impeller rotation if the flow is photographed using a full-scale model. However, flow analysis with greater precision and higher spatial resolution will be difficult because of the limits of photographing speed and optical magnification. In this case, a scale model makes it possible to increase the spatial and temporal resolution using the flow similarity law, but these types of models are expensive. Table 13.8 shows the properties of model scale.

The principle for a scale model is as follows (Stepanoff 1957). To be dynamically similar flow, the Reynolds number *Re*, which is the ratio of the inertial force to the viscous force,

$$Re = \frac{uR}{\nu} = \frac{2\pi R^2 N}{\nu} \tag{13.8}$$

and the flow rate coefficient ϕ, which is the geometry of the velocity triangle,

$$\phi = \frac{Q}{Au} = \frac{Q}{2\pi R^3 N} \tag{13.9}$$

coincide between a full-scale model and a scale model simultaneously, where R denotes the impeller radius, N the impeller rotational speed, ν the kinematic viscosity, Q the flow rate, A the representative area, and u the impeller tangential velocity. When the flow rate coefficient ϕ coincides between the two models, the head coefficient ψ

$$\psi = \frac{gH}{u^2} = \frac{gH}{R^2 N^2} \tag{13.10}$$

coincides between the two models. From Eq. (13.8), the impeller rotational speed in the scale model is given by

$$\frac{N_2}{N_1} = \left(\frac{R_1}{R_2}\right)^2 \times \frac{\nu_2}{\nu_1}. \tag{13.11}$$

where the subscript 1 indicates the value in the full-scale model, and the subscript 2 indicates the value in the scale model. From Eq. (13.9), the flow rate in the scale model is given by

$$\frac{Q_2}{Q_1} = \left(\frac{R_2}{R_1}\right)^3 \times \frac{N_2}{N_1} = \frac{R_2}{R_1} \times \frac{\nu_2}{\nu_1}. \tag{13.12}$$

From Eq. (13.10), the head in the scale model is given by

$$\frac{H_2}{H_1} = \left(\frac{R_2}{R_1}\right)^2 \times \left(\frac{N_2}{N_1}\right)^2. \tag{13.13}$$

From Eqs. (13.12) and (13.13), removing R gives

$$n_{S1} \equiv \frac{N_1\sqrt{Q_1}}{H_1^{3/4}} = \frac{N_2\sqrt{Q_2}}{H_2^{3/4}} \equiv n_{S2} \tag{13.14}$$

where n_s is defined as the specific speed. Therefore, to complete Eqs. (13.9) and (13.10), the specific speed n_s for the full-scale model and that for the scale model coincide.

2. Working fluid

Transparent fluids are used as the working fluid for the PIV measurement in the blood pump. Table 13.9 shows the physical properties of working fluids to simulate blood. These fluids include water, saline solution, glycerol solution, sodium iodide solution (Nishida et al. 1997), and a mixed solution of sodium iodide and glycerol

Table 13.9 Physical properties of working fluid to simulate blood (at room temperature, ≈20 °C)

	Specific gravity	(cSt)	Refractive index	(Advantage)	(Disadvantage)
Blood (37°C)	1.056	3 (or 3.5)	–		
Water	1	1	1.33	Easy handling	
Saline solution (5 wt% aq)	1.05	1. 03	N.A.	Easy handling, specific gravity matching	
Glycerol solution					
(40 wt% aq)	1.1	3.5	N.A.	Viscosity matching	
[100 %]	[1.26]	[1300]	[1.8]	[As a reference]	
Mixed solution of sodium iodide and glycerol					
(79 % saturated NaI aq + 20 % glycerin + 1%water)	1.75	3.8	1.49	Viscosity matching	Metal oxidation
(90 % saturated NaI aq +10 % glycerin)	1.87	3			
Sodium iodide solution (64 wt% aq)	1.9	1.8	1.49		Metal oxidation
Air	0.001	15	1	Easy handling	Visualization difficulty
Mineral oil	0.83	4.3–5.7	1.46		Flammable
Mixed solution of potassium thiocyanate and glycerol (56.4 KSCN w/o in 36 % glycerin aq)	1.43	3.5	1.49		Toxic

(Baldwin et al. 1994). Water, saline solution, and glycerol solution are easy to handle and are used to understand the overall flow. Saline solution is used to adjust the specific gravity, whereas glycerol solution is used to adjust the viscosity. However, because the refractive indices of these solutions are lower than that of acrylic resin (1.49), which is the general material for making pump models, the obtained images must be corrected to be quantified in the PIV measurement algorithm. It is not necessary to correct images if the refractive index of the working fluid coincides with that of acrylic resin. Sodium iodide solution (64 wt%, saturated) has the same refractive index as that of acrylic resin. Furthermore, a mixed solution of sodium iodide and glycerol has not only the same refractive index as acrylic resin but also the same kinematic viscosity as blood, which removes any translation of physical properties using the full-scale model. However, the sodium iodide solution is usually better because it achieves a lower flow velocity resulting from a lower kinematic viscosity of the working fluid to increase the temporal resolution. Table 13.10 shows the physical properties of human whole blood, water, and 64 wt% sodium iodide solution, and experimental setting values expressed as multipliers to acquire the similar flow using a 3× scale model versus the whole blood flow using a full-scale model.

3. Tracer particles

Various commercial tracer particles can be used for PIV measurements when the working fluid is water. However, the specific gravity of tracer particles should be large when the working fluid is 64 wt% sodium iodide solution, which has a specific gravity of 1.9. Silica particles have a specific gravity (1.9) identical to that of sodium iodide solution, and silver-coated hollow glass particles (specific gravity of 1.4) are commercially available. The silver-coated hollow glass particles are useful as tracer particles for the particle tracking method because of their good visibility.

4. Imaging system

Figure 13.17 shows the visualization regions in a centrifugal blood pump. Flow is periodic in a blood pump whether it is a continuous-flow or pulsatile-flow pump. Therefore, analyzing the flow in the blood pump is possible if a video camera can capture both continuous and periodic particle images. In our laboratory, the continuous particle images are firstly obtained to determine lens magnification and photographing speed. A large number of periodic particle images are then obtained by synchronizing constant impeller rotational phases. Finally, accuracy of the velocity measurement is improved by summation of the velocity data.

5. Image analysis

The PIV measurement method is generally classified into the particle tracking method and the correlation method, and both methods are used for the PIV measurement of the blood pump. In our laboratory, the four-frame particle tracking method (Kobayashi et al. 1989) is used to measure the flow in the vicinity of the wall surface with high spatial resolution. Figure 13.18 shows the flow visualization analysis results for various flows; the flow in the vicinity of the wall surface is important in blood pumps. The properties of both methods for the analysis of the flow in blood pumps are described as follows.

(a) Particle tracking method

In the particle tracking method, velocity is calculated by the movement of a particle's centroid. Thus, the spatial resolution of the particle tracking method is high; it comprises point measurements. This method is better for measuring the flow in the vicinity of the wall precisely because particles hardly exist

Table 13.10 Physical properties of human whole blood, water, and 64 wt% sodium iodide solution, and experimental setting values expressed as multipliers to acquire the similar flow using a 3× scale model versus the whole blood flow using a full-scale model

	Physical property		Multiple of the setting value to be similar flow				
Liquid	Density $\rho \times 10^3$ kg/m^3	Kinematic viscosity $\upsilon \times 10^{-6}$ m^2/s	Length L (arbitrarily)	Velocity U ($\propto \upsilon$/L)	Rotational speed n (U/L)	Flow rate Q (UL^2)	Pressure P (ρU^2)
Human whole blood (at 37°C)	1.056	3	1	1	1	1	1
Water (at 20°C)	1	1	3	1/9	1/27	1	1/85. 536
64 wt% NaI solution (at 20°C)	1. 9	1. 8	3	1/5	1/15	9/5	19/264

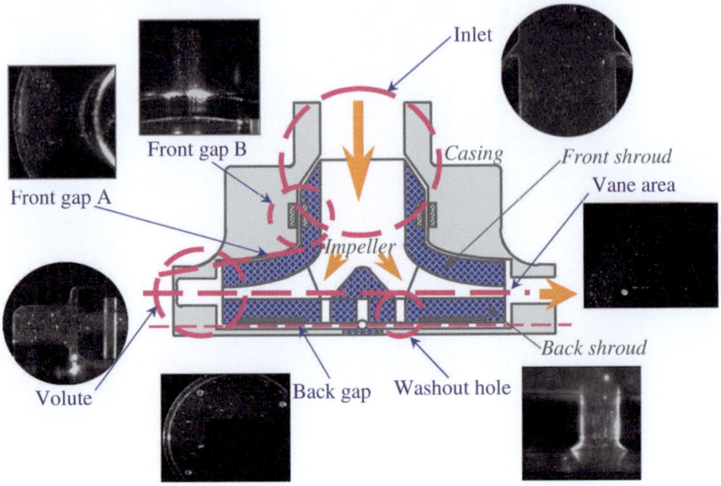

Fig. 13.17 Visualization regions in a centrifugal blood pump (Asztalos et al. 1999)

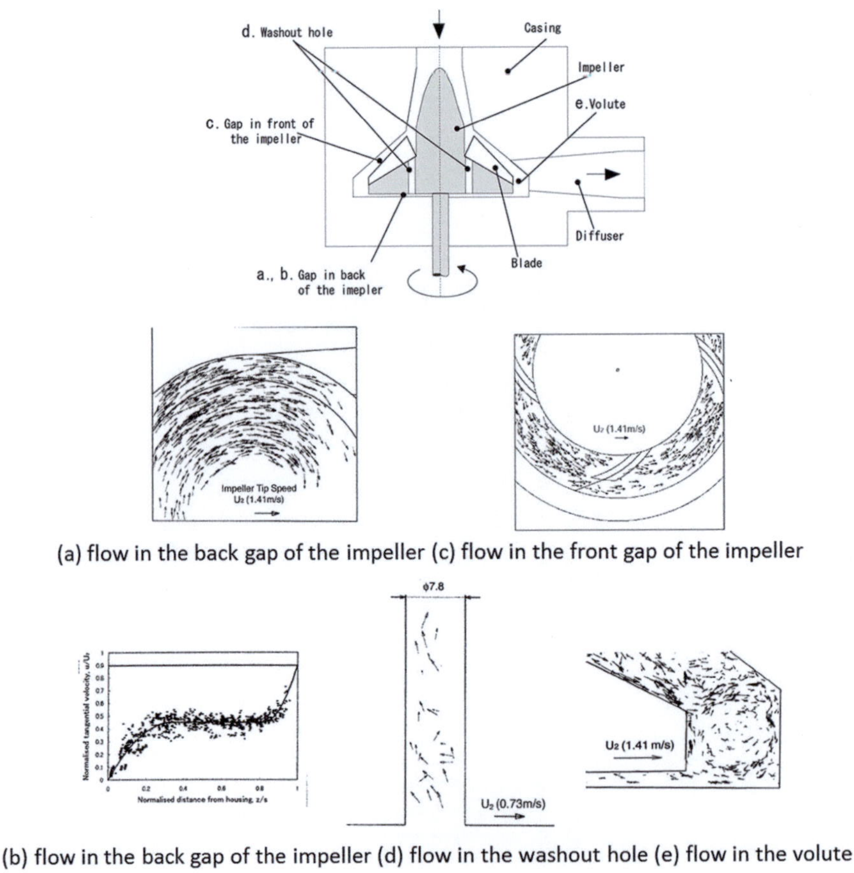

Fig. 13.18 Flow visualization analysis results for various flows

inside the flow, and the flow in the vicinity of the wall surface is important in the blood pump.

(b) Correlation method

In the correlation method, velocity is calculated as the average movement of particles in an interrogation area. Thus, the spatial resolution of the correlation method is lower than that of the particle tracking method and depends on the size of the interrogation area. However, the accuracy of the velocity is better than that of the particle tracking method because the velocity of the correlation method is an average velocity in the interrogation area.

6. Post analysis

The shear rate and the turbulence component are calculated from the velocity data obtained by the particle tracking method or the correlation method. The Reynolds stress and the shear stress are also calculated from the shear rate and the turbulence component.

13.6.2.2 PIV of Pump Performance

The important flow regions to be analyzed for the pump performance of blood pumps are the gap between impeller blades and near the cut-water, which locates from the impeller to the outlet. Figure 13.19 shows the example of the flow analysis of pump performance (Nishida et al. 1999). The flow in pumps with low pump performance sometimes causes a separation in the gap between the impeller blades and near the tongue. Thus, achieving smooth flow in a pump by improving the impeller geometry and outlet position results in a high-performance pump. However, the flow is rarely quantified for the pump performance because qualitative flow visualization is sufficient for observing the cause of the separation, as previously described.

13.6.2.3 PIV of Hemolysis

The amount of hemolysis is expressed as a function of the shear stress and the exposure time as shown in Eq. (13.4). High shear locates in the boundary layer in

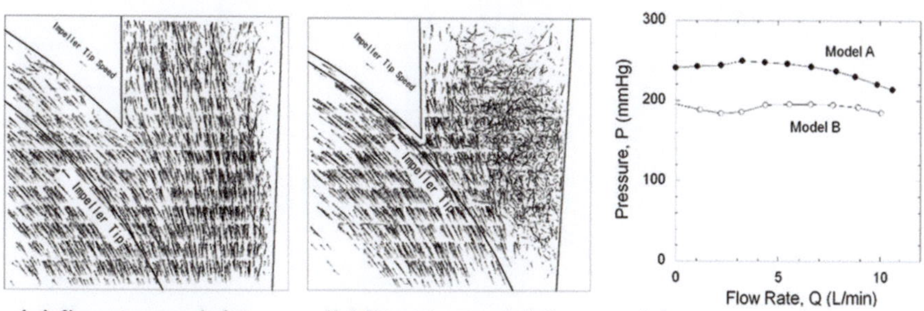

Fig. 13.19 Flow analysis of pump performance (Nishida et al. 1999)

the vicinity of the walls of the casing and the impeller in blood pumps where velocity is high or in the narrow gap. The thickness of the boundary layer is usually less than 100 μm in a centrifugal blood pump. The PIV measurement is a good method for quantifying the high shear in the boundary layer because the spatial resolution can be improved by adjusting the optical system. However, it is difficult to measure the exposure time for the shear, which makes it almost impossible to estimate the amount of hemolysis calculated by Eq. (13.4) experimentally. Figure 13.20 shows the relationship between high shear in the vicinity of the wall surface and hemolytic properties of tested pumps. At present, the PIV measurement is regarded as a reliable method to quantify the high shear in the boundary layer.

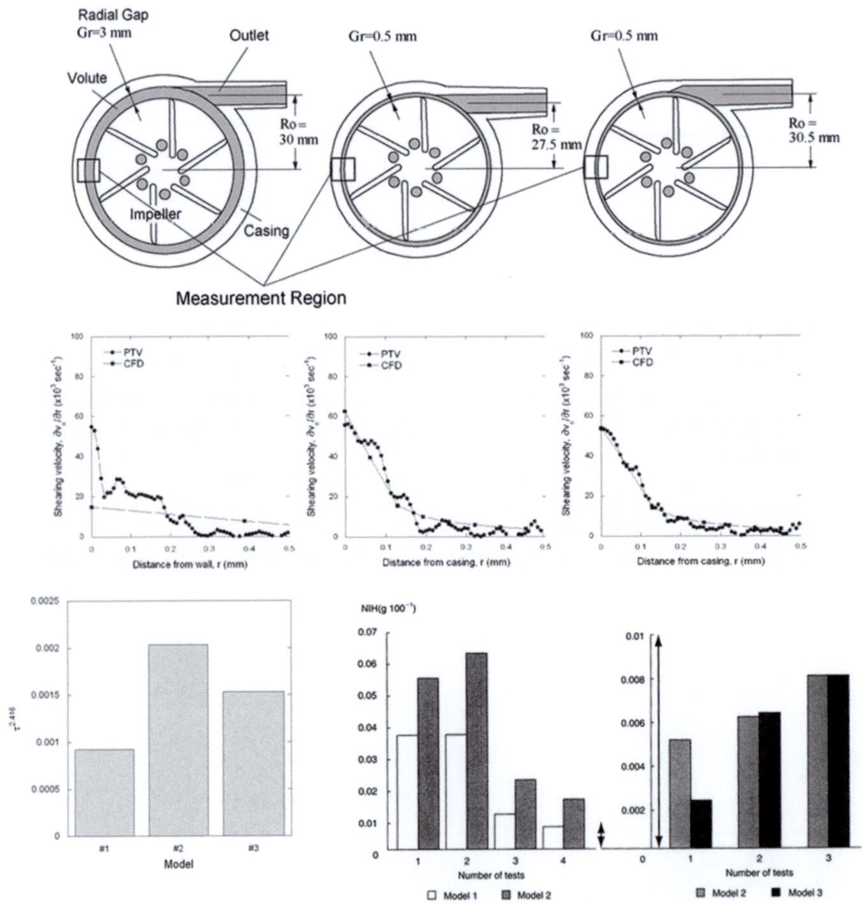

Fig. 13.20 Relationship between high shear in the vicinity of the wall surface and hemolytic properties of tested pumps (Nishida et al. 2006b). *Upper row*: pump geometries; *middle row*: shear rates; *lower row*: summations of shear stress $\tau^{2.416}$ and hemolysis test results)

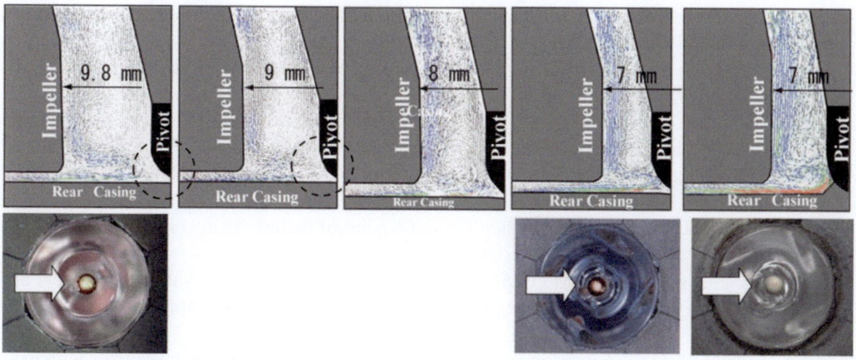

Fig. 13.21 Optimal design of the monopivot centrifugal pump using flow quantification and anti-thrombogenicity (Nishida et al. 1997). *Upper row*: PIV results; *lower row*: anti-thrombogenesis test results; *dotted circle*: stagnation region

13.6.2.4 PIV of Thrombus Formation

The mechanism of thrombus formation is more complex than that of hemolysis. The thrombus is generally considered to form in the low shear region, which is a region of flow stagnation, although it sometimes forms in the narrow gap and by hemolysis. Figure 13.21 shows the example of the optimal design of the blood pump using flow quantification and anti-thrombogenicity (Toyoda et al. 2002). If photographing particles is possible, the flow analysis of the stagnation region itself is not very difficult. However, it is usually difficult to photograph the flow in the stagnation region because it is located around the rotating axis.

13.7 Summary of This Chapter

This chapter presents current VADs and CPB pumps, their specifications, their classifications, and methods for designing and evaluating them, including flow analysis inside the pump to optimize the geometries. One particular VAD has exceeded 8 years of clinical durability and has provided a high quality of life for its wearers. In the future, VADs and CPB pumps will require enhanced performance and durability and will be adaptable to individual cases.

References

Adrian RJ, Westerweel J (2011) Particle image velocimetry. Cambridge University Press, New York

Affeld K et al (1976) New methods for the *in vitro* investigations of the flow patterns in artificial heart. Trans Am Soc Artif Intern Organs 22:460–467

Antaki JF et al (1995) Computational flow optimization of rotary blood pump components. Artif Organs 19(7):608–615

Apel J, Neudel F, Reul H (2001a) Computational fluid dynamics and experimental validation of a microaxial blood pump. ASAIO J 47(5):552–558

Apel J et al (2001b) Assessment of hemolysis related quantities in a microaxial blood pump by computational fluid dynamics. Artif Organs 25(5):341–347

Araki K, Taenaka Y, Masuzawa T et al (1993) A flow visualization study centrifugal blood pumps developed for long-term usage. Artif Organs 17(5):307–312

Arora D, Behr M, Pasquali M (2004) A tensor-based measure for estimating blood damage. Artif Organs 28(11):1002–1015

ASTM F1841-97 (1997) Standard practice for assessment of hemolysis in continuous flow blood pumps

Asztalos B, Yamane T, Nishida M (1999) Flow visualization analysis for evaluation of shear and recirculation in a new closed-type, monopivot centrifugal blood pump. Artif Organs 23 (10):939–946

Baldwin JT et al (1994) LDA measurements of mean velocity and Reynolds stress fields with in an artificial heart ventricle. J Biomech Eng 116:190–200

Blackshear PL, Dorman FD, Steinbach JH (1965) Some mechanical effects that influence hemolysis. Trans Am Soc Artif Intern Organs 11:112–120

Bludszweit C (1995) Three-dimensional numerical prediction of stress loading of blood particles in a centrifugal pump. Artif Organs 19(7):590–596

Burgreen GW, Antaki JF et al (2001) Computational fluid dynamics as a development tool for rotary blood pumps. Artif Organs 25(5):336–340

Burgreen GW et al (2004) Computational fluid dynamics analysis of a Maglev centrifugal left ventricular assist device. Artif Organs 28(10):874–880

Butler K, Thomas D, Antaki JF et al (1997) Development of the Nimbus/Pittsburgh axial flow left ventricular assist system. Artif Organs 21(7):602–610

Chan WK et al (2002) A computational study of the effects of inlet guide vanes on the performance of a centrifugal blood pump. Artif Organs 26(6):534–542

Chua LP et al (2002) Gap velocity measurements of a blood pump model. Artif Organs 26(8):682–694

Cooney DO (1976) Biomedical engineering principles. Dekker, New York

DeBakey ME (1934) A simple continuous flow blood transfusion instrument. New Orleans Med Surg J 87:386–389

Garon A, Farinas MI (2004) Fast three-dimensional numerical hemolysis approximation. Artif Organs 28(11):1016–1025

Gibbon JH Jr (1937) Artificial maintenance of circulation during experimental occlusion of pulmonary artery. Arch Surg 34:1150

Gibbon JH Jr, Miller BJ, Feinberg C (1953) An improved mechanical heart and lung apparatus. Med Clin N Am 37:1603

Giersiepen M, Wurzinger LJ et al (1990) Estimation of shear stress-related blood damage in heart valve prostheses: in vitro comparison of 25 aortic valves. Int J Artif Organs 13:300–306

Gobel C et al (2001) Development of the MEDOS/HIA DeltaStream extracorporeal rotary blood pump. Artif Organs 25(5):358–365

Golding LAR et al (1980) Chronic non-pulsatile blood flow in alive, awake animal 34-day survival. Trans Am Soc Artif Intern Organs 26:251–254

Goubergrits L, Affeld K (2004) Numerical estimation of blood damage in artificial organs. Artif Organs 28(5):499–507

Helmus JD, Brown CH (1977) Blood cell damage by mechanical forces. In: Hwang NHC, Normann NA (eds) Cardiovascular flow dynamics and measurement. University Park Press, Baltimore, pp 799–823

Heuser G (1980) A Couette viscometer for short time shearing of blood. Biorheology 17:17–24

Ikeda T, Yamane T et al (1996) A quantitative visualization study of flow in a scaled-up model of a centrifugal blood pump. Artif Organs 20(2):132–138

Imachi K, Mussivand T (2010) Outline of the International Organization for Standardization Standard for circulatory support devices. Artif Organs 34:695–698

ISO 14708–5 (2010) Implants for surgery: active implantable medical devices. Part 5: Circulatory support devices

ISO 5198 (1987) Centrifugal, mixed flow and axial flow pumps. Code for hydraulic performance tests: precision grade

Joyce DL, Joyce LD, Loebe M (eds) (2012) Mechanical circulatory support: principles and applications. McGraw-Hill, New York

Japan Society of Mechanical Engineers (ed) (1988) Fundamentals of computational fluid dynamics (in Japanese). Corona, Tokyo

Kawahito K, Nosé Y (1997) Hemolysis in different centrifugal pumps. Artif Organs 21:323–326

Kerrigan JP, Yamazaki K et al (1996) High-resolution fluorescent particle-tracking flow visualization within an intraventricular axial flow left ventricular assist device. Artif Organs 20 (6):534–540

Kobayashi T et al (1989) Velocity measurement of three -dimensional flow around rotating parallel disks by digital image processing. ASME FED 85:29–36

Kyo S (ed) (2014) Ventricular assist devices in advanced-stage heart failure. Springer, Heidelberg

Mitamura Y et al (2001) Prediction of hemolysis in rotary blood pumps with computational fluid dynamics analysis. J Congest Heart Fail Circ Support 1(4):331–336

Miyazoe Y et al (1998) Computational fluid dynamic analysis to establish design process of centrifugal blood pump. Artif Organs 22(5):381–385

Mizuguchi K et al (1995) Development of an axial flow ventricular assist device: in vitro and in vivo evaluation. Artif Organs 19(7):653–659

Nakamura S, Yano K (1999) Computational simulation of flows in an entire centrifugal heart pump. Artif Organs 23(6):572–575

Nakamura S et al (1999) Numeric flow simulation for an innovative ventricular assist system secondary impeller. ASAIO J 45(1):74–78

Nishida M, Yamane T et al (1997) Quantitative visualization of flow through a centrifugal blood pump: effect of washout holes. Artif Organs 21(7):720–729

Nishida M, Asztalos B, Yamane T et al (1999) Flow visualization study to improve hemocompatibility of a centrifugal blood pump. Artif Organs 23(8):697–703

Nishida M, Yamane T et al (2006a) Computational fluid dynamic analysis of the flow around the pivot bearing of the centrifugal ventricular assist device. JSME Int J Ser C 49(3):837–851

Nishida M et al (2006b) Quantitative Analysis of shearing velocity near the wall in a centrifugal blood pump, JSME Fluids Engineering Conference 2006 (CD-ROM): 812; 1–4

Oka S (1974) Biorheology (in Japanese). Shokabo, Tokyo

Pinotti M, Paone N (1996) Estimating mechanical blood trauma in a centrifugal blood pump: laser Doppler anemometer measurements of the mean velocity field. Artif Organs 20(6):546–552

Pinotti M, Rosa ES (1995) Computational prediction of hemolysis in a centrifugal ventricular assist device. Artif Organs 19(3):267–273

Pantalos GM et al (1998) Long-term mechanical circulatory support system reliability recommendation: American Society for Artificial Internal Organs and The Society of Thoracic Surgeons: long-term mechanical circulatory support system reliability recommendation. Ann Thorac Surg 66(5):1852–9.

Qian Y (2004) CFD application in Ventrassist implantable rotary blood pump design and validation. In: Proceedings, Bioengineering Conference Annual Meeting BED JSME, Kitakyushu, 22–23 January 2004, pp 259–260

Qian Y, Bertram CD (2000) Computational fluid dynamics analysis of hydrodynamic bearings of the VentrAssist rotary blood pump. Artif Organs 24(6):488–491

Rafferty EH et al (1968) Artificial heart II: application of nonpulsatile radially increasing pressure gradient pumping principle. Minn Med 51:191–193

Reul HM, Akdis M (2000) Blood pumps for circulatory support. Perfusion 15:295–311

Schima H et al (1992) Effect of stationary guiding vanes on improvement of the washout behind the rotor in centrifugal blood pumps. ASAIO J 38:220–224

Song X et al (2004) Quantitative evaluation of the blood damage in a centrifugal blood pump by computational fluid dynamics. J Fluids Eng 126(3):410–418

Stepanoff AJ (1957) Centrifugal and axial flow pumps, 2nd edn. Wiley, New York

Takatani S, Matsuda H, Hanatani A, Nojiri C, Yamazaki K, Motomura T, Ohuchi K, Sakamoto T, Yamane T (2005) Mechanical circulatory support devices (MCSD) in Japan: current status and future directions. J Artif Organs 8:13–27

Toyoda M et al (2002) Geometric optimization for non-thrombogenicity of a centrifugal blood pump through flow visualization. JSME Int J C45(4):1013–1019

Triep M et al (2006) Computational fluid dynamics and digital particle image velocimetry study of the flow through an optimized micro-axial blood pump. Artif Organs 30(5):384–391

Tsukamoto Y et al (2001) Computational fluid dynamics analysis for centrifugal blood pumps. J Congest Heart Fail Circ Support 1(4):337–343

Tsukiya T, Taenaka Y et al (2002) Improvement of washout flow in a centrifugal blood pump by a semi-open impeller. ASAIO J 48(1):76–82

Umezu M et al (1992) The effects of inter shapes of plastic connectors on blood in a extracorporeal circulation. In: Proceedings, 7th International Conference on Biomedical Engineering, Singapore, 2–4 December 1992, pp 197–199

Wells RE, Merrill EW, Gabelnick H (1962) Shear-rate dependence of viscosity of blood: interaction of red cells and plasma proteins. Trans Soc Rheol 6:19–24

Wood HG et al (1999) Numerical solution for blood flow in a centrifugal assist device. Int J Artif Organs 22(12):827–836

Wurzinger LJ et al (1985) Platelet and coagulation parameters following millisecond exposure to laminar shear stress. Thromb Haemost 54:381–386

Yamane T (2002) The present and future state of nonpulsatile artificial heart technology. J Artif Organs 5:149–155

Yamane T et al (2010) Japanese guidance for ventricular assist devices/total artificial hearts. Artif Organs 34:699–702

Innovations in Measuring Cellular Mechanics

14

Navid Bonakdar, Achim Schilling, Richard Gerum, José Luis Alonso, and Wolfgang H. Goldmann

Abstract
This article describes several novel mechanical methods for elucidating cellular responses to different types of mechanical loading (adhesive, pulling, pushing, shearing, and stretching forces). Understanding how cells deform and transmit stresses into the cell is important for gene expression, cytoskeletal remodeling, and focal adhesion reorganization and crucial for a variety of higher fundamental cell functions including cell division, motility, and differentiation. Introducing these unique methods of measuring and understanding cellular mechanics, therefore, provides a valuable platform for cell biology research.

Keywords
Magnetic tweezer • Magnetic twisting cytometry • Traction force microscopy • Cell poking • Plate rheometer • Nano-scale tracking

14.1 Introduction

14.1.1 Muscle Physiology, Biomechanics, and Biochemistry

Biomechanical studies of single cells and cell populations have been of interest for a while, and in recent years applied methods have experienced rapid development (Goldmann 2000, 2002; Mierke et al. 2008, 2010; Möhl et al. 2009; Lange

N. Bonakdar • A. Schilling • R. Gerum • W.H. Goldmann (✉)
Department of Physics, Biophysics Group, University of Erlangen-Nuremberg/Friedrich-Alexander-University (FAU), Henkestr. 91, 91052 Erlangen, Germany
e-mail: wgoldmann@biomed.uni-erlangen.de

J.L. Alonso
Department of Medicine, Mass. General Hospital/Harvard Medical School, Building 149, 13th Street, Charlestown, MA 02129, USA

K. Tanishita, K. Yamamoto (eds.), *Vascular Engineering*,
DOI 10.1007/978-4-431-54801-0_14

et al. 2013). Measuring cell mechanical properties in the field of muscle physiology, which has a century-old tradition, has contributed significantly to the current understanding of the mechanisms of muscle contraction (Huxley 1957; Ra et al. 1999). The dynamics and kinetics of biochemical reactions between the most important muscle contractile proteins, actin and myosin, were determined with high accuracy and temporal resolution based on relatively simple measurements of force and length changes of the skeletal muscle as determined by Hill (1965). The regular, almost crystalline, arrangement in the muscle contractile apparatus allowed the use of simple mathematical models to interpret the measurements obtained macroscopically in relation to molecular processes (Huxley 1957; Hill 1965; Kawai and Brandt 1980).

In nonmuscle cells, however, the conditions are more complex. Their diverse mechanical functions including migration, division, phagocytosis are determined by myosin motors, but the numerous regulatory molecules and structure-forming proteins of the cytoskeleton are also significantly involved (Hartmann and Spudich 2012). This complexity is often exasperated by the time-varying irregular network-like structures of the cytoskeleton. In nonmuscle cells, the assignment of mechanical measurements and biochemical processes are not clear, particularly the absence of macroscopic tissue that embeds cells. Nonetheless, tests can be carried out at the cellular level. In the following paragraphs, some important methods for measuring mechanical properties at the cellular level and some common conceptual models for their interpretation are listed.

14.1.2 Methods for Measuring Cell Mechanical Properties

James Crick and Arthur Hughes developed in the 1950s a method for quantitative measurements of rheological (mechanical) properties of individual cells under culture conditions (Crick and Hughes 1950). The method was based on microscopically small (about 5 μm) acicular ferromagnetic particles, which had been endocytosed or phagocytosed by cells. These particles were magnetized along their major axis and then subjected to a magnetic field in the plane perpendicular to the major axis, which results in a rotation in the direction of the magnetic field. The rotation angle and the rotation speed of the particles were recorded by a camera, and their movement gives information about the mechanical properties of the cell.

In the 1970s, Brain and Cohen replaced the complex evaluation of particle rotation by filming magnetometric measurements (Cohen et al. 1979), which was also used later by Valberg et al. (Valberg and Albertini 1985). Because of the limited sensitivity of the magnetometer, several thousand particles had to be added to the cell culture, and the behavior of individual cells could not be observed. However, the method was suitable for measurements in living animals, for instance, in determining the mechanical properties of alveolar macrophages. The limitation of the method for examining only endocytotic or phagocytotic cells was overcome in the 1990s by Wang et al. (1993), who used magnetic particles coated with peptides, proteins, or antibodies that bind to specific receptors on the cell surface. The restriction of measuring only the behavior of cell populations was resolved in 2001 by Fabry et al. (2001a, b) using computerized video analysis of the microscopic particle motion on individual cells.

Over the past 25 years, many other methods for measuring the mechanical properties of cells in culture have been developed. These include microindentation (cell poking) (Goldmann 2000; Zahalak et al. 1990), aspiration of cells into micropipettes (Evans and Yeung 1989; Merkel 2001), measurements of diffusive motion of intracellular particles or granuoles (Caspi et al. 2002), atomic force microscopy (Moy et al. 1994; Goldmann et al. 1998; Alcaraz et al. 2003), disk-microrheometry (Müller et al. 1991; Goldmann and Ezzell 1996; Thoumine and Ott 1997), and diffusion measurements of receptor-bound particles (beads) in optical traps (optical tweezer) (Choquet et al. 1997) or in magnetic field gradients (Alenghat et al. 2000). All these methods have their own advantages and limitations for specific applications. A selection of the methods is described below in detail: (i) nano-scale particle tracking, (ii) magnetic tweezer, (iii) rotation disc rheometer, (iv) magnetic twisting cytometry, (v) cell poking, and (vi) traction microscopy for the application in general biology research.

14.2 Methods

14.2.1 Nanoscale Particle Tracking

To describe active dynamic, mechanical operations such as cytoskeletal remodeling processes in cell migration, cell proliferation, adhesion, or contraction quantitatively, the method of nanoparticle tracking is ideally suited (An et al. 2004; Raupach et al. 2007; Metzner et al. 2007, 2010; Alonso and Goldmann 2012). The method is based on the assumption that a bead (<1 μm diameter) is firmly bound to the cytoskeleton via integrins and focal adhesions and moves when the actin cytoskeleton is spontaneously restructured, i.e., without the influence of (e.g., external magnetic) forces. The bead movement, therefore, reflects the extent and speed of cytoskeletal reorganization. The impromptu bead motion can be determined by subpixel arithmetic with nanometer accuracy (Fig. 14.1).

14.2.2 Magnetic Tweezer

When focal adhesion contacts of adherent cells are stimulated by external forces, they respond by reconstructing and reinforcing the contacts (focal adhesion strengthening) (Goldmann 2002; Choquet et al. 1997; Giannone et al. 2003; Deng et al. 2004). A controlled generation of such forces allows studying dynamic processes in great detail. For this purpose, a magnetic tweezer was developed with which forces of up to 10 nN on superparamagnetic beads of 4.5 μm diameter can be applied. The design of the magnetic tweezer was based on a prototype developed by

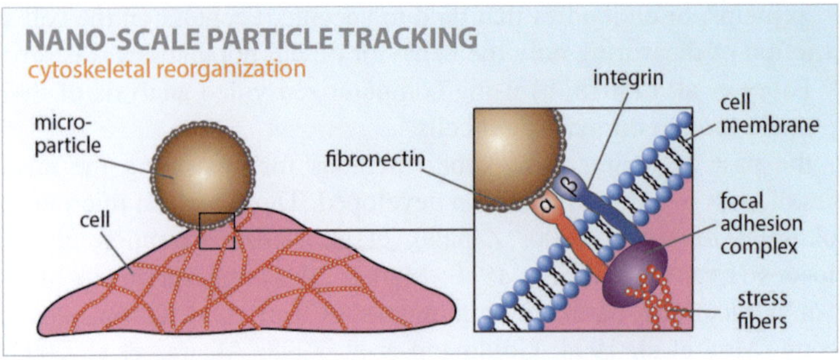

Fig. 14.1 Nano-scale particle tracking. Schematic representation of a < 1 μm RGD-coated polystyrene bead attached to integrins that span the cell membrane (*left*). Integrins are connected *via* focal adhesions to the actin cytoskeleton (inset, *right*)

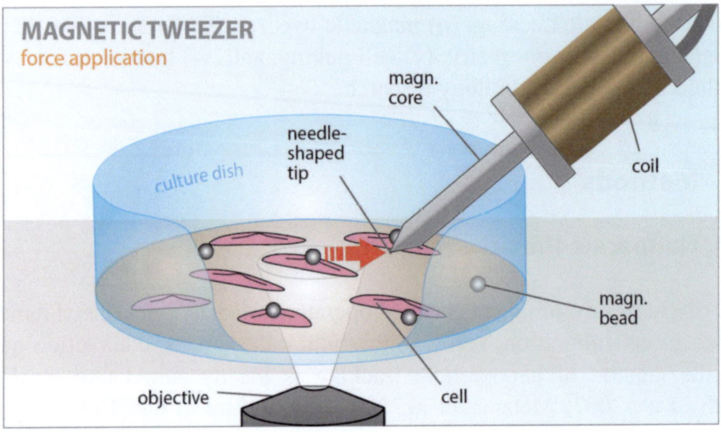

Fig. 14.2 Schematic image of the magnetic tweezer experimental setup. A high magnetic field gradient is generated by a needle-shaped high-permeability core of a solenoid attached to a micromanipulator. The gradient force generated by the magnetic tweezer acts on superparamagnetic beads coated with extracellular matrix proteins. Beads coated with extracellular matrix proteins are bound to the cell surface *via* integrin receptors, which connect the extracellular space with the intracellular cytoskeleton. The bead displacement is the basis for the calculation of cellular stiffness (Bonakdar et al. 2012, 2014, 2015)

Alenghat et al. (2000), which allowed force generation up to about only 1 nN. The latest generation of magnetic tweezers exerts forces up to 100 nN (Kollmannsberger and Fabry 2011).

Figure 14.2 shows the schematic image of a magnetic tweezer setup. The tweezer is composed of a soft, superparamagnetic iron alloy (HyMU80) core, which is tapered at one end, and a surrounding magnetic induction coil (copper wire of 0.5 mm diameter and 250 turns), which magnetizes the core with a maximum current of 3 A and a power up to 10 W. The current for the coil is generated by a microprocessor-controlled current source, which is connected to the

PC via a DA converter board. Near the core tip, the magnetic field shows a strong inhomogeneity, i.e., a high local field gradient. A superparamagnetic bead in close vicinity is magnetized and pulled in the direction of the strongest field gradient. By coating the beads with ligands, e.g., proteins of the extracellular matrix, the beads can adhere to specific receptors of the cell and transfer controlled forces. The lateral bead movement toward the needle tip allows the calculation of the mechanical properties of the cell (Alenghat et al. 2000). Because of higher forces per bead in the nano-Newton range compared to other methods (laser tweezer in pN range), the magnetic tweezer is a versatile tool that is suitable for the investigation of focal adhesion reinforcement and many other cellular processes (Kollmannsberger and Fabry 2007).

14.2.3 Rotation Disk Rheometer

Important advances have been made in recent years regarding the investigation of passive rheological properties of living cells occurring within short timescales. Rheological experiments that probe at the micron level have shown broad power-law frequency responses. In Fig. 14.3, we briefly describe the apparatus and details of the data analysis used by Müller et al. (1991). The apparatus consists of a cylindrical glass cuvette containing the cells, a metal disk on top of the cells, and two pairs of surrounding perpendicularly oriented magnetic coils. One of these serves to fix the orientation of the disk and the other (the deflection coil) to apply an

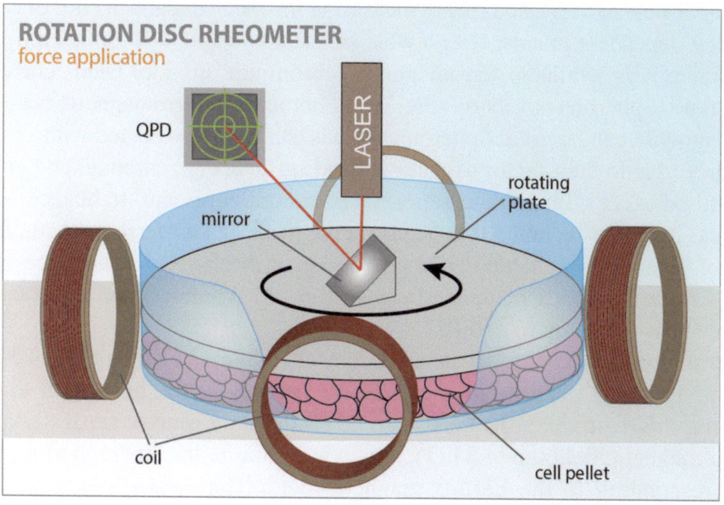

Fig. 14.3 A schematic representation of the rotation disc rheometer setup designed to perform cell rheology measurements. Two pairs of coils serve to orient the mirror on top of the metal plate and to oscillate the disc by a small amplitude. A laser beam hits the mirror and is reflected onto a quadrant photodiode. The rotation of the disc leads to a lateral deflection of the laser spot on the detector

oscillatory shear force to the cell pellet. The "in-phase" and "out-phase" components of the resulting rotational amplitude $\alpha(t)$ of the disk are analyzed as follows: The beam of a He–Ne laser in a direction vertical to the rotational axis of the disk is horizontally deflected by the mirror mounted on the disk, and its horizontal orientation is recorded by a gradient photodiode. The complex shear modulus $G^*(\varpi\Psi\psi\rho) = G'(\varpi\Psi\psi\rho) + G''(\varpi\Psi\psi\rho)$, where $G'(\varpi\Psi\psi\rho)$ is the frequency-dependent storage modulus and $G''(\varpi\Psi\psi\rho)$ is the loss modulus, was determined for a pellet of ~5.5×10^7 F9 wild-type and vinculin-deficient cells (Goldmann and Ezzell 1996). For further information regarding disk rheometers, refer to (Fernandez et al. 2007).

14.2.4 Magnetic Twisting Cytometry (MTC) with Optical Detection

14.2.4.1 Principle

The basic idea of the MTC is to manipulate living, adherent cells mechanically in a microrheometer to measure their mechanical properties. Similar to the conventional rheometer, in which the material to be examined is sheared between two plates, the cells are sheared between a rigid plate (disk) and magnetically rotated ("twisted") ferromagnetic beads adhered to the cells. From the relationship between the generated magnetic shear forces and the measured deformation of the cells, the constitutive mechanical properties of the living cells can be derived. The advantage of this method is the possibility (i) of coating the beads with different ligands and thus coupling them to specific receptors, (ii) of tracking several (up to 100) cells simultaneously per field of view, and (iii) of measuring the mechanical behavior of cells in a frequency-dependent manner over a wide frequency range (0.01–1000 Hz).

Commercially available ferromagnetic (chromium dioxide) beads coated with polystyrene (Spherotech, Libertyville, IL) or homemade ferromagnetic beads made from magnetite can be used. Spherotech beads can easily be coated with secondary antibodies (due to their factory-functionalized surface, e.g., carboxylated or coated with antibodies) or with proteins (fluorescent or nonfluorescent) to bind cell surface receptors. It should be noted that beads on the basis of low-ferromagnetic fraction (<40 vol.%) are only suitable for the production of mechanical shear forces of up to 50 Pa. Magnetite beads of defined sizes (1–6 μm diameter) can be obtained to produce shear forces up to 150 Pa.

After the beads are bound to the cells, they are magnetized by a short (10 μs) and strong (>150 mT) pulse from a permanent magnet in the horizontal direction. Following this procedure, the beads can be rotated in a weaker vertically aligned, uniform magnetic field (up to 8 mT), which is similar to the situation of a compass needle responding to the Earth's magnetic field. The torque T acting on each magnetic bead depends on the magnetization constant c of the beads and the rotation angle θ between the magnetic moment M and the magnetic field H according to the relation $\mathrm{T} = \mathrm{c} \times \mathrm{H} \times \cos\,\theta$ (Wang et al. 1993; Fabry et al. 2001a). The magnetization constant c is calibrated by measuring the rotational velocity of the beads in a viscous standard. For Spherotech beads, c is about

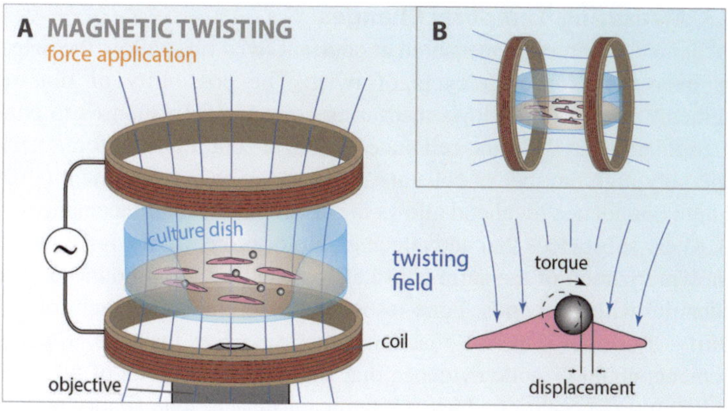

Fig. 14.4 Magnetic twisting device. Prior to experimentation, cells are incubated for 20 min with magnetite beads that are coated with, e.g., a synthetic peptide (RGD) that in turn connects the beads with the cellular actin cytoskeleton *via* integrin receptors. A brief (10 μs) but strong (>150 mT) magnetic pulse in the horizontal direction (parallel to the culture surface), using one pair of magnetic coils, is applied (*top, right*, **B**). Then after several seconds, a much weaker magnetic twisting field (up to 8 mT) in the vertical direction is applied (*left*, **A**). The vertical magnetic field produces a torque of the beads, which depends on the mechanical properties of the underlying cells in response to the deflection of the beads. The bead rotation (displacement) can be measured microscopically with high accuracy (*bottom, right*) (Goldmann et al. 2000)

5 Pa/mT and for magnetite beads 20 Pa/mT. With the usually small rotation angles of less than 5°, cos θ is negligible. The time characteristic of the homogeneous magnetic field is varied sinusoidally, which causes a rotation and movement of the beads with the same frequency along the direction of magnetization (Fig. 14.4).

The amplitude of the bead movement is typically around 100 nm but may vary depending on the cell type and other factors such as coating by up to two orders of magnitude. The bead movement can be directly visualized by a microscope with a CCD camera, and bead positions can be highly resolved with subpixel arithmetic. The round shape of the beads allows analysis via a simple center-of-mass algorithm. About 100 cells can be measured simultaneously, which ensures a high statistical confidence of the results. The image acquisition of the CCD camera is synchronized with the generation of the magnetic field. This is achieved by a phase-synchronous control of the current source and by means of a real-time trigger microcontroller of the camera (Fabry et al. 2001a).

14.2.4.2 Calculating Cell Mechanical Properties

Forced bead movement in the magnetic field gives rise to a deformation of the cell and of bead-associated intracellular structures. Within the cell, mechanical shear forces are generated opposite to the bead movement. Consequently, the mechanical properties can be derived from the relationship between the magnetically generated mechanical torque and bead movement. Due to periodic excitation, Fourier analysis can be used to calculate the mechanical properties, greatly simplifying the calculation and making it substantially insensitive to noise effects (Fabry et al. 2001a).

14.2.4.3 Measuring Temporal Changes

Cell mechanical parameters measured at constant twist rates allow the detection of changes over time for each cycle of twist. The possibility of time-resolved measurements using MTC allows measuring cell-specific responses to pharmacological stimulations in the same cell under the same conditions (before/after). Since a considerably high number of cells are measured in each experiment (~100), this gives a high confidence level and allows the detection of small mechanical changes (<5 %). Many substances that activate, e.g., motor proteins, or trigger polymerization/depolymerization of the actin cytoskeleton can lead to pronounced changes in cell mechanics within seconds. Time-resolved measurements are therefore essential to quantify dynamic (kinetic) reaction processes. For instance, time-resolved measurements could provide evidence that smooth muscle cells of the respiratory tract under culture conditions have a significant muscle tone (Fabry et al. 2001b) and that individual muscle cells show considerable heterogeneity regarding contractile behavior (Fabry et al. 2001b) or that mechanical stimulation of muscle cells leads to actin polymerization (Deng et al. 2004).

14.2.4.4 Measuring the Frequency Response

The constitutive mechanical properties of engineering materials are usually determined from the step or frequency responses of the material deformation. A similar approach has also been used in traditional muscle physiology. The frequency dependence of different types of cells using various agonists and different bead (coating) attachments over a large frequency range has been measured by MTC (Fabry et al. 2001a, 2003; Puig-de-Morales et al. 2004). The main aim of these studies was to make predictions about which molecular processes dominate in cell mechanics; i.e., dominant molecular processes should have characteristic relaxation times, which are reflected in a characteristic frequency response. This hypothesis could not be proven. Instead, a surprising power-law behavior was determined for cellular mechanical properties, for which no explanation in the literature of cell biology existed (Fabry et al. 2001a, 2003; Puig-de-Morales et al. 2004). The power-law behavior of cell mechanical properties can be described empirically by the so-called structural damping equation. This has been used for a systematic investigation of cell mechanics, in which cells were stimulated with contractile, relaxant, or cytoskeleton-changing agonists (histamine, bradykinin, ML-7, ML-9, BDM, wortmannin, N-formyl-methionyl-leucylphenylalanine (FMLP), N6, 2′-O-dibutyryladenosine 3′, 5′-cyclic monophosphate (DBcAMP), latrunculin A, jasplakinolide, and cytochalasin D) (Fabry et al. 2003; Puig-de-Morales et al. 2004). All mechanical cell responses were mainly determined by one parameter, while the parameters in the power-law remained largely unchanged (Fabry et al. 2003; Puig-de-Morales et al. 2004). This observation was confirmed in different cell types (macrophages, leukocytes, epithelial cells, endothelial cells, and fibroblasts) and with different ligations to the beads (e.g., AcLDL, urokinase, and different activating and nonactivating antibodies against various β-integrin subunits) (Fabry et al. 2003; Puig-de-Morales et al. 2004).

14.2.5 Cell Poking

The cell poking device was developed by Dr. Elliot Elson's group to measure the forces required to indent rigid surfaces, for instance, the cell membranes of circulating blood cells and adherent cells (Daily et al. 1984, Duszyk et al. 1989). Their aim was to record the dynamic response of the cell membrane to external signals in order to quantitatively investigate changes in cell shape and cytoskeletal organization associated with physiological processes. These researchers provided the theoretical basis for measuring elastic and viscous material properties resulting from tension generation at the lipid/protein interface, i.e., at the cell membrane-cytoskeleton connection (Zahalak et al. 1990). We have used this biophysical technique of cell indentation or poking to manipulate and probe cell structure and mechanics. The principle is shown in Fig. 14.5. A cell adhered to the bottom of a culture dish is indented by a glass stylus. In order to prevent attachment to the cell, the glass stylus is coated with BSA. With a high-speed video camera, the relaxation of the cell membrane after poking and retraction of the glass stylus is recorded and the relaxation rate determined from video analysis using an elastic response theory. This method has led to further insight into how cellular viscoelasticity is regulated based on the cell architecture, where mechanical, biological, and biochemical processes are tightly coupled. Better understanding of how cells control these functions will require further optimization of the present techniques and inclusion of methods for controlling and quantitating changes in the cell (Goldmann 2000).

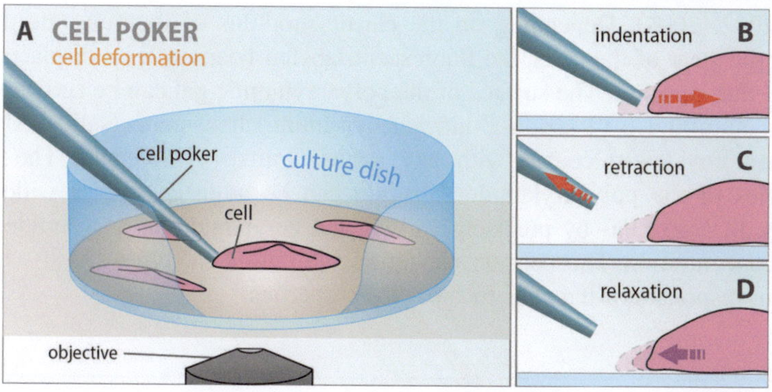

Fig. 14.5 Schematic representation of the cell poking device and detection by microscopy (**A**). The adherent cell is indented by a glass stylus that is coated with BSA to prevent its attachment to the cell (**B**). Images of the cell edge are taken after poking and immediate retraction of the glass stylus (**C**). The relaxation of the cell membrane after deformation is recorded by a high-speed video camera (**D**). The rate of relaxation is determined from the analysis of video frames using an elastic response theory (Goldmann 2000)

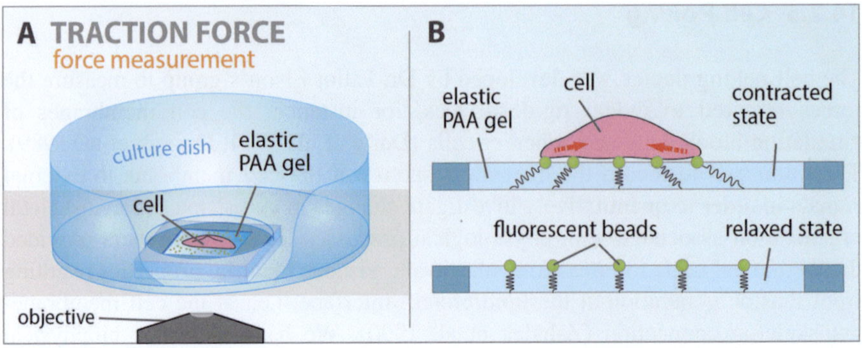

Fig. 14.6 Basic principle of traction force microscopy (**A**). Cells are seeded on an elastic polyacridamide (PAA) gel containing 200 nm diameter fluorescent beads embedded on the surface, which serve as markers to visualize any gel deformation due to cell contraction (**B**). The traction forces are calculated from the gel deformation; i.e., relaxed versus contracted state (as shear stress in Pa) caused by the cells (Bonakdar et al. 2014, 2015)

14.2.6 Fourier Transform Traction Force Microscopy

Adherent cells transfer forces to the extracellular matrix. When cells are cultured on a polyacrylamide elastic substrate coated with extracellular proteins in which fluorescent markers (e.g., 200 nm FluoSpheres) are embedded (Fig. 14.6), such forces can be optically determined (Deng et al. 2004; Butler et al. 2002; Stamenovic et al. 2002, 2004). Depending on the elastic modulus of the substrate and the contractile state of the cells, the fluorescent-labeled beads will be displaced from their resting position. The surface of the polyacrylamide gel can be activated with sulfosuccinimidyl 6 (4′-azido-2′-nitrophenyl-amino) hexanoate (sulfo-SANPAH) and then covalently coated with any adhesive matrix molecule. The elastic properties of the polyacrylamide substrate can be adjusted over a wide range (100 Pa to >50 kPa) by precisely varying the acrylamide/bis-acrylamide cross-linker concentration. This permits adaptation to different cell types and also induces different responses of the cells to substrate properties.

14.3 Method Application to Smooth Muscle Cells

To determine the contractile response of smooth muscle cells under culture conditions, the methods of traction force microscopy and optical magnetic twisting cytometry (MTC) can be applied. Traction force microscopy allows the direct determination of the contractile moment of individual cells, while MTC gives information about cell elasticity (G'). Classic muscle physiology predicts a linear relationship between contractile strength and elasticity of a muscle; for instance, when a muscle is activated, each individual actomyosin binding contributes to both the strength and the elasticity of the entire muscle. A linear relationship even at the cellular level has been reported (Butler et al. 2002; Stamenovic et al. 2004). Therefore, it seems reasonable to equate the cytomagnetometrically measured temporal changes of G' with a proportional change in force development. The advantage of this approach is the high statistical quality of cytomagnetometric measurements (up to 100 individual cells per measurement) and the high temporal resolution compared with traction force microscopic measurements. Interestingly, the linear relationship between cell contraction stress P and cell elasticity G' is lost when the mechanical properties are determined with higher frequencies of >1 Hz. Therefore, the deviation from the linear relationship between G′ and contractile force is related to the frequency dependence of G′. This relationship could possibly be a direct function of the power-law exponent of the acting cytoskeletal contractile force (prestress) (Stamenovic et al. 2004; Gardel et al. 2004).

To quantify cytoskeletal remodeling processes in smooth muscle cells, the nanoscale particle tracking method can also be applied (An et al. 2004). From the bead movement (expressed by the mean square displacement) within a defined time interval, the effective diffusion coefficient that increases with increasing speed of cytoskeletal remodeling can be determined. Cytoskeletal remodeling processes can be directly manipulated pharmacologically by using, for instance, cytochalasin D or latrunculin A (leading to actin depolymerization) or jasplakinolide (stabilizing the actin cytoskeleton). Cytoskeletal conversion processes can also be affected indirectly, for example, by stimulating cells with contractile (histamine or bradykinin) and relaxant (isoproteronol or DBcAMP) substances.

14.3.1 Relationship Between Contraction and Cytoskeletal Remodeling in Airway Smooth Muscle Cells

In human respiratory muscle cells, depending on the dynamics of the cytoskeleton remodeling induced pharmacologically over a wide range, the beads showed a more or less pronounced superdiffusive motion (An et al. 2004). Likely causes for this are active processes that keep the cytoskeletal matrix in a thermodynamic imbalance. The exact molecular details of the superdiffusive behavior are still unclear. It was shown, however, that metabolic, ATP-consuming processes influence spontaneous bead motion strongly; for instance, when the ATP turnover of cells was reduced with 2 mM deoxyglucose and 2 mM NaN_3 to about 5 % of baseline activity, the transition from subdiffusive to superdiffusive behavior was observed only at Δt of ~ 5 s (An et al. 2004).

Cytoskeletal remodeling processes can also be made directly visible, either by rhodamine-phalloidin staining of fixed cells (An et al. 2004; Deng et al. 2004) or with transfection of GFP-actin (Hu et al. 2003). Numerous experiments not only confirm the already suspected correlation between contraction and cytoskeletal remodeling in airway smooth muscle cells but also suggest a previously undescribed close temporal coupling of these two processes. Data showed under existing baseline conditions that the cytoskeleton is affected by the baseline tone of cells; i.e., the baseline cytoskeleton is too weak to transmit additional forces to the extracellular matrix resulting from myosin activation. For this, it needs a reinforcement of the existing cytoskeletal and adhesive structures. It is particularly surprising that the reinforcement and restructuring of the cytoskeleton can occur with the same speed as the contractile activation of muscle cells, namely, within a few seconds. Such a coupling between contraction and cytoskeletal remodeling has been described only for longer time intervals (Balaban et al. 2001; Gunst and Fredberg 2003).

The relationship between contractile forces and cell elasticity under steady state is well documented in smooth muscle cells. However, doubts about this relationship were reported (An et al. 2002). For example, myosin inhibitors such as 2,3-butanedione monoximes (BDM), wortmannin, or ML-7 are not able to completely block the increase in elasticity of cells after stimulation (Smith et al. 2007). A possible reason for this could be a myosin-independent actin polymerization after contractile stimulation (An et al. 2002; Mehta et al. 1998). Such a decoupling between actin polymerization and prestress, however, contradicts the well-established view of the cell as being a "stress-supported structure" (Wang et al. 2002). For example, a tight, timely coupling between the pharmacologically (BDM) modified prestress of fibroblasts and the size of focal adhesion contacts was observed (Balaban et al. 2001). An alternative, more trivial explanation for the increase in elasticity after contractile stimulation (despite myosin inhibition) could be an incomplete effect of the inhibitors. For a definite decision between these alternative hypotheses, the time course of the contractile force of the cell elasticity and the dynamic cytoskeletal reorganization together (preferably simultaneously) must be investigated.

Future experiments using airway smooth muscle cells should focus on the time course of contractile forces after stimulation at a time resolution of 1 s using traction force microscopy, on the elasticity using magnetic twisting cytometry, and on the dynamics of cytoskeletal reorganization using nanoscale particle tracking as well as live cell microscopy with fluorescently tagged actin, vinculin, and paxillin. Cells should be treated with myosin inhibitors such as wortmannin, BDM, or ML-7, or with actin-active substances such as cytochalasin D, latrunculin-A, or jasplakinolide. The chronological sequence of strength, elasticity, and cytoskeletal reorganization should be determined during pretreatment with these substances. These studies should be designed in a way that can provide clear answers to key open questions about smooth muscle contraction processes. In addition, future studies should provide answers to general questions about cell mechanics and mechanotransduction.

Acknowledgments We thank Dr. Ben Fabry for helpful discussions and Dr. Vicky Jackiw for proofreading the manuscript. This work was supported in part by grants from Deutscher Akademischer Austauschdienst (DAAD) and Deutsche Forschungsgemeinschaft (DFG).

References

Alcaraz J, Buscemi L, Grabulosa M, Trepat X, Fabry B, Farre R, Navajas D (2003) Microrheology of human lung epithelial cells measured by atomic force microscopy. Biophys J 84:2071–2079

Alenghat FJ, Fabry B, Tsai KY, Goldmann WH, Ingber DE (2000) Analysis of cell mechanics in single vinculin-deficient cells using a magnetic tweezer. Biochem Biophys Res Commun 277:93–99

Alonso JL, Goldmann WH (2012) Influence of divalent cations on the cytoskeletal dynamics of K562 cells determined by nano-scale bead tracking. Biochem Biophys Res Commun 421:245–248

An SS, Laudadio RE, Lai J, Rogers RA, Fredberg JJ (2002) Stiffness changes in cultured airway smooth muscle cells. Am J Physiol Cell Physiol 283:C792–C801

An SS, Fabry B, Mellema M, Bursac P, Gerthoffer WT, Kayyali US, Gaestel M, Shore SS, Fredberg JJ (2004) Role of heat shock protein 27 in cytoskeletal remodeling of the airway smooth muscle cell. J Appl Physiol 96:1701–1713

Balaban NQ, Schwarz US, Riveline D, Goichberg P, Tzur G, Sabanay I, Mahalu D, Safran S, Bershadsky A, Addadi L, Geiger B (2001) Force and focal adhesion assembly: a close relationship studied using elastic micropatterned substrates. Nat Cell Biol 3:466–472

Bonakdar N, Lautscham LA, Czonstke M, Koch TM, Mainka A, Jungbauer T, Goldmann WH, Schröder R, Fabry B (2012) Biomechanical characterization of a desminopathy in primary human myoblasts. Biochem Biophys Res Commun 419:703–707

Bonakdar N, Schilling A, Lennert P, Spörrer M, Gerum RC, Alonso JL, Goldmann WH (2014) Measuring mechanical properties in cells: three easy methods for biologists. Cell Biol Int 38:1227–1232

Bonakdar N, Schilling A, Spörrer M, Lennert P, Mainka A, Winter L, Walko G, Wiche G, Fabry B, Goldmann WH (2015) Determining the mechanical properties of plectin in mouse myoblasts and keratinocytes. Exp Cell Res 331:331–337

Butler JP, Tolic-Norrelykke IM, Fabry B, Fredberg JJ (2002) Traction fields, moments, and strain energy that cells exert on their surroundings. Am J Physiol Cell Physiol 282:C595–C605

Caspi A, Granek R, Elbaum M (2002) Diffusion and directed motion in cellular transport. Phys Rev E Stat Nonlinear Soft Matter Phys 66:011916

Choquet D, Felsenfeld DP, Sheetz MP (1997) Extracellular matrix rigidity causes strengthening of integrin- cytoskeleton linkages. Cell 88:39–48

Cohen D, Arai SF, Brain JD (1979) Smoking impairs long-term dust clearance from the lung. Science 204:514–517

Crick FHC, Hughes AFW (1950) The physical properties of cytoplasm. Exp Cell Res 1:37–80

Daily B, Elson EL, Zahalak GL (1984) Cell poking. Determination of the eleastic area compressibility modulus of the erythrocyte membrane. Biophys J 45:671–82

Deng L, Fairbank NJ, Fabry B, Smith PG, Maksym GN (2004) Localized mechanical stress induces time-dependent actin cytoskeletal remodeling and stiffening in cultured airway smooth muscle cells. Am J Physiol Cell Physiol 287:C440–C448

Duszyk M, Schwab B 3rd, Zahalak GL, Qian H, Elson EL (1989) Cell poking: quantitative analysis of indentation of thick viscoelastic layers. Biophys J 55:683–90

Evans E, Yeung A (1989) Apparent viscosity and cortical tension of blood granulocytes determined by micropipet aspiration. Biophys J 56:151–160

Fabry B, Maksym GN, Butler JP, Glogauer M, Navajas D, Fredberg JJ (2001a) Scaling the microrheology of living cells. Phys Rev Lett 87:148102

Fabry B, Maksym GN, Shore SA, Moore PE, Panettieri RA Jr, Butler JP, Fredberg JJ (2001b) Time course and heterogeneity of contractile responses in cultured human airway smooth muscle cells. J Appl Physiol 91:986–994

Fabry B, Maksym GN, Butler JP, Glogauer M, Navajas D, Taback AN, Millet EJ, Fredberg JJ (2003) Time scale and other invariants of integrative mechanical behavior in living cells. Phys Rev E 68:041914

Fernandez P, Heymann L, Ott A, Aksel N, Pullarkat PA (2007) Shear rheology of a cell monolayer. New J Phys 9:1–29

Gardel ML, Shin JH, MacKintosh FC, Mahadevan L, Matsudaira P, Weitz DA (2004) Elastic behavior of cross-linked and bundled actin networks. Science 304:1301–1305

Giannone G, Jiang G, Sutton DH, Critchley DR, Sheetz MP (2003) Talin1 is critical for force-dependent reinforcement of initial integrin-cytoskeleton bonds but not tyrosine kinase activation. J Cell Biol 163:409–419

Goldmann WH (2000) Mechanical manipulation of animal cells: cell indentation. Biotechnol Lett 22:431–435

Goldmann WH (2002) Mechanical aspects of cell shape regulation and signaling. Cell Biol Int 26:313–317

Goldmann WH, Ezzell RM (1996) Viscoelasticity in wild-type and vinculin-deficient (5.51) mouse F9 embryonic carcinoma cells examined by atomic force microscopy and rheology. Exp Cell Res 226:234–237

Goldmann WH, Galneder R, Ludwig M, Xu W, Adamson ED, Wang N, Ezzell RM (1998) Differences in elasticity of vinculin-deficient F9 cells measured by magnetometry and atomic force microscopy. Exp Cell Res 239:235–242

Goldmann WH, Alonso JL et al (2000) Cell shape control and mechanical signaling through the cytoskeleton. In: Carraway KL, Carraway CAC (eds) Cytoskeleton: signaling and cell regulation: Chapter 11. Oxford University Press, Oxford

Gunst SJ, Fredberg JJ (2003) The first three minutes: smooth muscle contraction, cytoskeletal events, and soft glasses. J Appl Physiol 95:413–425

Hartmann MA, Spudich JA (2012) The myosin superfamily at a glance. J Cell Sci 125:1627–1632

Hill AV (1965) Trails and trials in physiology. E. Arnold, London, pp 14–15

Hu S, Chen J, Fabry B, Numaguchi Y, Gouldstone A, Ingber DE, Fredberg JJ, Butler JP, Wang N (2003) Intracellular stress tomography reveals stress focusing and structural anisotropy in cytoskeleton of living cells. Am J Physiol Cell Physiol 285:C1082–C1090

Huxley AF (1957) Muscle structure and theories of contraction. Prog Biophys Biophys Chem 7:255–318

Kawai M, Brandt PW (1980) Sinusoidal analysis: a high resolution method for correlating biochemical reactions with physiological processes in activated skeletal muscles of rabbit, frog, and crayfish. J Muscle Res Cell Motil 1:279–303

Kollmannsberger P, Fabry B (2007) High-force magnetic tweezers with force feedback for biological applications. Rev Sci Instrum 78:114301

Kollmannsberger P, Fabry B (2011) Linear and nonlinear rheology of living cells. Annu Rev Mater Res 41:75–97

Lange J, Auernheimer V, Strissel PL, Goldmann WH (2013) Influence of focal adhesion kinase on the mechanical behavior of cell populations. Biochem Biophys Res Commun 436:246–251

Mehta D, Wang Z, Wu MF, Gunst SJ (1998) Relationship between paxillin and myosin phosphorylation during muscarinic stimulation of smooth muscle. Am J Physiol 274:C741–C747

Merkel R (2001) Force spectroscopy on single passive biomolecules and single biomolecular bonds. Phys Rep 346:343–385

Metzner C, Raupach C, Paranhos Zitterbart D, Fabry B (2007) Simple model of cytoskeletal fluctuations. Phys Rev E 76:021925

Metzner C, Raupach C, Mierke CT, Fabry B (2010) Fluctuations of cytoskeleton-bound microbeads – the effect of bead-receptor binding dynamics. J Phys Condens Matter 22:194105

Mierke CT, Kollmannsberger P, Paranhos-Zitterbart D, Smith J, Fabry B, Goldmann WH (2008) Mechano-coupling and regulation of contractility by the vinculin tail domain. Biophys J 94:661–670

Mierke CT, Kollmannsberger P, Zitterbart DP, Diez G, Koch TM, Marg S, Ziegler WH, Goldmann WH, Fabry B (2010) Vinculin facilitates cell invasion into three-dimensional collagen matrices. J Biol Chem 285:13121–13130

Möhl C, Kirchgessner N, Schäfer C, Küpper K, Born S, Diez G, Goldmann WH, Merkel R, Hoffmann B (2009) Becoming stable and strong: the interplay between vinculin exchange dynamics and adhesion strength during adhesion site maturation. Cell Motil Cytoskeleton 66:350–364

Moy VT, Florin EL, Gaub HE (1994) Intermolecular forces and energies between ligands and receptors. Science 266:257–259

Müller O, Gaub HE, Sackmann E (1991) Viscoelastic moduli of sterically and chemically cross-linked actin networks in the dilute to semidilute regime: measurements by an oscillating disk rheometer. Macromolecules 24:3111–3120

Puig-de-Morales M, Millet E, Fabry B, Navajas D, Wang N, Butler JP, Fredberg JJ (2004) Cytoskeletal mechanics in adherent human airway smooth muscle cells: probe specificity and scaling of protein-protein dynamics. Am J Physiol Cell Physiol 287:C643–C654

Ra HJ, Picart C, Feng H, Sweeney HL, Discher DE (1999) Muscle cell peeling from micropatterned collagen: direct probing of focal and molecular properties of matrix adhesion. J Cell Sci 112:1425–1436

Raupach C, Zitterbart DP, Mierke CT, Metzner C, Müller FA, Fabry B (2007) Stress fluctuations and motion of cytoskeletal-bound markers. Phys Rev E Stat Nonlinear Soft Matter Phys 76:011918

Smith LA, Aranda-Espinoza H, Haun JB, Dembo M, Hammer DA (2007) Neutrophil traction stresses are concentrated in the uropod during migration. Biophys J 92:L58–L60

Stamenovic D, Mijailovich SM, Tolic-Norrelykke IM, Chen J, Wang N (2002) Cell prestress. II. Contribution of microtubules. Am J Physiol Cell Physiol 282:C617–C624

Stamenovic D, Suki B, Fabry B, Wang N, Fredberg JJ (2004) Rheology of airway smooth muscle cells is associated with cytoskeletal contractile stress. J Appl Physiol 96:1600–1605

Thoumine O, Ott A (1997) Time scale dependent viscoelastic and contractile regimes in fibroblasts probed by microplate manipulation. J Cell Sci 110:2109–2116

Valberg PA, Albertini DF (1985) Cytoplasmic motions, rheology, and structure probed by a novel magnetic particle method. J Cell Biol 101:130–140

Wang N, Butler JP, Ingber DE (1993) Mechanotransduction across the cell surface and through the cytoskeleton. Science 260:1124–1127

Wang N, Tolic-Norrelykke IM, Chen J, Mijailovich SM, Butler JP, Fredberg JJ, Stamenovic D (2002) Cell prestress. I. Stiffness and prestress are closely associated in adherent contractile cells. Am J Physiol Cell Physiol 282:C606–C616

Zahalak GI, McConnaughey WB, Elson EL (1990) Determination of cellular mechanical properties by cell poking, with an application to leukocytes. J Biomech Eng 112:283–294

Innovation of Vascular Engineering by Mechanomedicine

15

Ken Takahashi and Keiji Naruse

Abstract

This article describes how physical forces contribute to development, physiology, and pathology of vascular cells, focusing on endothelial cells and vascular smooth muscle cells. Based on these basic understandings of the mechanobiology, we discuss mechanomedicine, an application of the mechanobiology to medicine. Basic knowledge about cellular responses, such as cellular signal transduction pathway, gene expression, and cytoskeletal remodeling, to mechanical stimuli is important for understanding the pathology of vascular diseases including atherosclerosis. Introducing the knowledge of the vascular mechanobiology will not only contribute to the development of regenerative medicine using pluripotent stem cells but also provide a way to prevent diseases caused by thromboembolisms, such as myocardial and cerebral infarctions.

Keywords

Vascular cell • Endothelial cell • Vascular smooth muscle cell • Angiogenesis • Regenerative medicine • Mechanobiology • Mechanical stimulus

15.1 Introduction

Mankind has made remarkable advances in machine technology, which have enabled the realization of the creation of not only devices that act on the external world but also artificial organs that complement or replace the diminished functions of organs in our bodies. Functions of machine technology-derived organs (e.g.,

K. Takahashi • K. Naruse (✉)
Graduate School of Medicine, Dentistry, and Pharmaceutical Sciences, Okayama University, 2-5-1 Shikata-cho, Kita-ku, 700-8525 Okayama, Japan
e-mail: knaruse@md.okayama-u.ac.jp

K. Tanishita, K. Yamamoto (eds.), *Vascular Engineering*,
DOI 10.1007/978-4-431-54801-0_15

artificial heart and eyes) are amazing, but they do not fully replace the biological functions intrinsic to organs in the human body. Currently, regenerative medicine seeks to overcome this problem through the use of biotechnologically grown tissues, with the aim of developing artificial organs with more complete body functions without causing foreign body reactions. Such technologies are adding new pages to the history of medicine.

An important factor in regenerative medicine is the formation of blood vessels to supply nutrients to grown tissues. For example, even when a layer of cardiomyocytes grows to 1 mm in thickness, necrosis occurs rapidly in the internal cells if capillaries do not supply nutrients. This is confounded by the fact that thicker myocardial tissues are required to recover the contractile force of the heart weakened by heart failure.

Normal blood vessels are constantly under the action of mechanical stimuli due to the pulse wave of blood flow (i.e., stretch, pressure, and shear). These mechanical stimuli play an essential role in the formation of the vascular structure and, ultimately, its function. Therefore, understanding of the vascular tissue's responses to such mechanical stimuli is required in regenerative medicine to facilitate its proper construction for the sufficient delivery of nutrients to the tissue.

Health and illness are two sides of the same coin. Mechanosensitivity is the property of biological tissues to respond to mechanical stimuli and is essential to maintain healthy blood vessels, as described below. Disorders of the mechanosensitive systems lead to certain pathological conditions. Furthermore, mechanosensitive responses of the blood vessels are important in relation to the induction of challenging problems, such as arteriosclerosis and inflammation.

In this chapter, we first describe the biological responses of the vascular tissue to mechanical stimuli. We then outline the involvement of these mechanosensitive responses in the occurrence of vascular disease. Finally, we review the means for the induction of vascular tissue formation using the mechanosensitive responses of vascular cells or their progenitors.

15.2 Mechanosensitive Systems of the Blood Vessel

From the interior to the exterior, the blood vessel comprises endothelial cells, extracellular matrix (ECM), smooth muscle cells, and the outer membrane. Of these, endothelial cells and smooth muscle cells detect and respond to mechanical stimuli, such as shear stress, stretch, and pressure. These stimuli are converted to chemical reactions within the cell, which exert crucial effects on both the morphologies and functions of the endothelial cells and smooth muscle cells constituting the blood vessel.

15.2.1 Endothelial Cells

Endothelial cells in the innermost layer of the blood vessel form a single layer ranging in thickness from 0.3 to 1 μm (Pries et al. 2000). This single endothelial cell layer forms the interface between the blood and the blood vessel and encompasses a surface area of 350 m^2 (Pries et al. 2000). This thin layer also exerts critical influences on the functional regulation of the blood vessel with a wall thickness of up to 2–3 mm (Li et al. 2004) as well as disease progression.

15.2.1.1 Mechanosensitive Systems of Endothelial Cells

Endothelial cells are equipped with a variety of factors to detect mechanical stimuli (Ando and Yamamoto 2013; Zaragoza et al. 2012; Chatzizisis et al. 2007). The long protruding primary cilium (Egorova et al. 2012), glycocalyx, G-protein coupled receptors (GPCRs) (Chachisvilis et al. 2006), ion channels (Ranade et al. 2014), receptor tyrosine kinases (RTKs), and caveolae are present at the cell's apical surface in contact with the bloodstream. In addition, the lipid bilayer constituting the cell membrane is considered to demonstrate mechanosensitivity (Chien 2003; Teng et al. 2014). Proteins of known three-dimensional structure are shown in Fig. 15.1.

Three types of junctional structures are formed between a pair of adjacent endothelial cells, namely, tight, gap, and adherens junctions. The adherens junction has a complex structure composed of cadherin, beta-catenin, and alpha-catenin and is involved in sensing shear stress. Mechanical stimuli induce structural changes in alpha-catenin and render the molecule capable of binding to certain proteins such as vinculin (Gulino-Debrac 2013). Cadherin forms homodimers between adjacent endothelial cells and contributes to binding between these endothelial cells. Platelet endothelial cell adhesion molecule-1 (PECAM-1) is another protein with a function similar to cadherin. On mechanical stimulation, PECAM-1 activates the downstream phosphatidylinositol 3-kinase (PI3K), which results in intracellular signal transduction that facilitates focal adhesion growth and hardening of the cytoskeleton (Collins et al. 2012).

On the ECM side of the cell, an integrin protein complex is responsible for mechanosensitive responses. Integrin binds to ECM outside the cell and serves to fix the cells, similar to the function of an anchor. Inside the cell, it binds to the cytoskeleton to create the tension therein and sense the resistance of the cell scaffold. This mechanism plays an important role in the differentiation of stem cells into specific tissues such as bones and nerves. A protein complex including integrin that is formed at the cell attachment is known as adhesome, and it contains nearly 200 different proteins, including talin, vinculin, and zyxin (Zaidel-Bar and Geiger 2010).

When mechanical stimuli are detected by these mechanosensors, intracellular signaling by Ca^{2+} (Naruse et al. 1998) and NO, in addition to the PI3K-Akt and MAPK pathways, occurs leading to activation of nuclear transcription factors.

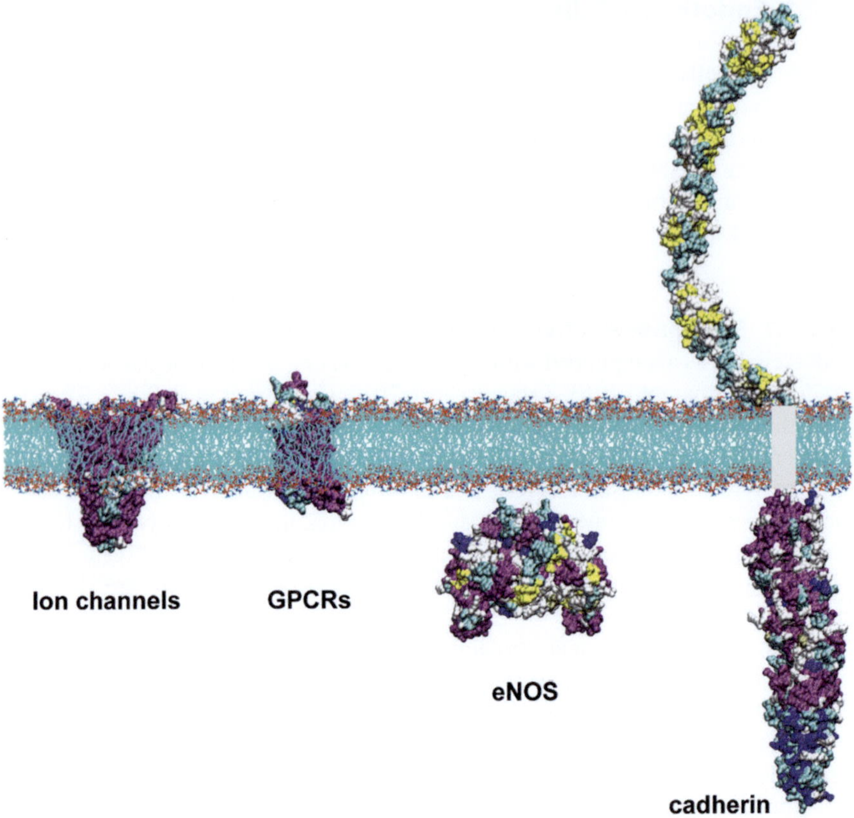

Fig. 15.1 Three-dimensional structure of mechanosensitive molecules. From *left to right*, TREK1 channel (PDB ID: 4TWK), adenosine A2A receptor (PDB ID: 2YDO), eNOS oxygenase domain (3EAH), and cadherin extracellular (PDB ID: 1L3W) and intracellular (1I7X) domains

15.2.1.2 Mechanosensitive Response of Endothelial Cells

As described above, vascular endothelial cells sense mechanical stimuli using a diverse range of factors, and mechanical stimuli induce a variety of responses to control vascular functions as follows:

1. Within a few minutes of sensing stimuli, endothelial cells change their alignment to adjust their long axis direction to mirror that of the shear stress (Ladoux and Nicolas 2012; Ives et al. 1986). This movement is mediated by reconfiguration of cytoskeletal actin filaments and is a highly dynamic phenomenon involving deconstruction of the existing configuration and rearrangement of the filaments in a new direction. Endothelial cells also respond to a stretch stimulus by orienting themselves vertically with respect to the direction of the stimulus (Ives et al. 1986). This response has been suggested to involve an ion channel

as the mechanosensor and to occur through Ca^{2+}-dependent intracellular signaling (Naruse et al. 1998).

2. Endothelial cells adjust their own rigidity depending on the composition of the ECM scaffold, a mechanism involving PECAM-1 (Collins et al. 2014). In fact, the vascular ECM varies in composition depending on the location, and collagen is the major protein in the aortic arch. Therefore, endothelial cells may be presumed to function by appropriately adjusting their own rigidity depending on the location of the blood vessel.
3. Furthermore, endothelial cells regulate proliferation of vascular smooth muscle cells depending on the mechanical strain. This regulation requires the glycosaminoglycans perlecan and heparan sulfate, which are glycocalyx components (Baker et al. 2008). The role of glycocalyx as a mechanosensor has been discussed above.
4. Both shear stress and stretch induce angiogenesis, although their underlying mechanisms are different (Milkiewicz et al. 2007). Hypoxia-induced factors (HIF)-1α and HIF-2α promote angiogenesis and are expressed at elevated levels in endothelial cells under hypoxic conditions. Their expression levels are increased by stretch, whereas they are reduced by shear stress. Angiogenesis is also affected by the rigidity of the endothelial cell scaffold, a mechanism that has been suggested to involve the mechanosensitive protein zyxin, which forms a complex with integrin at focal adhesion sites (Abbey and Bayless 2014).
5. Cellular mechanosensitivity is critically involved in the differentiation of embryonic stage tissues. In angiogenesis, the mechanosensitive ion channel Piezo 1 senses shear stress, thus inducing normal vascular remodeling. Mice lacking this gene die at the midgestation embryo stage due to abnormalities in vascular remodeling (Ranade et al. 2014).

Mechanical stimuli induce various cellular responses, as described above. These responses accompany changes in gene expression levels. For example, shear stress alters the expression levels of hundreds of genes in endothelial cells (Ando and Yamamoto 2011). Recently, emerging data have shown that miRNA expression is also involved in mechanosensitive responses. Specifically, shear stress regulates cellular redox and inflammation states via changing miRNA expression levels in endothelial cells (Marin et al. 2013).

15.2.2 Vascular Smooth Muscle Cells

Vascular smooth muscle cells undergo mechanical stimulation by the pulse wave of blood flow. Pressure in the vessel lumen affects the degree of vasoconstriction (Bevan and Laher 1991), which is considered a regulatory mechanism for local blood flow. Mechanical stimuli alter gene expression in smooth muscle cells, thereby exerting effects on their proliferation, apoptosis, migration, and remodeling (Qiu et al. 2014).

15.2.2.1 Mechanosensitive Systems of Vascular Smooth Muscle Cells

Like endothelial cells, vascular smooth muscle cells utilize GPCRs, ion channels, and RTKs as mechanosensors on the cell membrane. Caveolin is associated with smooth muscle caveolae and is believed to detect stretch through complex formation with RTK or integrin, in addition to regulating smooth muscle cell proliferation (Sedding and Braun-Dullaeus 2006). The membrane protein Notch has also been considered to be a mechanosensor of vascular smooth muscle cells, as described below (Qiu et al. 2014).

Smooth muscle cells, which are normally lined by endothelial cells in the blood vessel, receive a relatively small level of shear stress from the interstitial fluid flow. However, once the layer of endothelial cells is removed (e.g., as occurring in arteriosclerosis), smooth muscle cells are directly exposed to the shear stress from blood flow. The shear stress load on pulmonary artery smooth muscle cells elevates the intracellular Ca^{2+} level via the TRPM7 and TRPV4 ion channels (Song et al. 2014), thus leading to smooth muscle contraction. TRPM7 and TRPV4 ion channels are known to be expressed at increased levels in patients with idiopathic pulmonary hypertension. Similar examples include pressure-responsive contraction of renal artery smooth muscle cells, which is mediated by the DEG/ENaC channel as a mechanosensor (Jernigan and Drummond 2005).

15.2.2.2 Mechanosensitive Response of Vascular Smooth Muscle Cells

As seen in endothelial cells, smooth muscle cells are oriented perpendicularly with respect to the direction of stretch. This phenomenon requires Notch3, a protein expressed on the cell membrane (Zhu et al. 2011). The Notch signaling pathway plays an important role in determining the cell fate at the embryogenesis stage and is envisioned to be an important target of vascular engineering. Cyclic stretch load results in cell cycle entry in smooth muscle cells (Sedding et al. 2003). This phenomenon is a possible cause of vascular proliferative pathologies.

Given that even a single example of a mechanosensor includes many types, mechanosensitive systems in the body are not only left largely to be elucidated in terms of their functions but are also assumed to include more types yet to be identified. Furthermore, cellular responses to mechanosensor-transmitted information also remain to be elucidated, although gene expression- and miR-mediated mechanisms are being increasingly revealed.

15.3 Pathological Conditions Due to Vascular Mechanosensitive Systems

Mechanical force causes pathological conditions of the blood vessel. However, these conditions do not occur because of physical damage to the blood vessel as a result of mechanical force. Conversely, arteriosclerosis lesions frequently develop when shear stress is small. Shear stress due to blood flow and stretch of the vessel wall caused by pulsation elicit biological responses in vascular endothelial cells and smooth muscle cells. In other words, changes in the mechanical environment activate the program of vascular pathology progression.

15.3.1 Arteriosclerosis

Arteriosclerosis is well known as a common cause of ischemic heart disease, which causes heart attack. However, a little-known fact is that mechanosensitivity of the blood vessel causes arteriosclerosis. Shear stress applied to vascular endothelial cells is highly related to atherosclerosis pathology (Chatzizisis et al. 2007). Hydrodynamic shear stress tends to be ≤ 10 dyn/cm^2 at the exterior of a branching point and at the interior of a curved portion of the blood vessel, where atherosclerotic plaques frequently develop. Mechanosensors enabling the detection of shear stress in endothelial cells in arteriosclerosis are the same as those described in Sect. 15.2.1.1.

A pulsatile stress of 12 dyn/cm^2 and 1 Hz promotes phosphorylation of sirtuin (SIRT) 1 by Ca^{2+}/calmodulin-dependent protein kinase kinase (CaMKK) β. SIRT1 prevents the formation of atherosclerosis through the enhancement of the antioxidation and anti-inflammatory activities of endothelial cells (Wen et al. 2013). Similarly, the expression level of the antioxidant enzyme peroxiredoxin (PRX) 1 increases when vascular endothelial cells are subjected to a shear stress of 15 dyn/cm^2. However, this phenomenon does not occur when the cells are subjected to an oscillatory stress of ± 5 dyn/cm^2 and 1 Hz (Mowbray et al. 2008).

Cathepsins are cysteine proteases capable of degrading elastin and collagen, which are found at high levels in arterial ECM. Expression of cathepsin S and K subtypes is virtually undetectable in healthy arteries, but the expression can be detected in the blood vessels of patients with arteriosclerosis (Sukhova et al. 1998). Although vascular endothelial cells express cathepsin K at relatively high levels under an oscillatory stress of ± 5 dyn/cm^2 and 1 Hz, the expression level is suppressed under a shear stress of 15 dyn/cm^2 (Platt et al. 2007).

Arteriosclerosis is a process involving inflammation. When blood flow is artificially disturbed by applying a ligature to the mouse common carotid artery (disturbed blood flow, d-flow), vascular endothelial cells produce and release miR-712, a type of miR produced in a mechanosensitive manner. In endothelial cells, this miR not only enhances vascular permeability through activation of the ECM-degrading proteins MMP, ADAM, and ADAMTS4, but it also promotes inflammation by allowing monocytes to adhere to the endothelium (Son et al. 2013). Furthermore, disturbed blood flow inhibits expression of antiatherogenic genes via excessive methylation of their promoter region, thus promoting the development of arteriosclerosis (Dunn et al. 2014).

Although a pulsatile flow to endothelial cells induces antioxidative, anti-inflammatory gene expression, an oscillatory flow induces pro-oxidative, proinflammatory gene expression (Marin et al. 2013). Small, wavy shear stress changes are found at sites prone to developing disturbed blood flow and atherosclerosis. Interestingly, a specific frequency component of shear stress changes has been suggested to control expression of the proinflammatory phenotypes in endothelial cells through PECAM-1 as a mechanosensor (Feaver et al. 2013).

15.3.2 In-Stent Restenosis

Therapeutic options for arteriosclerosis-induced narrowing of blood vessel include percutaneous transluminal coronary angioplasty (PTCA), in which the affected blood vessel is dilated by inflating a balloon therein followed by stent placement. These treatments expose the blood vessels to strong mechanical stimuli and cause undesirable biological reactions (Mitra and Agrawal 2006; Best et al. 1999). Apoptosis of vascular smooth muscle cells is caused by application of a few atmospheres of pressure with a balloon (Perlman et al. 1997; Han et al. 1995), while stenting elicits apoptosis and proliferation of these cells (Kollum et al. 1997). Pressure from a balloon or stent induces programmed cell death, apoptosis, rather than physically destroying the cells.

15.4 Mechanomedicine: Application of the Mechanobiology to Medicine

One of the practical applications of the mechanobiology to medicine would be generation of blood vessels. To regenerate tissue or organs such as the heart or skin, it is necessary to create proper vascular structure to nurture cells. Capillaries, the smallest blood vessels, are a dynamic tissue that progresses and regresses in response to mechanical stimuli. For example, exercise induces angiogenesis, which is caused by raised levels of shear stress on the vessel wall (Egginton 2011). Increased blood flow induces intussusceptive angiogenesis, a process of capillary splitting by intraluminal longitudinal divide (Prior et al. 2004). Shear stress controls angiogenic sprouting of endothelial cells (Galie et al. 2014). These responses are thought to regulate vascular structure to be maintained adequately. Combining this knowledge of the mechanobiology and engineering techniques, it would be possible to control the formation of vascular structure.

Another application of the mechanobiology is to predict prognosis of artery disease. As we discussed in Sect. 15.3.1, shear stress plays an important role in the development of atherosclerosis. We outline this interesting technology below.

15.4.1 Induction of Vasculogenesis by Mechanical Stimuli

Mesenchymal stem cells (MSCs) and induced pluripotent stem cells (iPSCs) are candidates to construct vascular tissues in regenerative medicine. Recently, methods for differentiation of these stem cells into vascular tissues have been actively studied. Certain substances (such as transforming growth factor-β1 (TGFβ1), vascular endothelial growth factor (VEGF), or fibroblast growth factor (FGF)) are added to induce the differentiation of stem cells into vascular tissue (Wang et al. 2014; Morin and Tranquillo 2013). Although particular substances are required to induce stem cell differentiation, it is also true that mechanical stimuli play a crucial role in formation of vascular structure (Egginton 2011). Indeed, stem cells differentiate depending on the rigidity of the surrounding environment (Delaine-Smith and Reilly 2013; Engler et al. 2006). Mechanical cues, including surface property on which cells attach, determine the fate of multipotent stem cells to be either cardiogenic or endothelial (Kshitiz et al. 2014). Examples of two-dimensional and three-dimensional culture systems for inducing vasculogenesis by mechanical stimuli are outlined below.

15.4.1.1 Two-Dimensional Culture Systems

Examples of mechanical stimuli in two-dimensional culture systems are shown in Fig. 15.2. ESCs differentiate into vascular smooth muscle cells on application of cyclic strain. These cells are oriented in the vertical direction with respect to the strain, as seen in adult vascular smooth muscle cells. Activation of platelet-derived growth factor receptor (PDGFR) β is involved in differentiation of ESCs into vascular smooth muscle (Shimizu et al. 2008). PDGFRβ is a type of RTK whose expression has been shown to be increased by stretch in an experiment using rabbit pulmonary arterial tissue (Tanabe et al. 2000).

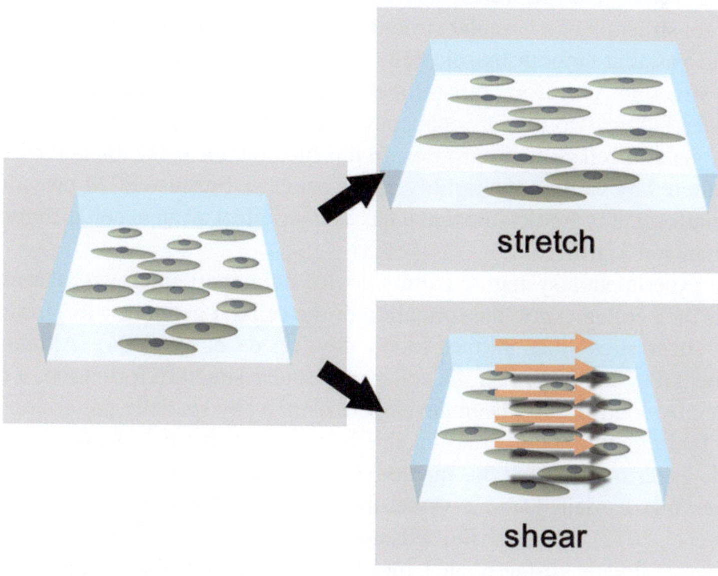

Fig. 15.2 Mechanical stimuli to the tissue. Stretch is elongation of the tissue (shown here is unidirectional). Shear stress is the force parallel to the tissue plane per unit area

In the above example, mechanical stimulation of the stem cells led to their differentiation into vascular lineage cells. Examples are also available for differentiation of stem cells into vascular lineage cells indirectly promoted by mechanical stimuli. Bone marrow-derived blood mononuclear cells are known to differentiate into endothelial cells. Although the direct application of stretch to these EC progenitor cells does not induce their differentiation into the endothelium, stretch applied indirectly to cardiovascular cells affects differentiation into the endothelium (Wang et al. 2004). Specifically, application of cyclic stretch to coronary artery smooth muscle cells results in suppressed EC production.

In addition, there are examples in which mechanical stimuli appear to alter ECM synthesis in stem cells and thereby affect their morphogenesis. For example, the mRNA expression level of the ECM component collagen I increased when strain was applied to MSCs (Heo et al. 2011). Further, the mRNA level of lysyl oxidase (LOX), which cross-links collagen fibers, was also increased. These findings suggest that strain affects ECM formation of MSCs and regulates morphogenesis.

A method of varying flow rates of the culture medium is often used to apply shear stress to cells (Ando and Yamamoto 2011; Ueda et al. 2004). In contrast, application of stretch often involves a method of stretching the silicon film on which the cells are cultured. Modes of stretch application include biaxial stretch, in which the membrane is stretched concentrically, and uniaxial stretch (Naruse et al. 1998; Ando and Yamamoto 2011), in which the membrane is stretched linearly in a single axis direction. Mechanical stimuli can be applied quantitatively by controlling the degree, velocity, and frequency of the stretch with a computer.

15.4.1.2 Three-Dimensional Culture Systems

For the construction of a vascular structure able to carry blood, vascular endothelial cells and vascular smooth muscle cells are required to form a three-dimensional column structure. Therefore, three-dimensional culture with an ECM protein gel is used instead of the two-dimensional culture environment with flasks or petri dishes used for general cell culture purposes. In this method, the mechanical environment can be altered by varying the rigidity of the gel used. Because ECM provides cells with a platform, it is predictable that it has a major effect on stem cell differentiation (Kleinman and Martin 2005).

In an experimental system to culture bovine pulmonary microvascular endothelial cells on a collagen gel, the formation of the cellular network was promoted by loading shear stress that mimics blood flow (Ueda et al. 2004). Although this experimental system relies on the cells for vascular structure formation, a method to artificially control it has been developed, namely, by spatially patterning the gel rigidity by irradiating photocrosslinkable hydrogels with a two-photon laser (West 2011). In experiments using this method, HUVEC and 10 T1/2 cells formed tubular structures that hardened into a vessel-like structure along the pattern of the gel (Culver et al. 2012). Although the cells used in these studies were not stem cells but, rather, were already differentiated into the blood vessel, the results from these studies are significant in that they have demonstrated the concept of promoting vascular formation in three-dimensional culture by applying a mechanical stimulus.

In recent years, constructing the vascular structure from stem cells has also been studied actively. Chen et al. constructed a vessel-like structure from human blood-derived endothelial colony-forming cells and bone marrow-derived MSCs *in vitro* using a methacrylate hydrogel as a culture substrate. This study succeeded in anastomosis of the construct implanted into immunodeficient mice with the host blood vessels (Chen et al. 2012). This study suggests the presence of a compressive modulus effectively promoting angiogenesis.

We have described computer-controlled application of stretch stimuli to two-dimensionally cultured cells in Sect. 15.4.1.1. Based on this method, our group has established a method of applying stretch stimuli in the three-dimensional culture environment using a self-assembling hydrogel peptide (Nagai et al. 2012).

Once vascular mechanosensitive systems are clarified and engineering technologies are developed to apply the obtained findings, the outcomes would not be limited to devising therapeutic means for refractory diseases such as arteriosclerosis. At that point, we may anticipate breakthroughs in regenerative medicine that surpass the existing artificial organs and organ transplantation.

15.4.2 Prediction of Prognosis of Artery Disease Using Mechanobiology

Atherosclerosis, a leading cause of death across the world, is called a silent killer. Low shear stress on arterial walls modulates gene expression of endothelial cells, inducing local lipid accumulation, inflammation, oxidative stress, matrix breakdown, and eventually formation of an early atherosclerotic plaque (Chatzizisis et al. 2007). Rupture of an enlarged plaque will obstruct the blood vessel and cause ischemia, which may cause death of the tissue. Widespread disturbance of blood flow at the coronary artery may culminate in severe functional disorder of the heart. Therefore, it would be considerably meaningful if we can predict the prognosis of the atherosclerosis.

"Vascular profiling" is a technique to predict progression of vascular pathology using structural information of blood vessel and fluid dynamics (Stone et al. 2003). Three-dimensional structure of the coronary artery of an individual patient can be reconstituted using intravascular ultrasound and coronary angiography. Shear stress on the arterial wall can be accurately predicted by fluid dynamics simulation. Role of local hemodynamic characteristics in coronary plaque progression was determined by the Prediction of Progression of Coronary Artery Disease and Clinical Outcome Using Vascular Profiling of Shear Stress and Wall Morphology (PREDICTION) Study (Stone et al. 2012). This clinical study concluded that low local endothelial shear stress provides prediction to identify plaques that develop progressive enlargement. Future development of the vascular profiling technology may provide prevention of diseases caused by thromboembolic events, such as myocardial and cerebral infarctions.

References

Abbey CA, Bayless KJ (2014) Matrix density alters zyxin phosphorylation, which limits peripheral process formation and extension in endothelial cells invading 3D collagen matrices. Matrix Biol 38:36–47

Ando J, Yamamoto K (2011) Effects of shear stress and stretch on endothelial function. Antioxid Redox Signal 15:1389–1403

Ando J, Yamamoto K (2013) Flow detection and calcium signalling in vascular endothelial cells. Cardiovasc Res 99:260–268

Baker AB, Ettenson DS, Jonas M, Nugent MA, Iozzo RV et al (2008) Endothelial cells provide feedback control for vascular remodeling through a mechanosensitive autocrine TGF-beta signaling pathway. Circ Res 103:289–297

Best PJ, Hasdai D, Sangiorgi G, Schwartz RS, Holmes DR Jr et al (1999) Apoptosis. Basic concepts and implications in coronary artery disease. Arterioscler Thromb Vasc Biol 19:14–22

Bevan JA, Laher I (1991) Pressure and flow-dependent vascular tone. FASEB J 5:2267–2273

Chachisvilis M, Zhang YL, Frangos JA (2006) G protein-coupled receptors sense fluid shear stress in endothelial cells. Proc Natl Acad Sci U S A 103:15463–15468

Chatzizisis YS, Coskun AU, Jonas M, Edelman ER, Feldman CL et al (2007) Role of endothelial shear stress in the natural history of coronary atherosclerosis and vascular remodeling: molecular, cellular, and vascular behavior. J Am Coll Cardiol 49:2379–2393

Chen YC, Lin RZ, Qi H, Yang Y, Bae H et al (2012) Functional human vascular network generated in photocrosslinkable gelatin methacrylate hydrogels. Adv Funct Mater 22:2027–2039

Chien S (2003) Molecular and mechanical bases of focal lipid accumulation in arterial wall. Prog Biophys Mol Biol 83:131–151

Collins C, Guilluy C, Welch C, O'Brien ET, Hahn K et al (2012) Localized tensional forces on PECAM-1 elicit a global mechanotransduction response via the integrin-RhoA pathway. Curr Biol 22:2087–2094

Collins C, Osborne LD, Guilluy C, Chen Z, O'Brien ET 3rd et al (2014) Haemodynamic and extracellular matrix cues regulate the mechanical phenotype and stiffness of aortic endothelial cells. Nat Commun 5:3984

Culver JC, Hoffmann JC, Poche RA, Slater JH, West JL et al (2012) Three-dimensional biomimetic patterning in hydrogels to guide cellular organization. Adv Mater 24:2344–2348

Delaine-Smith RM, Reilly GC (2013) Mesenchymal stem cell responses to mechanical stimuli. Muscles Ligaments Tendons J 2:169–180

Dunn J, Qiu H, Kim S, Jjingo D, Hoffman R et al (2014) Flow-dependent epigenetic DNA methylation regulates endothelial gene expression and atherosclerosis. J Clin Invest 124:3187–3199

Egginton S (2011) In vivo shear stress response. Biochem Soc Trans 39:1633–1638

Egorova AD, van der Heiden K, Poelmann RE, Hierck BP (2012) Primary cilia as biomechanical sensors in regulating endothelial function. Differentiation 83:S56–S61

Engler AJ, Sen S, Sweeney HL, Discher DE (2006) Matrix elasticity directs stem cell lineage specification. Cell 126:677–689

Feaver RE, Gelfand BD, Blackman BR (2013) Human haemodynamic frequency harmonics regulate the inflammatory phenotype of vascular endothelial cells. Nat Commun 4:1525

Galie PA, Nguyen DH, Choi CK, Cohen DM, Janmey PA et al (2014) Fluid shear stress threshold regulates angiogenic sprouting. Proc Natl Acad Sci U S A 111:7968–7973

Gulino-Debrac D (2013) Mechanotransduction at the basis of endothelial barrier function. Tissue Barriers 1, e24180

Han DK, Haudenschild CC, Hong MK, Tinkle BT, Leon MB et al (1995) Evidence for apoptosis in human atherogenesis and in a rat vascular injury model. Am J Pathol 147:267–277

Heo SJ, Nerurkar NL, Baker BM, Shin JW, Elliott DM et al (2011) Fiber stretch and reorientation modulates mesenchymal stem cell morphology and fibrous gene expression on oriented nanofibrous microenvironments. Ann Biomed Eng 39:2780–2790

Ives CL, Eskin SG, McIntire LV (1986) Mechanical effects on endothelial cell morphology: in vitro assessment. In Vitro Cell Dev Biol 22:500–507

Jernigan NL, Drummond HA (2005) Vascular ENaC proteins are required for renal myogenic constriction. Am J Physiol Renal Physiol 289:F891–F901

Kleinman HK, Martin GR (2005) Matrigel: basement membrane matrix with biological activity. Semin Cancer Biol 15:378–386

Kollum M, Kaiser S, Kinscherf R, Metz J, Kubler W et al (1997) Apoptosis after stent implantation compared with balloon angioplasty in rabbits. Role of macrophages. Arterioscler Thromb Vasc Biol 17:2383–2388

Kshitiz, Afzal J, Kim DH, Levchenko A (2014) Concise review: mechanotransduction via p190RhoGAP regulates a switch between cardiomyogenic and endothelial lineages in adult cardiac progenitors. Stem Cells 32:1999–2007

Ladoux B, Nicolas A (2012) Physically based principles of cell adhesion mechanosensitivity in tissues. Rep Prog Phys 75:116601

Li AE, Kamel I, Rando F, Anderson M, Kumbasar B et al (2004) Using MRI to assess aortic wall thickness in the multiethnic study of atherosclerosis: distribution by race, sex, and age. AJR Am J Roentgenol 182:593–597

Marin T, Gongol B, Chen Z, Woo B, Subramaniam S et al (2013) Mechanosensitive microRNAs-role in endothelial responses to shear stress and redox state. Free Radic Biol Med 64:61–68

Milkiewicz M, Doyle JL, Fudalewski T, Ispanovic E, Aghasi M et al (2007) HIF-1alpha and HIF-2alpha play a central role in stretch-induced but not shear-stress-induced angiogenesis in rat skeletal muscle. J Physiol 583:753–766

Mitra AK, Agrawal DK (2006) In stent restenosis: bane of the stent era. J Clin Pathol 59:232–239

Morin KT, Tranquillo RT (2013) In vitro models of angiogenesis and vasculogenesis in fibrin gel. Exp Cell Res 319:2409–2417

Mowbray AL, Kang DH, Rhee SG, Kang SW, Jo H (2008) Laminar shear stress up-regulates peroxiredoxins (PRX) in endothelial cells: PRX 1 as a mechanosensitive antioxidant. J Biol Chem 283:1622–1627

Nagai Y, Yokoi H, Kaihara K, Naruse K (2012) The mechanical stimulation of cells in 3D culture within a self-assembling peptide hydrogel. Biomaterials 33:1044–1051

Naruse K, Yamada T, Sokabe M (1998) Involvement of SA channels in orienting response of cultured endothelial cells to cyclic stretch. Am J Physiol 274:H1532–H1538

Perlman H, Maillard L, Krasinski K, Walsh K (1997) Evidence for the rapid onset of apoptosis in medial smooth muscle cells after balloon injury. Circulation 95:981–987

Platt MO, Ankeny RF, Shi GP, Weiss D, Vega JD et al (2007) Expression of cathepsin K is regulated by shear stress in cultured endothelial cells and is increased in endothelium in human atherosclerosis. Am J Physiol Heart Circ Physiol 292:H1479–H1486

Pries AR, Secomb TW, Gaehtgens P (2000) The endothelial surface layer. Pflugers Arch 440:653–666

Prior BM, Yang HT, Terjung RL (2004) What makes vessels grow with exercise training? J Appl Physiol (1985) 97:1119–1128

Qiu J, Zheng Y, Hu J, Liao D, Gregersen H et al (2014) Biomechanical regulation of vascular smooth muscle cell functions: from in vitro to in vivo understanding. J R Soc Interface 11:20130852

Ranade SS, Qiu Z, Woo SH, Hur SS, Murthy SE et al (2014) Piezo1, a mechanically activated ion channel, is required for vascular development in mice. Proc Natl Acad Sci U S A 111:10347–10352

Sedding DG, Braun-Dullaeus RC (2006) Caveolin-1: dual role for proliferation of vascular smooth muscle cells. Trends Cardiovasc Med 16:50–55

Sedding DG, Seay U, Fink L, Heil M, Kummer W et al (2003) Mechanosensitive p27Kip1 regulation and cell cycle entry in vascular smooth muscle cells. Circulation 108:616–622

Shimizu N, Yamamoto K, Obi S, Kumagaya S, Masumura T et al (2008) Cyclic strain induces mouse embryonic stem cell differentiation into vascular smooth muscle cells by activating PDGF receptor beta. J Appl Physiol (1985) 104:766–772

Son DJ, Kumar S, Takabe W, Kim CW, Ni CW et al (2013) The atypical mechanosensitive microRNA-712 derived from pre-ribosomal RNA induces endothelial inflammation and atherosclerosis. Nat Commun 4:3000

Song S, Yamamura A, Yamamura H, Ayon RJ, Smith KA et al (2014) Flow shear stress enhances intracellular Ca2+ signaling in pulmonary artery smooth muscle cells from patients with pulmonary arterial hypertension. Am J Physiol Cell Physiol 307:C373–C383

Stone PH, Coskun AU, Kinlay S, Clark ME, Sonka M et al (2003) Effect of endothelial shear stress on the progression of coronary artery disease, vascular remodeling, and in-stent restenosis in humans: in vivo 6-month follow-up study. Circulation 108:438–444

Stone PH, Saito S, Takahashi S, Makita Y, Nakamura S et al (2012) Prediction of progression of coronary artery disease and clinical outcomes using vascular profiling of endothelial shear stress and arterial plaque characteristics: the PREDICTION study. Circulation 126:172–181

Sukhova GK, Shi GP, Simon DI, Chapman HA, Libby P (1998) Expression of the elastolytic cathepsins S and K in human atheroma and regulation of their production in smooth muscle cells. J Clin Invest 102:576–583

Tanabe Y, Saito M, Ueno A, Nakamura M, Takeishi K et al (2000) Mechanical stretch augments PDGF receptor beta expression and protein tyrosine phosphorylation in pulmonary artery tissue and smooth muscle cells. Mol Cell Biochem 215:103–113

Teng J, Loukin S, Anishkin A, Kung C (2014) The force-from-lipid (FFL) principle of mechanosensitivity, at large and in elements. Pflugers Arch 467:27–37

Ueda A, Koga M, Ikeda M, Kudo S, Tanishita K (2004) Effect of shear stress on microvessel network formation of endothelial cells with in vitro three-dimensional model. Am J Physiol Heart Circ Physiol 287:H994–H1002

Wang C, Jiao C, Hanlon HD, Zheng W, Tomanek RJ et al (2004) Mechanical, cellular, and molecular factors interact to modulate circulating endothelial cell progenitors. Am J Physiol Heart Circ Physiol 286:H1985–H1993

Wang Y, Qian DJ, Zhong WY, Lu JH, Guo XK et al (2014) TGF-beta1 induces the formation of vascular-like structures in embryoid bodies derived from human embryonic stem cells. Exp Ther Med 8:52–58

Wen L, Chen Z, Zhang F, Cui X, Sun W et al (2013) Ca2+/calmodulin-dependent protein kinase kinase beta phosphorylation of Sirtuin 1 in endothelium is atheroprotective. Proc Natl Acad Sci U S A 110:E2420–E2427

West JL (2011) Protein-patterned hydrogels: customized cell microenvironments. Nat Mater 10:727–729

Zaidel-Bar R, Geiger B (2010) The switchable integrin adhesome. J Cell Sci 123:1385–1388

Zaragoza C, Marquez S, Saura M (2012) Endothelial mechanosensors of shear stress as regulators of atherogenesis. Curr Opin Lipidol 23:446–452

Zhu JH, Chen CL, Flavahan S, Harr J, Su B et al (2011) Cyclic stretch stimulates vascular smooth muscle cell alignment by redox-dependent activation of Notch3. Am J Physiol Heart Circ Physiol 300:H1770–H1780

Integrated Vascular Engineering: Vascularization of Reconstructed Tissue

16

Ryo Sudo, Seok Chung, Yoojin Shin, and Kazuo Tanishita

Abstract

In this chapter, we describe culture methods to construct microvascular networks as well as approaches to integrating capillary networks with 3D epithelial tissue-engineered constructs. First, culture models of microvascular networks such as *in vitro* angiogenesis and vasculogenesis models are introduced. Using these culture models, the roles of endothelial cells (ECs), such as endothelial tip, stalk, and phalanx cells, are demonstrated. Additionally, regulatory factors, including both biochemical and biophysical factors, are discussed in the context of 3D capillary formation, including the process of vascular development, growth, and maturation. Next, we focus on the use of microfluidics technologies for investigating capillary morphogenesis. Examples of 3D capillary formation assays with growth factor gradients and different extracellular matrix materials are described. Cocultures of ECs and the other cell types in microfluidic devices are also introduced to show the potential of microfluidic vascular formation models. The vascularization of constructed tissues is discussed from the viewpoints of horizontal and vertical approaches for combining capillary structures and epithelial tissues *in vitro*. Finally, the concept of integrated vascular engineering and future perspectives are discussed.

R. Sudo (✉) • K. Tanishita
Department of System Design Engineering, Keio University, 3-14-1 Hiyoshi, Kohoku-ku, Yokohama 223-8522, Japan
e-mail: sekim@kansai-u.ac.jp

S. Chung
School of Mechanical Engineering, Korea University, 145 Anam-ro, Seongbuk-gu, Seoul 02841, Korea

Y. Shin
Department of Mechanical Engineering, Massachusetts Institute of Technology, 77 Massachusetts Avenue, Cambridge, MA 02139, USA

K. Tanishita, K. Yamamoto (eds.), *Vascular Engineering*,
DOI 10.1007/978-4-431-54801-0_16

Keywords
Angiogenesis • Vasculogenesis • Microfluidic device • Vascularization

16.1 In Vitro Culture Models of Microvascular Networks

16.1.1 General Introduction for In Vitro Capillary Formation

Blood vessels are tubular structures through which blood passes to provide oxygen and nutrients to individual cells within tissues and organs. Because oxygen and nutrients are the most important fundamental factors for vital activity, blood vessels are essential tissues in the body. The luminal surface of blood vessels is covered by endothelial cells (ECs). Thus, we need to understand how ECs form blood vessels for the development of vascular tissue engineering. In particular, in the context of tissue engineering for three-dimensional (3D) tissues and organs, construction of microvascular networks, rather than large blood vessels, has a high priority.

There are two processes of microvessel formation *in vivo*, referred to as "angiogenesis" and "vasculogenesis" (Risau 1997). Angiogenesis is the formation of new capillaries branching from preexisting blood vessels. To investigate this angiogenic process, an *in vitro* 3D angiogenesis model has been developed. In this model, hydrogels, such as collagen gel and fibrin gel, are formed in a culture dish, and ECs are seeded on the surface of the 3D hydrogel scaffold. ECs grow on the gel surface and finally form a confluent monolayer. When the cells are cultured with no growth factors, they maintain a confluent monolayer, which is similar to the ECs covering the luminal surface of blood vessels *in vivo* (step 1, Fig. 16.1a). However, when ECs are cultured and supplemented with angiogenic growth factors, such as vascular endothelial growth factor (VEGF) and basic fibroblast growth factor (bFGF), some ECs penetrate into the underlying gel to form vascular sprouts, which is the beginning of vascular formation (step 2, Fig. 16.1a). These vascular sprouts gradually extend and branch into the gel and finally develop into 3D capillary-like structures (step 3, Fig. 16.1a).

The other process of microvessel formation, vasculogenesis, is the process of vascular formation in the absence of preexisting blood vessels, which occurs in the developmental process and the process of adult vascular growth (Risau and Flamme 1995). First, cells from the mesoderm form aggregates that are called blood islands. The peripheral cells of the blood islands are angioblasts, which are precursors of ECs and have not yet formed a lumen. Then, angioblasts differentiate into ECs while fusion of blood islands results in the formation of a primary capillary plexus, which, subsequently, generates primitive blood vessels.

This vasculogenesis process can also be reproduced in an in *vitro* 3D vasculogenesis model. In this model, ECs are mixed in hydrogel prepolymer solution. The prepolymer solution is then poured into a culture dish and incubated for gelation, resulting in a hydrogel with embedded individual ECs (step 1, Fig. 16.1b). ECs embedded in the hydrogel, such as a collagen gel or fibrin gel, extend in 3D and connect with each other, while some cells form vacuoles within their cytoplasm (step

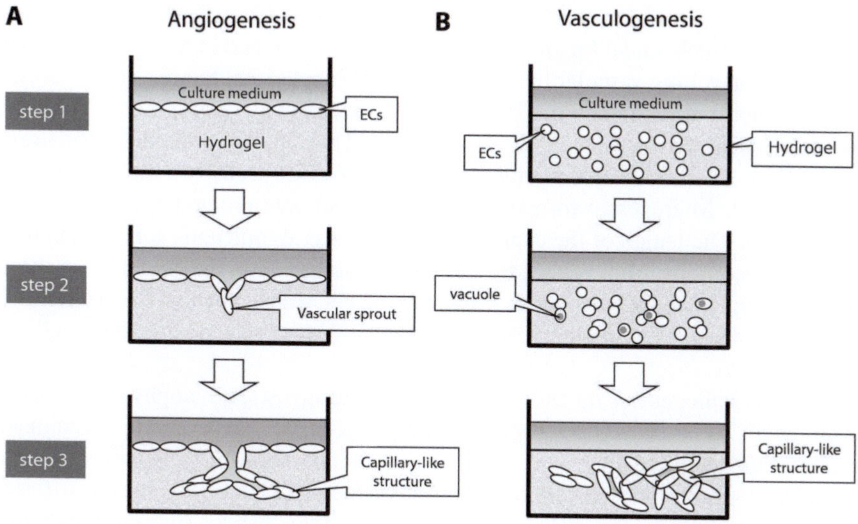

Fig. 16.1 *The process of microvascular network formation in angiogenesis and vasculogenesis culture models.* (**a**) In the angiogenesis model, ECs form a confluent monolayer covering the surface of the hydrogel. Some cells within the monolayer form vascular sprouts, which finally develop into three-dimensionally extending capillary-like networks. (**b**) In the vasculogenesis model, individual ECs are embedded in a hydrogel. These cells extend within the hydrogel and connect with each other. Some cells start to form vacuoles within their cytoplasm. These cells finally connect and develop into capillary-like structures with partial lumens

2, Fig. 16.1b). These vacuoles, formed in individual cells, subsequently fuse with each other, resulting in capillary-like structures with partial lumens (step 3, Fig. 16.1b).

Control of microvascular network formation is important not only in physiological but also in pathological conditions. For example, the formation of microvascular networks is required to promote 3D tissue formation in the field of tissue engineering and regenerative medicine. In contrast, it is also important to inhibit the formation of microvascular networks in terms of cancer therapy because *inhibition* of vascular formation is an important strategy to treat cancer. Thus, *in vitro* culture models of both angiogenesis and vasculogenesis are used to investigate how to control the formation of microvascular networks.

16.1.2 In Vitro Angiogenesis and Vasculogenesis Models

In vitro culture models of angiogenesis and vasculogenesis play important roles in investigating the mechanisms of vascular generation, growth, stabilization, and pathophysiology because cells in these models can be cultured under controlled conditions, which are independent from the complex cellular microenvironments *in vivo*.

Numerous *in vitro* angiogenesis and vasculogenesis models have been used to reveal molecular mechanisms and cell-to-cell interactions for constructing microvascular networks by ECs in 3D extracellular matrix (ECM) environments, focusing on biochemical factors (Davis et al. 2002; Newman et al. 2011). Another aspect to controlling vascular formation is that of biophysical factors, such as flow-induced mechanical signals and ECM stiffness. Ueda et al. (2004) investigated the effects of shear flow in a 3D angiogenesis model. They applied a laminar shear stress of 0.3 Pa to the surface of a confluent monolayer of ECs seeded on the surface of a collagen gel. Microvessel formation within the gel was promoted by shear flow application. The length of the capillary networks was significantly longer than that in static conditions. Microvessel formation was promoted in the gel whereas shear flow was applied to the cells on the surface of the gel, suggesting that the behavior of the cells *on* the gel is important to understand capillary morphogenesis of the cells *in* the gel.

A 3D vasculogenesis model can also be used to investigate biophysical factors, such as ECM stiffness, on capillary morphogenesis. For example, Sieminski et al. (2004) cultured ECs in floating or constrained collagen gels with different collagen concentrations to control matrix stiffness. EC morphogenesis differed depending on the matrix stiffness. Additionally, their results suggested that the relative magnitudes of cellular force generation due to the different matrix stiffness regulated capillary morphogenesis.

16.1.3 Endothelial Tip, Stalk, and Phalanx Cells

In the process of angiogenesis, ECs form vascular sprouts and gradually develop into branching networks with continuous lumens. Although vascular structures are formed by a single cell type, the EC, three different EC phenotypes are involved in the process of angiogenesis: endothelial tip, stalk, and phalanx cells (Hellström et al. 2007; Tammela et al. 2008; Carmeliet et al. 2009; Mazzone et al. 2009). Thus, we need to understand the characteristics of each EC phenotype and consider them in controlling the construction of microvascular networks *in vitro*.

Endothelial tip cells are the cells that locate at the tip of branching vascular networks. These cells extend filopodia to sense the surrounding microenvironment, such as growth factor gradients and ECM signaling molecules and stiffness, leading to the extension of vascular networks through tip cell migration. Thus, tip cells play an important role in determining the direction in which the capillary extends, while the cells rarely proliferate. In the context of vascular tissue engineering, the control of endothelial tip cells is important in guiding the direction of capillaries.

Endothelial stalk cells are the cells that follow the leading tip cells. Stalk cells proliferate to fill the elongating stalk in capillary networks. Although endothelial tip cells have no lumen, stalk cells create a lumen in the process of vascular formation. The assignment of ECs into the tip and stalk cells is regulated by endothelial cell-cell communication via Notch receptors and their transmembrane ligands, the delta-like ligand 4 (‘Dll4’) (Carmeliet et al. 2009). When ECs are activated by cytokines, such as VEGF, Dll4 expression is induced in the cells. The cells that strongly express Dll4 become endothelial tip cells. Dll4 expressed in tip cells activates

Notch in neighboring cells, which induce stalk cells. This Dll4/Notch signaling from tip to stalk cells downregulates expression of VEGFR2 in stalk cells.

Endothelial phalanx cells are the third phenotype, which are located far from the tip cells. These ECs form a monolayer with a cobblestone appearance, which also resembles a phalanx, a Greek military formation. Phalanx cells form continuous and tightly packed monolayers as a quiescent endothelium (Mazzone et al. 2009).

16.1.4 Regulatory Factors in Vascular Formation

Advances in angiogenesis and vasculogenesis studies have revealed that both biochemical and biophysical factors regulate vascular formation. *In vivo*, ECs are exposed directly to blood flow, which supplies biophysical factors directly and indirectly. Blood flow delivers soluble factors, such as cytokines and growth factors, produced in different tissues. ECs are not only exposed to blood flow but also exposed to mural cells and ECM, such as basement membrane matrices, which also give rise to biophysical and biochemical factors.

The effects of biophysical factors, such as shear flow, interstitial flow, and ECM stiffness, have been investigated using *in vitro* models. For example, shear stress has been recognized as a promoter of angiogenesis (Ueda et al. 2004; Kang et al. 2008; Kaunas et al. 2011). In addition to shear flow, interstitial flow also plays important roles in capillary morphogenesis (Ng et al. 2004; Helm et al. 2005; Semino et al. 2006; Hernández Vera et al. 2009). ECM stiffness is also important in the regulation of angiogenesis. Yamamura et al. (2007) revealed that ECs cultured in rigid and flexible gels formed 3D networks via different processes; cells formed dense, thin networks in a flexible gel, while thicker and deeper networks were formed in a rigid gel. Sieminski et al. (2004) also reported that the relative magnitudes of cellular force generation and ECM stiffness modulated capillary morphogenesis *in vitro*.

Biochemical factors, such as growth factors, cytokines, chemokines, and their concentration gradients, have also been investigated in the context of vascular formation using *in vitro* angiogenesis and vasculogenesis models. VEGF and bFGF are major growth factors promoting vascular formation (Montesano et al. 1986). Phorbol myristate acetate (PMA), a tumor promoter that increases the production of collagenase and plasminogen activator, is also a promoter of angiogenesis. Montesano and Orci (1985) revealed that bovine microvascular ECs grown on collagen gels invaded the underlying collagen gel and formed extensive capillary networks when they were treated with PMA, whereas control ECs were confined to the surface of the gels.

Not only proangiogenic factors but also antiangiogenic factors have been identified, such as endostatin, tissue inhibitor of matrix metalloproteinase 3 (TIMP-3; Anand-Apte et al. 1997), and leukemia inhibitory factor (LIF; Pepper et al. 1995). Some of these factors are summarized in Table 16.1. Transforming growth factor β (TGFβ) has been reported to have biphasic effects on vascular formation; low concentrations (0.1-1 ng/mL) promoted angiogenesis while higher concentrations (5-10 ng/mL) inhibited capillary lumen formation (Pepper

Table 16.1 Pro- and antiangiogenic factors

Proangiogenic factors	Antiangiogenic factors
PMA (Montesano and Orci 1985)	Thrombospondin (Good et al. 1990)
bFGF (Montesano et al. 1986)	Platelet factor-4 (Maione et al. 1990)
Sodium orthovanadate (Montesano et al. 1988)	Angiostatin (O'Reilly et al. 1994)
VEGF (Bikfalvi et al. 1991; Pepper et al. 1992)	LIF (Pepper et al. 1995)
Fibroblast-derived factors (Montesano et al. 1993)	Endostatin (O'Reilly et al. 1997)
Membrane-type-matrix metalloproteinases (MT-MMPs; Lafleur et al. 2002)	Chondromodulin (Hiraki et al. 1997)
Sphingosine-1-phosphate (Bayless and Davis 2003)	TIMP-3 (Anand-Apte et al. 1997)

et al. 1993). In addition to individual factors, their combined effects are also important. For example, Vernon and Sage (1999) demonstrated that VEGF and bFGF have synergistic proangiogenic effects using an *in vitro* culture model. Newman et al. (2011) also demonstrated the importance of growth factor combinations. They showed that a combination of angiopoietin-1 (Ang-1), angiogenin, hepatocyte growth factor (HGF), TGFα, and tumor necrosis factor (TNF) could substitute for the effects of fibroblast-derived factors on sprout formation whereas no protein alone could function in this way. They also identified fibroblast-derived factors required for lumen formation, which were collagen I, procollagen C endopeptidase enhancer 1, "secreted protein acidic and rich in cysteine" (SPARC), the TGFβ-induced protein "ig-h3," and insulin growth factor-binding protein 7 (Newman et al. 2011). The combined effects of multiple biochemical and biophysical factors need further clarification in future studies.

16.2 Microfluidic Devices for the Culture of Microvascular Networks

16.2.1 Endothelial Monolayers in Microfluidic Devices

This section provides an introduction to the use of microfluidic devices to study microvessel formation, which deals with the culture of ECs or monolayers in microfluidic channels. After EC loading, the cells adhere to the channel surface and form a confluent endothelial monolayer. This mimics the intima of the endothelium, the blood contact surface formed by a single layer of ECs. Mechanotransduction is an important issue for such a monolayer; signal transduction is regulated by the shear stress of blood flow.

Many studies since the 1990s have reported remodeling and alignment of ECs in monolayers due to shear stress gradients (Tardy et al. 1997). Another role for shear stress, in changing endothelial barrier function and permeability, has also been reported (Seebach et al. 2000; Colgan et al. 2007). This process involves cytoskeletal remodeling and molecular expression within ECs. Precise regulation based on fluidic shear stress types – laminar, turbulent, or pulsatile – has been an important challenge, attracting many interesting microfluidic approaches.

16.2.1.1 Endothelial Monolayer on Microfluidic Channel Surface

A realistic *in vivo*-like microenvironment of the endothelial monolayer would include shear stress, physiological pressure, and stretch. A microfluidic device has been used to culture endothelial monolayers on a stretchable and thin planar membrane (Estrada et al. 2011) (Fig. 16.2a). The membrane was made on the bottom of a microfluidic channel, forming a concave shape under pressure, representing a segment of a blood vessel wall. Precision regarding the physical stimuli can be achieved by a computer-controlled microcirculatory support system integrated with microfluidic devices (Song et al. 2005). It has microfluidic valve and pump units, and the advantages achieved by the integration show a good correlation between levels of shear stress that should be applied to the ECs and pump capacity. The tiny volumes of microfluidic channels have made it challenging to connect "conventional" pumps and microscale channels in parallel. Fluctuation, even due to tube compliance between a pump and a microfluidic device, can make the precise regulation of microfluidic shear stress inefficient. Other types of microfluidic units have been reported enabling pulsatile or oscillatory regulation of shear stress to mimic the *in vivo* situation (Shao et al. 2009) (Fig. 16.2b). Another study included sensor integration in microfluidic channels to precisely measure flow rate, pressure, and shear stress around an endothelial monolayer (Liu et al. 2013). The studies showed an important advantage of the microfluidic strategy: the potential for parallelization and increased throughput over conventional assays. It is also clear that the endothelial monolayer can discriminate pulsatile environments from a static flow condition. However, as yet, the response and mechanism(s) involved cannot be investigated fully, even using microfluidic

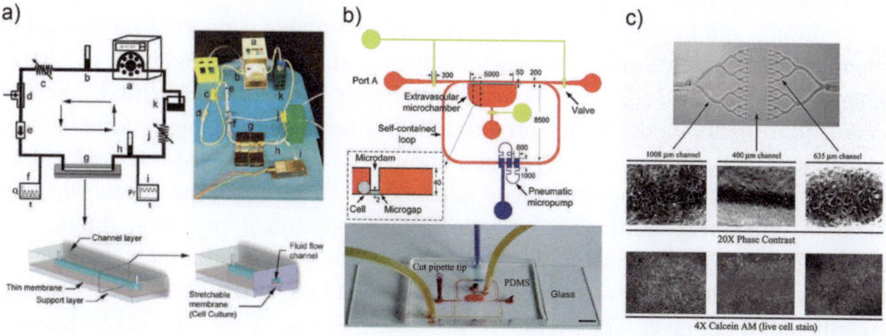

Fig. 16.2 *Formation of an endothelial monolayer and tube-like endothelium in the microfluidic channel.* (**a**) EC culture model integrated with flow loop. (**b**) Integrated microfluidic chip exposed to a pulsatile and oscillatory shear stress. (**c**) Reconstituted functional endothelialized microvascular networks with circular cross sections

approaches. This is because of the complexity of pulsatile flow patterns and also the scale of microfluidic devices.

16.2.1.2 Endothelial Monolayer as Part of In Vitro Blood Vessels

The mechanotransduction studies with shear stress mentioned above revealed interesting features of ECs and/or endothelial monolayers as the interface of a blood vessel and the blood flow within. ECs are reorganized and remodeled actively by shear stress and circulating cytokines, and these effects can be explored precisely with microfluidic regulation. Other factors in recapitulating the functionality of blood vessels include the permeability of the endothelial monolayer, the tube-like morphological characteristics, and interactions between ECs and other neighboring cells. These factors characterize many functions of blood vessels as an important transport system of blood throughout the body (Srigunapalan et al. 2011). When an endothelial monolayer forms in microfluidic channels, the cells cover every surface inside the microfluidic channels, in general, and form a conformal tube-like endothelium. Microfabrication techniques can draw two-dimensional (2D) pipelines and complex networks, with square, rectangular, or hemi- or fully circular cross sections (Borenstein et al. 2010) (Fig. 16.2c). Cell-cell interactions and tube-like morphological characteristics can also be examined with various antibodies and signaling molecules to test the effects of drugs and to study the interaction between different human vascular cell types (ávan der Meer et al. 2013).

However, these models lack the possibility of exploring one of the most important features of blood vessels: the barrier function for oxygen, nutrients, molecules, and/or cells. The barrier is semiselective and can control the passage of many materials into or out of blood. However, molecules cannot pass through an endothelial monolayer cultured on a channel surface. Layers *beyond* the endothelium are needed to support molecular transport through the monolayer. In blood vessels *in vivo*, layers with smooth muscle cells (SMCs), fibroblasts, and ECM are located outside the endothelial monolayer. They support the monolayer, providing rigidity for the vascular tube structure.

Hydrogels mimicking ECM have been used as suitable supports for endothelial monolayers in microfluidic channels (Wong et al. 2012; Zheng C et al. 2012; Zheng Y et al. 2012; Bischel et al. 2013; Baker et al. 2013). With ECM, the endothelial monolayers finally form endothelial tubes mimicking endothelium in blood vessels. Type I collagen hydrogel, fibrin, and Matrigel have been used widely in many applications to form tubes attached to ECM. They are perfusable by medium containing nutrients, oxygen, proteins, chemical reagents, drugs, and cells. Microfabrication techniques can also make planar structures behave as complex 2D networks, with tube-like cross sections of ECM attached to the endothelial monolayer (Zheng C et al. 2012; Zheng Y et al. 2012; Bischel et al. 2013) (Fig. 16.3). Not only permeability but also other important factors regarding the physics of blood vessels, thrombosis and angiogenesis, can be analyzed *in vitro*. Compared with "traditional" *in vitro* assays with ECs, this is physically relevant to *in vivo* vascular networks, with liquid flow inside and ECM materials outside (ávan der Meer et al. 2013). A gradient of diffusive molecules

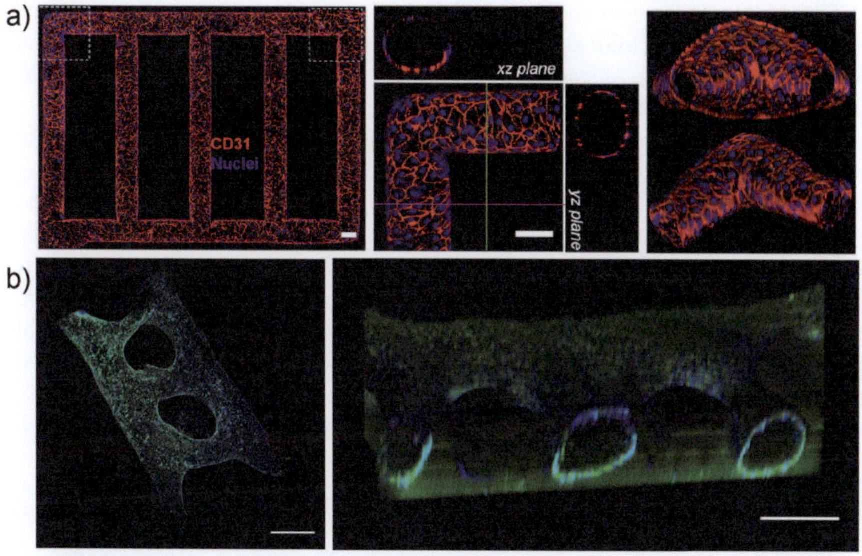

Fig. 16.3 *Microvessel formation in a microfluidic system incorporating ECM hydrogel.* Microfluidic-based reconstitution of 3D lumens with circular cross-sectional geometries within ECM hydrogel

three-dimensionally within ECM can be applied to the formed 3D endothelial tubes, to mimic the guidance provided by the molecules in many processes within development, angiogenesis, and cancer (Baker et al. 2013). The applied molecules, distributed spatiotemporally within 3D ECM, and their diffusion patterns, are affected by the tubes. At the same time, the diffusion patterns also affect the tubes, guiding locations and driving morphogenetic events, such as angiogenesis.

16.2.2 3D Capillary Formation Assays Using Microfluidic Devices

16.2.2.1 3D Angiogenesis Assay

The microfluidic assays described can be applied to the basic morphogenetic analysis of blood vessels; however, they lack any ability to induce fully functional structures due to the high stiffness of the ECM used to support the fluidic channel structure (Borenstein et al. 2010) and limited ECM space around the endothelial tubes (ávan der Meer et al. 2013). Hydrogel-incorporating microfluidic devices have been developed to investigate the morphogenetic nature of blood vessels, angiogenesis (Chung et al. 2009b; Shin et al. 2011, 2012; Jeong et al. 2011a, b; Song and Munn 2011), and vasculogenesis (Kim et al. 2013; Whisler et al. 2014; Chen et al. 2013). Angiogenesis denotes the formation of new blood vessels from preexisting vessels, and vasculogenesis is the process that occurs when there are no preexisting vessels. Angiogenesis is a normal and vital process in growth and development. It is also a fundamental step in the transition of tumors (Penn

2008). Many mechanical and chemical factors regulating angiogenesis have been revealed, including VEGF and angiopoietins. For diseases due to poor or abnormal vascularization, such as cardiovascular diseases and tumors, angiogenesis has already emerged as an effective therapeutic target.

When ECs are seeded in a hydrogel incorporating microfluidic channels, the seeded cells attach to the hydrogel and form an endothelial monolayer. Type I collagen, Matrigel, and fibrin are used widely to mimic the ECM surrounding blood vessels. Chung et al. (2009b) developed an optimized protocol to induce sprouting angiogenesis into type I collagen hydrogel and confirmed its 3D tube-like structure by confocal microscopy (Fig. 16.4a). They found that a VEGF gradient induced sprouting angiogenesis formation but also that additional signals are probably necessary to determine which segments sprout, which dilate, and which remain quiescent. Combined treatment with angiogenic factors and VEGF can induce a complex and integrated network of blood vessels, angiopoietin (Shin et al. 2011),

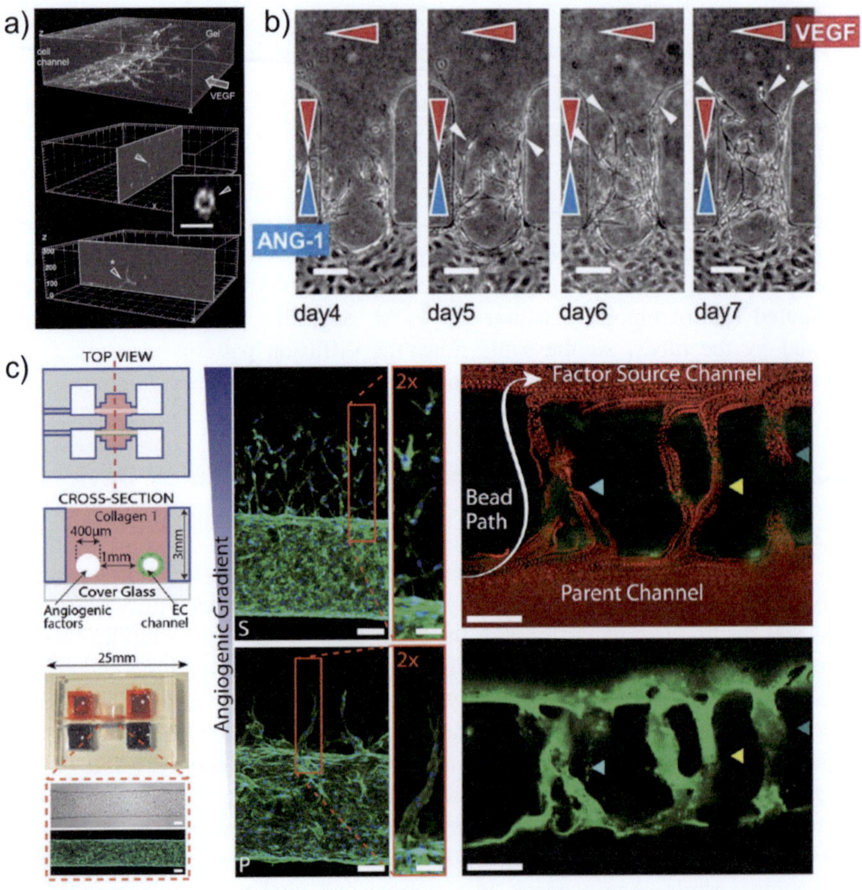

Fig. 16.4 *3D sprouting angiogenesis in a microfluidic system.* **(a)** **(b)** **(c)** Formation of 3D capillary sprouting into ECM hydrogel under gradients of various angiogenic factors

and sphingosine (Nguyen et al. 2013) (Fig. 16.4b, c). Microfluidic schemes enable *in situ* monitoring of collective and complex angiogenic procedures with enhanced views. Many interesting phenomena involved in 3D angiogenesis have been reported, including 3D cooperative migration of tip cells (with filopodia-like protrusions) and stalk cells (with apical-basal polarity), lumen formation, and the connection and division of capillary branches with perfusion inside the lumens. Anti- or proangiogenic drugs can be evaluated (Nguyen et al. 2013; Lee et al. 2014), as can angiogenic inhibitors and matrix metalloproteinase inhibitors.

Another study on the combined effects of growth factors and fluid shear stress also revealed the important nature of capillary angiogenesis (Shin et al. 2011). Endothelial monolayers are always under fluidic shear stress and interstitial (leaking) flow, and these effects, combined with various growth factors, are now under investigation. It also makes the assay more quantitative and robust if these issues are considered. Studies on local and global gradient formation, 3D diffusion, convection, and interactions with cells (barrier function and materials consumption) can provide a basis for understanding the true distribution of the many factors involved and how they interact with blood vessels (Jeong et al. 2011b; Song and Munn 2011; Helm et al. 2005). Local gradients of the factors regulate angiogenic structures but are also interrupted by endothelial monolayers. When the monolayer prohibits the molecule(s) from passing through, a rapid change in concentration occurs across the monolayer. When the cells in the monolayer take up the molecule (s), a local gradient forms around the cells. Studies should also consider the synergistic interaction of molecule distributions under convection by blood flow and interstitial flow and interactions with cells and endothelial monolayers. It is still a very challenging issue to understand the true mechanism of angiogenesis.

Coculturing with stromal fibroblast, pericytes, or SMCs (Kim et al. 2013; Mack et al. 2009) has also been investigated to confirm these cells' roles in adaptive remodeling of capillary structures. Mack et al. (2009) found that KLF2-expressing ECs cocultured with SMCs showed significantly reduced SMC migration. The cells form vascular smooth muscle, the majority of the wall of blood vessels outside the endothelial monolayer, to support the lumen structure and stabilize ECs. Coculturing enables a deep investigation of complex interactions between blood vessels and neighboring cells in development and many diseases, which cannot be achieved readily with conventional cell culture assays.

16.2.2.2 3D Vasculogenesis Assay

The hydrogel-incorporating vasculogenesis assays developed have a similar structure to angiogenesis assays. However, vasculogenesis assays usually involve culturing ECs embedded in ECM hydrogel. Before the development of hydrogel-incorporating microfluidic devices, the differentiation of angiogenesis and vasculogenesis procedures in conventional *in vitro* assays was almost impossible, while current vasculogenesis assays can form very complex, highly connected, functional, and perfusable 3D microvascular networks (Kim et al. 2013) (Fig. 16.5a). The assays show strong barrier function and long-term stability of the networks. Precise control of growth factors and coculturing with stromal

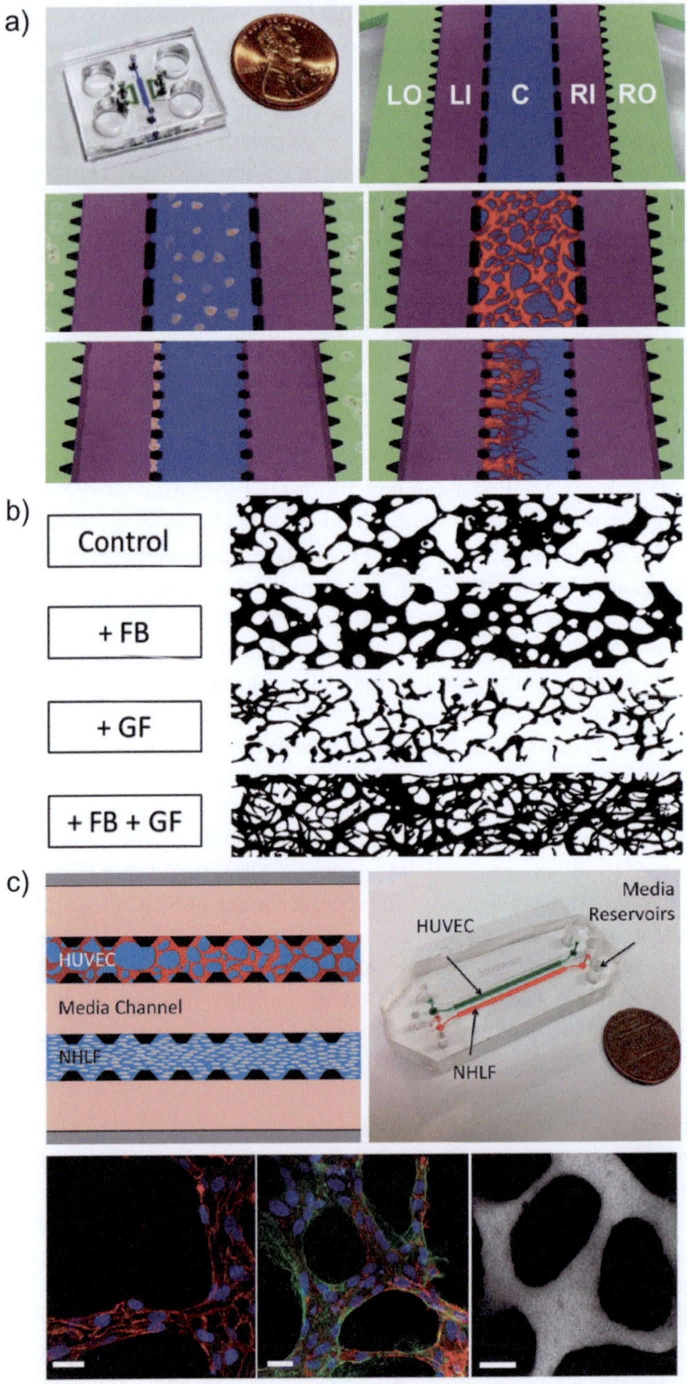

Fig. 16.5 *Vasculogenesis in a microfluidic system.* (**a**) (**b**) (**c**) 3D formation of microvascular network within ECM hydrogel in coculture with stromal cells such as fibroblasts

fibroblasts, SMCs, or pericytes help to form stable but complex vascular structures (Whisler et al. 2014) (Fig. 16.5b). These assays are also used for tumor transendothelial studies, intra- and extravasation (Chen et al. 2013) (Fig. 16.5c). This will be discussed in detail later.

16.2.3 Vascular Formation Under a Gradient of Growth Factors in Microfluidic Devices

The generation and precise control of biochemical factor distributions is one of the most useful features of microfluidic devices. Concentration gradients formed around cells induce 2D chemotactic responses of ECs. ECs exposed to growth factor gradients displayed an increased number of filopodia and an asymmetric distribution of filopodia, with enhanced numbers of protrusions present along the leading edge (Shamloo et al. 2008). The absolute concentration of VEGF enhanced the total number of filopodia, while the steepness of the VEGF gradient induced filopodia localization, cell polarization, and subsequent directed migration.

A study on chemotactic responses in three-dimensionally aligned endothelial monolayer-induced angiogenic responses of ECs, in ECMs with varying densities and stiffnesses, has been reported. The mechanical properties of ECMs could regulate angiogenic sprout and lumen formation when combined with gradients of vascular growth factors (Shamloo and Heilshorn 2010). Microfluidic chips enable the precise regulation of gradients and *in situ* investigations and monitoring of complex 3D cellular morphogenesis. They also enable quantitative studies on chemical release systems to deliver growth factors, therapeutic agents, and drugs. Kim et al. (2012) evaluated the proangiogenic potential of factors secreted from encapsulated fibroblast cells on an endothelial monolayer. An extensive network of functional lumen structures was made *in vitro*, perfused with blood cells.

The true distribution of a chemical gradient around an endothelial monolayer is not easy to define because many factors are involved synergistically, as described in Section 16.2.2 (Jeong et al. 2011b; Song and Munn 2011; Helm et al. 2005). Pioneering works to precisely estimate actual distributions of chemical gradients by computational modeling have been reported, with experimental verification by monitoring 3D sprouting angiogenesis from an endothelial monolayer (Jeong et al. 2011b). Moreover, molecular diffusion perturbed by the endothelium helps to regulate the degree of gradient around the endothelial monolayer. The orchestration of multiple growth factors (e.g., VEGF and angiopoietins) has also been simulated, with experimental confirmation by 3D cooperative migration of tip and stalk cells and the lumen structures formed (Shin et al. 2011).

16.2.4 ECM Materials for Microfluidic Angiogenic Studies

Various ECM materials, including type I collagen, fibrin, hyaluronic acids (HA), Matrigel, fibronectin, and gelatin, have been used in angiogenic studies as a base for

Table 16.2 ECM materials for angiogenic studies

ECMs	Applications	References
Type I collagen	Basic ECM for angiogenic sprouting, angiogenesis, blood vessel formation, thrombosis study, formation of brain blood barrier model, and cancer metastasis study	Zheng C et al. (2012), Zheng Y et al. (2012), Nguyen et al. (2013), Sudo et al. (2009), and Chrobak et al. (2006)
Fibrin	Basic ECM for angiogenic sprouting, angiogenesis, vasculogenesis, and cancer metastasis study	Chen et al. (2013), Yeon et al. (2012), and Carrion et al. (2010)
HA-collagen	Basic ECM for cell migration	Jeong et al. (2011a)
Matrigel	Reducing cell movement in angiogenesis, blood vessel formation, and cancer metastasis study	Chaw et al. (2007), Dai et al. (2011), and Bischel et al. (2013)
Fibronectin	Enhancing cell movement in angiogenesis, cell migration, and cell adhesion	Shamloo et al. (2011) and Young et al. (2007)
Gelatin	Sacrificial element for blood vessel formation	Golden and Tien (2007) and Chung et al. (2010)

ECs (Table 16.2). Matrigel has been widely accepted as an ECM in 2D angiogenic studies, because ECs form tube-like "2.5-dimensional" structures on Matrigel (Soriano et al. 2004; Kieda et al. 2006). However, the structure formed is far from that of the *in vivo* vasculature. It grows on the ECM (Matrigel), with biased polarity, of the top to the medium and bottom to the ECM, which is the opposite of the *in vivo* vasculature, which has ECM outside containing blood flow inside.

Pioneering microfluidic works incorporating ECM have enabled the study of angiogenic responses in ECM. The vasculature in the ECM was made by combined effects of collective and cooperative migration of ECs, with active remodeling of the ECMs by soluble proteases and matrix metalloproteinases (MMPs). Type I collagen and fibrin are the major ECMs used for microfluidic angiogenesis and vasculogenesis studies. In some studies, the ECMs have been modified with other ECM components, such as laminin and fibronectin, to mimic the ECM composition of specific target organs. For example, the ECM of adult brain tissue has a unique composition. It has lecticans, proteoglycans with a lectin domain and a hyaluronic acid-binding domain. Hyaluronic acid and tenascin family adhesive/antiadhesive proteins are also abundant (Ruoslahti 1996). Brain also has high concentrations of matrix proteins common in other tissues, such as laminin and collagen, but only near the brain vasculature. The unique composition of brain tissue has trophic effects on specific brain cells and in diseases. ECM-incorporating microfluidic studies enable the exploration of the unique effects of cellular microenvironment on specific cells, precisely mimicking *in vivo* tissue physiology (Shin et al. 2014).

16.3 Coculture of ECs and the Other Cell Types in Microfluidic Devices

16.3.1 Using ECs as a Vascular Microenvironment for Other Cell Types: Mesenchymal Cells

16.3.1.1 Proangiogenic Effects of Fibroblasts

Fibroblasts are a type of mesenchymal cell; they play important roles in the synthesis and maintenance of ECM. Since it was first reported that factors produced by mouse embryo 3 T3 fibroblasts stimulated the proliferation of cultured ECs (Birdwell et al. 1977), fibroblasts have been cocultured with ECs to promote angiogenesis in 3D culture models. Montesano et al. (1993) investigated the effect of various fibroblasts, such as Swiss 3 T3, BALB 3 T3, NIH 3 T3 cells, and epithelial cells, such as MDCK and intestine 407 cells, on the angiogenic responses of bovine microvascular ECs. Their results revealed that only Swiss 3 T3 fibroblasts, among the various fibroblasts and epithelial cells tested, were able to induce an angiogenic response when cocultured with the ECs in their 3D angiogenesis model. This proangiogenic effect of fibroblast-derived factors was also used in recent *in vitro* angiogenesis models created in microfluidic devices (Yeon et al. 2012; Kim et al. 2012, 2013; Moya et al. 2013). Fibroblasts are known to produce various proangiogenic soluble factors, including VEGF, bFGF, IL-8, TGFα, TGFβ, HGF, angiogenin, Ang-1, TNF, and PDGF (Mansbridge et al. 1999; Hartlapp et al. 2001; Hurley et al. 2010; Newman et al. 2011). In addition to the growth factors secreted by fibroblasts, fibroblast-derived matrix proteins also play important roles in angiogenic responses (Berthod et al. 2006). Newman et al. (2011) identified collagen I, procollagen C endopeptidase enhancer 1, SPARC, ig-h3, and insulin growth factor-binding protein 7, which were expressed in fibroblasts and were essential for EC lumen formation.

16.3.1.2 Capillary Stabilization by Pericytes or SMCs

ECs have been cocultured with fibroblasts because of their ability to secrete various proangiogenic factors. In the context of vascular tissue engineering, not only the formation but also the stabilization of capillary networks is important to construct long-lasting capillary networks *in vitro*. Other types of mesenchymal cells, such as SMCs and pericytes, may be expected to interact with ECs because they localize around blood vessels *in vivo*. Pericytes wrap themselves around small vessels, such as capillaries <10 μm in diameter, as well as microvessels, 10-100 μm in diameter, while SMCs form many layers around larger blood vessels (Wanjare et al. 2013). These perivascular cells have different effects on angiogenesis compared with fibroblasts. SMCs and pericytes inhibit EC proliferation and angiogenic responses in cocultures as well as *in vivo* (Orlidge and D'Amore 1987; Wanjare et al. 2013; Crocker et al. 1970; Shepro and Morel 1993; Díaz-Flores et al. 2009). Activated TGFβ was the key factor for the inhibitory effect of pericytes on EC growth in EC-pericyte coculture (Antonelli-Orlidge et al. 1989). Saunders et al. (2006) found that EC-derived TIMP-2 and pericytes-derived TIMP-3 coregulated capillary tube stabilization. Pericytes-derived TIMP-3 inhibited vascular sprout formation by ECs in an *in vitro* angiogenesis model, while TIMP-3 contributed to capillary tube stabilization in an *in vitro* vasculogenesis model. Numerous studies suggest that Ang-1/Tie2 signaling between pericytes and ECs is required for vessel stabilization (Armulik et al. 2005).

16.3.1.3 Use of Stem Cell Source for EC-Pericyte Coculture

Mesenchymal stem cells (MSCs) have also been used for coculture with ECs. Because it is difficult to establish primary culture of pericytes, MSCs have been used to derive pericytes. Au et al. (2008) demonstrated that MSCs derived from human bone marrow differentiated into perivascular cells, which stained positively for α-smooth muscle actin (α-SMA), a pericyte marker, *in vivo* when coimplanted with ECs. Although capillary structures quickly regressed when only human umbilical vein endothelial cells (HUVECs) were implanted, pericytes differentiated from MSCs stabilized capillary structures when coimplanted with ECs, resulting in the maintenance of functional capillary structures for >130 days *in vivo*. Furthermore, these capillary structures constricted in response to stimulation by endothelin-1, an endogenous vasoconstrictive peptide.

Differentiation of MSCs into pericytes has also been confirmed in recent studies using *in vitro* culture models. Carrion et al. (2010) performed HUVEC-MSC coculture in an *in vitro* vasculogenesis model created in a microfluidic device and demonstrated that MSCs localized around the periphery of the vessel structure formed by HUVECs. These MSCs became positive for α-SMA and eventually wrapped themselves around the forming vessels. Yamamoto et al. (2013) reported that HUVEC-MSC direct contact was required for the differentiation of MSCs into pericytes in 2D culture. In this study, HUVEC-MSC coculture was also performed in a microfluidic device. Results showed that MSCs attenuated vascular sprout formation of HUVECs in the early stage of 3D angiogenesis as well as the extension of microvascular networks in the later stage, suggesting that MSCs have stabilization effects on the formation of microvascular networks. Jeon et al. (2014) also reported that HUVEC-MSC direct contact was required for the differentiation of MSCs into pericytes in 3D culture. In this study, HUVEC-MSC coculture was performed in an *in vitro* vasculogenesis model created in a microfluidic device. MSC-derived α-SMA-positive cells colocalized with HUVECs and wrapped around the newly formed microvascular networks, which resulted in a smaller vessel diameter versus HUVECs alone. According to the recent advances in *in vitro* culture systems mentioned here, there is increasing evidence that HUVEC-MSC coculture is promising for the development of functional, physiological, and long-lasting microvascular networks *in vitro*.

16.3.2 Using ECs to Create a Vascular Microenvironment for Other Cell Types: Neural Stem Cells

Cellular microenvironments include signaling molecules, small molecules, neighboring cells, ECMs, small organs, and mechanical stimuli. These provide structural and trophic support for cells, regulating their life cycle. The vascular microenvironment, one of the cellular microenvironments formed by blood vessels, supplies neighboring cells with nutrients, oxygen, and other soluble factors and actively regulates their life cycle. For example, neural stem cells (NSCs) reside in a brain vascular microenvironment. Brain vessels play an important role in maintaining an appropriate balance between NSC self-renewal and differentiation (Shin et al. 2014). NSCs in the subventricular zone maintain contact with the brain vasculature, maintaining its multipotency by self-renewal. Brain tissue damage by, for example, ischemic stroke, neurodegenerative diseases, and brain tumors, promotes homing and integration of NSCs. The NSCs integrated into the damaged brain tissue choose the proper differentiation lineage into neurons, astrocytes, or oligodendrocytes. In the process, the vascular microenvironment plays an essential role, and a study of the microenvironments may facilitate basic research on the development of potential therapies for degenerative disorders.

Shin et al. (2014) developed a microfluidic assay to reconstitute NSC-vascular microenvironments with a 3D brain vasculature and ECMs. The reconstituted vascular microenvironment regulated self-renewal, proliferation, and colony formation of NSCs and suppressed neuronal generation. It also promoted NSC differentiation into astrocytes and oligodendrocytes. Cell culture on a microfluidic scale was shown to emphasize cell-cell communications, by enrichment of secreted factors in the tiny microscale environment. It dramatically enhanced the proximity effect on NSCs from the brain vasculature, pushing NSCs to differentiate into astrocytes to form a brain-specific vascular structure, the blood-brain barrier (Shin et al. 2014).

Not only NSCs but also the other stem cells require an assay for assessing vascular microenvironments. Many *in vivo* models from mouse and Drosophila have helped to explore the role of stem cell microenvironments on stem cell maintenance and activity. However, details of morphogenetic processes of stem cells under specific microenvironments cannot be systematically monitored *in vivo*, raising the need for an *in vitro* assay regulating the microenvironmental factors. Another challenge involves demonstrating the combined effects of multiple microenvironmental factors. For example, low oxygen tension induces angiogenesis by increased expression of various angiogenic factors. The expressed angiogenic factors not only affect ECs but also stem cells and other stromal cells. Newly formed vasculature also affects the previously low oxygen tension and other stromal cells. These cell-cell and cell-microenvironment interactions are complex, affecting each other in a very complicated manner. Reconstitution of each microenvironmental factor and their combinations markedly increases the chances of developing new knowledge.

16.3.3 Using Other Cell Types to Interact with ECs: Immune and Cancer Cells

Cancer occurs by genetic damage to normal cells, leading to uncontrolled proliferation, resulting in a primary heterogenic tumor. The tumor induces angiogenesis and recruits new blood vessels to enhance its nutrient and oxygen supply. Some cancer cells acquire the ability to invade neighboring ECMs and/or tissues, and finally intravasate into lymphatic or blood vessels. They become circulating tumor cells and extravasate to secondary tumor sites. These extravasated tumor cells may die or become dormant due to the different microenvironment of the secondary site from that of the primary site, but sometimes proliferate and form micrometastases, resulting in secondary tumor site(s). Proliferation in the secondary tumor site also requires angiogenic responses from neighboring blood vessels. Angiogenesis is thus an essential process for a cancer to acquire malignancy by metastasis; it supplies the tumor site with nutrient and oxygen, and also works as a major route for tumor cells to escape from the original site.

Cancer angiogenesis is thus a promising drug target; suppressing it could starve tumor cells and minimize the chances of metastasis. The permeability and 3D structures of blood vessels are related to the metastasis procedure; they also work as a route for drug delivery, directly targeting tumor cells. The whole process of a cancer acquiring malignancy by metastasis has been addressed and defined as described above, but there is still a need for appropriate *in vitro* assays to investigate the individual steps involved in metastasis.

16.3.3.1 Remote Interaction Between Tumor Cells and ECs: Angiogenic Response

A microfluidic platform was developed to evaluate and quantify capillary growth from an endothelial monolayer, when placed in coculture with physiologically relevant cell types, including cancer cells and SMCs (Chung et al. 2009a) (Fig. 16.6a). Three-dimensional capillary structures were grown into type I collagen ECM, under a precisely controlled coculture microenvironment. Breast cancer cells were shown to induce capillary formation, while SMCs suppressed it. A similar interaction between endothelial cells and tumor cells was also realized on the 2D surface of a microfluidic array (Zheng C et al. 2012; Zheng Y et al. 2012). The device offered 16 parallel arrays to study interactions between cells. It successfully provided an accurate spatiotemporal microenvironment in cocultures of multiple types of cells (Zheng C et al. 2012; Zheng Y et al. 2012) (Fig. 16.6b).

16.3.3.2 Remote Interaction Between Tumor Cells and ECs: Antiangiogenic Response

It is now widely recognized that cancer cells induce angiogenesis (Folkman 1971). However, antiangiogenic responses were also observed in *in vitro* coculture models. Kalchman et al. (2013) reported the coculture of HUVECs and HepG2, a hepatocellular carcinoma cell line. HUVECs and HepG2 cells were cultured in separate microfluidic channels. When HUVECs were cultured alone, they migrated

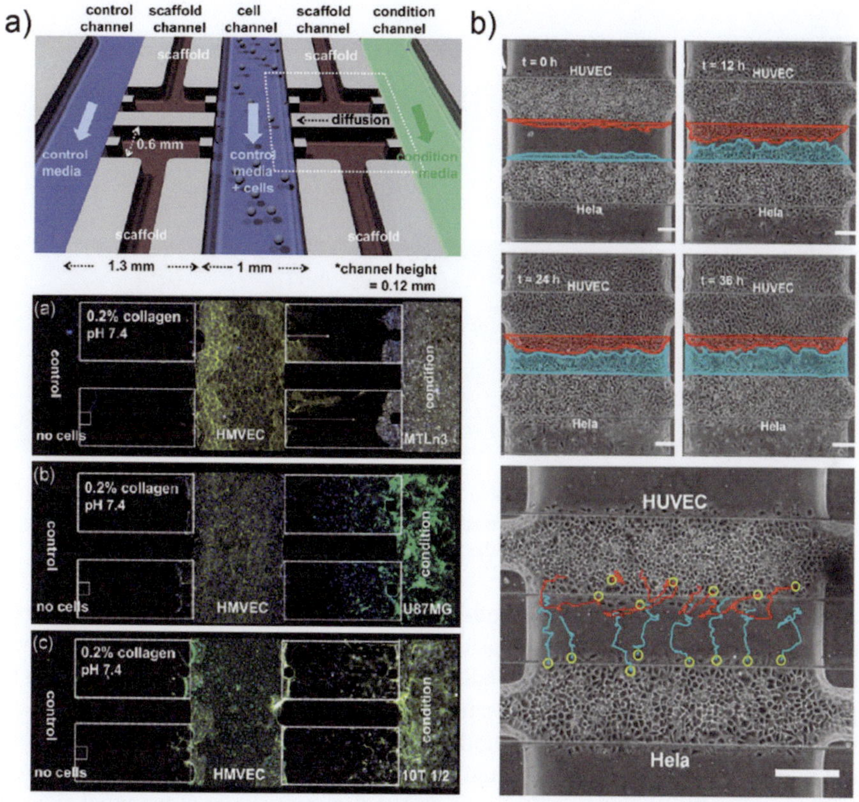

Fig. 16.6 *Interactions between endothelial cells and cancer cells in a microfluidic system.* (**a**) Cancer angiogenesis within a microfluidic system incorporating ECM hydrogel. (**b**) A coculture assay for the quantitative study of the interaction between endothelial cells and cancer cells

into collagen gel and formed microvascular networks. However, when HUVECs were cultured with HepG2, the formation of vascular sprouts was completely inhibited and HUVECs formed monolayers (Fig. 16.7). Instead, HUVECs induced the formation of protrusions by HepG2, indicating that there are interactions between HUVEC and HepG2, although no angiogenic response was observed. These results suggest that tumor-induced angiogenesis can be confirmed in culture models, depending on the cell lines used. Further investigations are needed to clarify interactions between cancer cells and ECs in terms of angiogenesis.

16.3.3.3 Direct Interaction Between Tumor Cells and ECs: Intravasation and Extravasation

Many microfluidic devices have been used to investigate intravasation, entry into the bloodstream, or extravasation, movement out of the blood stream, of tumor cells. A microfluidic device reconstituting the 3D tumor-vascular interface,

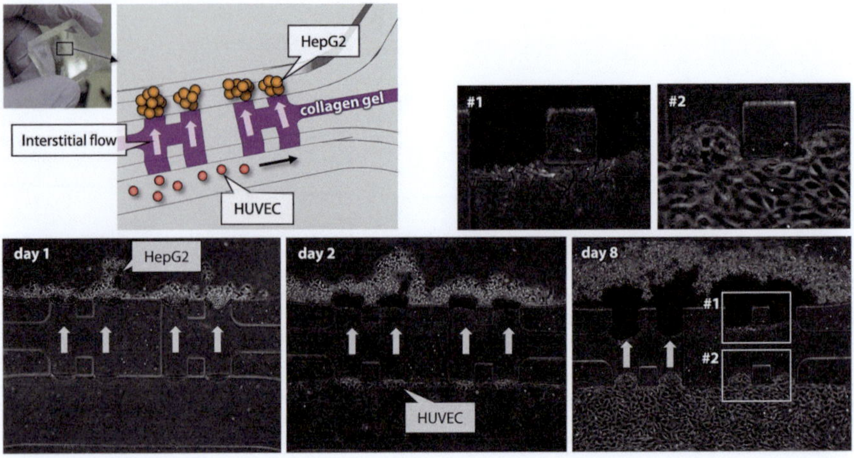

Fig. 16.7 *HUVEC-HepG2 coculture in a microfluidic device*. HepG2 cells were cultured in interstitial flow conditions. HUVECs were added to the other side on day 2. Phase-contrast images #1 and 2 correspond to box # 1 and 2 in the image of day 8

allowing real-time monitoring of cell-cell interactions and precise quantification of endothelial barricr function, such as endothelial permeability has been developed (Zervantonakis et al. 2012) (Fig. 16.8a). The endothelial monolayer was shown to pose a barrier to tumor cell intravasation that could be regulated by factors present in the tumor microenvironment. Another microfluidic device provided an endothelial monolayer culture in a microfluidic device to study the intravascular adhesion of circulating breast cancer cells on an endothelial monolayer at potential sites of metastasis (Song et al. 2009). Spatially restricted stimulation of an endothelial monolayer with the chemokine CXCL12 from the basal side mimicked the organ-specific localization and polarization of chemokines and other signaling molecules, and was shown to affect cancer cell adhesion on the endothelial monolayer.

Another study dealt with the effects of CXCL12 on tumor cell aggregates near the endothelial monolayer (Zhang et al. 2012) (Fig. 16.8b). CXCL12 stimulation damaged the integrity of the endothelial monolayer and attracted cancer cells transmigrating through the monolayer. Thus, the evaluation of antimetastasis agents by inhibiting transendothelial invasion or the adhesion of cancer cells on the endothelial monolayer can be performed with such microfluidic devices. A similar device was developed to mimic tumor cell extravasation, transmigrating across an endothelial monolayer and migrating into a hydrogel, to model the extracellular space (Jeon et al. 2013). Precise measurement of endothelial permeability and *in situ* monitoring of cancer cell transmigration can help in understanding permeability changes during tumor cell intra/extravasation events.

A tumor extravasation study in a microfluidic device was enhanced dramatically by reconstituting complex 3D vascular networks (Chen et al. 2013) (Fig. 16.8c).

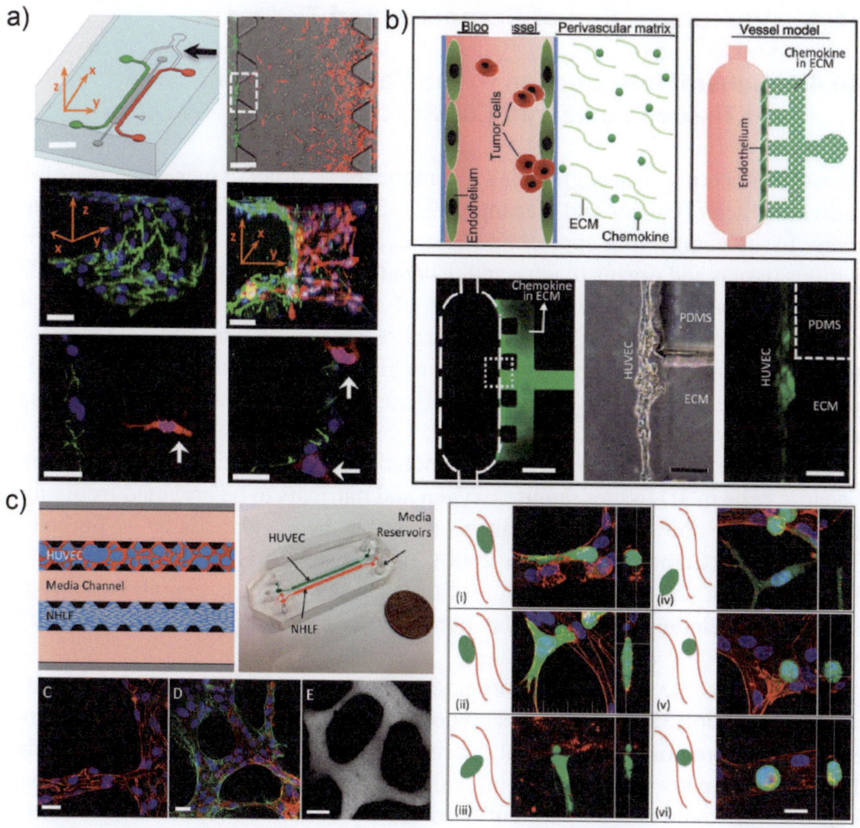

Fig. 16.8 *Intravasation and extravasation in an in vitro model.* (**a**) Microfluidic tumor-vascular interface model for intravasation. (**b**) Blood vessel formation for the study of transendothelial migration of cancer cells. (**c**) Microfluidic microvascular network for the study of tumor cell extravasation

The network was made by vasculogenesis in a fibrin ECM, having tight endothelial junctions with low vessel permeability. Extravasation efficiency was increased by inflammatory cytokine stimulation and trapping cancer cells in complex junction structures. Dynamic interactions of tumor cells and ECs were monitored: protrusion formation, extrusion of the cell body through a very small opening in the endothelial barrier, and nuclei transmigration. These methods can be applied not only to tumor cell metastasis studies but also to the spatiotemporal analysis of the inflammatory response to characterize transendothelial migration behaviors of neutrophils under the influence of various inflammatory stimuli (Han et al. 2012).

16.4 Vascularization of Reconstructed Tissues

16.4.1 General Introduction to the Vascularization of Tissue-Engineered Constructs

Tissue engineering is an engineering technique to construct tissues and organs from individual cells (Langer and Vacanti 1993). Since the concept was first reported, many tissue engineering approaches have been studied with a view to producing various tissues and organs *in vitro* and *in vivo* (Khademhosseini et al. 2009; Griffith and Naughton 2002). The development of tissue engineering has allowed the construction of 2D tissues, such as skin substitutes and cornea, and 3D avascular tissues, such as cartilage. Although these engineered tissues have been assessed in clinical trials, construction of 3D vital organs, such as liver, heart, and pancreas, remains a key challenge in this field.

In contrast to 2D tissues, 3D organs include abundant microvascular networks, because cells in 3D tissues require oxygen and nutrients to be provided from blood flow. Because the diffusion distance of oxygen in a living tissue is limited to 100-200 μm (Carmeliet and Jain 2000), microvessels in 3D tissues have to be located within the interval of this critical distance. Thus, complex 3D tissue structures, including microvascular networks, make it difficult to construct 3D vital organs.

To construct 3D vital organs, formation of microvascular networks is important, as is constructing 3D epithelial tissues composed of parenchymal cells. In previous studies, the formation processes of the microvascular networks and the 3D epithelial tissues have been investigated individually. However, in recent studies, not only constructing the microvessels but also incorporating the microvessels within engineered 3D tissues, referred to as vascularization, has become increasingly important. Thus, this vascularization of tissue-engineered constructs is a key challenge in achieving 3D thick tissues *in vitro* (Rouwkema et al. 2008; Lovett et al. 2009; Auger et al. 2013).

16.4.2 Coculture of ECs and Hepatocytes in a Microfluidic Device: A Horizontal Approach

To achieve the vascularization of tissue-engineered constructs, it is essential to investigate the heterotypic interaction between microvascular ECs that form the microvascular networks and host organ cells (*e.g.*, hepatocytes) used to construct the 3D tissues. Liver is one of the most highly vascularized organs and is composed mainly of functional epithelial cells, the hepatocytes. Because hepatocytes assemble into organized 3D tissues *in vivo*, various 3D hepatocyte culture models have been developed (Sudo et al. 2005; Sudo 2014; Hwa et al. 2007; Toh et al. 2007). Culturing hepatocytes in a 3D configuration is useful to maintain differentiated hepatic functions *in vitro*. However, the current techniques fail to incorporate microvascular networks within the 3D hepatocyte tissues. Vascular tissue engineering has developed separately from liver tissue engineering. The formation of microvascular networks has been investigated with various *in vitro* 3D angiogenesis models, and the molecular mechanism of capillary formation has been studied (Davis et al. 2002; Vailhé et al. 2001). However, little is known concerning the interaction between the resulting capillary structures and other epithelial cell types, such as hepatocytes. Thus, there is a need to establish a coculture system where microvascular ECs and the other parenchymal cell types can interact in a controlled microenvironment with physiological realistic topology and cell organization (*e.g.*, capillary networks formed by ECs and 3D structures, composed of hepatocytes). Because 3D tissue formation and vascularization can be achieved in 3D tissues *in vivo*, there is an increasing demand for *in vitro* models that capture the physiological complexity such as controllable biochemical (*e.g.*, local chemotactic gradient and ECM properties) and biophysical factors (*e.g.*, interstitial fluid flow, shear flow, and mechanical properties of scaffold materials) (Griffith and Swartz 2006; Yamada and Cukierman 2007). Although *in vitro* models that incorporate these features are difficult to create, recent advances in microfluidic technologies (Kim et al. 2007; Khademhosseini et al. 2006) have allowed the development of such an *in vitro* culture model.

A microfluidic platform for the vascularization of 3D tissue-engineered constructs was reported, integrating knowledge from microfluidics, *in vitro* angiogenesis models, and liver tissue engineering (Sudo et al. 2009). The microfluidic device was made of polydimethylsiloxane (PDMS) and a cover glass. There were two microfluidic channels and an intervening collagen gel (Fig. 16.9a). To investigate the interaction between hepatocytes and ECs, hepatocytes were seeded into a microfluidic channel, while ECs were seeded into the channel of the other side. This microfluidic design allowed locating hepatocytes and ECs in close proximity, separated by collagen gel, the width of which was 750 μm (Fig. 16.9b).

First, hepatocytes were seeded in one of the two microfluidic channels and cultured in interstitial flow conditions (day 0, Fig. 16.10). This interstitial flow application during day 0-1 in the beginning of the culture induced 3D tissue formation by the hepatocytes. Hepatocytes maintained a 3D tissue structure even

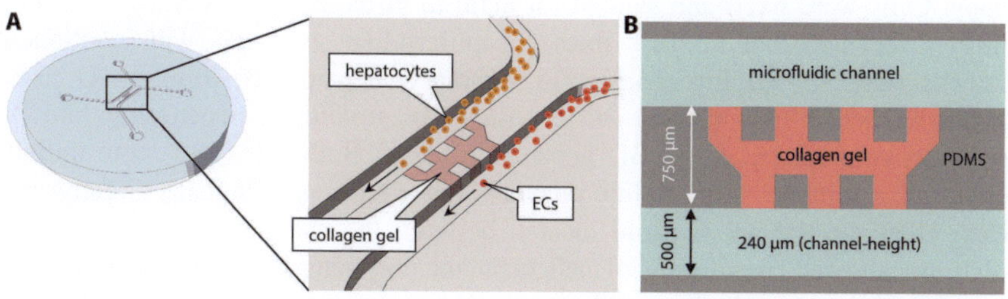

Fig. 16.9 *Microfluidic device for coculture of hepatocytes and microvascular ECs.* (**a**) A schematic image of a microfluidic device made of PDMS and a cover glass. There are two microfluidic channels and an intervening collagen gel. Hepatocytes were seeded in the microfluidic channel while ECs were seeded on the other side for a coculture. (**b**) Enlarged image of collagen gel scaffold in a microfluidic device

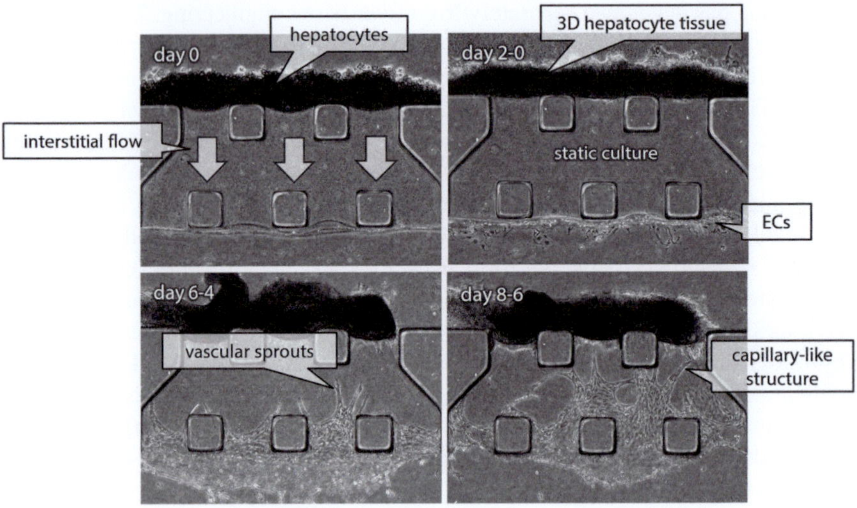

Fig. 16.10 *Phase-contrast images from cocultured hepatocytes and microvascular ECs in a microfluidic device*. Day 0: hepatocytes were seeded into a microfluidic channel and cultured in interstitial flow conditions for 1 day. Day 2–0: Hepatocytes form 3D tissues. ECs were added to the other side of the microfluidic channel, and coculture was performed in static conditions to enhance interactions between hepatocytes and ECs. Day 6–4: ECs formed vascular sprouts in the collagen gel. Day 8–6: Vascular sprouts extended toward the 3D hepatocyte tissues and developed into capillary-like structures

when cultured in static conditions going forward from day 1. ECs were then added to the other side to start the coculture (day 2–0, Fig. 16.10). Although rat ECs failed to form capillary-like structures in the absence of hepatocytes, they formed vascular sprouts extending in the collagen gel only in the case of coculture (day 6–4, Fig. 16.10). Vascular sprouts subsequently extended toward 3D hepatocyte tissues and developed into capillary-like structures (day 8–6, Fig. 16.10). Phase-contrast images of the capillary-like structures formed in the coculture suggested luminal spaces (day 10–8, Fig. 16.11), which was investigated further by confocal microscopy. Cells were fixed and stained for actin to visualize 3D structures. Confocal images revealed that rat ECs formed capillary-like structures with continuous lumens (day 12–10, Fig. 16.11). In summary, hepatocytes formed 3D tissue structures while ECs formed 3D capillary-like structures, penetrating into the collagen gel adjacent to the 3D hepatocyte tissue. Both cell types were initially cultured in separate microfluidic channels but could communicate directly with each other via the gel scaffold located between the channels. The microfluidic platform allowed monitoring of cell behavior and heterotypic interactions. This experimental approach showed the potential for constructing vascularized organs *in vitro*. Further investigations are needed to achieve vascularized liver tissues.

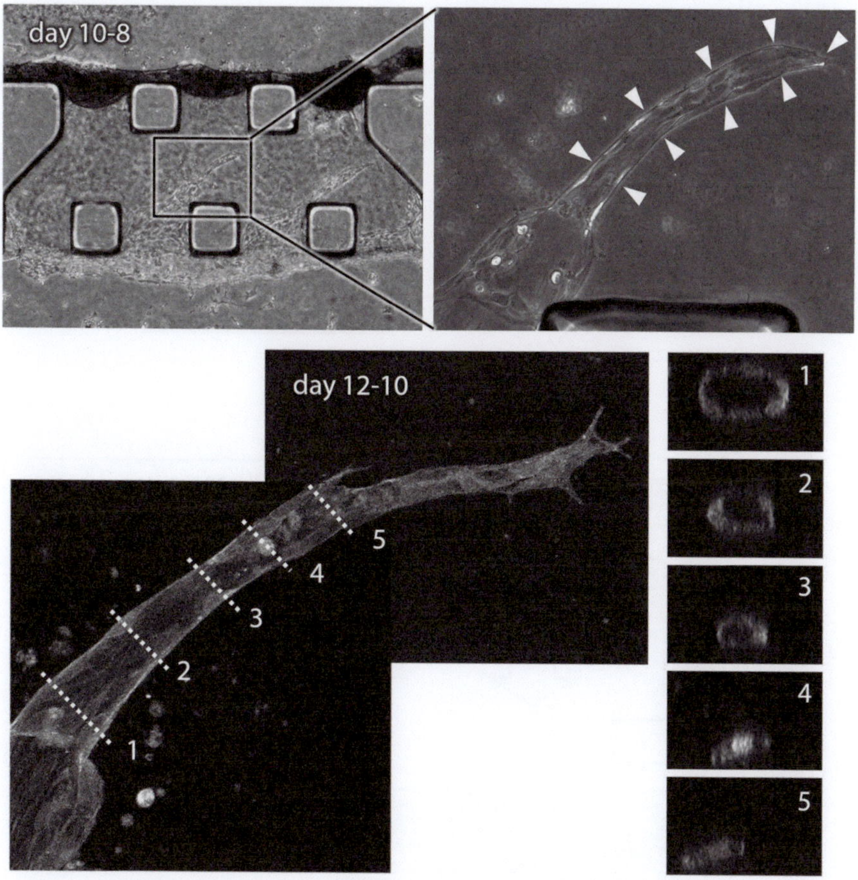

Fig. 16.11 *Capillary-like structures formed in hepatocyte-EC coculture*. Phase-contrast images taken on day 10–8 showed that ECs formed capillary-like structures extending toward the 3D hepatocyte tissues (arrowheads). Day 12–10: Cells were fixed on day 12–10 and stained for actin. A projection image was reconstructed from *z*-stack confocal images. Images #1-5 are cross section images corresponding to the dotted lines #1-5

16.4.3 Coculture of ECs and Hepatocytes Using a Microporous Membrane: A Vertical Approach

Another approach for constructing vascularized tissues is a method using microporous membranes. The microfluidic approach mentioned in Section 16.4.2 is a method to combine liver and vascular tissues horizontally. However, liver and vascular tissues can also be combined vertically, by stacking microporous membranes.

Heterotypic interactions between hepatocytes and ECs were investigated in a coculture model using microporous membranes. First, small hepatocytes (SHs), hepatic progenitor cells, were cultured on microporous membranes. SHs grew on

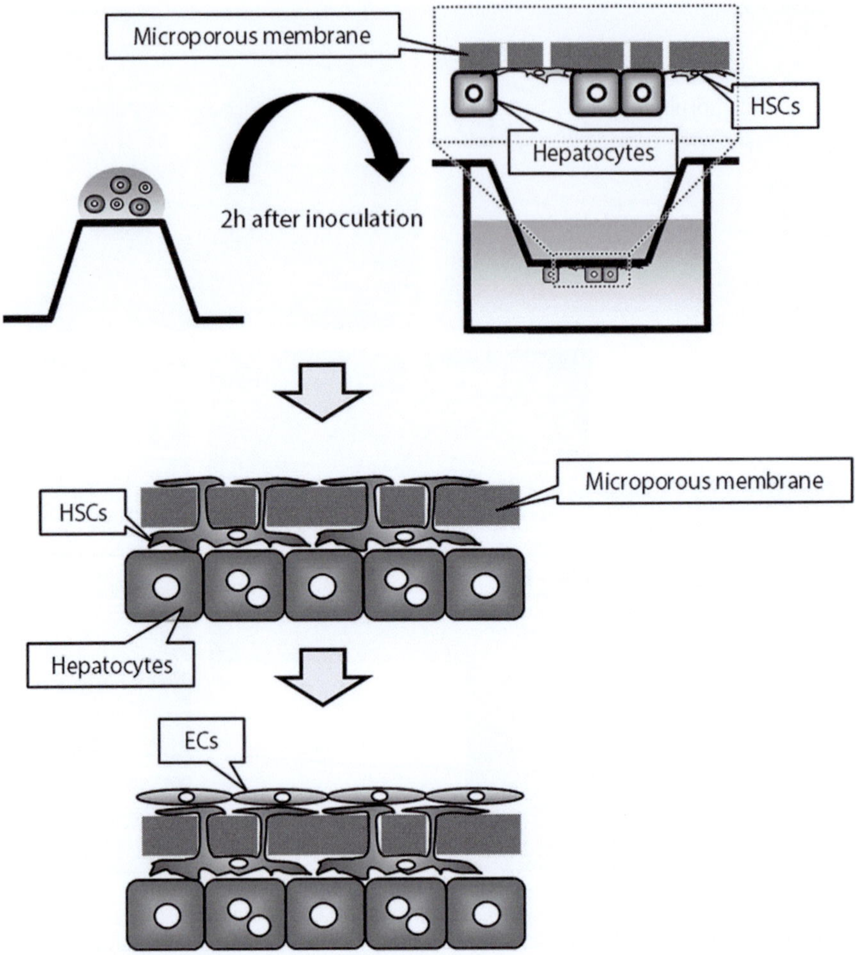

Fig. 16.12 *A triculture model using a microporous membrane*. First, SHs and HSCs were seeded on the bottom surface of a microporous membrane. The membrane was turned over in the holding well after 2 h. The SHs proliferated and formed colonies within 2 weeks in culture, while HSCs located between the SH colonies and the membrane. ECs were seeded onto the top surface of the membrane on day 14, resulting in the HSC-mediated layered architecture of SHs and ECs

the membranes, and the membranes were stacked onto other membranes to construct 3D stacked structures (Sudo et al. 2005). This 3D culture model was extended to coculture with ECs and hepatic stellate cells (HSCs), resulting in a 3D triculture model, established using microporous membranes, to create the functional unit of proximal layers of hepatocytes, HSCs, and ECs (Kasuya et al. 2011) (Fig. 16.12). HSC behavior was controlled by the membrane pore size, which was key for achieving proximal cell layers. With a specific pore size, the HSCs intercalated between layers of hepatocytes and ECs. When only cytoplasmic processes of quiescent HSCs were adjacent to ECs, while the HSC bodies remained on the side of the hepatocytes, the ECs changed morphologically and were capable of long-term survival. HSCs mediated communication between hepatocytes and ECs in terms of EC morphogenesis. This triculture model allowed investigating the roles of HSCs as both facilitators and integrators of cell-cell communication between hepatocytes and ECs and is useful for investigating heterotypic cellular communication *in vitro*. However, in terms of vascularization of the hepatocyte tissues, ECs in the triculture model formed monolayers rather than capillaries. Thus, next, this triculture model was modified to have ECs forming capillary-like structures.

To modify the hepatocyte-HSC-EC triculture model, the effects of direct HSC-EC contacts on EC capillary morphogenesis were investigated using the triculture model where HSC behavior was controlled spatially to achieve HSC-mediated proximal layers of hepatocytes and ECs. EC capillary morphogenesis was induced by overlaying Matrigel on an EC layer (Fig. 16.13). Direct HSC-EC contacts inhibited EC capillary morphogenesis, suggesting that the HSC-EC contacts may be an important factor in capillary formation (Kasuya et al. 2012a). Next, the hypothesis that, in addition to spatial control, temporal control of HSC behavior is also important in achieving capillary morphogenesis in the triculture was tested. ECs responded to the induction of capillary morphogenesis before the formation of direct HSC-EC contacts, while the ECs remained to form monolayers when capillary morphogenesis was induced after the HSC-EC contacts were established (Kasuya et al. 2012a) (Fig. 16.13). When capillary morphogenesis was achieved in the triculture, HSCs tended to preferably localize near the preformed capillary-like structures, resulting in reconstruction of liver sinusoid-like structures. In these structures, maturation of the hepatocytes was induced. These findings indicated that control, both spatially and temporally, of HSC behavior is a key engineering strategy for the vascularization of engineered liver tissue *in vitro*.

In liver sinusoids, HSCs line the outer surface of microvessels to form a functional unit with endothelia and hepatocytes. To reconstruct a functional liver tissue *in vitro*, recapturing the HSC-lined sinusoidal structures is essential. We previously demonstrated capillary formation in triculture where a Matrigel angiogenesis assay and SH-HSC coculture were combined using polymeric microporous membranes. However, the large thickness and low porosity of the membranes limited the heterotypic cell-cell interactions that are essential in forming HSC-lined structures. In addition to the achievement of capillary formation in the triculture, we previously established a stack-up culture method for reconstructing 3D hepatic tissues using thin and highly porous poly (D,L-lactide-co-glycolide acids) (PLGA) membranes (Kasuya et al. 2012b). We thus focused on the effective use of the PLGA microporous membranes in a hepatocyte-HSC-EC triculture to reconstruct the HSC-lined liver sinusoidal structures *in vitro*. First, the formation of EC capillary-like structures was induced on Matrigel-coated microporous PLGA membranes (Fig. 16.14). Then, the membranes were stacked on hepatic organoids

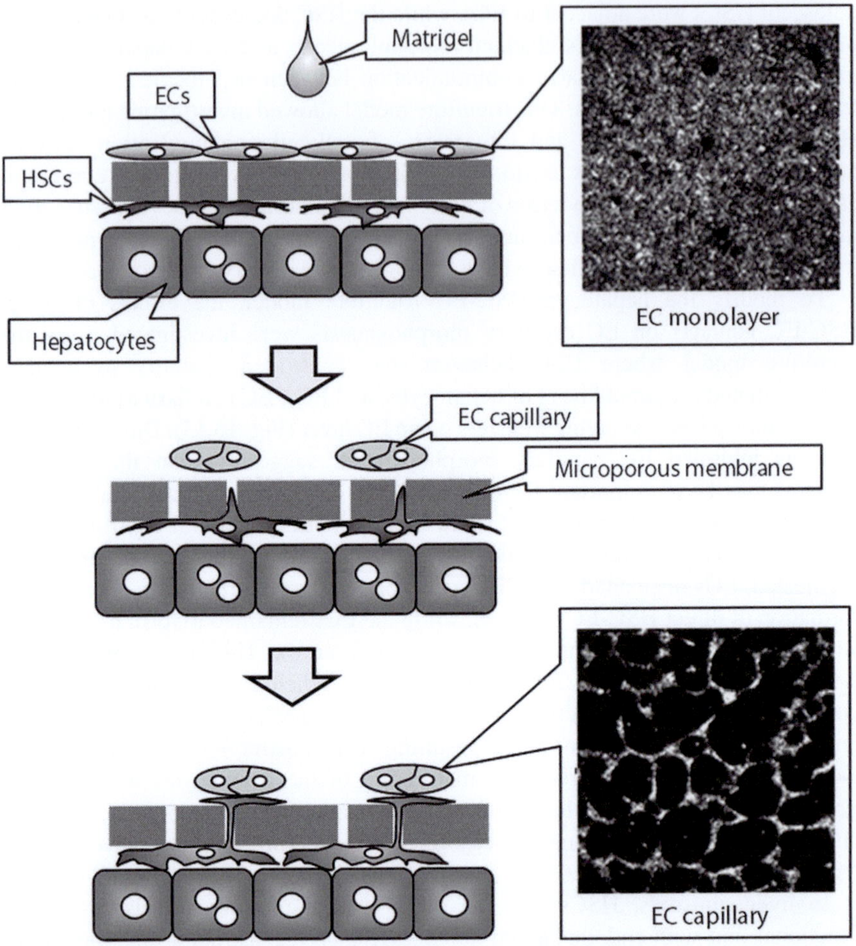

Fig. 16.13 *A triculture model with modification of capillary morphogenesis*. First, a SH-HSC-EC triculture model was constructed, in which ECs formed a confluent monolayer. Then, capillary morphogenesis was induced by overlaying Matrigel on the confluent EC monolayer. HSCs were recruited by the capillary-like structures, resulting in a triculture model where HSCs mediated an SH layer and capillary-like structures

composed of SHs and HSCs. When membranes with optimized configurations were used, HSCs selectively migrated to the EC capillary-like structures. This process was mediated, in part, by platelet-derived growth factor (PDGF) signaling. In addition, these HSCs firmly lined the outer surface of the EC capillary-like structures with long cytoplasmic processes (Kasuya et al. 2015) (Fig. 16.14). In the HSC-lined sinusoidal tissues, SHs retained higher levels of differentiated functions compared with those without ECs. This model provides a basis for the construction of functional, thick, vascularized liver tissues *in vitro*.

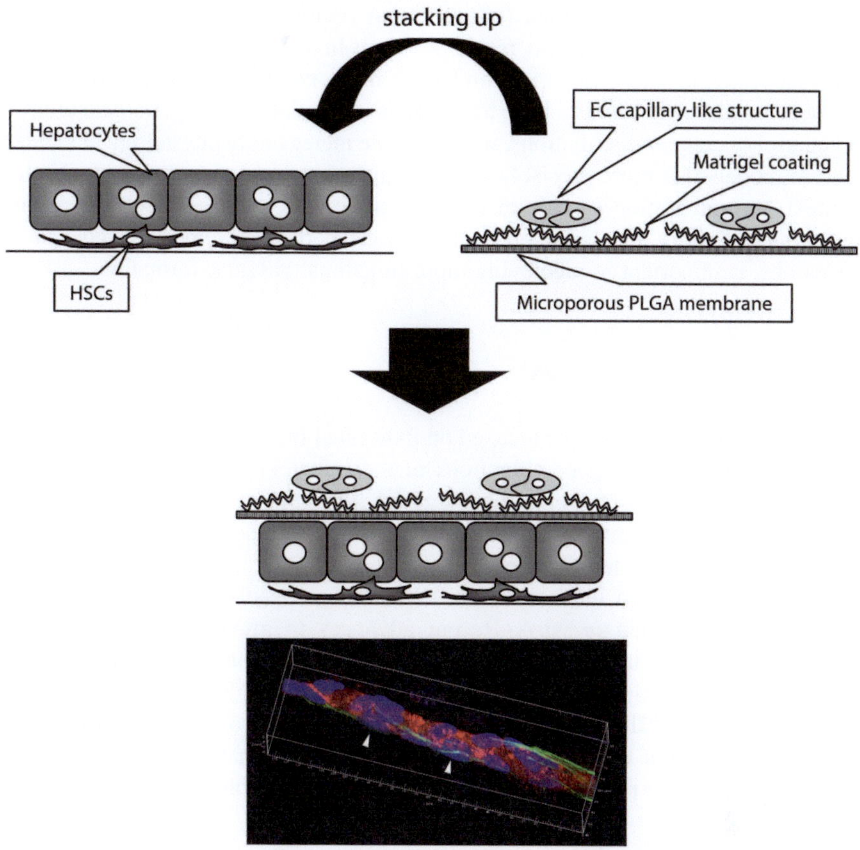

Fig. 16.14 *A triculture model using a biodegradable, PLGA microporous membrane*. SHs, including the HSC fraction, were cultured on a culture dish, while ECs were cultured on a Matrigel-coated microporous PLGA membrane to allow them to form capillary-like structures. Next, the membrane was stacked on the top of SH colonies grown in a culture dish, resulting in a 3D stacked-up triculture. Initially, HSCs were located between the SH layer and culture dish. However, HSCs passed through the microporous membrane and finally covered the outer surface of the capillary-like structures

16.5 Summary

16.5.1 Integrated Vascular Engineering

In this chapter, we introduced *in vitro* culture models for constructing microvascular networks, which are angiogenesis and vasculogenesis models. Because the control of microvascular formation is important both in the context of tissue engineering and cancer therapy, numerous studies have been performed to clarify the mechanism as to how we can construct or regress the microvasculature using these *in vitro* models. Although "conventional" culture models are still useful in investigating vascular formation, recent advances in microfluidic technologies have allowed studying vascular formation under more controlled culture conditions. Because microfluidic devices have advantages in regulating the spatial and temporal control of cellular distributions, we can create increasingly physiological *in vitro* culture models that more precisely mimic capillary morphogenesis *in vivo*. In this context, the integration of culture microenvironments, such as biochemical and biophysical factors and cell-to-cell interactions both with temporal and spatial variations, is important for achieving more functional vascular formation.

16.5.2 Future Perspectives

Because previous studies have focused on individual factors, such as one biochemical or biophysical factor, combinations of these factors need to be investigated further, which may result in the discovery of nonlinear, synergistic, and system-level effects. Heterotypic cellular interactions are also important aspects. Because microvessels always interact with other tissues, ECs interact with other epithelial tissues and cells. Such interactions will be characterized once vascularized 3D tissues are constructed *in vitro*. Recent advances in microfluidic technologies have opened a new field exploring multiple organ-level functions, often referred to as "human-on-a-chip" or "organs-on-a-chip" (Zhang et al. 2009; Luni et al. 2014; Bhatia and Ingber 2014; Huh et al. 2011). Because organs are connected with the vascular networks of our body, methods for vascular formation described in this chapter also have the potential to be integrated into human-on-a-chip or organs-on-a-chip devices.

Acknowledgements This work was partially supported by Grants-in-Aid for Scientific Research (25282135, 25560208, 25249018) from the Japan Society for Promotions of Science, and by the Human Resources Program in Energy Technology of the KETEP grant from the Ministry of Trade, Industry & Energy, Republic of Korea. (No. 20124010203250).

References

Anand-Apte B, Pepper MS, Voest E, Montesano R, Olsen B, Murphy G, Apte SS, Zetter B (1997) Inhibition of angiogenesis by tissue inhibitor of metalloproteinase-3. Invest Ophthalmol Vis Sci 38(5):817–823

Antonelli-Orlidge A, Saunders KB, Smith SR, D'Amore PA (1989) An activated form of transforming growth factor beta is produced by cocultures of endothelial cells and pericytes. Proc Natl Acad Sci U S A 86(12):4544–4548

Armulik A, Abramsson A, Betsholtz C (2005) Endothelial/pericyte interactions. Circ Res 97 (6):512–23

Au P, Tam J, Fukumura D, Jain RK (2008) Bone marrow-derived mesenchymal stem cells facilitate engineering of long-lasting functional vasculature. Blood 111(9):4551–4558

Auger FA, Gibot L, Lacroix D (2013) The pivotal role of vascularization in tissue engineering. Annu Rev Biomed Eng 15:177–200 ávan der Meer AD, Dijke P, den Berg A (2013) Three-dimensional co-cultures of human endothelial cells and embryonic stem cell-derived pericytes inside a microfluidic device. Lab Chip 13:3562–3568

Baker B, Trappmann B, Stapleton SC, Toro E, Chen CS (2013) Microfluidics embedded within extracellular matrix to define vascular architectures and pattern diffusive gradients. Lab Chip 13:3246–3252

Bayless KJ, Davis GE (2003) Sphingosine-1-phosphate markedly induces matrix metalloproteinase and integrin-dependent human endothelial cell invasion and lumen formation in three-dimensional collagen and fibrin matrices. Biochem Biophys Res Commun 312 (4):903–913

Berthod F, Germain L, Tremblay N, Auger FA (2006) Extracellular matrix deposition by fibroblasts is necessary to promote capillary-like tube formation in vitro. J Cell Physiol 207 (2):491–498

Bhatia SN, Ingber DE (2014) Microfluidic organs-on-chips. Nat Biotechnol 32(8):760–772

Bikfalvi A, Sauzeau C, Moukadiri H, Maclouf J, Busso N, Bryckaert M, Plouet J, Tobelem G (1991) Interaction of vasculotropin/vascular endothelial cell growth factor with human umbilical vein endothelial cells: binding, internalization, degradation, and biological effects. J Cell Physiol 149(1):50–59

Birdwell CR, Gospodarowicz D, Nicholson GL (1977) Factors from 3T3 cells stimulate proliferation of cultured vascular endothelial cells. Nature 268(5620):528–531

Bischel LL, Young EW, Mader BR, Beebe DJ (2013) Tubeless microfluidic angiogenesis assay with three-dimensional endothelial-lined microvessels. Biomaterials 34(5):1471–1477

Borenstein JT et al (2010) Functional endothelialized microvascular networks with circular cross-sections in a tissue culture substrate. Biomed Microdevices 12:71–79

Carmeliet P, Jain RK (2000) Angiogenesis in cancer and other diseases. Nature 407(6801):249–257

Carmeliet P, De Smet F, Loges S, Mazzone M (2009) Branching morphogenesis and antiangiogenesis candidates: tip cells lead the way. Nat Rev Clin Oncol 6(6):315–326

Carrion B et al (2010) Recreating the perivascular niche ex vivo using a microfluidic approach. Biotechnol Bioeng 107:1020–1028

Chaw K, Manimaran M, Tay E, Swaminathan S (2007) Multi-step microfluidic device for studying cancer metastasis. Lab Chip 7:1041–1047

Chen MB, Whisler JA, Jeon JS, Kamm RD (2013) Mechanisms of tumor cell extravasation in an in vitro microvascular network platform. Integr Biol 5:1262–1271

Chrobak KM, Potter DR, Tien J (2006) Formation of perfused, functional microvascular tubes in vitro. Microvasc Res 71:185–196

Chung S et al (2009a) Cell migration into scaffolds under co-culture conditions in a microfluidic platform. Lab Chip 9:269–275

Chung S, Sudo R, Zervantonakis IK, Rimchala T, Kamm RD (2009b) Surface-treatment-induced three-dimensional capillary morphogenesis in a microfluidic platform. Adv Mater 21:4863–4867

Chung S, Sudo R, Vickerman V, Zervantonakis IK, Kamm RD (2010) Microfluidic platforms for studies of angiogenesis, cell migration, and cell–cell interactions. Ann Biomed Eng 38:1164–1177

Colgan OC et al (2007) Regulation of bovine brain microvascular endothelial tight junction assembly and barrier function by laminar shear stress. Am J Physioly-Heart Circ Physiol 292:H3190–H3197

Crocker DJ, Murad TM, Geer JC (1970) Role of the pericyte in wound healing. An ultrastructural study. Exp Mol Pathol 13(1):51–65

Dai X et al (2011) A novel in vitro angiogenesis model based on a microfluidic device. Chin Sci Bull 56:3301–3309

Davis GE, Bayless KJ, Mavila A (2002) Molecular basis of endothelial cell morphogenesis in three-dimensional extracellular matrices. Anat Rec 268(3):252–275

Díaz-Flores L, Gutiérrez R, Madrid JF, Varela H, Valladares F, Acosta E, Martín-Vasallo P, Díaz-Flores L Jr (2009) Pericytes. Morphofunction, interactions and pathology in a quiescent and activated mesenchymal cell niche. Histol Histopathol 24(7):909–969

Estrada R et al (2011) Endothelial cell culture model for replication of physiological profiles of pressure, flow, stretch, and shear stress in vitro. Anal Chem 83:3170–3177

Folkman J (1971) Tumor angiogenesis: therapeutic implications. N Engl J Med 285 (21):1182–1186

Golden AP, Tien J (2007) Fabrication of microfluidic hydrogels using molded gelatin as a sacrificial element. Lab Chip 7:720–725

Good DJ, Polverini PJ, Rastinejad F, Le Beau MM, Lemons RS, Frazier WA, Bouck NP (1990) A tumor suppressor-dependent inhibitor of angiogenesis is immunologically and functionally indistinguishable from a fragment of thrombospondin. Proc Natl Acad Sci U S A 87 (17):6624–6628

Griffith LG, Naughton G (2002) Tissue engineering – current challenges and expanding opportunities. Science 295(5557):1009–1014

Griffith LG, Swartz MA (2006) Capturing complex 3D tissue physiology in vitro. Nat Rev Mol Cell Biol 7(3):211–24

Han S, Yan JJ, Shin Y, Jeon JJ, Won J, Jeong HE, Kamm RD, Kim YJ, Chung S (2012) A versatile assay for monitoring in vivo-like transendothelial migration of neutrophils. Lab Chip 12:3861–3865

Hartlapp I, Abe R, Saeed RW, Peng T, Voelter W, Bucala R, Metz CN (2001) Fibrocytes induce an angiogenic phenotype in cultured endothelial cells and promote angiogenesis in vivo. FASEB J 15(12):2215–2224

Hellström M, Phng LK, Hofmann JJ, Wallgard E, Coultas L, Lindblom P, Alva J, Nilsson AK, Karlsson L, Gaiano N, Yoon K, Rossant J, Iruela-Arispe ML, Kalén M, Gerhardt H, Betsholtz C (2007) Dll4 signalling through Notch1 regulates formation of tip cells during angiogenesis. Nature 445(7129):776–780

Helm CL, Fleury ME, Zisch AH, Boschetti F, Swartz MA (2005) Synergy between interstitial flow and VEGF directs capillary morphogenesis in vitro through a gradient amplification mechanism. Proc Natl Acad Sci U S A 102(44):15779–15784

Hernández Vera R, Genové E, Alvarez L, Borrós S, Kamm R, Lauffenburger D, Semino CE (2009) Interstitial fluid flow intensity modulates endothelial sprouting in restricted Src-activated cell clusters during capillary morphogenesis. Tissue Eng A 15(1):175–185

Hiraki Y, Inoue H, Iyama K, Kamizono A, Ochiai M, Shukunami C, Iijima S, Suzuki F, Kondo J (1997) Identification of chondromodulin I as a novel endothelial cell growth inhibitor. Purification and its localization in the avascular zone of epiphyseal cartilage. J Biol Chem 272 (51):32419–32426

Huh D, Hamilton GA, Ingber DE (2011) From 3D cell culture to organs-on-chips. Trends Cell Biol 21(12):745–754

Hurley JR, Balaji S, Narmoneva DA (2010) Complex temporal regulation of capillary morphogenesis by fibroblasts. Am J Physiol Cell Physiol 299(2):C444–C453

Hwa AJ, Fry RC, Sivaraman A, So PT, Samson LD, Stolz DB, Griffith LG (2007) Rat liver sinusoidal endothelial cells survive without exogenous VEGF in 3D perfused co-cultures with hepatocytes. FASEB J 21:2564–2579

Jeon JS, Zervantonakis IK, Chung S, Kamm RD, Charest JL (2013) In vitro model of tumor cell extravasation. PLoS One 8:e56910

Jeon JS, Bersini S, Whisler JA, Chen MB, Dubini G, Charest JL, Moretti M, Kamm RD (2014) Generation of 3D functional microvascular networks with human mesenchymal stem cells in microfluidic systems. Integr Biol (Camb) 6(5):555–63

Jeong GS, Kwon GH, Kang AR, Jung BY, Park Y, Chung S, Lee SH (2011a) Microfluidic assay of endothelial cell migration in 3D interpenetrating polymer semi-network HA-Collagen hydrogel. Biomed Microdevices 13(4):717–723

Jeong GS, Han S, Shin Y, Kwon GH, Kamm RD, Lee SH, Chung S (2011b) Sprouting angiogenesis under a chemical gradient regulated by interactions with an endothelial monolayer in a microfluidic platform. Anal Chem 83(22):8454–8459

Kalchman J, Fujioka S, Chung S, Kikkawa Y, Mitaka T, Kamm RD, Tanishita K, Sudo R (2013) A three-dimensional microfluidic tumor cell migration assay to screen the effect of anti-migratory drugs and interstitial flow. Microfluid Nanofluid 14:969–981

Kang H, Bayless KJ, Kaunas R (2008) Fluid shear stress modulates endothelial cell invasion into three-dimensional collagen matrices. Am J Physiol Heart Circ Physiol 295(5):H2087–H2097

Kasuya J, Sudo R, Mitaka T, Ikeda M, Tanishita K (2011) Hepatic stellate cell-mediated three-dimensional hepatocyte and endothelial cell triculture model. Tissue Eng A 17(3–4):361–370

Kasuya J, Sudo R, Mitaka T, Ikeda M, Tanishita K (2012a) Spatio-temporal control of hepatic stellate cell-endothelial cell interactions for reconstruction of liver sinusoids in vitro. Tissue Eng A 18(9–10):1045–1056

Kasuya J, Sudo R, Tamogami R, Masuda G, Mitaka T, Ikeda M, Tanishita K (2012b) Reconstruction of 3D stacked hepatocyte tissues using degradable, microporous poly(d, l-lactide-co-glycolide) membranes. Biomaterials 33(9):2693–2700

Kasuya J, Sudo R, Masuda G, Mitaka T, Ikeda M, Tanishita K (2015) Reconstruction of hepatic stellate cell-incorporated liver capillary structures in small hepatocyte tri-culture using microporous membranes. J Tissue Eng Regen Med 9(3):247–56

Kaunas R, Kang H, Bayless KJ (2011) Synergistic regulation of angiogenic sprouting by biochemical factors and wall shear stress. Cell Mol Bioeng 4(4):547–559

Khademhosseini A, Langer R, Borenstein J, Vacanti JP (2006) Microscale technologies for tissue engineering and biology. Proc Natl Acad Sci U S A 103:2480–2487

Khademhosseini A, Vacanti JP, Langer R (2009) Progress in tissue engineering. Sci Am 300 (5):64–71

Kieda C et al (2006) Suppression of hypoxia-induced HIF-1α and of angiogenesis in endothelial cells by myo-inositol trispyrophosphate-treated erythrocytes. Proc Natl Acad Sci 103:15576–15581

Kim L, Toh YC, Voldman J, Yu H (2007) A practical guide to microfluidic perfusion culture of adherent mammalian cells. Lab Chip 7:681–694

Kim C, Chung S, Yuchun L, Kim MC, Chan JK, Asada HH, Kamm RD (2012) In vitro angiogenesis assay for the study of cell-encapsulation therapy. Lab Chip 12(16):2942–2950

Kim S, Lee H, Chung M, Jeon NL (2013) Engineering of functional, perfusable 3D microvascular networks on a chip. Lab Chip 13:1489–1500

Lafleur MA, Handsley MM, Knäuper V, Murphy G, Edwards DR (2002) Endothelial tubulogenesis within fibrin gels specifically requires the activity of membrane-type-matrix metalloproteinases (MT-MMPs). J Cell Sci 115(Pt 17):3427–3438

Langer R, Vacanti JP (1993) Tissue engineering. Science 260(5110):920–926

Lee H, Kim S, Chung M, Kim JH, Jeon NL (2014) A bioengineered array of 3D microvessels for vascular permeability assay. Microvasc Res 91:90–98

Liu M-C et al (2013) Electrofluidic pressure sensor embedded microfluidic device: a study of endothelial cells under hydrostatic pressure and shear stress combinations. Lab Chip 13:1743–1753

Lovett M, Lee K, Edwards A, Kaplan DL (2009) Vascularization strategies for tissue engineering. Tissue Eng B Rev 15(3):353–370

Luni C, Serena E, Elvassore N (2014) Human-on-chip for therapy development and fundamental science. Curr Opin Biotechnol 25:45–50

Mack PJ et al (2009) Biomechanical regulation of endothelium-dependent events critical for adaptive remodeling. J Biol Chem 284:8412–8420

Maione TE, Gray GS, Petro J, Hunt AJ, Donner AL, Bauer SI, Carson HF, Sharpe RJ (1990) Inhibition of angiogenesis by recombinant human platelet factor-4 and related peptides. Science 247(4938):77–79

Mansbridge JN, Liu K, Pinney RE, Patch R, Ratcliffe A, Naughton GK (1999) Growth factors secreted by fibroblasts: role in healing diabetic foot ulcers. Diabetes Obes Metab 1(5):265–279
Mazzone M, Dettori D, Leite de Oliveira R, Loges S, Schmidt T, Jonckx B, Tian YM, Lanahan AA, Pollard P, Ruiz de Almodovar C, De Smet F, Vinckier S, Aragonés J, Debackere K, Luttun A, Wyns S, Jordan B, Pisacane A, Gallez B, Lampugnani MG, Dejana E, Simons M, Ratcliffe P, Maxwell P, Carmeliet P (2009) Heterozygous deficiency of PHD2 restores tumor oxygenation and inhibits metastasis via endothelial normalization. Cell 136:839–851
Montesano R, Orci L (1985) Tumor-promoting phorbol esters induce angiogenesis in vitro. Cell 42 (2):469–477
Montesano R, Vassalli JD, Baird A, Guillemin R, Orci L (1986) Basic fibroblast growth factor induces angiogenesis in vitro. Proc Natl Acad Sci U S A 83(19):7297–7301
Montesano R, Pepper MS, Belin D, Vassalli JD, Orci L (1988) Induction of angiogenesis in vitro by vanadate, an inhibitor of phosphotyrosine phosphatases. J Cell Physiol 134(3):460–6
Montesano R, Pepper MS, Orci L (1993) Paracrine induction of angiogenesis in vitro by Swiss 3T3 fibroblasts. J Cell Sci 105(Pt 4):1013–1024
Moya ML, Hsu YH, Lee AP, Hughes CC, George SC (2013) In vitro perfused human capillary networks. Tissue Eng C Methods 19(9):730–737
Yamada KM, Cukierman E (2007) Modeling tissue morphogenesis and cancer in 3D. Cell 130 (4):601–10
Newman AC, Nakatsu MN, Chou W, Gershon PD, Hughes CC (2011) The requirement for fibroblasts in angiogenesis: fibroblast-derived matrix proteins are essential for endothelial cell lumen formation. Mol Biol Cell 22(20):3791–3800
Ng CP, Helm CL, Swartz MA (2004) Interstitial flow differentially stimulates blood and lymphatic endothelial cell morphogenesis in vitro. Microvasc Res 68(3):258–264
Nguyen D-HT et al (2013) Biomimetic model to reconstitute angiogenic sprouting morphogenesis in vitro. Proc Natl Acad Sci 110:6712–6717
O'Reilly MS, Holmgren L, Shing Y, Chen C, Rosenthal RA, Moses M, Lane WS, Cao Y, Sage EH, Folkman J (1994) Angiostatin: a novel angiogenesis inhibitor that mediates the suppression of metastases by a Lewis lung carcinoma. Cell 79(2):315–328
O'Reilly MS, Boehm T, Shing Y, Fukai N, Vasios G, Lane WS, Flynn E, Birkhead JR, Olsen BR, Folkman J (1997) Endostatin: an endogenous inhibitor of angiogenesis and tumor growth. Cell 88(2):277–285
Orlidge A, D'Amore PA (1987) Inhibition of capillary endothelial cell growth by pericytes and smooth muscle cells. J Cell Biol 105(3):1455–1462
Penn JS (2008) Retinal and choroidal angiogenesis. Springer Netherlands
Pepper MS, Ferrara N, Orci L, Montesano R (1992) Potent synergism between vascular endothelial growth factor and basic fibroblast growth factor in the induction of angiogenesis in vitro. Biochem Biophys Res Commun 189(2):824–831
Pepper MS, Vassalli JD, Orci L, Montesano R (1993) Biphasic effect of transforming growth factor-beta 1 on in vitro angiogenesis. Exp Cell Res 204(2):356–363
Pepper MS, Ferrara N, Orci L, Montesano R (1995) Leukemia inhibitory factor (LIF) inhibits angiogenesis in vitro. J Cell Sci 108(Pt 1):73–83
Risau W (1997) Mechanisms of angiogenesis. Nature 386(6626):671–674
Risau W, Flamme I (1995) Vasculogenesis. Annu Rev Cell Dev Biol 11:73–91
Rouwkema J, Rivron NC, van Blitterswijk CA (2008) Vascularization in tissue engineering. Trends Biotechnol 26(8):434–441
Ruoslahti E (1996) Brain extracellular matrix. Glycobiology 6:489–492
Saunders WB, Bohnsack BL, Faske JB, Anthis NJ, Bayless KJ, Hirschi KK, Davis GE (2006) Coregulation of vascular tube stabilization by endothelial cell TIMP-2 and pericyte TIMP-3. J Cell Biol 175(1):179–191
Seebach J et al (2000) Endothelial barrier function under laminar fluid shear stress. Lab Investig 80:1819–1831

Semino CE, Kamm RD, Lauffenburger DA (2006) Autocrine EGF receptor activation mediates endothelial cell migration and vascular morphogenesis induced by VEGF under interstitial flow. Exp Cell Res 312(3):289–298

Shamloo A, Heilshorn SC (2010) Matrix density mediates polarization and lumen formation of endothelial sprouts in VEGF gradients. Lab Chip 10:3061–3068

Shamloo A, Ma N, Poo M-M, Sohn LL, Heilshorn SC (2008) Endothelial cell polarization and chemotaxis in a microfluidic device. Lab Chip 8:1292–1299

Shamloo A, Xu H, Heilshorn S (2011) Mechanisms of vascular endothelial growth factor-induced pathfinding by endothelial sprouts in biomaterials. Tissue Eng A 18:320–330

Shao J et al (2009) Integrated microfluidic chip for endothelial cells culture and analysis exposed to a pulsatile and oscillatory shear stress. Lab Chip 9:3118–3125

Shepro D, Morel NM (1993) Pericyte physiology. FASEB J 7(11):1031–1038

Shin Y et al (2011) In vitro 3D collective sprouting angiogenesis under orchestrated ANG-1 and VEGF gradients. Lab Chip 11:2175–2181

Shin Y, Han S, Jeon JS, Yamamoto K, Zervantonakis IK, Sudo R, Kamm RD, Chung S (2012) Microfluidic assay for simultaneous culture of multiple cell types on surfaces or within hydrogels. Nat Protoc 7(7):1247–1259

Shin Y et al (2014) Reconstituting vascular microenvironment of neural stem cell niche in three-dimensional extracellular matrix. Adv Healthcare Mater 3:1457–1464

Sieminski AL, Hebbel RP, Gooch KJ (2004) The relative magnitudes of endothelial force generation and matrix stiffness modulate capillary morphogenesis in vitro. Exp Cell Res 297 (2):574–584

Song JW, Munn LL (2011) Fluid forces control endothelial sprouting. Proc Natl Acad Sci 108:15342–15347

Song JW et al (2005) Computer-controlled microcirculatory support system for endothelial cell culture and shearing. Anal Chem 77:3993–3999

Song JW et al (2009) Microfluidic endothelium for studying the intravascular adhesion of metastatic breast cancer cells. PLoS One 4:e5756

Soriano JV et al (2004) Inhibition of angiogenesis by growth factor receptor bound protein 2-Src homology 2 domain bound antagonists. Mol Cancer Ther 3:1289–1299

Srigunapalan S, Lam C, Wheeler AR, Simmons CA (2011) A microfluidic membrane device to mimic critical components of the vascular microenvironment. Biomicrofluidics 5:013409

Sudo R (2014) Multiscale tissue engineering for liver reconstruction. Organogenesis 10 (2):216–224

Sudo R, Mitaka T, Ikeda M, Tanishita K (2005) Reconstruction of 3D stacked-up structures by rat small hepatocytes on microporous membranes. FASEB J 19:1695–1717

Sudo R, Chung S, Zervantonakis IK, Vickerman V, Toshimitsu Y, Griffith LG, Kamm RD (2009) Transport-mediated angiogenesis in 3D epithelial coculture. FASEB J 23(7):2155–2164

Tammela T, Zarkada G, Wallgard E, Murtomäki A, Suchting S, Wirzenius M, Waltari M, Hellström M, Schomber T, Peltonen R, Freitas C, Duarte A, Isoniemi H, Laakkonen P, Christofori G, Ylä-Herttuala S, Shibuya M, Pytowski B, Eichmann A, Betsholtz C, Alitalo K (2008) Blocking VEGFR-3 suppresses angiogenic sprouting and vascular network formation. Nature 454(7204):656–660

Tardy Y, Resnick N, Nagel T, Gimbrone M, Dewey C (1997) Shear stress gradients remodel endothelial monolayers in vitro via a cell proliferation-migration-loss cycle. Arterioscler Thromb Vasc Biol 17:3102–3106

Toh YC, Zhang C, Zhang J, Khong YM, Chang S, Samper VD, van Noort D, Hutmacher DW, Yu H (2007) A novel 3D mammalian cell perfusion-culture system in microfluidic channels. Lab Chip 7:302–309

Ueda A, Koga M, Ikeda M, Kudo S, Tanishita K (2004) Effect of shear stress on microvessel network formation of endothelial cells with in vitro three-dimensional model. Am J Physiol Heart Circ Physiol 287(3):H994–H1002

Vailhé B, Vittet D, Feige JJ (2001) In vitro models of vasculogenesis and angiogenesis. Lab Investig 81:439–452
Vernon RB, Sage EH (1999) A novel, quantitative model for study of endothelial cell migration and sprout formation within three-dimensional collagen matrices. Microvasc Res 57 (2):118–133
Wanjare M, Kusuma S, Gerecht S (2013) Perivascular cells in blood vessel regeneration. Biotechnol J 8(4):434–447
Whisler JA, Chen MB, Kamm RD (2014) Control of perfusable microvascular network morphology using a multiculture microfluidic system. Tissue Eng C Methods 20(7):543–552
Wong KH, Truslow JG, Khankhel AH, Chan KL, Tien J (2012) Artificial lymphatic drainage systems for vascularized microfluidic scaffolds. J Biomed Mater Res A 101:2181–2190
Yamamoto K, Tanimura K, Mabuchi Y, Matsuzaki Y, Chung S, Kamm RD, Ikeda M, Tanishita K, Sudo R (2013) The stabilization effect of mesenchymal stem cells on the formation of microvascular networks in a microfluidic device. J Biomech Sci Eng 8(2):114–128
Yamamura N, Sudo R, Ikeda M, Tanishita K (2007) Effects of the mechanical properties of collagen gel on the in vitro formation of microvessel networks by endothelial cells. Tissue Eng 13(7):1443–1453
Yeon JH, Ryu HR, Chung M, Hu QP, Jeon NL (2012) In vitro formation and characterization of a perfusable three-dimensional tubular capillary network in microfluidic devices. Lab Chip 12:2815–2822
Young EW, Wheeler AR, Simmons CA (2007) Matrix-dependent adhesion of vascular and valvular endothelial cells in microfluidic channels. Lab Chip 7:1759–1766
Zervantonakis IK et al (2012) Three-dimensional microfluidic model for tumor cell intravasation and endothelial barrier function. Proc Natl Acad Sci 109:13515–13520
Zhang C, Zhao Z, Abdul Rahim NA, van Noort D, Yu H (2009) Towards a human-on-chip: culturing multiple cell types on a chip with compartmentalized microenvironments. Lab Chip 9 (22):3185–3392
Zhang Q, Liu T, Qin J (2012) A microfluidic-based device for study of transendothelial invasion of tumor aggregates in realtime. Lab Chip 12:2837–2842
Zheng C et al (2012) Quantitative study of the dynamic tumor–endothelial cell interactions through an integrated microfluidic coculture system. Anal Chem 84:2088–2093
Zheng Y et al (2012) In vitro microvessels for the study of angiogenesis and thrombosis. Proc Natl Acad Sci 109:9342–9347

Novel Technology to Assay the Multicellular Network: On-Chip Cellomics Technology

17

Kenji Yasuda

Abstract

A series of studies aimed at developing methods and technologies of analyzing epigenetic information in cells and in those networks, as well as that of genetic information, was examined to expand our understanding of how living systems are determined. Technologies of analyzing epigenetic information was developed starting from the twin complementary viewpoints of cell regulation as an "algebraic" system (emphasis on temporal aspects) and as a "geometric" system (emphasis on spatial aspects). Exploiting the combination of latest microfabrication technologies and measurement technologies, which we call on-chip cellomics technology, we can select, control, and reconstruct the environments, interaction of single cells and cell networks from "algebraic" and "geometric" viewpoints. In this chapter, our developed technolgoeis and some results for spatial viewpoint of epigenetic information as a part of a series of cell-network-based "geometric" studies of celluler systems in our research groups are summarized and reported. The knowlege and technolgies acquired from these viewpoints may lead to the use of cells that fully control practical applications like cell-network-based drug screening and the regeneration of organs from cells.

Keywords

On-chip cellomics • Epigenetic information • Community effect • Cell network • Microfabrication

K. Yasuda (✉)
Institute of Biomaterials and Bioengineering, Tokyo Medical and Dental University, 2-3-10 Kanda-Surugadai, Chiyoda, Tokyo 101-0062, Japan
e-mail: yasuda.bmi@tmd.ac.jp

K. Tanishita, K. Yamamoto (eds.), *Vascular Engineering*,
DOI 10.1007/978-4-431-54801-0_17

17.1 Introduction

After the dramatic progress of life science from the twentieth century, a great challenge remains in the epigenetic information, such as how the living system of cells can reserve and inherit the acquired epigenetic information among neighboring cells and between generations. As we move into the postgenomics/proteomics period, such a complementarity of genetic information will become more apparent. The cells in a group are individual entities, and differences arise even among cells with identical genetic information that have grown under the same conditions. These cells respond differently to perturbations (Spudich and Koshland 1976). Why and how do these differences arise? To understand the rules underlying possible differences occurring in cells, we need to develop methods of simultaneously evaluating both the genetic and epigenetic information (Fig. 17.1). In other words, if we are to understand whole information of living systems, we need to analyze their epigenetic information such as adaptation processes, populatin effects (a group of same cells), and community effects (a group of different cells), and the meaning of spatial cell network patterns. We thus started a series of studies focusing on developing a system that could be used to evaluate the epigenetic information in cells by continuously observing specific examples and their interactions under controlled conditions. The importance of understanding epigenetic information is expected to become apparent in cell-based biological and medical fields like cell-based drug screening and the regeneration of organs from stem cells, fields where phenomena cannot be interpreted without taking epigenetic factors into account.

As cells are minimum units in genetic and epigenetic information, we must analyze their epigenetic information starting from the twin complementary

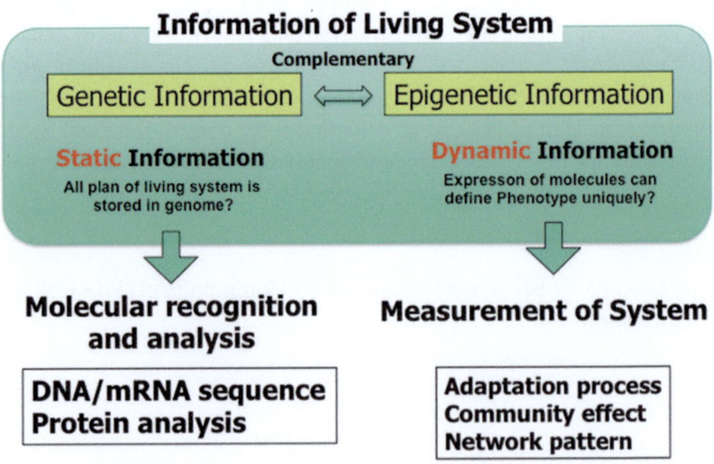

Fig. 17.1 *Epigenetic information: complementary to genetic information.* As epigenetic information is based on the functional pathway information, it is sometimes quite difficult to measure using the conventional destructive molecule-based analysis. To measure the epigenetic information, re-constructive approach of cell network model with non-destructive long-term measurement is one of the powerful methods

viewpoints of cell regulation as being an "algebraic" system (with emphasis on temporal aspects; adaptation among generation) and as a "geometric" system (with emphasis on spatial aspects; spatial pattern-dependent community effect) using identified single cells and their groups. We thus commenced a series of studies to analyze epigenetic information among neighboring generations of cells, and in the spatial structures of a cell network in order to expand our understanding of how the fates of living systems are determined. As shown in Fig. 17.2, the strategy behind

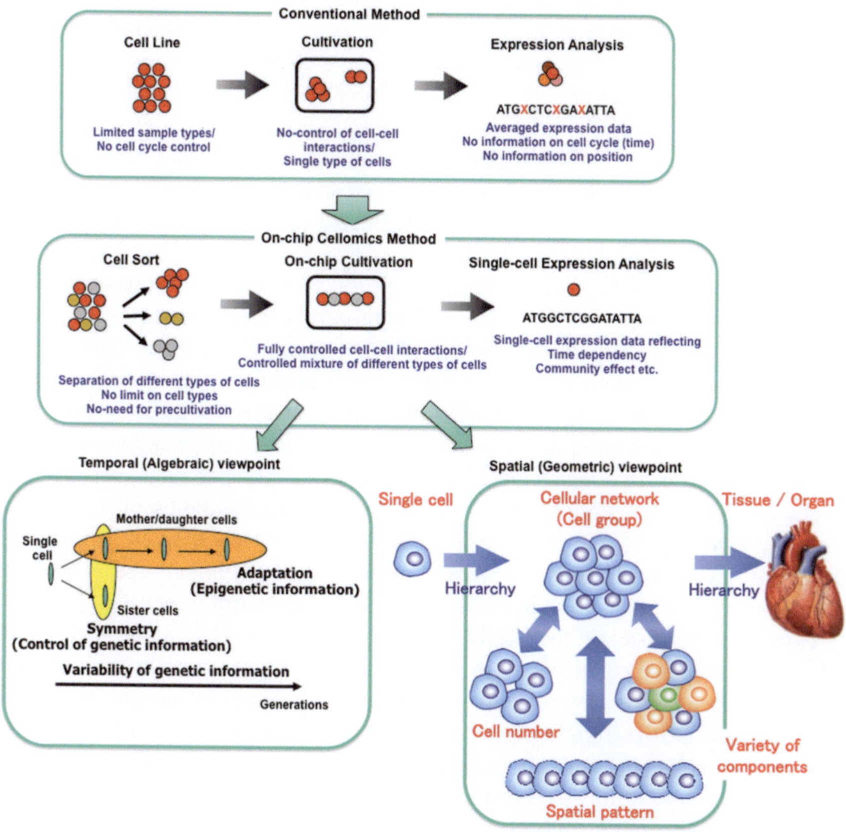

Fig. 17.2 *On-chip cellomics analysis. Aim of single-cell-based analysis for multicellular system: Temporal (algebraic) aspect and spatial (geometric) aspect.* In geometric viewpoint, two viewpoints of analysis also should be done; one is population/community size dependence and the other is spatial (network) pattern dependence of group of cells. In the conventional cell-based studies, cell lines were usually used for acquiring the same type of cells, and then cultivated them in the cultivation dish without any control of their population or no formation of community with other cell types, and finally they were analyzed as a group regardless of their differences of cell cycles regardless of their possible differences. In contrast, on-chip cellomics technology adopted the new three step strategies: first, the cells were adopted from the community using non-destructive cell sorting procedure, then the cells were cultivated in a mocrochambers, in which cell network formation and medium environment was controlled, and finally the genome/proteome measurement in each cell was measured

our on-chip microfabrication methods – we call them "on-chip cellomics technologies" – is constructive, involving three steps, which we describe as "on-chip cellomics." First, we purify cells from tissue singly in a nondestructive manner using several technologies such as digestible DNA-aptamer labeling and cell collection, ultrahigh-speed camera-based real-time cell sorting, or noninvasive handling of cells by acoustic radiation force (Yasuda 2000). We then cultivate and observe cells under fully controlled conditions (*e.g.*, cell population, network patterns, or nutrient conditions) using an on-chip single-cell cultivation chip (Inoue et al. 2001a, b, 2004; Wakamoto et al. 2001, 2003; Hattori et al. 2003; Matsumura et al. 2003a, b; Takahashi et al. 2003; Umehara et al. 2003) or an on-chip agarose microchamber system exploiting photothermal etching technology, which can control the microstructure of microchambers even during cell cultivation (Moriguchi et al. 2002, 2004; Kojima et al. 2003, 2005; Hattori et al. 2004; Sugio et al. 2004; Suzuki et al. 2004a, b). Finally, we undertake single-cell-based genome/proteome analysis through a set of nanoprobe and adaptive electron microscopy or a single-cell-based 3 min ultrahigh-speed polymerase chain reaction (PCR) measurement technology (Yasuda et al. 2000).

The advantage of our on-chip cellomics approach is that as it is a reconstructive approach of the simplified artificial minimum cell network model on a chip; it removes the complexity in underlying physicochemical reactions that are not always completely understood and for which most of the necessary variables cannot be measured. Moreover, this approach shifts the view of cell regulatory processes from a basic chemical ground to a paradigm of the cell as an information-processing unit working as an intelligent machine capable of adapting to changing environmental and internal conditions. This is an alternative representation of the cell and can bring new insights into cellular processes. Thus, models derived from such a viewpoint can directly help in more traditional biochemical and molecular biological analyses that assist in our understanding of control in cells.

The main purpose of the study was to develop on-chip single-cell-based cultivation and analysis systems to monitor dynamic processes in the cell. We have used these systems to extend ideas from the genetic to the genetic-epigenetic network in investigating topics like the population effect and the community effect of cells, and network pattern formation in cell groups. After sufficient experimental observations, we can understand the role of epigenetic information in modeling more complex signaling cascades. This field has almost been entirely monopolized by physicochemical models, which provide a good standard for comparison, evaluation, and development with our approach. The ultimate aim of our study is to provide a comprehensive understanding of living systems as products of both genetic and epigenetic information. It would permit us to describe the phenomena occurring in cell systems sufficiently well to be able to interpret and control them.

17.2 Techniques Developed for Non-Invasive Cell Identification, Separation, and Purification

Cell identification, separation, and purification are essential to life science research and practical applications such as cell-based regenerative medicine and drug screening. Efficient and rapid purification of cells has been accomplished with techniques such as fluorescence-activated cell sorting (FACS) (Bonner et al. 1972; Hulett et al. 1973; Herzenberg and Sweet 1976; Ogawa et al. 1994; Assenmacher et al. 1995; Johnson et al. 2007), magnetic-activated cell separation (MACS) (Owen and Sykes 1984; Imamura et al. 1995; Zborowski and Chalmers 2005, 2011), automated single-cell sorting using dual-beam optical trapping (Grover et al. 2001), differential adhesion cell sorting (Hsu et al. 2008), and disposable microfabricated fluorescence-activated cell sorting (μFACS) (Azuma et al. 2007; Liu et al. 2011; Yoon et al. 2012). Recent advances in optofluidic cytometry, in which optics and microfluidics are used together to create novel functionalities on a small chip, have made it very promising for lab-on-a-chip flow cytometry (Hatayama et al. 1994; Yasuda et al. 2006).

In the conventional measurement methods, particularly flow cytometry and FACS, the target cells are labeled with fluorescent dyes using specific antibodies or other molecules, and the optical signals they create (scattered angular light or emitted fluorescence) are detected and analyzed to identify individual cells or biological samples. However, these conventional methods can damage cells during droplet generation, and signal detection based on nondirect scattering in principle contains the risk of a low cell recognition rate. Furthermore, if cells are to be used for cultivation after purification, the damage to them caused by sorting should be minimized. For example, MACS separates cells by incubating them with magnetic particles coated with antibodies against a particular surface antigen. While this method is simple and economical, removing the antibodies attached to the cell surfaces is difficult, and these antibodies can cause changes in the target cell properties.

There are several approaches proposed to overcoming this problem with conventional cell labeling at present. One is to use optical microscope images instead of fluorescent intensities to identify the target cells in the limited fields, which is a next generation of FACS technology for expansion of limitation of conventional FACS technology. With this morphological approach, there is no staining with fluorescence dyes. Several studies have shown that target cells can be identified using optical images without any labeling, both *in vitro* (Schonbrun et al. 2012) and *in vivo* (Golan et al. 2012). However, the appeared morphological characteristics in target cells are usually too vague to be used for general approach to identifying specific cells compared to the molecular biomarkers.

The second approach is to use degradable cell labeling to particular molecular biomarkers appeared on the surface of target cells, which enables the cells to be restored to an intact condition. For example, Yamaguchi and his colleagues developed a technique for selecting nonadherent cells that uses photocleavable poly(ethylene glycol)-coated lipids. However, those techniques still have not reached to the universal technologies for identifying and recollection of single cells noninvasively without any

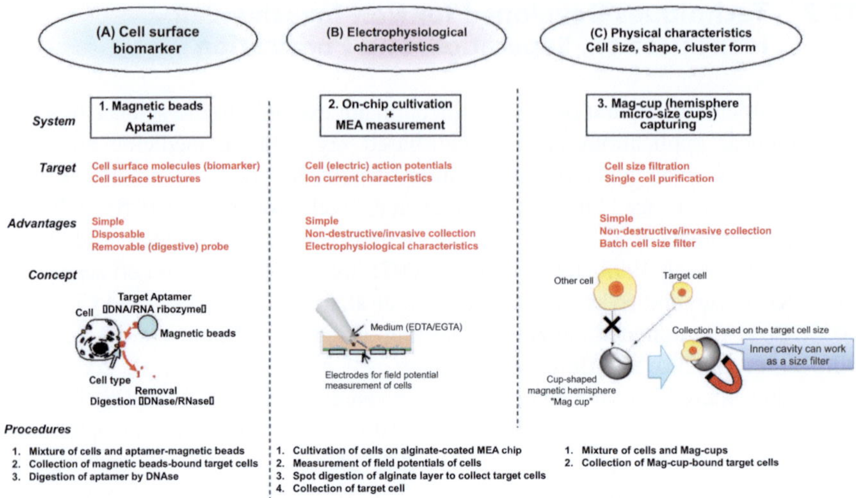

Fig. 17.3 *Three techniques developed for nondestructive and noninvasive cell sorting:* (**A**) magnetic bead acquisition using DNA aptamer; (**B**) on-chip cultivation using MEA measurement on alginate layer, and (**C**) specific size filtration of cells using hemisphere microsize magnetic cup, "mag-cup," capturing

remaining fragments of target probes yet. Moreover, there are no combined technologies of noninvasive electrophysiological measurement technology, and non-invasive cell collection technology was proposed because of the difficulties of their combination before we propose the new approach using digestive alginate gel-coated external microelectrode array measurement (Terazono et al. 2012).

Hence, in this subsection, as shown in Fig. 17.3, we introduce three new approaches we have developed for noninvasive and nondestructive cell identification, separation, and purification that work at the single-cell level and that are simple and safe enough for use in biological research and clinical diagnostics, i.e., (1) digestive DNA aptamer and magnetic beads for noninvasive cell surface recognition, (2) digestive polyalginate coated multielectrode arrary (MEA) for noninvasive identification of cells' electrophysiological characteristics, and (3) superparamagnetic hemisphere microsize cups for selective collection of particular size of cells. We also introduce briefly an advance of flow cytometer exploiting multi-imaging and real-time cell shape recognition.

17.2.1 Magnetic Bead Acquisition Using DNA Aptamer

The separation of uniform phenotype cells without changing their native state is important both for postgenome cell-based studies and practical applications of phenotype cells for regenerative medicine. With conventional antibody-based labeling and separation techniques, strong, often poorly reversible conjugation between antibody and antigen molecules on the cell surface may affect the function of the targeted cells and their interaction efficiency with other cells (Trickett and Kwan 2003). To overcome this problem, we have developed nucleotide-based probes capable of binding to target cell antigens (Gopinath 2007). Aptamers are single-strand DNA (ssDNA), RNA, or modified nucleic acids. Systematic evolution of ligands using exponential enrichment (SELEX) procedures for molecular evolution and aptamer engineering has been reported (Ellington and Szostak 1990; Tuerk and Gold 1990). The development of ssDNA aptamers for specific binding to live cells (cell-SELEX) expanded the potential of ligands for use in cell separation and labeling (Herr et al. 2006; Ohuchi et al. 2006; Shangguan et al. 2006). However, the problem of removing the aptamers from the target cells and restoring the cells to their original unlabeled state remains. To address this problem, we have added a DNAse digestion procedure after target cell collection. This procedure removes the attached DNA aptamers, which are tagged with magnetic beads, enabling reversible labeling of the cells (Fig. 17.4A).

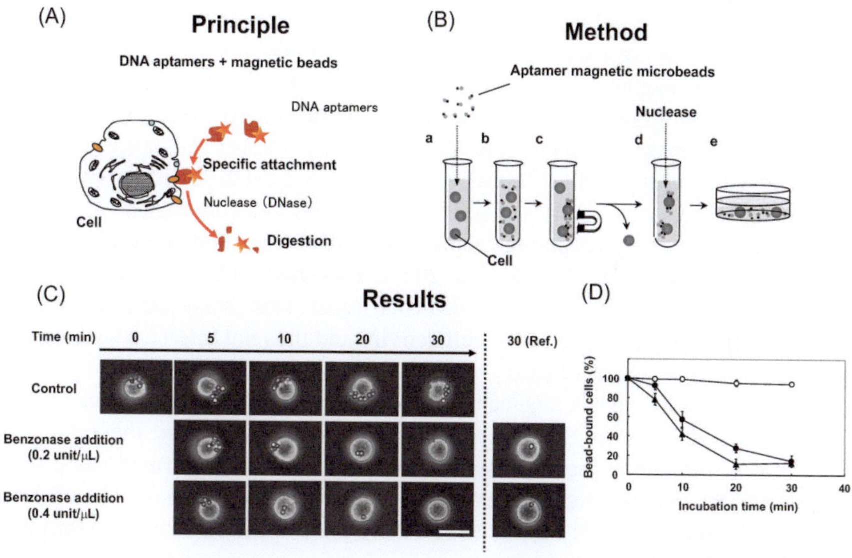

Fig. 17.4 *Magnetic beads and DNA aptamer*. (**A**) Principle of DNA aptamer cell sorting. DNA aptamer probes are attached to biomarkers on cell surface; target cells are then collected using DNA aptamers; and finally attached DNA aptamers are digested and removed by DNase after target cells are collected. (**B**) Protocols for separating cells using aptamer-conjugated magnetic microbeads and for detaching beads using nuclease. In (*a*, *b*) cells are attached to aptamer-magnetic beads in test tube, (*c*) unbound cells are extracted using a magnet, (*d*) bead-bound cells are incubated with nuclease, and (*e*) bead-bound cells are counted. (**C**) Phase-contrast optical microscope images of bead detachment during incubation with Benzonase nuclease. *Upper row*: control (absence of Benzonase). *Middle row*: 0.2 unit/μL Benzonase. *Lower row*: 0.4 unit/μL Benzonase. Right reference microphotographs show nonspecifically attached magnetic beads that could not be removed even after incubation for 30 min in 0.2 or 0.4 unit/μL Benzonase. Bar = 20 μm. (**D**) Growth rate (normalized by that on day 0) of separated cells after nuclease digestion. *Filled circles* and *filled triangles*: 0.2 or 0.4 unit/μL Benzonase-treated cells, respectively. *Open circles*: untreated cells. *Error bars* indicate S.D. for three measurements (Anzai et al. 2007)

Our technique uses high affinity ssDNA aptamers specific to surface-exposed target molecules on cells. The target cells are purified by first attaching aptamer-conjugated magnetic microbeads to them, then magnetically separating these labeled target cells from the unlabeled cells, and finally detaching the microbeads from them by nuclease digestion.

We tested our cell separation technique by using cultured human leukemic lymphoblast (CCRF-CEM) cells (Fig. 17.4C). Cells were dispersed in media buffer with aptamer-magnetic beads. After the suspension was incubated with gentle mixing, the bead-bound cells were remained using a magnet and rinsed with fresh media buffer to remove the unbound and low-affinity cells.

The separated target cells were incubated at 37 °C with gentle mixing in media buffer containing DNase (Benzonase® nuclease), and the time course of the detachment is shown in Fig. 17.4C and plotted in Fig. 17.4D, where 100 % represents the number of cells with beads attached at 0 min. During incubation in the presence of 0.2 or 0.4 unit/μl Benzonase, the percentage of bead-bound cells decreased with time and reached an equilibrium level within about 30 min for both concentrations. In the absence of Benzonase, the percentage of bead-bound cells did not change.

The potential damage to the cells due to the nuclease digestion process was evaluated by examining the growth rate of the separated cells. After digestion with 0.4 unit/μl Benzonase for 20 min, the aptamer-detached target cells were collected, the buffer was exchanged three times by centrifugation, and the collected cells were cultured for 5 days while untreated cells were cultured as a control. The number of cells was counted using a counting chamber. The treated and untreated cells grew at the same rate for the 5 days (Anzai et al. 2007). This shows that the nuclease had no obvious effect on the growth of the detached cells. This means that the method should not change the nature of the cell activity and thus not interfere with further applications of the separated cells.

Another noninvasive labeling technique uses ssDNA aptamer-conjugated Q-dots (Terazono et al. 2010a). Binding of the fluorescent aptamers to the cells was evaluated by measuring the fluorescence intensity and was confirmed using flow cytometry. The aptamers were removed in ~10 min by DNase digestion. Incubation of cells with aptamers or with nucleases resulted in no apparent damage to the cells and did not affect their growth rates.

These two aptamer-based techniques for cell identification and separation are promising alternatives to conventional antibody-based techniques, particularly for noninvasive reversible cell labeling and cell separations when it is important to maintain native cell activity. The digestion of the probe aptamers that attach the magnetic beads and Q-dots to the cells could help solve the problem of attached antibodies affecting cell efficacy. The results obtained also indicate the potential of multiple-step separation using a series of stepwise different aptamer-magnetic bead or Q-dot separations for precise refinement of target cells when there are multiple target molecules on the cell surface. The potential of multistep separation method using our digestible DNA aptamers should thus contribute to the rigorous purification of subpopulations and more detailed analysis of phenotypic differences in uniform cell types.

17.2.2 On-Chip Cultivation and Selective Collection of Single Cells on Alginate Layer

Alginate has advantages for single-cell-level spot removal of the cultivation matrix. This matrix has two states, a calcium conjugated gel state and a calcium chelated sol state, and cells can be cultivated on the gel state matrix layer. Sodium alginate is gelled by replacing sodium ions with calcium ions and is changed into a sol state by chelating the calcium ions from gel in micrometer resolution with spot introduction of a calcium chelate buffer such as ethyleneglycol bis tetraacetic acid (EGTA).

In our on-chip cellomics techniques, we use a newly developed complementally another functional-identification-based cell collection/purification procedure that includes spot melting of the calcium conjugated alginate gel and the noninvasive microelectrodes for external field potential measurement of cells. The target cells like neurons and cardiomyocytes are nondestructively collected at the single-cell level after identification of their phenotypes on the basis of their geometry or electrophysiological characteristics by using a multielectrode-equipped cultivation dish.

Figure 17.5 shows an example of cell purification procedure: (1) First, coating sodium alginate on a microelectrode equipped culture dish using a spin-coater, drying the coating in air, and creating calcium alginate gel by applying $CaCl_2$, as shown, (2) next, type IV collagen and sodium alginate were mixed at a ratio of one part to two, and coated on the microelectrode equipped culture dish with a spin-coater, the coating was dried in air, and the mixed layer of sodium alginate and collagen on the calcium alginate layer was gelled by applying $CaCl_2$, (3) once the chip was prepared, cells were cultivated on the sheet by covering it with cultivation medium containing ethylenediaminetetraacetic acid (EDTA) or EGTA, (4) after cell growth was confirmed or the electrophysiological characteristics of each neuron were identified, the medium was spot loaded at the target neuron using a microcapillary, (5) then a micrometer-sized portion of the calcium alginate layer around the target neuron was immediately melted and dispersed, and the target neuron was released from the bottom layer of the chip, (6) finally, the released cells were collected with a micropipette (Fig. 17.5B).

Figure 17.5C shows time course images of the collection of human embryonic stem (hES) cell-derived cardiomyocytes. In this example, all steps, from cell release (a) to released target cell collection (f), were performed quickly (within 2 min). Then, within only a few minutes after the collected cells were placed in another cultivation dish, recultivation started. The collected cells were attached and adhered to the collagen-coated cultivation dish, and started to beat. In contrast, cells enzymatically released from the culture dish with trypsin or collagenase did not beat or adhere again because of the damage caused by the digestion of the cell surface proteins by the enzymes. That is, our collection technique did not cause any apparent damage to the cell surface proteins of the released target cells, so the cells were able to begin adhering and beating again within a few minutes.

Extracellular field potential measurement using a multi-microelectrode array aligned on the bottom surface of a cultivation dish was used to noninvasively identify the electrophysiological characteristics of the target cells. The

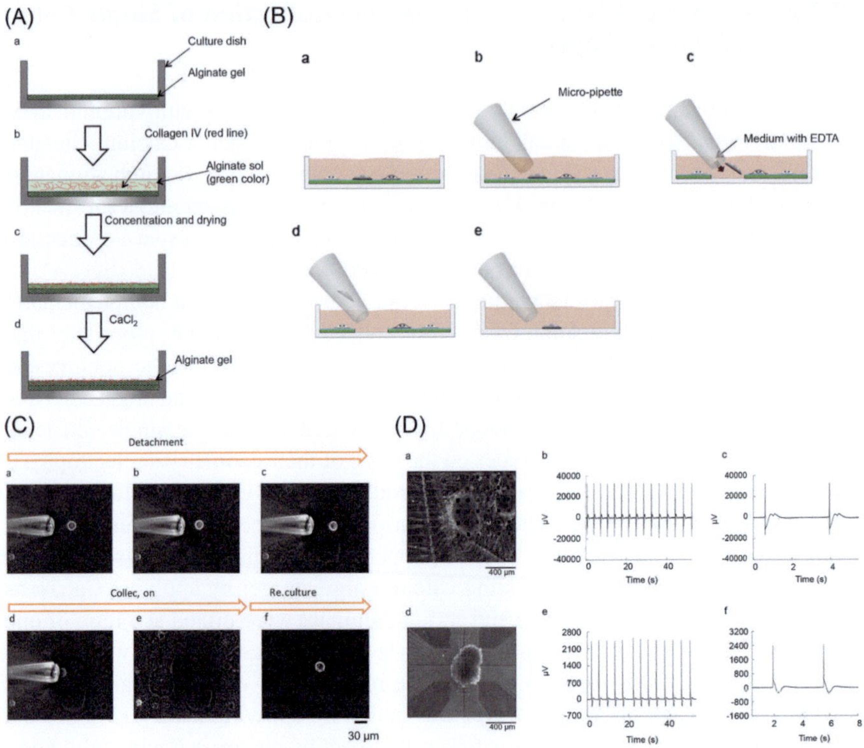

Fig. 17.5 *Digestible alginate layer for on-chip cultivation and selective collection of single cells.* (**A**) Alginate gel layer fabrication procedure. First, a calcium alginate layer was formed in two culture dishes (*a*). For the cardiomyocyte culture dish, a mixture containing a low concentration of alginate and collagen type IV extracellular matrix was layered onto the alginate sheet (*b*). The solution was then concentrated and dried (*c*), gelled with $CaCl_2$, and washed three times (*d*). (**B**) Selective cell collection procedure. First, cells are precultured in an alginate-coated culture dish (*a*). Then, the alginate layer is locally solated from gel by applying EDTA-containing culture medium with a micropipette (*b–c*), and cells cultured on the alginate are released from the layer and can be collected (*d*). In Finally, cells collected with the micropipette were recultured in the other culture dish (*e*). (**C**) Series of micrographs of cell collection procedure: from preculture, detachment, and collection to reculturing of single cardiomyocytes derived from human ES cells. During solation process, the cardiomyocytes stopped beating when the EDTA buffer was applied. However, after reculturing, cardiomyocytes began to beat again within a few minutes. (**D**) Results of electrophysiological measurement of cardiomyocyte cells before and after recultivation. Micrographs of cardiomyocyte clusters derived from human ES cells before and after detachment, with their recorded extracellular action potentials. Both before and after detachment, action potentials were regular and had normal waveforms

electrophysiological characteristics of human cardiomyocyte clusters differentiated from hES cells cultivated on a calcium alginate layer were measured with the microelectrode array through this calcium alginate layer. As shown in Fig. 17.5D, the extracellular action potential amplitude of the cardiomyocyte clusters before detachment was weak. This is because the alginate sheet was a spacer between the electrodes and cluster. Nevertheless, the waveform before detachment was similar to the one after detachment and recultivation. After detachment, the cardiomyocyte cluster beat regularly, as before detachment. This indicates that the detachment was nondestructive and did not disturb the physiological function of the cardiomyocytes. This technique thus enables target cells to be acquired on the basis of the invisible electrophysiological indices of the cells.

17.2.3 Hemisphere Superparamagnetic Microcups (Mag-Cups) for Cell Size Filtration

Size separation is one of the most important cell purification technologies. The membrane filtration is the most popular and widespread size separation technology; however, it has several problems in cell damage and purification throughput.

To overcome existing problems, we have developed a new method of size-specific filtration of target cells exploiting precisely size-controlled hemispherical superparamagnetic microparticles and depletion effect on concave microstructure of them (Kim et al. 2014a, b, c).

Nanoparticles having characteristic superparamagnetic cup-shaped hemispherical conformations (referred to as "mag-cups," hereafter) were fabricated as following procedures: First, desired sizes of polystyrene nanospheres were placed on a flat substrate as a monolayer to use templates for the mag-cups, and desired elements like Au, Ag, or Ni were coated on the spheres by thermal evaporation. Next, polystyrene spheres were removed from the element-coated particles by organic decomposition. Finally, mag-cups having the same diameters with polystyrene templates composed of the evaporated elements were obtained (Kim et al. 2011, 2014a, b, c).

As shown in Fig. 17.6, a multilayered structure of a 2-nm nickel layer inserted between 15-nm silicon dioxide layers was formed on polystyrene cast spheres by vapor deposition. The thickness of nickel layers and silicon dioxide layers were strictly chosen to maintain superparamagnetic properties for mag-cups. The standard target cells (CCRF-CEM, 12 ± 2 μm) were mixed with a set of different sizes of the fabricated magcups, and we confirmed that the cells were captured in the mag-cups having cavities larger than 15 μm in diameter, and then gathered by magnetic force. The collected cells were grown in a culture medium without any damage.

The results showed that (1) target cells were captured into the cavity of mag-cups and the conjugates were collected with magnetic force, (2) inner diameter of the cup has a function of cell-size filtration, and (3) the depletion effect is one driving force for the acquisition of target particles of closer size of mag-cup's concave microstructures than smaller size of cells. The results indicated that this method is quick, simple, and noninvasive size separation of target cells.

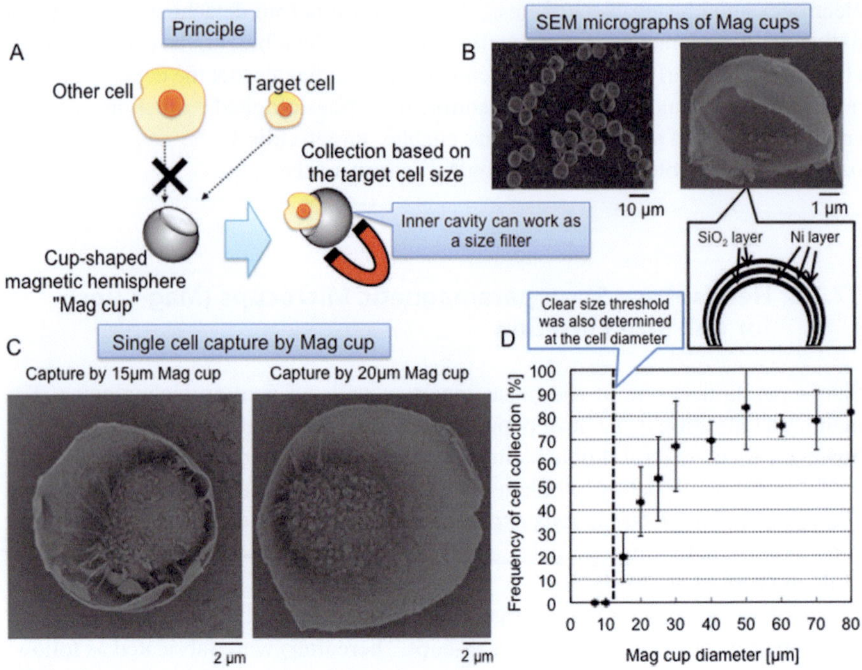

Fig. 17.6 *Cell size filtration with hemisphere superparamagnetic microcups (Mag-cups).* (**A**) Principle of size filtration procedure using Mag-cups. (**B**) Scanning electron microscope (SEM) images of Mag-cups. To maintain the superparamagnetic properties of Ni, the 2-nm thin Ni layers are separated by SiO_2 layers. (**C**) Examples of SEM images of the single cells captured by Mag-cups. (**D**) Mag-cup size (diameter) dependence for 10 μm cell collection (Kim et al. 2014a, b, c)

17.2.4 On-Chip Imaging Cell Sorter for Identification and Separation of Cells Based on Imaging Biomarkers

As described above, imaging cell sorter has a potential to identify individual cells or biological samples based on both the physical properties (e.g., size, shape) and biochemical properties (e.g., cell cycle distribution, DNA contents) of cells; we call them "imaging biomarkers." Information about the cells of interest is obtained optically in a nondestructive and quantitative manner (Hayashi et al. 2011; Nomura et al. 2011a, b; Yasuda et al. 2013).

As a step toward achieving a cell sorter system for "imaging biomarkers," we have developed an on-chip high-speed multi-imaging cell-sorting system for label-free collection of target cells (Takahashi et al. 2004; Hattori and Yasuda 2010). The developed system has five core features (Fig. 17.7): (1) fully automated sorting and collection of samples using a 3D micropipetting device, (2) arrangement of the entire fluidic system in a disposable plastic chip, enabling biological samples to be lined up in a laminar flow using hydrodynamic focusing, (3) direct observation-based cell identification using specific image indexes and an optical system comprising phase-contrast/fluorescence simultaneous multi-imaging microscopy and

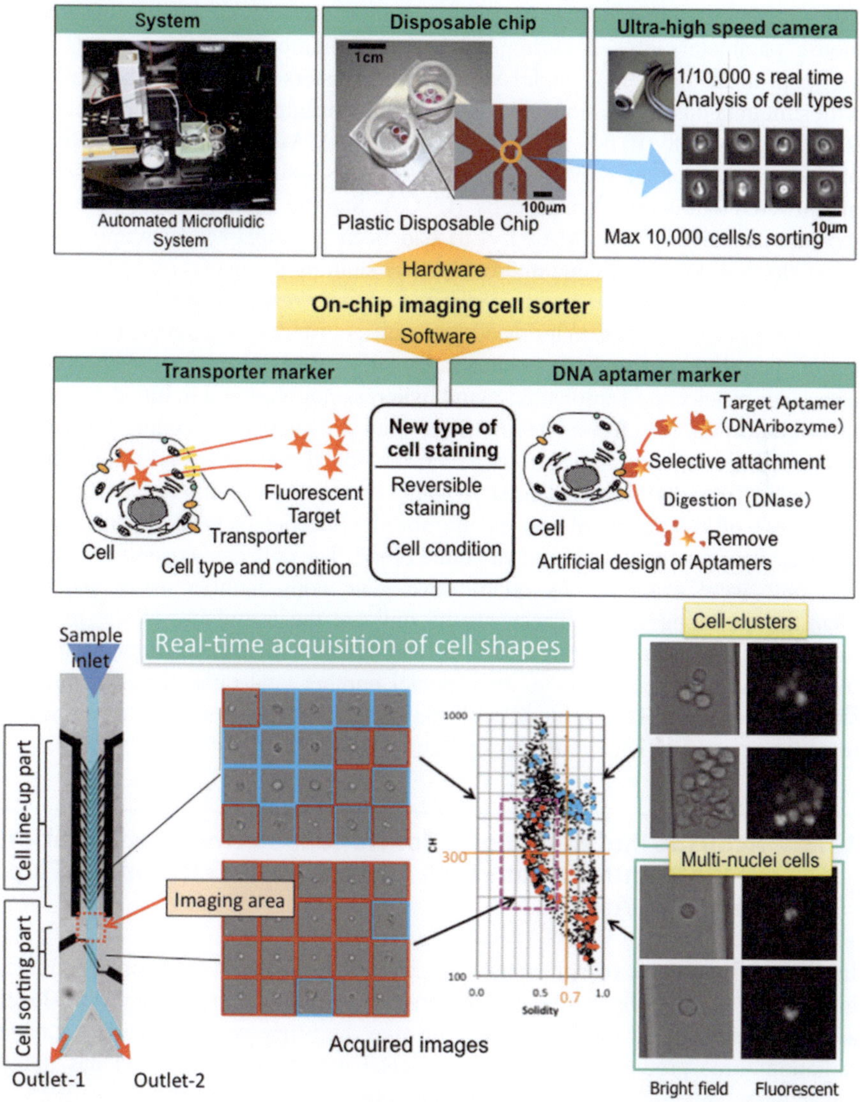

Fig. 17.7 *Imaging cell sorter system for identifying and separation of different shaped cells*. Imaging cell sorter consists of three parts; Automated microfluidic system, plastic disposable chip, and ultrahigh-speed camera. The target cells were recognized by the images of cell shapes and also identified by the reversible cell staining representing their conditions like transporter marker and DNA aptamers. Using imaging cell sorter, the clustered cells are easily distinguished and can be applied for label-free circulating tumor cell cluster identification and collection (Kim et al. 2014a, b, c)

real-time image processing, (4) nondestructive wider dynamic range sorting using mild electrostatic force in a laminar flow (agarose gel electrodes are used to prevent electrode loss and electrolysis bubble formation), and (5) contamination-free recultivation of collected target cells in a microreservoir.

Using the on-chip multi-imaging flow cytometry system, we can obtain morphometric parameters of cell clusters such as cell number, perimeter, total cross-sectional area, number of nuclei, and size of clusters as "imaging biomarkers," with simultaneous acquisition and analysis of both bright-field (BF) and fluorescent (FL) images at 200–10,000 frames per second (fps). As an example of application of "imaging biomarker"-based cell purification, this system can be used for identity of clustered circulating tumor cells (CTCs). Sample blood of rats in which a prostate cancer cell line (MAT-LyLu) had been preimplanted was applied to a microchannel on a disposable microchip after staining the nuclei using fluorescent dye for their visualization, and the acquired images were measured and compared with those of healthy rats. In terms of the results, clustered cells having (1) cell area larger than 200 μm^2 and (2) nucleus area larger than 90 μm^2 were specifically observed in cancer cell-implanted blood, whereas not observed in blood of healthy rats. In addition, (3) clusters having more than 3 nuclei were specific for cancer-implanted blood, and (4) a ratio between the actual perimeter and the perimeter calculated from the obtained area, which reflects a shape distorted from ideal roundness, of less than 0.90 was specific for all clusters having more than 3 nuclei and was also specific for cancer-implanted blood. The collected clusters larger than 300 μm^2 were examined by quantitative gene copy number assay and were identified as being CTCs. These results indicate the usefulness of the imaging biomarkers for characterizing clusters, and all of the four examined imaging biomarkers, cluster area, nuclei area, nuclei number, and ratio of perimeter, can identify clustered CTCs in blood with the same level of preciseness using multi-imaging cytometry (Kim et al. 2014a, b, c).

17.3 Acoustic Radiation Force for Nondestructive Manipulation of Bioresources from DNA Molecules to Cells for On-Chip Cellomics Assay

In diagnostic analysis of epigenetic cell characteristics, a method for noncontact handling of single cells can be incorporated in a fully automated single-cell-level analysis system providing contamination-free, real-time measurement. Acoustic radiation force produces one of the universal noncontact forces and has many advantages in terms of handling biomaterials for mixing, concentrating, fractionalizing, arrangement, and separation (Fig. 17.8). It can, for example, be used to form a hyperbolic potential field with a span in the order of millimeters, and it can continuously separate different materials by exploiting the competition between acoustic radiation force and electrostatic force. The use of acoustic radiation force for handling biomaterials has several benefits because (1) it is a noncontact force, (2) it allows a simple and compact design without moving parts, (3) it is suitable for continuous driving for concentrating, separating, and mixing, and (4) it is more economical than other noncontact handling methods.

In this subsection, a brief summary of acoustic radiation force as one of the desirable fundamental technologies for on-chip cellomics studies.

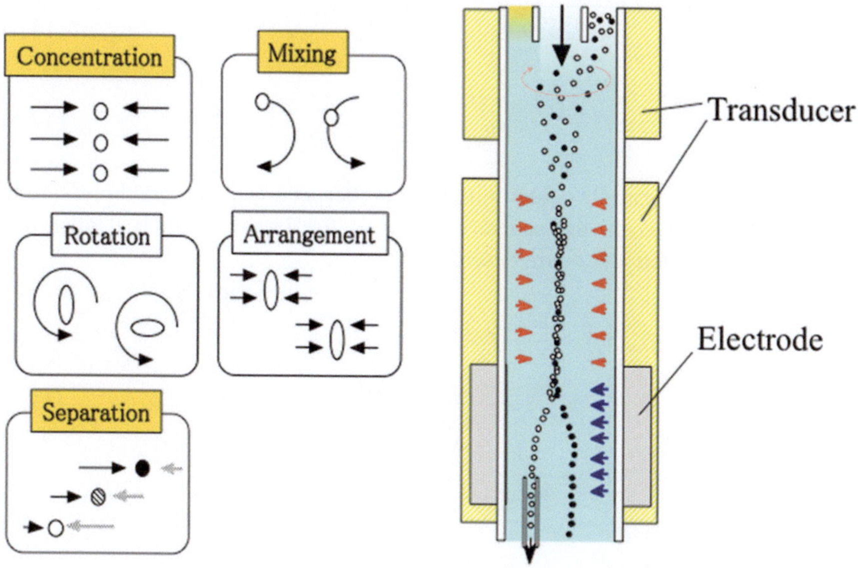

Fig. 17.8 Applications of acoustic radiation force in microfluidic pathways

17.3.1 Principles

17.3.1.1 Acoustic Radiation Force on Particles

According to the theory of acoustic radiation pressure, (King 1935; Yoshioka and Kawashima 1955; Gorkov 1962; Nyborg 1967; Kozuka et al. 1995; Yoshioka et al. 1997) the acoustic radiation force on a particle, F_{ac}, in an ultrasound field is given as

$$F_{ac} = V[B\nabla\langle K_a\rangle - (1-\gamma)\nabla\langle P_a\rangle] + \Delta \tag{17.1}$$

$$B = \frac{3(\rho-\rho_0)}{2\rho+\rho_0} \cong \frac{\rho}{\rho_0} - 1, \tag{17.2}$$

where V is the volume of the particle, $<K_a>$ is the time mean value of kinetic energy, $<P_a>$ is the time mean value of potential energy, ρ_0 is the density of the medium, ρ is the density of the particle, γ is the compressibility of the particle relative to that of the medium [i.e., $\gamma =$ (compressibility of particle, β)/(compressibility of medium, β_0)], and Δ is a value negligible under the standing wave condition. When an ultrasound wave with peak velocity u_0, wave number k, and angular frequency ω is introduced into the chamber, which is closed at one end by a rigid wall ($x = L$) and at the other end by a flat vibrating wall ($x = 0$), the standing wave generated in the chamber is described as

$$u(x,t) = |u_0| \cdot \frac{\sin[k(L-x)]}{\sin(kL)} \cdot e^{j\omega t}, \tag{17.3}$$

$$p(x,t) = -j\rho_0 c_0 \cdot |u_0| \cdot \frac{\cos[k(L-x)]}{\sin(kL)} \cdot e^{j\omega t}, \tag{17.4}$$

where $u(x, t)$ is the particle velocity in the chamber, $p(x, t)$ is the sound pressure in the chamber, and c_0 is the sound velocity in the liquid medium. Under this condition, $<K_a>$ and $<P_a>$ are described as

$$\begin{aligned} \langle K_a(x) \rangle &= \frac{1}{2}\rho_0 \cdot \langle u^2 \rangle \\ &= \frac{\rho_0 \cdot |u_0|^2 \cdot \sin^2[k(L-x)]}{4\sin^2(kL)} \\ &= \frac{\rho_0 \cdot |u_0|^2}{8\sin^2(kL)} \cdot (1 - \cos[2k(L-x)]), \end{aligned} \tag{17.5}$$

$$\begin{aligned} \langle P_a(x) \rangle &= \left\langle \int -p dv \right\rangle = \frac{1}{2} \cdot \frac{\langle p \rangle^2}{K_0} \\ &= \frac{1}{2} \cdot \frac{\langle p \rangle^2}{\rho_0 \cdot c_0^2} \\ &= \frac{\rho_0 \cdot |u_0|^2}{8\sin^2(kL)} \cdot (1 + \cos[2k(L-x)]), \end{aligned} \tag{17.6}$$

where K_0 is the bulk modulus of the medium and c_0 is assumed to be equal to $(K_0/\rho_0)^{1/2}$. Substituting Eqs. (17.5) and (17.6) into Eq. (17.1), the force generated in the standing wave is described as

$$\langle F_{ac}(x) \rangle = -V \cdot \rho_0 \cdot |u_0|^2 \cdot kB \cdot \left[1 + \frac{(1-\gamma)}{B}\right] \cdot \frac{\sin[2k(L-x)]}{2\sin^2(kL)} + \Delta. \tag{17.7}$$

The above equations clearly show that the strength of the acoustic radiation force is determined by the magnitude of the gradient of the total energy distribution, which consists of kinetic energy and elastic energy as shown in Fig. 17.9.

17.3.1.2 Concentration Efficiency of Particles in the Acoustic Standing Wave

Chemical potential, μ, of a solution containing particles subjected to an external conservative force such as acoustic radiation force F_{ac} ($= -\nabla U$, U the potential energy corresponding to the force F_{ac}) is written as $u = u_0 + (N_A k_B) T_A \ln c + N_A U$, where u_0 is a constant, N_A is Avogadro's constant, k_B is the Boltzmann constant, T_A is the absolute temperature, and c is the particle concentration in the liquid. Applying this chemical potential to Fick's diffusion law, the particle distribution in the external conservative force is given as the following diffusion equation:

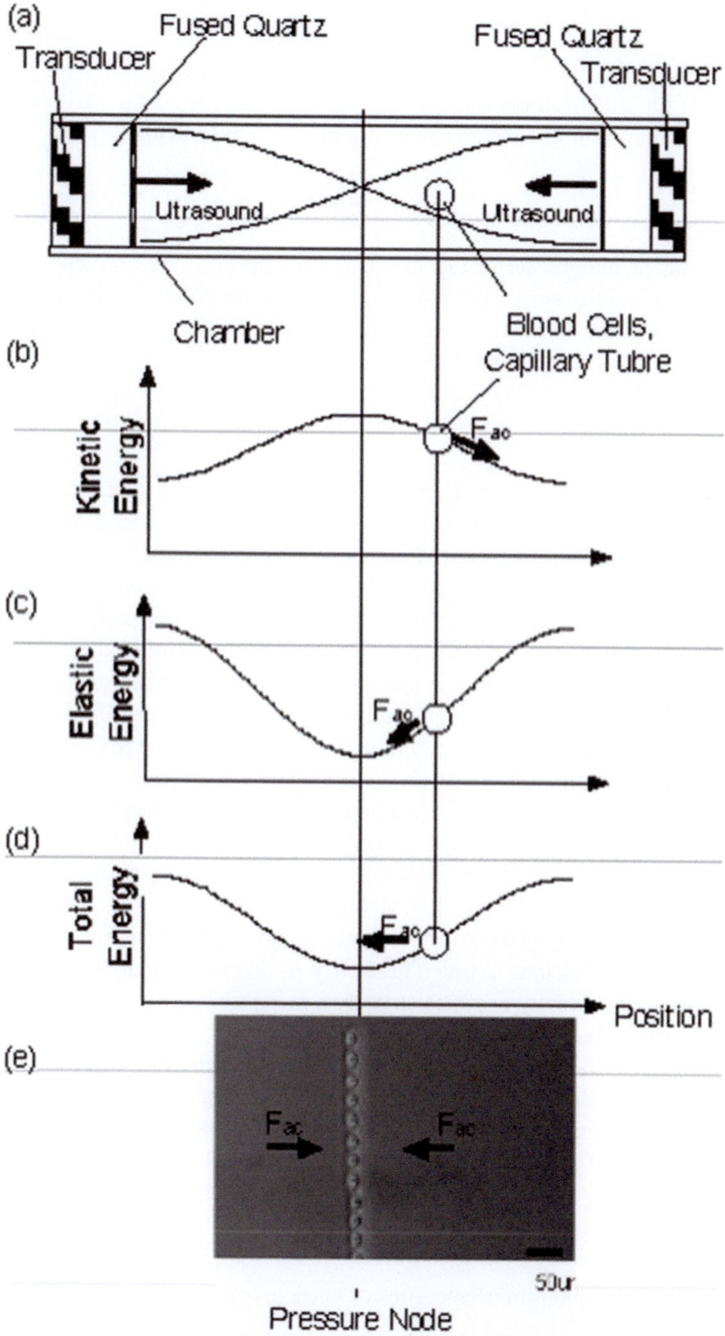

Fig. 17.9 *Principle of acoustic radiation force acting on particles.* (**a**) Schematic of acoustic microchamber, (**b**) spatial distribution of kinetic energy, (**c**) elastic energy, (**d**) total energy, and (**e**) migrograph of lined-up polystyrene spheres on the pressure node

$$\frac{\partial c}{\partial t} = D\Delta c - \operatorname{div}(\mu c F_{ac}), \tag{17.8}$$

where D is the diffusion coefficient and μ is the mobility of the particle. For the particle distribution in the steady state, Eq. (17.8) is rewritten as

$$D\frac{dc}{dx} = \mu c F_{ac}. \tag{17.9}$$

The constant of integration in Eq. (17.9) is zero because the gradient of particle concentration is zero where the external conservative force F_{ac} is zero. Solving Eq. (17.9) gives the relation between ultrasound frequency f, irradiated acoustic energy E_{ac}, and particle radius R as

$$f^2 \cdot E_{ac} \propto \frac{1}{R^5} \tag{17.10}$$

so that the concentration efficiency y of particles remains equal. As shown in Fig. 17.10, the half-width of particle distribution of the 5-μm polystyrene spheres in the 500-kHz standing wave is 50.9 μm, which is 10 times larger than the radius of the particles.

The above theoretical result indicates that, first, higher-frequency ultrasound improves the concentration efficiency and, second, the smaller particles can hardly be handled by acoustic radiation force. Our previous experimental results showed that the apparent size of particles is the sum of the actual particle size and the shell of viscous fluid surrounding the particle and that the minimum size of particles which acoustic radiation force can handle is smaller than we expected (Fig. 17.11) (Yasuda and Kamakura 1997).

When the size dependency of maximum velocity of polystyrene spheres in the acoustic radiation force field was evaluated, both the size dependency of the maximum velocity and of the estimated acoustic radiation force (mean ± standard error, $n = 15$) are shown in Fig. 17.11. Although v_{max} decreased with decreasing particle size when the particles were larger than 2 μm, it increased when the particles were smaller than 2 μm. The theory of acoustic radiation force predicts that the maximum velocity of particles decreases in proportion to the square of their radii as shown in the dashed line for $\delta = 0$ μm.

One possible explanation of this discrepancy is the limit of the theory, since the theory assumes particles suspended in a perfect fluid. But we can also attribute the discrepancy to the existence of the shell with thickness δ surrounding the particle in Yosioka's theory. If we regard the apparent size of particles in the acoustic radiation force field R' as

$$R' = (R + \delta), \tag{17.11}$$

we can rewrite the maximum velocities of particles as

$$v \propto \frac{R'^3}{R} = \frac{(R + \delta)^3}{R} \tag{17.12}$$

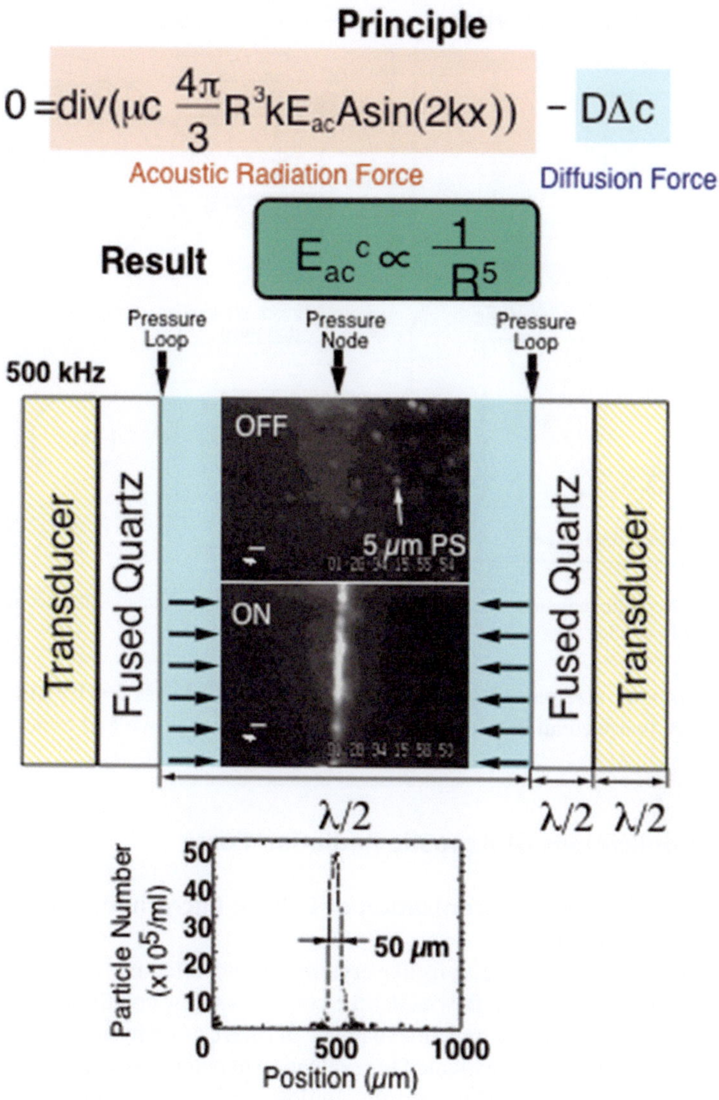

Fig. 17.10 Principle of acoustic concentration [*PS* polystyrene sphere, *k* wave number of ultrasound, E_{ac} energy density of ultrasound]

and the acoustic radiation force as

$$F_{ac} \propto (R+\delta)^3 \to \delta^3(R \to 0). \quad (17.13)$$

Applying Eq. (17.12), we can easily explain the increase of $v_{\max}$ with decreasing particle size. As shown in Fig. 17.4, when δ is 1.5 μm (broken line in figure), tendency of experimental results fits the line calculated by Eq. (17.12).

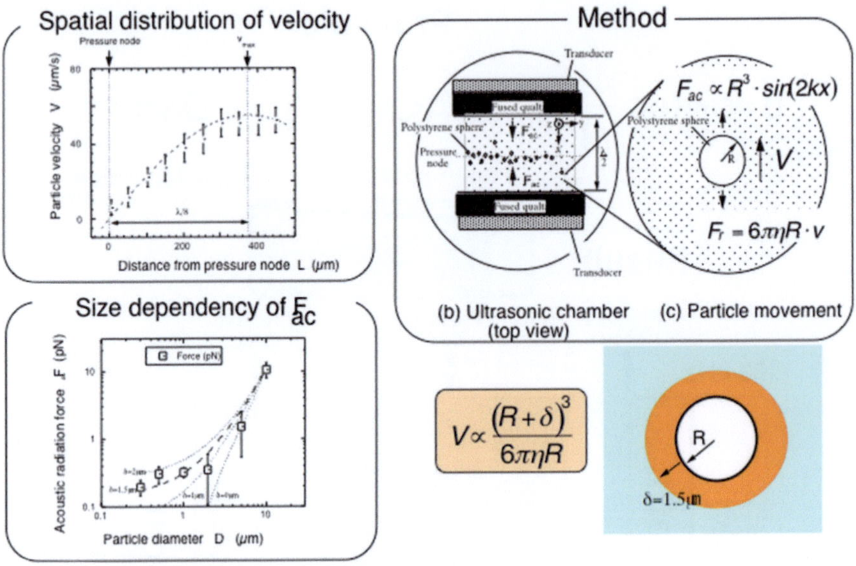

Fig. 17.11 Acoustic radiation force on micrometer-sized particles' existence of viscous shell on the surface of particles

If submicrometer-size particles can hold a shell with a thickness $\delta = 1.5$ μm, there is a possibility that we can handle particles much smaller than we had expected.

17.3.2 Applications of Acoustic Radiation Force

17.3.2.1 Continuous Concentration of Cells in Microfluidic Stream: Erythrocytes

Using acoustic radiation force, cells like erythrocytes are effectively concentrated at the center of the microfluidic flow under the continuous flow condition (Fig. 17.12) (Yasuda 1997). However, the potential damage of cells caused by acoustic vibration of medium buffer should be evaluated for evaluation of potential usage for living cell handling. Hence, we have examined the elution of potassium ion, iron ion, and lactate dehydrogenase (LDH) from erythrocyte cells quantitatively after irradiation of acoustic radiation force. For the experiment, we removed more than 60 % of dissolved gas from medium buffer to prevent cavitation formation. As shown in the graph in Fig. 17.12, after 2.5 min of irradiation (required time for erythrocyte concentration), no obvious elutions of potassium ion, iron ion, and LDH were observed. And even after 10 min irradiation, no apparent elution was observed. In contrast, when we kept dissolved gas in the buffer, the cavitation formation was observed just after acoustic radiation force irradiation and all of potassium ion, iron ion, and LDH were eluted from erythrocytes. The results indicate the importance of preservation of cavitation formation as far as we use acoustic radiation force.

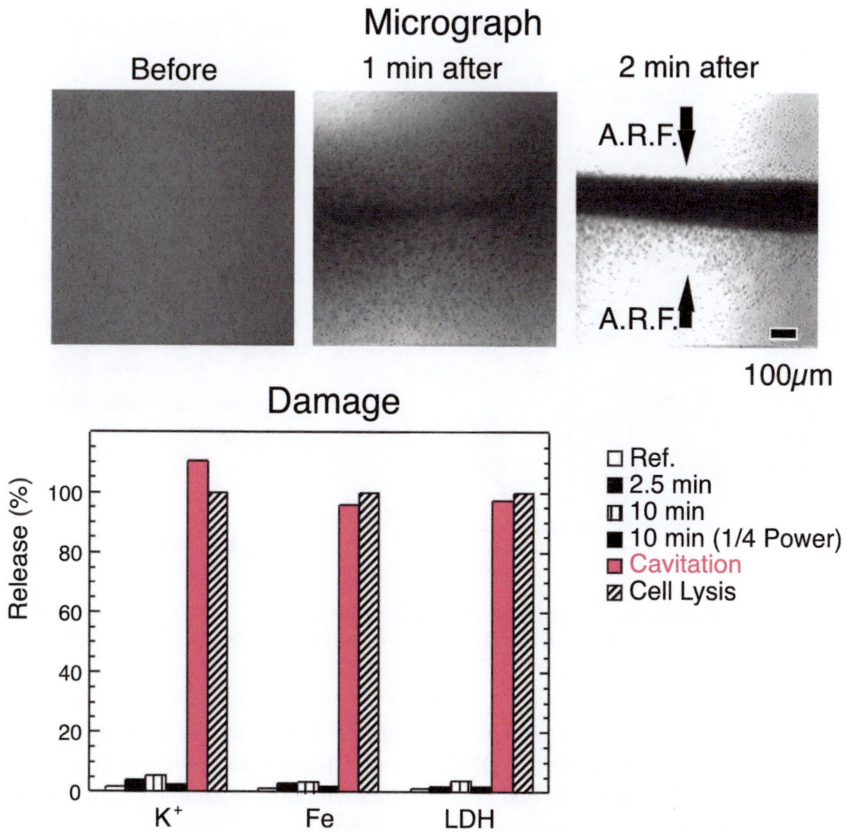

Fig. 17.12 Acoustic radiation force on erythrocytes in the microfluidic flow and quantitative evaluation of micrometer-sized particles' existence of viscous shell on the surface of particles

17.3.2.2 Concentration and Gathering of Dissolved DNA Fragments

As described in Eq. (17.7), acoustic radiation force decreased drastically when the target particle size decreased. Hence, the invisible small particles like DNA fragments are expected not to be able to be captured by acoustic radiation force. However, when we irradiated acoustic radiation force into the DNA fragment solution more than 80 % of ethanol was applied into the buffer (Yasuda et al. 1996a) (Fig. 17.13). As shown in micrographs in Fig. 17.13, after acoustic radiation force irradiation started, gathering of DNA fragment at the pressure node (center of micrograph) has started and large cluster of DNA was formed and trapped at the pressure node. It should be because the medium condition was optimized for DNA gathering and size dependence of vibration of DNA fragments, i.e., the different-sized DNA fragment clusters collided with each other and larger clusters were captured at the pressure node of acoustic standing wave. After trapping of DNA clusters, we also evaluated the potential damage caused by acoustic radiation

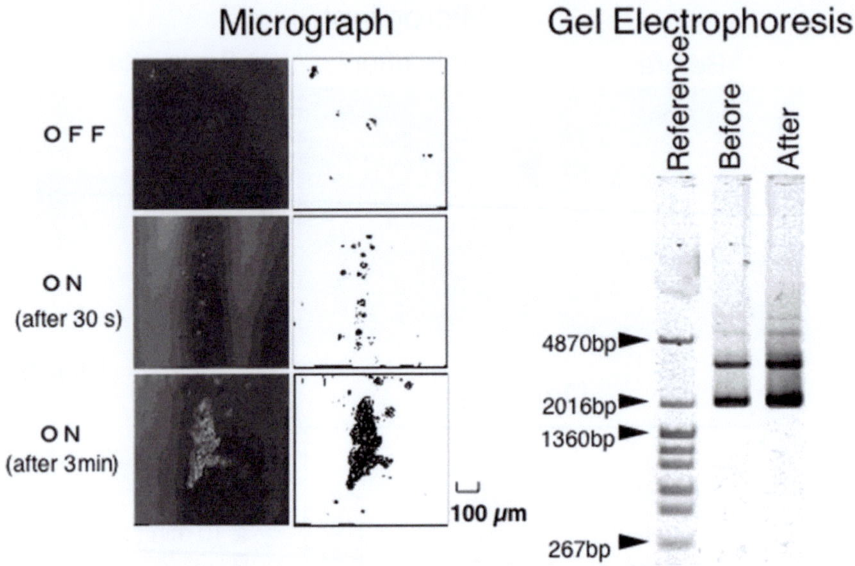

Fig. 17.13 Concentration of DNA fragments in acoustic radiation force field and quantitative evaluation of possible damage of DNA caused by acoustic radiation force

force like fragmentation using gel electrophoresis, and found that no fragmentation was observed from the gathered DNA clusters. The above results indicated that even DNAs could be handled and concentrated within microfluidic pathways, and no DNA fragmentation occurred after DNA trapping.

17.3.2.3 Improvement of Concentration Efficiency by Superposition of Higher Harmonics of Acoustic Standing Waves

The concentration efficiency of cells in the acoustic standing wave was determined by the gradient of acoustic radiation force field. Hence, the acoustic radiation force around the potential minimum was not strong because the shape of acoustic radiation force field is sinusoidal. To overcome above problems and increase the efficiency of concentration, we have examined the improvement acoustic radiation force around the pressure node (cell gathering place) exploiting the superposition of higher harmonics of fundamental standing acoustic waves (Yasuda 1997) (Fig. 17.14). As shown in the micrographs of Fig. 17.14a–d, the erythrocytes were gathered at the pressure node increasing the intensity of acoustic pressure (Fig. 17.14b, c; from 60 V_{pp} to 110 V_{pp}), and the concentration was improved when superposition method was applied and same peak voltage was applied (Fig. 17.14c, d; same 110 V_{pp}). The results above indicated that superposition method is effective to improve concentration efficiency with lower peak acoustic pressures.

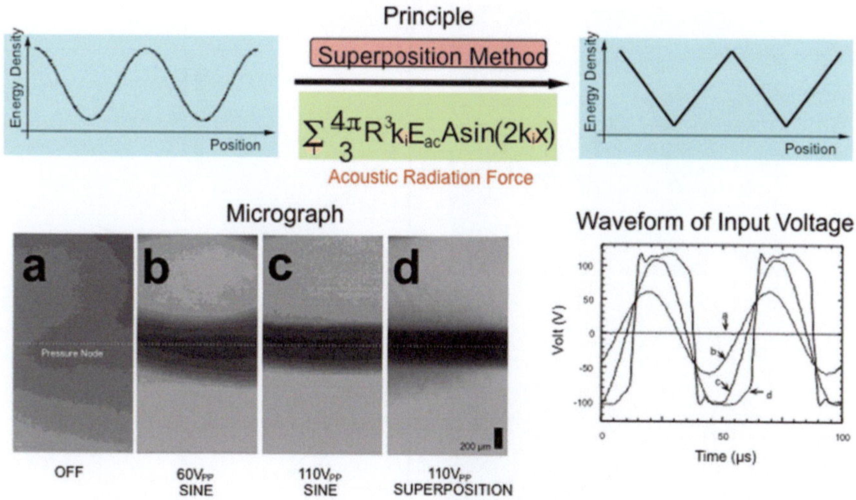

Fig. 17.14 Concentration of erythrocytes in microfluidic stream with superposition of higher harmonics of fundamental acoustic standing waves. (**a**) Before irradiation strated, (**b**) 60V sine wave, (**c**) 110V sine wave, and (**d**) 110V superposition wave irradiated

17.3.3 Manipulation and Position Fixation of Microparticles Using Acoustic Radiation Force

Applying acoustic radiation force on the small particles dispersed on the plate, we also can manipulate the particles as shown in Fig. 17.15 (Yasuda 2000). First, when we irradiated a pair of ultrasound waves from opposite ends of the place, dispersed polystyrene spheres were gathered around the pressure node of standing waves (Fig. 17.15a). Next, when we increased the irradiation power of acoustic waves, the slope of acoustic potential well increased, and finally the spheres were lined up at the pressure node (Fig. 17.15b). Then, when the frequency of acoustic ultrasounds changed, the place of pressure node plane was changed (Fig. 17.15c), and finally particles were shifted to the new pressure node line keeping their lined-up shapes (Fig. 17.15d). It is also one of the potential usages of acoustic radiation force as acoustic tweezers even within microfluidic spaces where optical tweezers cannot be used.

17.3.4 Mixing of Two Laminar Flows Within a Microfluidic Pathway Using Acoustic Radiation Force

The microfluidic flow chamber with a pair of 3.5-MHz PZT transducers, one on each side of the chamber (Fig. 17.16). As shown in Fig. 17.16a, when samples such as blood cells were introduced into the laminar flow in the chamber, the boundary between the region containing particles and particle-free region could be observed.

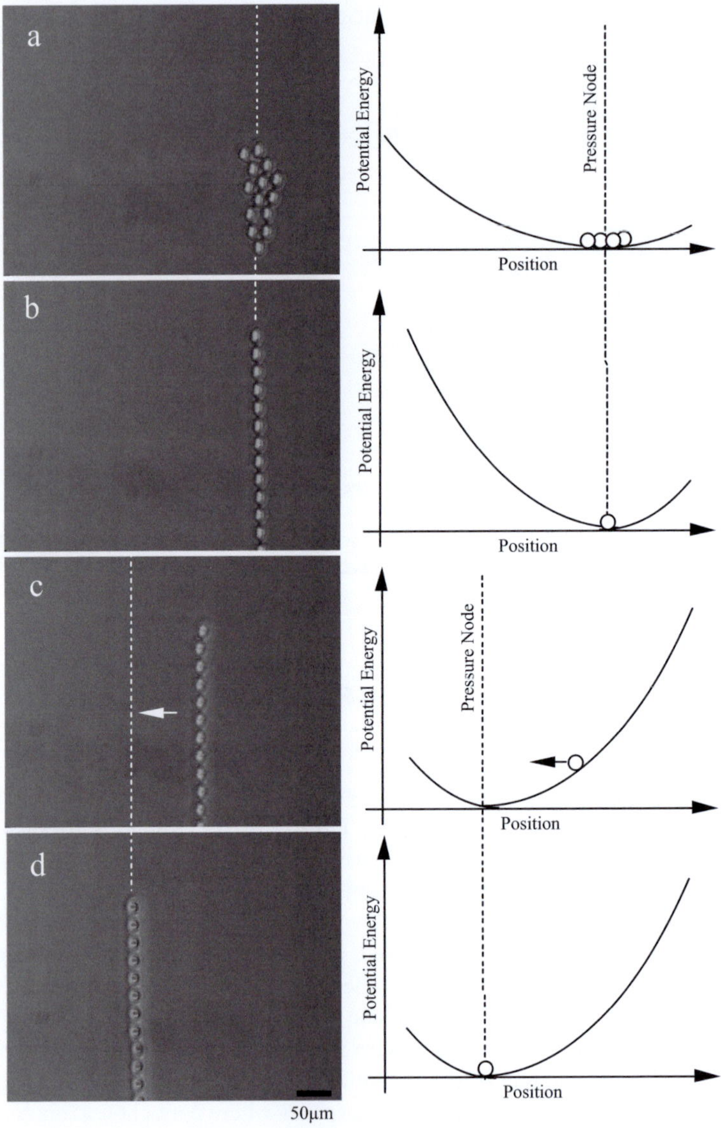

Fig. 17.15 Manipulation of polystyrene spheres set on the plate using acoustic radiation force. (**a**) Weak standing wave, (**b**) stronger standing wave, (**c**) pressure node shift, and (**d**) lined-up polystyrene spheres followed

When the chamber was excited by ultrasound, blood cells started to spread into the whole of microfluidic flow (Fig. 17.16c). When the wavelength λ of 3.5-MHz ultrasound in water is 0.43 mm, the width of the chamber having single pressure node (i.e., 0.215 mm) is 1.5 mm and thus equal to 7/2 λ. As shown in Fig. 17.16c, incoherent ultrasound wave was introduced into the microfluidic pathway; the

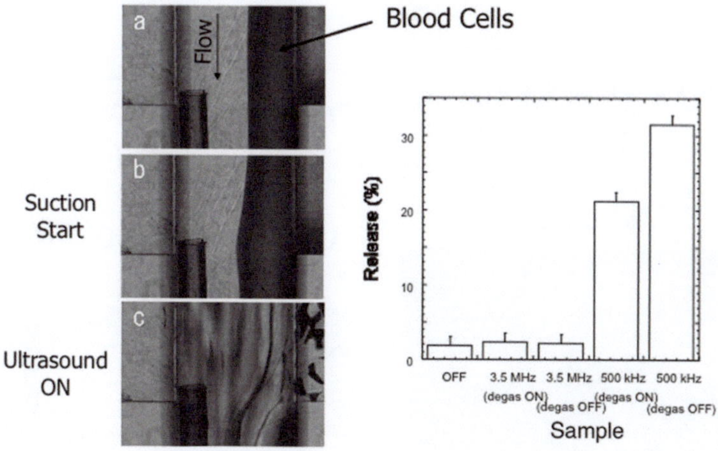

Fig. 17.16 Mixing and suctioning of erythrocytes in one of two laminar flows using acoustic radiation force. (**a**) Before ultrasond irradiation, (**b**) suctioning started, and (**c**) ultrasond irradiation started

boundary of two laminar flows disappeared and mixed the cells in the left side laminar flow into all over the flow.

We also have checked the leakage of intracellular components from erythrocytes by the ultrasound mixing under 3.5-MHz (20 V_{pp}) ultasound. The result indicates that there is no significant leakage of insights such as potasium under 3.5-MHz ultrasound. In contrast, in the 500-kHz ultrasound condtion, erythrocytes released more than 20 % of the intracellular component after suctioning. The damage of erythrocytes was thought to be related with cavitation generation in the water solution under the mixing condition. Though the degas process should be necessary for preventing the cavitation generation under ultrasound irradiation, we must know how much dissolved oxygen is required for keeping biomaterials alive. Cells eventually die without oxygen, and the death of the cells results in the leakage of intracellular components. For practical use of this method for biomaterials, we strongly recommend the use of MHz-order ultrasound.

17.3.5 Particle Separation Using Ultrasound and Electrostatic Force

For separation of different types of particles within microfluidic pathways, exploiting the competition of two different noncontact forces like acoustic radiation force and electrostatic force is one of the candidates. As shown in Fig. 17.17, we can set a pair of electrodes and acoustic transducers on the outer wall of microfluidic chamber. And when we applied both forces simultaneously, the polystyrene spheres having different diameters were shifted depending on their sizes (Yasuda et al. 1996b).

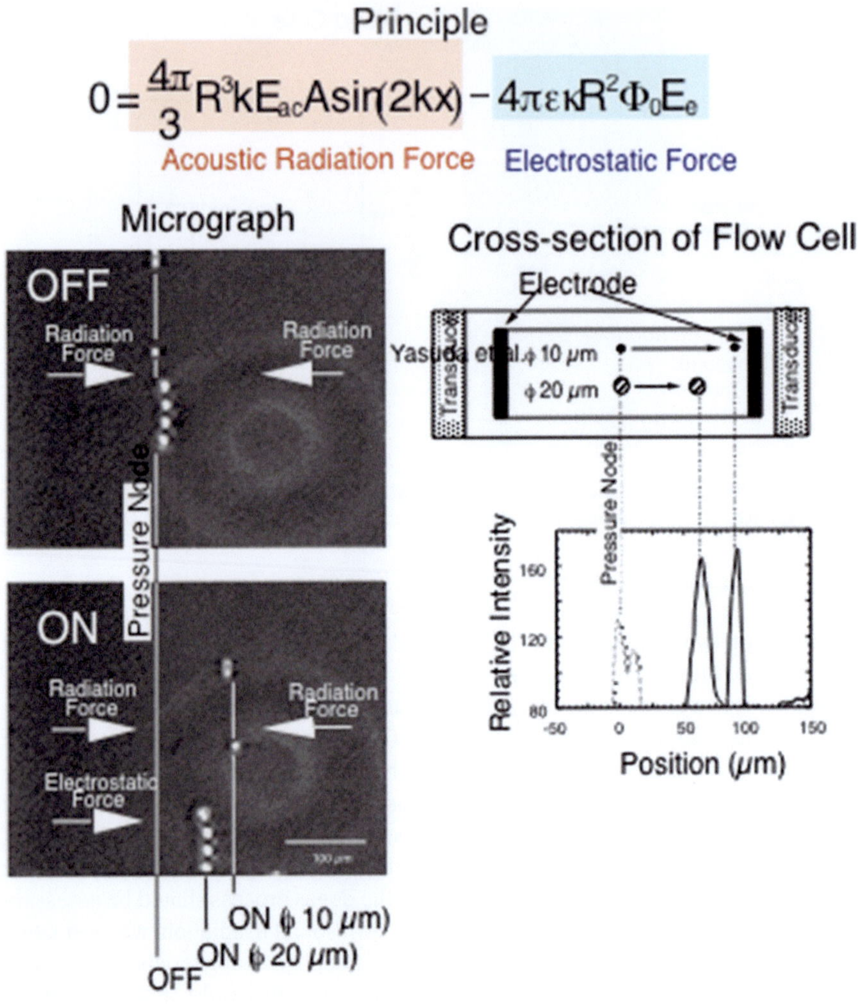

Fig. 17.17 Particle separation using ultrasound and electrostatic force

In this subsection, a series of fundamental usage of acoustic radiation force examined by our team was introduced. As explained above, the advantages of acoustic radiation force in microfluidic fabrications are (1) a noncontact force spread whole area of microfluidic area, (2) a simple and compact design without moving parts, (3) suitable for continuous driving for concentrating, separating, and mixing, and (4) more economical than other noncontact handling methods. We expect that more microfluidic devices exploiting above advantages of ultrasounds will be developed practical and useful in biological research and clinical diagnostics.

17.4 Photothermal Etching Method: Flexible Change of Cellular Network Formation of Cultivated Cells on a Chip

17.4.1 Photohermal Etching on Alginate Layer for Cell Network Formation Control

Flexible change of microstructures of cell-to-cell interactions or cell-network shapes on a chip during cultivation is necessary for reconstructive approach of cell-network model studies. To accomplish this requirement, we have developed a photothermal etching method (Moriguchi et al. 2002, 2004; Hattori et al. 2004) with agarose-microchamber cell cultivation system (Fig. 17.18). This is the area-specific melting of agarose microchambers by spot heating using a focused laser beam of 1480 nm, which has an absorbance to water, and of a thin layer made of a light-absorbing material such as chromium with a laser beam of 1064 nm, which is permeable to water. For phase-contrast microscopy and μm-scale photothermal etching, three different wavelengths (visible light for observation, and 1480-nm/1064-nm infrared lasers for spot heating) were used simultaneously to observe the positions of the agar chip surface and to melt a portion of the agar in the area being heated. Using the photothermal etching method, we can form microstructures within the agarose layer on the chip by melting a portion of the agarose layer at the spot of a focused infrared laser beam as described above. This method can be

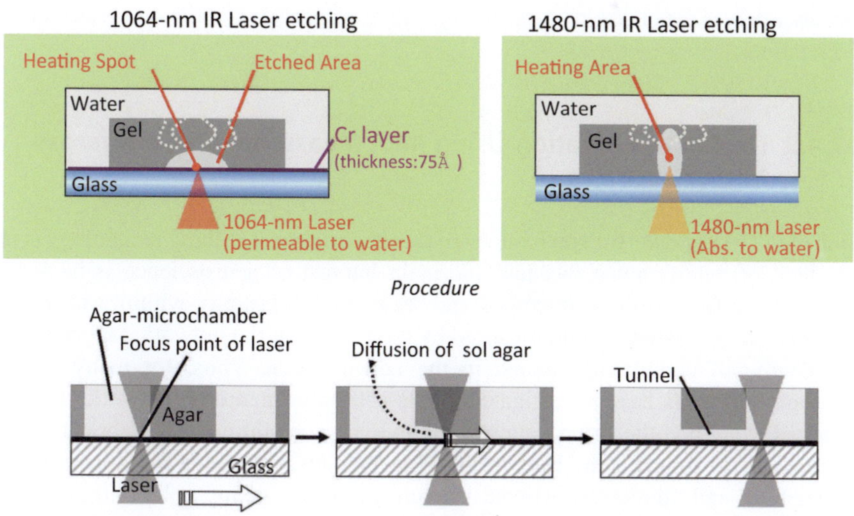

Fig. 17.18 *Photothermal etching method.* Using two types of focused infrared (IR) lasers, the thin layer of low-melting point agar on the chip was selectively melted. As a1064-nm IR laser has no absorption to water, only a portion of agar near to the absorption layer is heated and melted to change gel state to sol state. In contrast, as a 1480-nm IR laser has absorption to water, all the agar on the light pathway is heated and melted. The changed sol state agar is dispersed into the agar gel and holes or tunnels are formed in the agar layer

applied even during cultivation, so we can change the network pattern of nerve cells during cultivation by adding microchannels between two adjacent microchambers in a step-by-step fashion (Sugio et al. 2004; Suzuki et al. 2005).

We used this new type of noncontact three-dimensional photothermal etching for the agar-microetching exploiting the characteristics of the two different infrared laser beam wavelengths (1480 and 1064 nm). As the 1480-nm infrared beam was absorbed by water and agar gel, the agar gel in the 1480-nm infrared light pathway was heated and completely melted. As the 1064-nm infrared beam, on the other hand, did not have this absorbance, the agar melted just near the thin chromium layer, which absorbed the beam.

Using this noncontact etching, we could easily produce microstructures such as holes and tunnels within only a few minutes. As we can see from Fig. 17.18, the laser melted the agar as follows: (a) when a focused 1064-nm infrared laser beam was focused on the chromium layer on the glass slide, the agar at the focal point near the chromium layer started to melt. (b) When the focused beam was moved parallel to the chip surface, a portion of the agar at the heated spot melted and diffused into the water through the agar mesh. (c) After the heated spot had been moved, a tunnel was created at the bottom of the agar layer. (d) However, when a 1480-nm infrared laser beam was focused on the agar glass slide, the agar in the light path started to melt. (e) When the focused beam was moved parallel to the chip surface, a portion of the agar in the light path melted and diffused into the water. (f) Finally, after the heated spot had been moved, a hole was created on the glass slide. Figure 17.19 also shows an example of the features of photothermal etching: the laser power dependence of its etching width in agar layer, indicating their linear correlation.

17.4.2 On-Chip Cultivation Using MEA Measurement on Agarose Microstructures

One of the most powerful applications of this photothermal etching technology is in the field of neuroscience, because the main interest of neuroscience is how the epigenetic information is processed and recorded as plasticity within a neuronal network pattern, what might be caused by the change in the network pattern or by the degree of complexity related to the network size. Thus, for many years, neurophysiologists have investigated single-cell-based neuronal network cultivation and examined the firing patterns of single neurons through the fabrication of cultivation substrates using microprinting techniques, patterning on silicon oxide substrates, and three-dimensional structures made using photolithography. Although these conventional microfabrication techniques provide structures with fine spatial resolution, it is still hard to make flexible microstructures with simple steps or to change their shape during cultivation since the shape is usually unpredictable and only defined during cultivation.

Applying stepwise photothermal etching technology to an agarose microchamber (AMC) array during cultivation, we have developed a method of

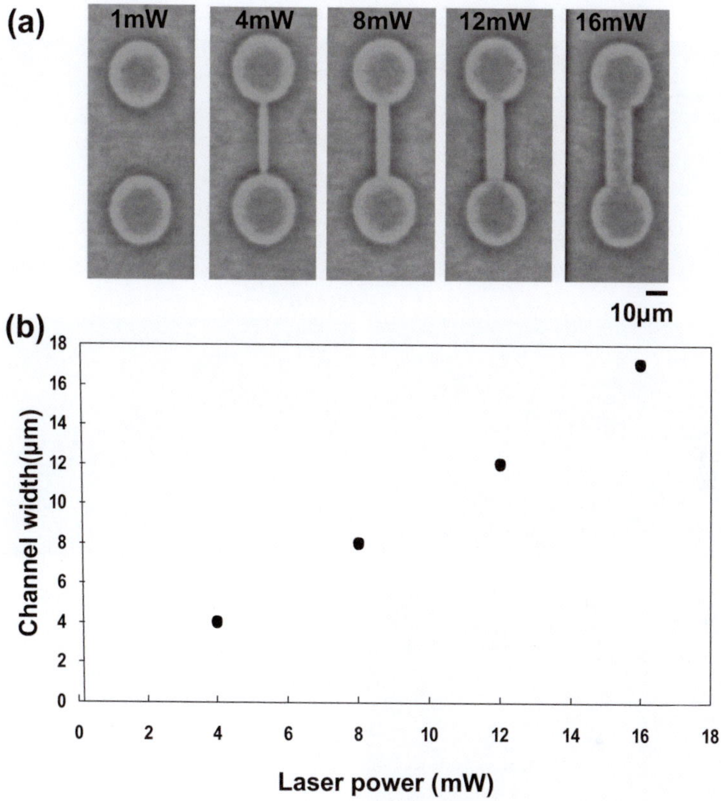

Fig. 17.19 *Laser power dependence of melting width of agar layer in photothermal etching method.* (**a**) Micrographs of IR laser power dependence of agar layer melting. (**b**) Laser power dependence of melted channel width

controlling the topography in the direction of synaptic connections in the network patterns of a living neuronal network. In contrast to the conventional methods for orientation control of connections, this method can allow the full control of direction in which axons and dendrites were elongated to be flexibly controlled by melting the narrow micrometer-order grooves (microchannels) in steps through photothermal etching where a portion of the agarose layer was melted with the 1064-nm infrared laser beam.

Figure 17.20 shows an example of this procedure. The micrographs are phase-contrast images of the growth of single hippocampal cells in the microchambers. When cultivation started, single cells were placed into the AMCs. Then, first single neurites, which are axons, elongated from the neurons into the microchannels. After we confirm that the elongation of first neurites (axons) were guided to the desired pathway sufficiently stable, additional photothermal etchings were performed to connect two adjacent AMCs. Then, a series of the additional photothermal etching for guiding second neurites, which are dendrites, to other neurons. Finally, all

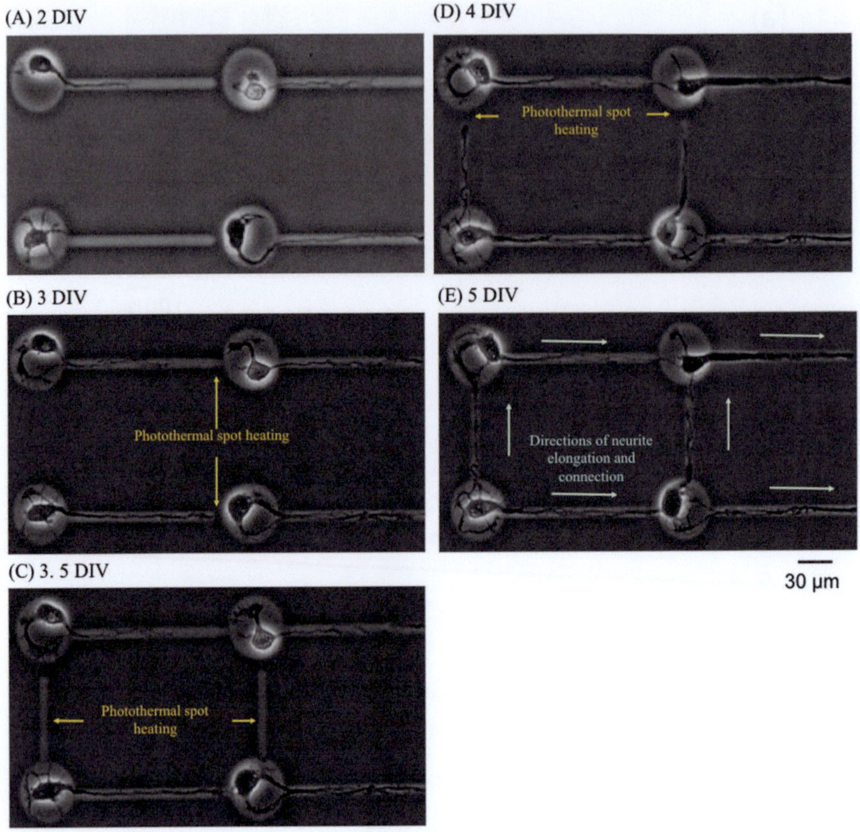

Fig. 17.20 *Stepwise direction control of axon and dendrite connections of neurons using photothermal spot heating*. (**A**) Micrograph of single neurons cultivated in microchambers fabricated in agarose gel layer 2 days after cultivation started (2 days in vitro, DIV). A neurite, which was destined to become an axon, elongated from each neuron and was guided into the single narrow channel connected to the microchamber. (**B**) After 3 DIV, photothermal spot heating was used to connect the first narrow channels and neighboring microchambers to enable the direction of the first neurite elongation to be controlled. (**C**) After 3.5 DIV, after confirmation of connections between first neuritis and neighboring neurons, second channels were fabricated to guide second neurites to this newly added channel. (**D**) After 4 DIV, after confirmation of successful guiding of the second neurites into the second channels, additional photothermal spot heating was used to connect these second channels and another neighboring microchamber to form another connection between second neurites and other neighboring neurons. (**E**) After 5 DIV, a fully direction-controlled neuronal network had been fabricated on the chip

neurons retained their shapes and were connected by the precisely defined neurites, axons, and dendrites with their full control of directions (Suzuki et al. 2004a, b, 2005; Suzuki and Yasuda 2007a, b).

Figure 17.21 is an example of AMC technology for multielectrode array (MEA) measurements. The problem of conventional MEA measurement was how to

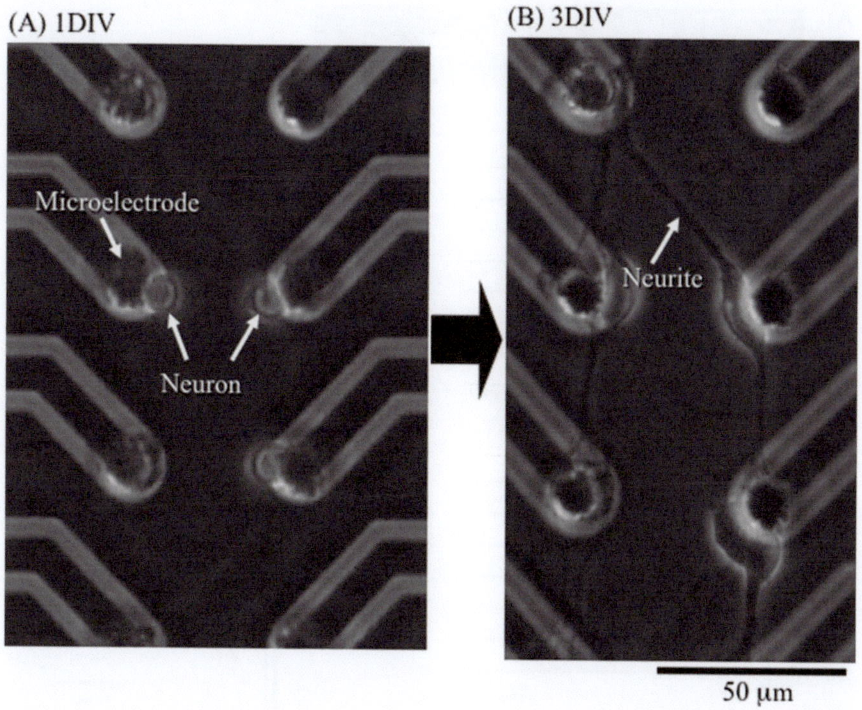

Fig. 17.21 *Strict control of positions of neurons and directions of neuritis using agarose microstructures on a microelectrode array chip.* (**A**) Micrograph of single neurons cultivated in microchambers fabricated in agarose gel layer on the microelectrode array chip (1 DIV). (**B**) After 3 DIV, the neurons were connected through the narrow channels in agarose layer

control the positions of cells to meet the positions of microelectrodes. When we combine these two technologies, the single neurons are fixed at the position of single microelectrodes, and their neurite connections also can be controlled.

Exploiting this combined technology, we also have examined the response of tetanic stimulation on single neuron within lined-up eight cells, as shown in Fig. 17.22. The result showed (1) the tetanic stimulation is successfully recorded into the whole lined-up cells even if they are small eight-cell single network, (2) the hysteresis disappeared gradually, and finally disappeared 24 h after the stimulation, (3) the waveforms of response of neurons are similar in such a single lined-up network (Suzuki and Yasuda 2007a, b).

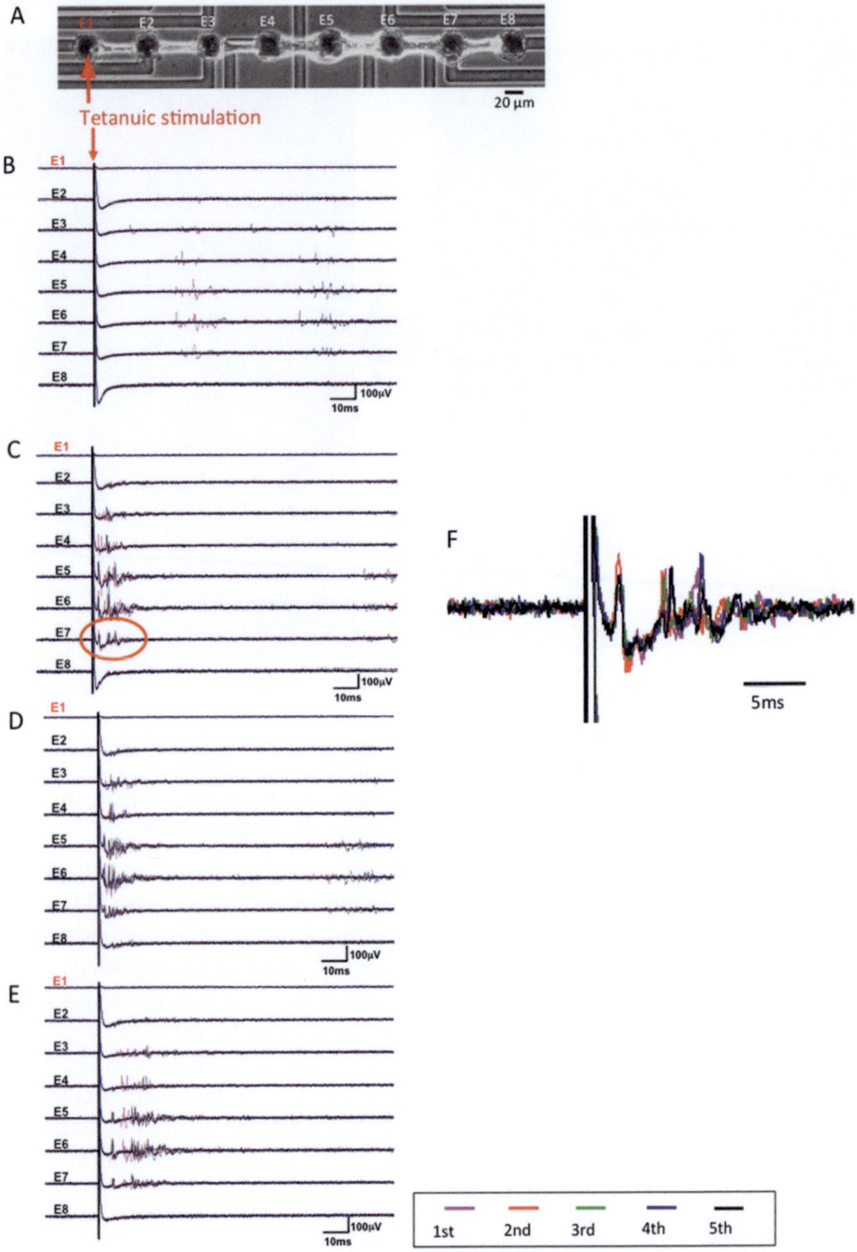

Fig. 17.22 *Tetanic stimulation of single neuron on the strict position controlled lined-up neurons using agarose microstructures on a microelectrode array chip*. (**A**) Micrograph of lined-up eight rat hippocampal cells cultivated in microchambers fabricated in agarose gel layer on the microelectrode array chip. (**B**) Before tetanic stimulation, (**C**) 30 min after the stimulation, (**D**) 6 h after the stimulation, (**E**) 24 h after the stimulation. (**F**) Magnified waveforms of eight neurons' responses for comparison of five time pulse stimulation responses

17.4.3 Community Effect of Cells for Their Stabilization and Synchronization

In the former subsection, we described the ability of agarose microstructures for controlling the cell arrangement, especially for neuronal network studies. In this subsection, we introduce another example of applications of this technology for finding the community effect of cardiomyocyte beating (Kojima et al. 2003, 2004, 2005, 2006).

First, we have investigated the roles of the beat rate and beat-rate fluctuation of isolated single cardiomyocytes on the reestablishment of synchronous beating by analyzing the changes of beating rates and their fluctuations before and after the synchronization of two cardiomyocytes through narrow channels with initially different rhythms (e.g., Fig. 17.23A, B). As a result, we found three types of synchronization; i.e., (1) the beating of the two cardiomyocytes synchronized at the faster of the two initial rates, whereas with the beating fluctuation of the lower of the two initial rates (Fig. 17.23C), (2) the beating of the two cells synchronized at the lower of the two initial rates, and with the fluctuation of the lower of the two initial rates (Fig. 17.23D), and (3) the synchronization at neither of the initial rates of single cardiomyocyte and with fluctuation of smaller of the initial fluctuations (Fig. 17.23E).

The interbeat intervals of 14 cell pairs before and after synchronization are listed in Table A of Fig. 17.24. Five of the cell pairs synchronized at the initial rate of the faster cell, two of the pairs synchronized at the initial rate of the slower cell, and the other seven pairs synchronized at a rate other than one of the initial rates. In Table B of Fig. 17.24, the fluctuation data for the 14 cell pairs whose rate data is listed in Table A is grouped according to the change of the fluctuation before and after synchronization. Thirteen pairs synchronized with a fluctuation equal to or less than the initial fluctuation of the slower member of the pair, and one pair synchronized with a fluctuation larger than that of either of the two initial fluctuations.

These results suggest that the fluctuation of reestablished synchronous beating by isolated cardiomyocytes is influenced more strongly by the fluctuation of the initial fluctuation of the beat rates of the isolated cardiomyocytes than the rate of the reestablished synchronous beating is influenced by the initial beat rates of the isolated cardiomyocytes. It might therefore be thought that a cardiomyocyte whose beat rate fluctuates less than that of another cardiomyocyte entrains the beating rhythm of that cardiomyocyte, but we observed one pair of cells in which this did not happen (Group H). This indicates that the influence of a single cell is still not strong enough to account for the process of entrainment in heart tissue.

Figure 17.25 also describes the community size effect of cardiomyocyte network on its beating stability. In this work, we explored the relation between entrainment and community size by examining the synchronization process of a cardiomyocyte network formed by the interaction of single cardiomyocytes cultured in a 3×3 grid of agarose microchambers with connecting microchannels. After nine isolated cells were cultured in the nine-chamber agarose microcultivation chip for 24 h, we started to measure the synchronization process continuously and found that when an isolated single cell came into contact with another cell and formed a two-cell network (Fig. 17.25 top and middle), these two cells synchronized at the initial rate of the

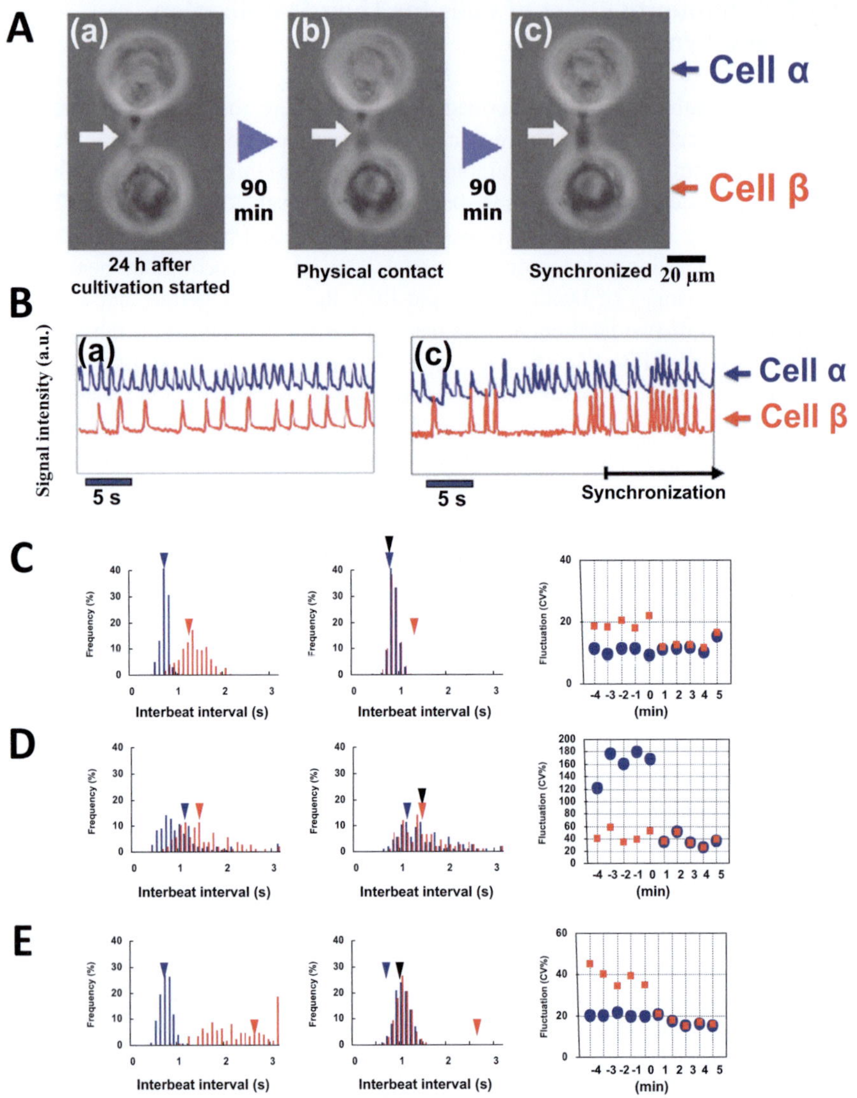

Fig. 17.23 *Synchronization of two cardiomyocytes.* (**A**) Micrographs of two cardiomyocytes under isolated conditions (*a*), just after connection (*b*), and synchronization started (*c*). (**B**) Beating waveforms at (*a*) and (*c*) in Fig. (a). (**c–e**) (*left graph*, and *center graph*) Beating frequency spectrum before and after synchronization, respectively; distribution of interbeat intervals of two cardiomyocytes and the change of the mean value of beating rhythm fluctuation at intervals of 1 min measured for 5 min before and after synchronization. *Blue* and *red triangle* shows the mean values before synchronization, and the *black triangle* shows the mean value for the two cells after synchronization. (*right graph*) The change of the mean value of beating rhythm fluctuation [CV%: coefficient of variation (100 × standard deviation/mean beat rate)] at intervals of 1 min measured for 5 min before and after synchronization. *Blue circles* and *red squares* show the corresponding mean values of beating rhythm fluctuation for 1 min. Three types of synchronization tendencies were described: (**C**) synchronization to faster beating cell, (**D**) synchronization to slower beating cell, (**E**) synchronization with new beating frequency

A. Three types of synchronization from the viewpoint of frequency (s) (n = 14)

Synchronize to faster beating cell			Synchronize to slower beating cell			Forming new beating interval		
Cell A	Cell B	Sync	Cell A	Cell B	Sync	Cell A	Cell B	Sync
0.64	1.23	0.76	1.10	1.40	1.40	0.64	2.7	0.94
0.93	1.01	0.83	0.59	0.63	0.62	0.84	1.77	1.10
0.74	1.13	0.82				0.56	1.21	0.89
0.87	1.43	0.86				0.71	0.92	0.81
0.94	2.18	0.89				0.56	1.21	0.90
						0.53	1.06	0.74
						0.43	3.10	1.00

B. Two types of synchronization from the viewpoint of fluctuation (CV%)

Fluctuation (CV%) decrease				Fluctuation (CV%) increase			
Before		After		Before		After	
25.1	12.3	12.3	Sync to faster beating cell	88.7	26.7	78.1	Forming new beating interval
20.8	16.4	8.9	Sync to faster beating cell				
46.5	19.7	15.9	Sync to faster beating cell				
117	19.7	18.9	Sync to faster beating cell				
164	14.7	13.2	Sync to faster beating cell				
149	41.2	41.7	Sync to slower beating cell				
16.4	11.6	10.7	Sync to slower beating cell				
42.9	20.1	17.3	Forming new beating interval				
19.7	11.7	10.9	Forming new beating interval				
29.0	17.9	18.1	Forming new beating interval				
20.5	17.5	12.9	Forming new beating interval				
29.3	17.9	18.7	Forming new beating interval				
22.8	21.7	11.8	Forming new beating interval				

Fig. 17.24 *Tendency of synchronization of two cardiomyocytes*. (**A**) Three types of synchronization of two cardiomyocytes from the viewpoint of beating intervals. (**B**) Two types of synchronization from the viewpoint of beating stability (fluctuation of beating)

first cell and with a fluctuation decreased from the initial fluctuation of the first cell. When the whole nine cells came into contact and formed the nine-cell network (Fig. 17.25 bottom), it synchronized at a rate equal to the initial rate of the first cell and with a fluctuation decrease. These results suggest that the beating rhythm of a single cardiomyocyte tends to entrain to the rhythm of the cell network and the strength of this tendency increases with the size of the network. Therefore it is thought that the fluctuation of the rate at which a network of cardiomyocytes beats decreases as

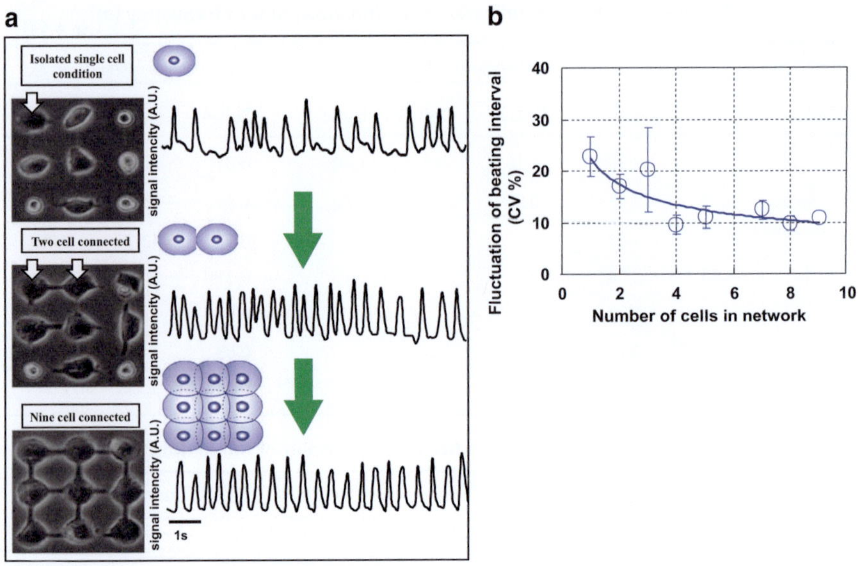

Fig. 17.25 *Tendency of synchronization of two cardiomyocytes.* (**a**) Three types of synchronization of two cardiomyocytes from the viewpoint of beating intervals. (**b**) Two types of synchronization from the viewpoint of beating stability (fluctuation of beating)

the size of the network increases. The tendency of the synchronization above was simply explained by saying that the synchronization of two cardiomyocytes was caused by the more unstable cell (the one with the more variable beating intervals) following the more stable cell. Such fluctuation reduction tendency was more obvious when the number of cardiomyocytes in the network increased, and we call this phenomenon as "community effect" of synchronization (Kaneko et al. 2007).

Using the agarose microchambers, we also can examine the spatial arrangement dependence of synchronization stability of cardiomyocyte networks. As shown in Fig. 17.26, we can arrange the cardiomyocytes in three types of shapes, linear shape, radial shape, and lattice shape, and compared their tendency of beating stabilization against cell numbers. The results indicated that there are no apparent relationships between the cell number and their shapes and that the most important index for stabilization of cell beating is not geometry of cells but cell numbers.

We also have examined the contribution of fibroblasts for synchronization of cardiomyocytes. We connected two cardiomyocytes through single fibroblast and synchronized them as shown in Fig. 17.27a, b, and used this heterogeneous cardiomyocyte-fibroblast coupling to examine the tendency of the stability of interbeat intervals and beating rhythm fluctuation of two cardiomyocytes through a fibroblast before and after their synchronization.

The first type of synchronization was the tendency of fluctuation reduction caused by synchronization, which is the same tendency seen in a network formed by the direct connection of two cardiomyocytes. As shown in Fig. 17.27C, D, in this type, the two cells having different interbeat intervals before synchronization had

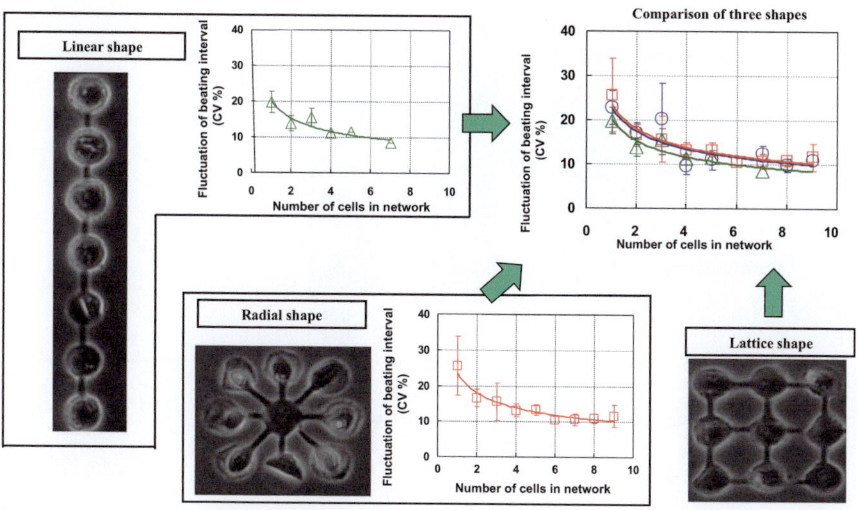

Fig. 17.26 *Effect of spatial arrangement of caridiomyocytes on cell number dependence of beating stability*. Three types of spatial arrangements (i.e., lattice shape, linear shape, and radial shape) were compared to check the contribution difference on their beating stability

made a synchronized interbeat interval of only within a second after synchronization (e.g., Fig. 17.27B). The fluctuation of synchronized network became almost equal or smaller than either of the two initial fluctuations.

In contrast, the second type was the tendency of fluctuation increase caused by synchronization, which was not occurring in the cardiomyocyte network (Fig. 17.27E). In this case, the two cardiomyocytes having two different interbeat intervals before synchronization had a mean interbeat interval after synchronization, and the fluctuation of the synchronized network was greater than that of the cell that had the lower fluctuation before the synchronization.

Our agarose microchamber system lets us regulate the cell type and community size of cultured cells at the single-cell level. This could not be done when using the conventional cell cultivation method, so the prolific growth of cardiac fibroblasts made it hard to culture only cardiomyocytes and investigate the properties of a single cell in a group of cells. By using single-cell-based cultivation, we were able to investigate ways in which the fluctuation of the rates at which cardiomyocytes affects the reestablishment of synchronized beating. The results of the present experiments can be summarized as follows:

1. When two isolated, independently beating cardiomyocytes came into contact, they tended to beat synchronously at a rate that fluctuated no more than did that of the cell whose beat rate fluctuated less than did that of the other cell.
2. When initially isolated cardiomyocytes formed a network, it tended to entrain to their rhythm the beating rhythm of single cells whose beating rhythm fluctuated more than did that of the network.

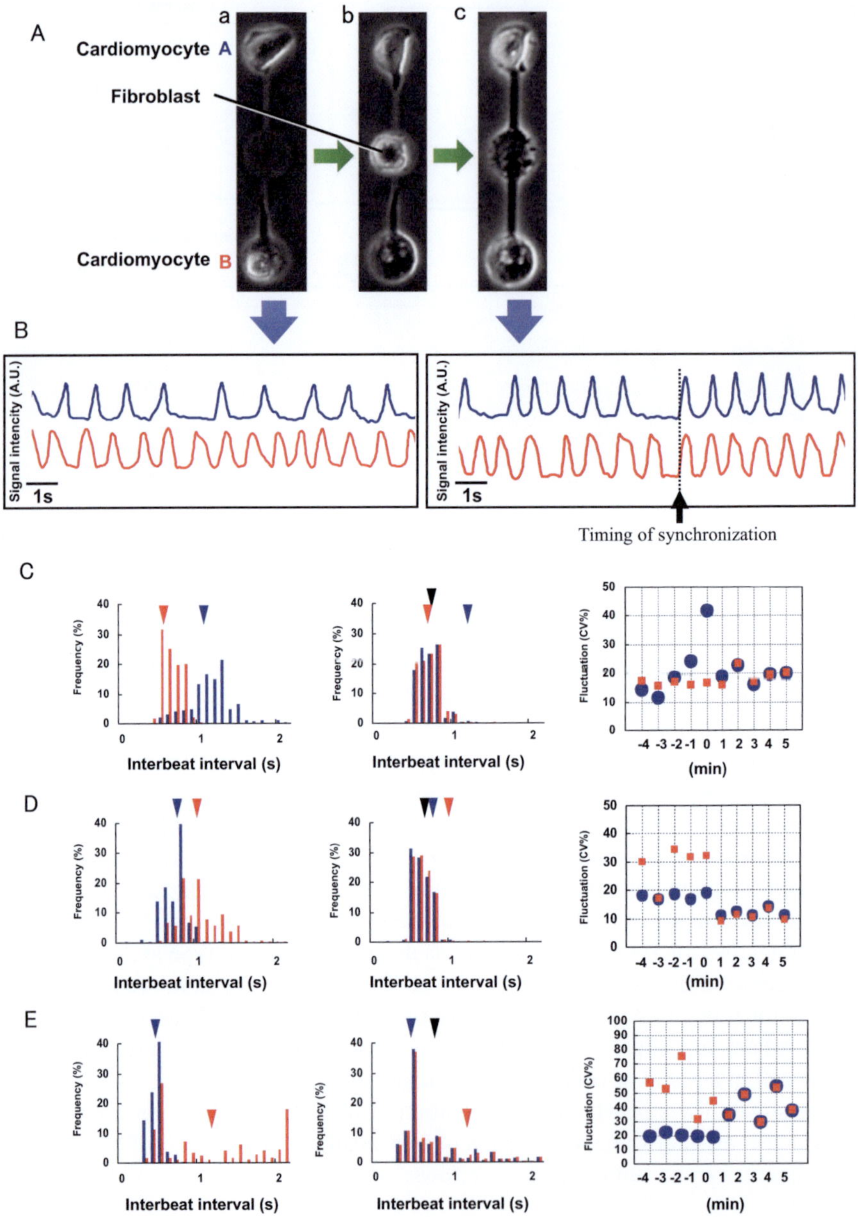

Fig. 17.27 Synchronization of two cardiomyocytes through fibroblast. (**A**) Micrographs of two cardiomyocytes under isolated conditions (*a*), a fibroblast was added between two cardiomyocytes (*b*), and two cardiomyocytes connected through fibroblast and synchronization started (*c*). (**B**) Beating waveforms at (*a*) and (*c*) in Fig. (A). (**C–E**) Three types of synchronization tendencies. Beating frequency spectrum before (*left graph*) and after (*right graph*) synchronization, and their beating fluctuation (*right graph*). (**C**) Synchronization to faster and more stable beating cell; (**D**) Synchronization and creating new beating intervals with contribution of beating stability; (**E**) Synchronization with new beating frequency, however, beating fluctuation increased

3. The entrainment activity of cell networks increased with their size.
4. Spatial arrangement difference does not affect the synchronization manner of cardiomyocytes, and only the cell number is ruling for their synchronization tendency.
5. The interbeat intervals after the synchronization of two cardiomyocytes connected by a fibroblast is not the same as that after the synchronization of two cardiomyocytes directly connected to each other, and the tendency of community effect seems to be suppressed when the cardiomyocytes are heterogeneously coupled through a fibroblast.

They might indicate that unstable isolated cardiomyocytes reestablish a cell network that beats stably and synchronously. A novel finding of this study is that a cardiomyocyte network containing only a few cells acquires to form a stable rhythm. And moreover, once the cell or the cell network acquired the stable beating, the additional attachment of an unstable cell can synchronize to the stable cell or cell network following their stable beating intervals. This phenomenon also suggests that the factor of stability is very important to determine the fate of the beating frequency of the network after the connection of unstable cells.

17.5 *Quasi In Vivo* Model of Stem-Cell-Derived Cardiomyocytes for Predictive Cardiotoxicity

17.5.1 Origin of Lethal Arrhythmia in Heart: How to Measure Cardiotoxicity *In Vitro*?

The establishments of human induced pluripotent stem (hiPS) cells and of hiPS-derived cardiomyocytes (hiPS-CMs) confer significant advantages to generate panels of cell lines that more closely reflect the genetic diversity of a population, such as familial cardiomyopathy, familial lethal arrhythmias, and congenital heart diseases. However, to date, there has been no report in the conventional *in vitro* caordiotoxicity assay on the difference of the cell responses depending on the difference of their spatial arrangements, which is important to predict the risk of the compounds quantitatively.

For drug discovery and toxicology, International Conference on Harmonisation (ICH) regulation S7B request a pair of in vitro channel measurement based study and in vivo animal QT prolongation test. Although *in vitro* patch-clamp analysis of cells is a well-accepted way to investigate the action potential parameters, it is just single-cell-based study and it is impossible to measure the abnormal cell-to-cell conductivity, which is one of the origins of ventricular tachycardia (VT) including torsade de pointes (TdP), a potentially life-threatening arrhythmia because it may lead to ventricular fibrillation, asystole, and sudden death. On the other hand, animal test is expensive and time consuming, and the high amounts of test sample necessary to perform an experiment make it unsuitable for screening purposes. Hence, cell-network-based drug screening assay has a potential being suited for drug discovery and toxicology as a minimum model of the ventricle. However it has a problem, for

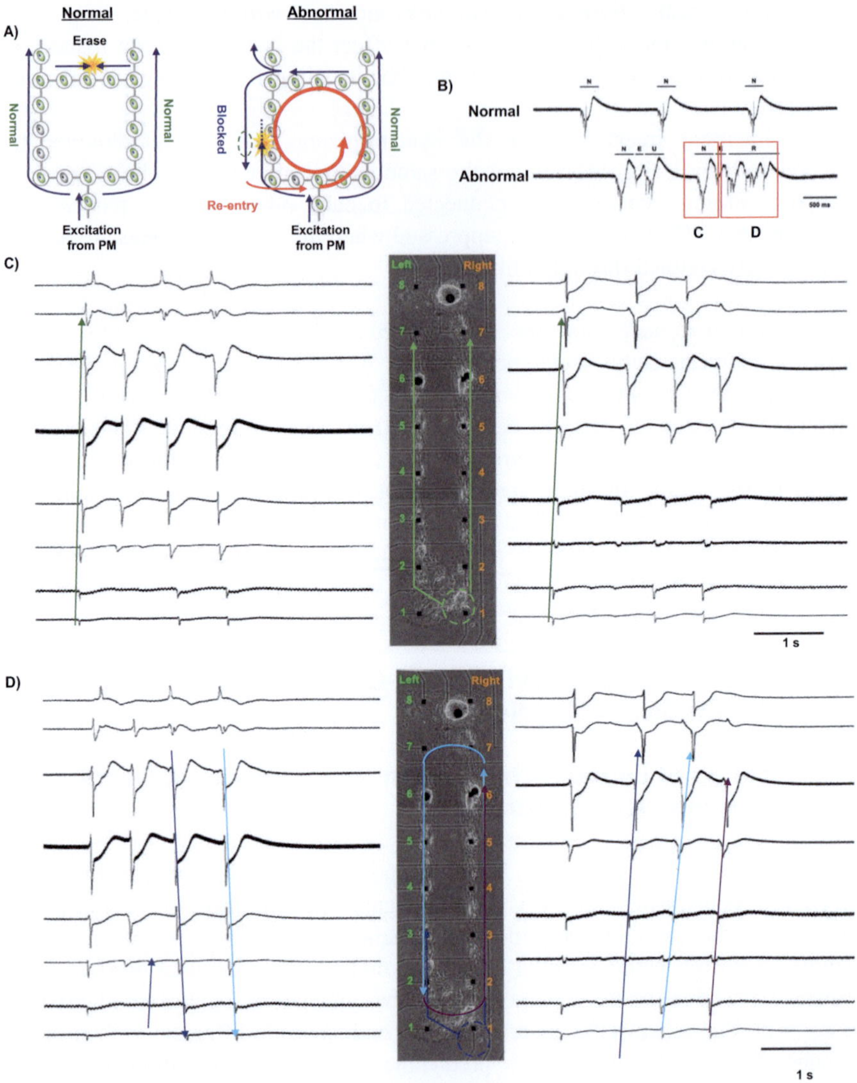

Fig. 17.28 *Conductance abnormality as an origin of cardiotoxicity.* (**A**) Schematic drawing explaining conductance abnormality occurred in cardiomyocyte network. (**B**) Convoluted external field potential waveforms of multielectrodes representing electrocardiograph (ECG) like information, which represents the abnormality of each cell and their conductance. (**C**) Normal propagation of depolarization from pace-making area in loop-shaped cariomyocyte network, which is on-chip *in vitro* re-entry model. (**D**) Abnormal propagation from pace-making area and circulating as re-entry

example, the difference of results occurred in cell-based screening using cultivated cell even using same cells in the networks.

Figure 17.28 is one example *of in vitro* model of the origin of lethal arrhythmia, such as micro reentry caused by abnormal conductance of excitation in cardiomyocyte network in ventricle tissue (Fig. 17.28A). To examine the summed field potential (FP) signals in the lined-up cardiomyocyte cell network (Fig. 17.28B), we have fabricated an on-chip closed circuit lined-up cell network on an MEA chip as shown in Fig. 17.28C, D. The microelectrodes (L1-L8, R1-R8) were set in the corridors to enable the recording of the FPs of cardiomyocytes. During the normal propagation of beating, as shown in Fig. 17.28C, the beating started from the pacemaking cardiomyocytes (PM) at R1, propagated along both corridors, and disappeared at L8 and R8, where the two propagating beating signals met and canceled each other. However, even in the same cell network, once the irregular propagation of beating signals occurred, as shown in Fig. 17.28D, the asymmetric propagation of beating in the left corridor and right corridor resulted in the circulation of beating in the closed circuit caused by the failure of signal cancelation, which is a characteristic of ventricular tachyarrhythmia.

The above results indicate the importance of detecting the propagation manner of beating, because we cannot predict tachyarrhythmias only from the AP profiles of single cells (and single-point FP profiles without propagation measurement), i.e., the observation of propagation between neighboring cells is not equivalent to the observation of particular cells' depolarization. It should also be noted that the coordinated arrangement of cells used to regulate the propagation manner of beating is important for acquiring the shape of coordinated quasi *in vivo* electrocardiogram (ECG) signals, because the roughly dispersed cells on the cultivation chip can make random connections, which might give unexpected circulation in beating propagation.

Figure 17.29 shows the relationships between electrophysiological profiles, for example, conventional *in vitro* action potentials (APs) (Fig. 17.29A), *in vivo* ECGs (Fig. 17.29D), and the FPs acquired from cardiomyocytes and on-chip cardiomyocyte cell networks (Fig. 17.29C, D, respectively). The AP of cardiomyocytes is their electrical membrane potential, which is generated by special types of voltage-gated

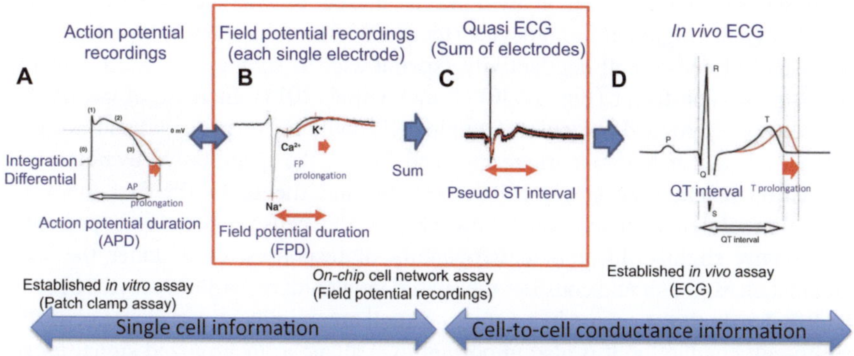

Fig. 17.29 *Relationship of electrophysiological signals of cardiomyocytes from single cell viewpoint to heart organ viewpoint.* (**A**) Action potential (AP) of (isolated) single cells. (**B**) External field potential (FP) of single cells or small cell clusters. (**C**) Convoluted external field potential signals of cardiomyocyte network, which includes ion-channel information of each cell and conductance of neighboring cells. (**D**) Electrocardiograph (ECG) signals from heart organ

ion channels embedded in a cell's plasma membrane. As shown in Fig. 17.29A, the first sharp rise in voltage corresponds to the influx of sodium ions, whereas the two decays correspond to the sodium-channel inactivation and the repolarizing flux of potassium ions. The second small rise after the first decay results from the influx of potassium upon the opening of voltage-sensitive calcium channels. On the other hand, the extracellular FP of cardiomyocytes (Fig. 17.29B) is the electrical ion current flux of sodium, potassium, and calcium ions from cardiomyocyte cells, i.e., the relationship between the AP and FP of cardiomyocytes is the derivative/integral of fluxes of ions.

As shown in Fig. 17.29D, an ECG is a set of transthoracic electrical signals of the heart muscle, which depolarize during each heartbeat, externally recorded by skin electrodes. During each heartbeat, a healthy heart will have an orderly progression in a wave of depolarization, which is triggered by the cells in the sinoatrial node, spreads out through the atrium, passes through intrinsic conduction pathways, and then spreads throughout the ventricles. A typical ECG tracing of the cardiac cycle (heartbeat) consists of a P wave, a QRS complex, and a T wave. As the ST segment represents the period when the ventricles are depolarized and the T wave represents the repolarization (or recovery) of the ventricles, the ECG signals of the ST period (ST segment and following T wave) consist of FP signals, which propagate in the ventricles. Hence, the convolution of the propagating FP signals in the lined-up cardiomyocyte cell network in the ventricles (Fig. 17.29C) should represent the characteristics of a piece of ventricular tissue, i.e., quasi *in vivo* ECG signals in the ST period of the ECG waveform.

17.5.2 On-Chip Quasi *In Vivo* Predictive Cardiotoxicity Measurement (I): Temporal Fluctuation Measurement of Field Potential Duration (FPD) of Single Cardiomyocytes

As described above, lethal arrhythmia is caused by the increase of response uncertainty of single cardiomyocytes (temporal aspect; strategy I) as a triggering factor and of cell-to-cell conductivity (spatial aspect; strategy II) as an enhancement/suppression factor (Fig. 17.30) (Kaneko et al. 2014). Increase of uncertainty of electrophysiological response of single cells could be the principal and essential origin of lethal arrhythmia triggering, and hence, the quantitative evaluation of fluctuation potential of single cells is critical and should be the first index for prediction of lethal arrhythmia. However, as the cells in the tissue present a functionally slightly different heterogeneity of responses, even under the same circumstances, we should consider the spatial viewpoint of cardiomyocytes because the lethal arrhythmia occurs in a tissue, i.e., cell community, not in single cells. For cell-to-cell conduction it is also important to evaluate asynchronized signal propagation in the cardiomyocyte network in heart tissues. Those community effects of cardiomyocytes could be the second origin of arrhythmia in addition to the first origin, temporal fluctuation of repolarization time of single cells after depolarization. The second origin could have either enhancing or suppressing roles for lethal arrhythmia occurrence. For example, the heterogeneity of cardiomyocyte

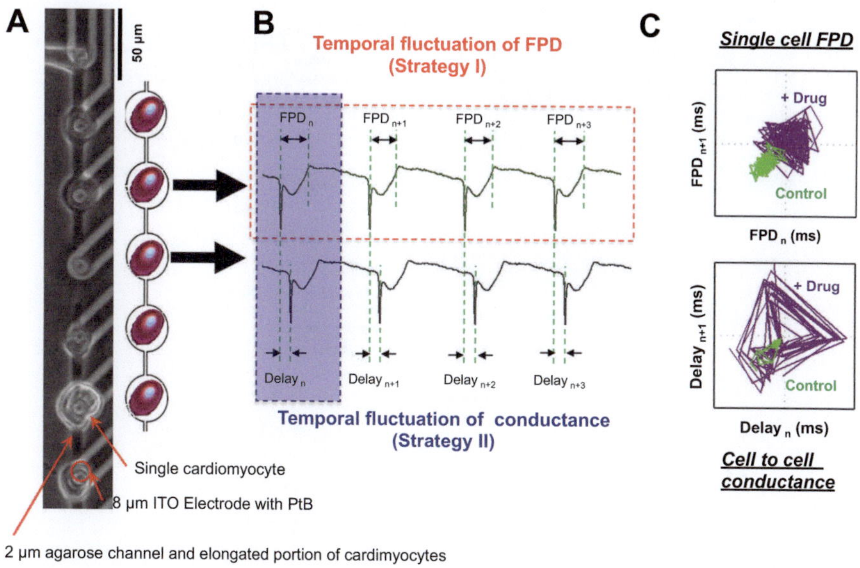

Fig. 17.30 *Two viewpoints of in vitro predictive cardiotoxicity measurement.* (**A**) Micrograph and schematic drawing of cardiomyocyte network arrangement. Human iPS derived cardiomyocytes are arranged linearly on the microelectrodes on a chip. (**B**) Temporal fluctuation of prolongation of field potential duration (FPD) of single cells (Strategy I), and temporal fluctuation of conductance between neighboring cells (Strategy II) can be measured simultaneously. (**C**) Acquired temporal fluctuation signals of single cell FPD and cell-to-cell conductance are plotted in the logistic maps and evaluate their uncertainty as a quantitative risk value of lethal arrhythmia of heart

functional characteristics has a potential to enhance the occurrence of lethal arrhythmia because of the different responses of neighboring cells from the first origin. In contrast, the community effect of cardiomyocytes has also a potential to suppress the occurrence of lethal arrhythmia by the enhancement of synchronization tendencies with suppression of fluctuation in cell groups as demonstrated in the former section. Hence, one promising way to predict ventricular arrhythmia at the *in vitro* level is to measure the first origin of arrhythmia, i.e., temporal fluctuation of repolarization time of single cells, and then, measure the second origin, i.e., spatial fluctuation of conductivity of neighboring cardiomyocytes using spatially arranged hCMs.

In this subsection, we describe the measurement technology of temporal aspect (strategy I) and its results. To study the temporal increase of uncertainty (fluctuation) of human cardiomyocyte (hCM) response, we have modified the developed on-chip cell network cultivation system, in which extracellular signals (field potentials: FP) of hCMs can be measured using a multielectrode array (MEA), and spatial arrangement control of cells can be performed using agarose microstructures designed on MEA chip (Fig. 17.31A). For the temporal fluctuation measurement, recorded FP duration (FPD) was defined as the time interval between

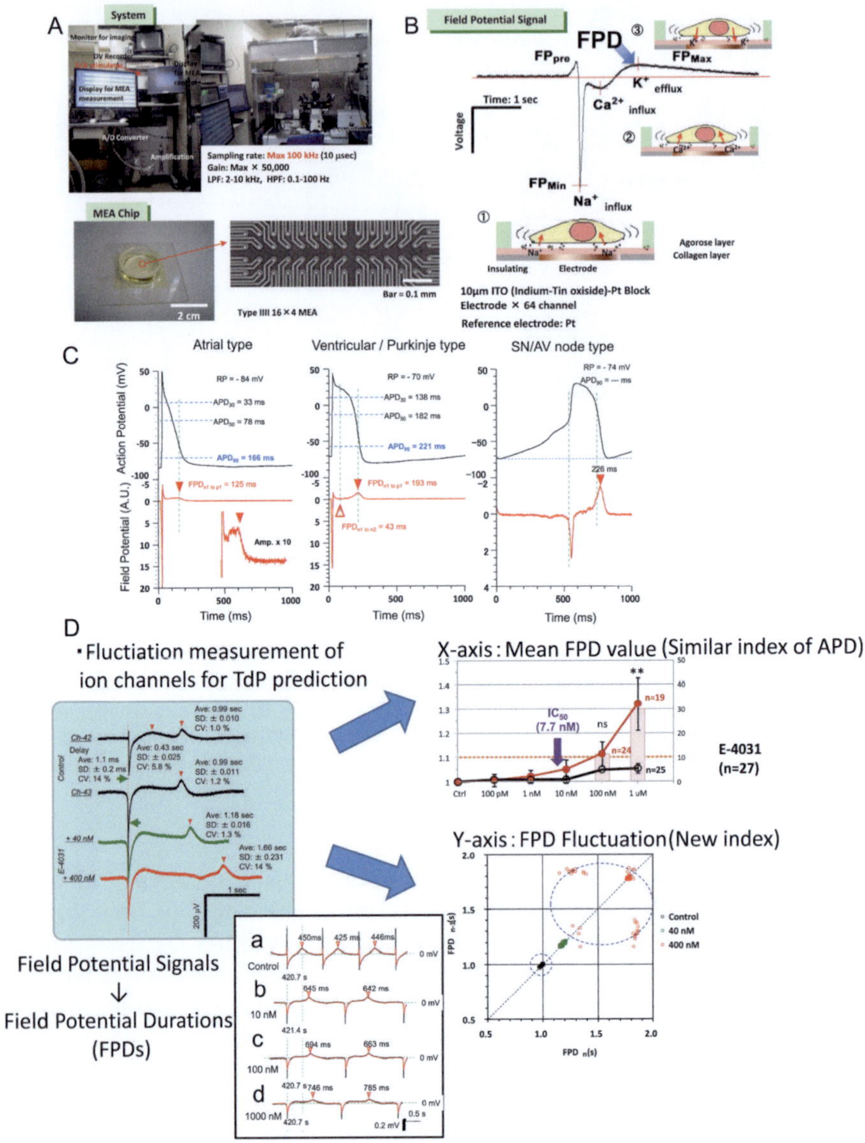

Fig. 17.31 *Method for predicting cardiotoxicity from the viewpoint of temporal fluctuation of field potential duration (FPD)*. (**A**) Multielectrode measurement system and a multielectrode array (MEA) chip. (**B**) Schematic drawing explaining the meaning of external field potential (FP) waveform. FP represents an inward/outward ion current through cardiomyocyte's cell membrane. First, inward current is observed by quick opening of Sodium ion channel, followed by inward Calcium ion channel and outward Potassium ion current. The peak of Potassium current is defined as field potential duration (FPD), and its prolongation is caused by the inhibition of Potassium ion channels by toxic compounds. (**C**) Relationship of action potentials (AP) and field potentials (FP). Position of FPD represents 90 % of action potential duration (APD_{90}). (**D**) Schematic drawing of cardiotoxicity evaluation procedure of temporal fluctuation signals of cardiomyocyte FPDs (Strategy I).

the initial field potential deflection and the peak of depolarization (Fig. 17.31B), and their waveforms were defined by the difference of cardiomycoytes such as atrial, ventricule, and node types (Fig. 17.31C). To evaluate the cardiotoxicity, we applied two dimensional indices, the conventional mean values of FPD prolongations as X axis and the new fluctuation values of FPDs as the short-term variability (STV) of FPD in Y axis, which is defined as the mean distance of points perpendicular to the line of identity in the Poincaré plot, calculated ($STV = \Sigma|Dn{+}1 - Dn|/[n \times \sqrt{2}]$, where *Dn* represents the FPD of *n*-th beating) (Fig. 17.31D).

The left graphs in Fig. 17.32A show the compound concentration dependence of FPD prolongation ratio (mean values of FPDs in a 5 min recording). For the interpretation of the FPD data, we set 10 % prolongation of FPD as risk judgment, which met to the value of prolongation, 10 %, in conventional APD measurements. Among the positive compounds (category I), FPD prolongation showed the almost similar ability to detect QT risk (prolongation), which was consistent to the clinical results, whereas hERG inhibition ratio was not always representing the clinical results (e.g., DL-sotalol). Next, among the false negative compounds (category II), some QT risks of the APD_{pm} negative compounds (astemizole, bepridil, paroxetine, and thioridazine) were correctly predicted by FPD prolongation measurement. Third, among the false positive compounds (category III), FPDs were consistent with the clinical results.

The results indicate that the FPD prolongation measurement of hCMs can predict the lethal arrhythmia risks in these four categories with a precision better than conventional *in vitro* measurements. However, no obvious advantage of this replacement was observed since no perfect improvement of false negative/positive problems was measured. In addition, the result of famotidine in the FPD measurement raised concerns related to the advantage of FPD measurement against conventional methods.

Hence, we added the second index on the first index of averaged values of APD/FPD prolongations; that is, temporal fluctuation of hCM responses, because one of the origins of VT occurrence is uncertainty of cell responses such as fluctuation of repolarization time after depolarization.

The center graphs in Fig. 17.32A show the drug concentration dependence of STV_{FPD} changes in the above four compound categories. Using the standard positive compound E-4031, we defined that the border of risk in STV_{FPD} was 1.9, i.e., when STV increased over 1.9 from the STV_{FPD} results of the control, we consider that VT risk appeared. The right graphs in Fig. 17.32A indicate the relationship of FPD prolongation and STV_{FPD} increase. As shown for the compounds in the categories I, and II, FPD and STV_{FPD} correlated linearly for QT prolonging compounds. Moreover,

Fig. 17.31 (continued) The conventional index of the mean value of FPD prolongation is set as X axis (representing APD_{90}), and the new index, short-term variability (STV) of temporal fluctuation of FPD is set as Y axis to evaluate the risk of compound two-dimensionally

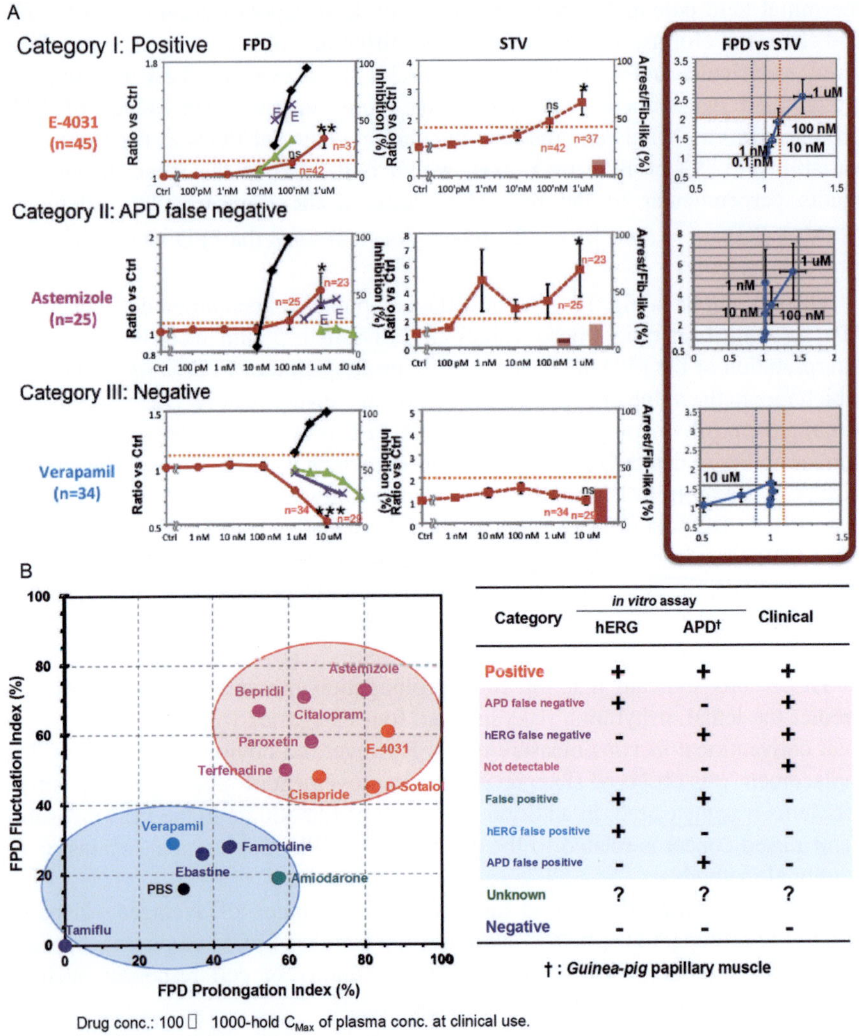

Category	*in vitro* assay hERG	APD†	Clinical
Positive	+	+	+
APD false negative	+	-	+
hERG false negative	-	+	+
Not detectable	-	-	+
False positive	+	+	-
hERG false positive	+	-	-
APD false positive	-	+	-
Unknown	?	?	?
Negative	-	-	-

† : *Guinea-pig* papillary muscle

Fig. 17.32 *Results of in vitro predictive cardiotoxicity measurement from the viewpoint of temporal FPD fluctuation.* (**A**) Typical screening results of positive (category I), APD false negative (category II), and negative (category III) compounds. Conpound concentration dependence of FPD (*left graph*), STV (*center graph*), and FPD-STV relation (*right graph*) are plotted. (**B**) Two-dimensional (2D) plotting of FPD prolongation (X-axis) and FPD fluctuation (STV; Y-axis). Although the conventional one-dimensional evaluation (X-axis only) cannot distinguish overlapping of false negative/positive compounds evaluation, the 2-D evaluation can categorize the difference of positive/negative risks of compounds

STV_{FPD} measurement predicted the risk of the compounds in the category II, which FPD measurement couldn't predict their risks, except for terfenadine. Figure 17.32B shows the summary of the two-dimensional results of FPD and STV_{FPD} of compounds, indicating the more precise and correct answers of false negative/positive compounds by adding the new Y-axis.

17.5.3 On-Chip Quasi *In Vivo* Predictive Cardiotoxicity Measurement (II): Temporal Fluctuation Measurement of Cell-to-Cell Conductance

A *quasi in vivo* cardiac toxicity assay is a new in vitro cell network assay technology platform to bridge the gap between conventional *in vitro* single-cell-based studies and *in vivo* human clinical settings in terms of cardiac toxicity of new chemical entities from spatial aspect (strategy II) for drug development. Potential advantages of the *quasi in vivo* assay are to predict lethal arrhythmia (torsades de pointes (TdP)/ ventricular tachyarrhythmia (VT)/ventricular fibrillation (Vf)) by evaluation of spatial cell-to-cell conductance fluctuation using the on-chip cell network loop, which can choose different conductance pathways of human cardiomyocytes among neighboring circulations (Nomura et al. 2011a, b).

As shown in Fig. 17.33, the ventricular electrocardiogram (ECG)-like field potential data, which includes both the repolarization and the conductance abnormality, was acquired from the self-convoluted extracellular field potentials (FPs) of

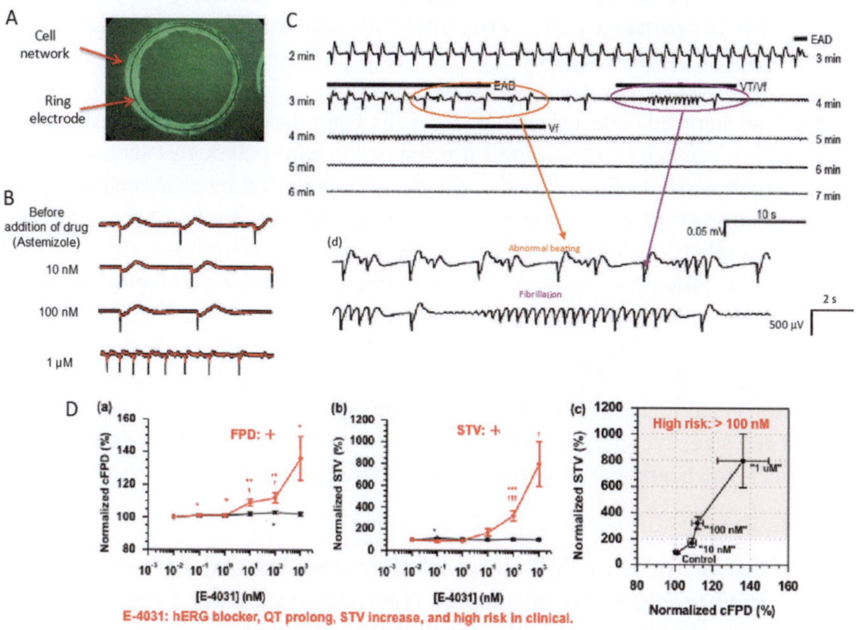

Fig. 17.33 *Abnormal beating and fibrillation measurement of circuit-type cardiomyocyte network using a ring-shaped single microelectrode caused by lethal arrhythmia compounds (Astemizole and E-4031).* (**A**) Micropgraph of circuit-type cardiomyocyte network on a 1-mm diameter *ring-shape* microelectrode. (**B**) Positive compound (Astemizole) concentration dependence of FP waveform change. (**C**) Time course FP (quasi *in vivo* ECG of ST interval) profiles after addition of 1 μM Astemizole. (**D**) Positive compound (E-4031) concentration dependence of FPD prolongation (*a*), short-term variability (STV) of FPD (*b*), and FPD-STV correlation (*c*), which are similar results of single-cell-based screening in Fig. 17.32a

a lined-up cardiomyocyte network on a circle-shaped microelectrode in an agarose microchamber (Fig. 17.33A). When positive compound (Astemisol) applied to the closed-loop cardiomyocyte network, QT-prolongation-like self-convoluted FP duration prolongation was according to the increase of Astemizole concentration (Fig. 17.33B). Moreover self-convoluted FP profile of normal beating changed into an early after-depolarization (EAD) like waveform, and then showed ventricular tachyarrhythmias and ventricular fibrilations (VT/Vf) (Fig. 17.33C). When we applied another lethal arrhythmia positive compound (E-4031), QT-prolongation-like self-convoluted FP duration prolongation and its fluctuation also increased according to the increase of E-4031 concentration (Fig. 17.33D).

The results indicate that the ECG-like convoluted FP waveforms of the quasi *in vivo* ring-shaped cell network assay has the strong potential to predict lethal arrhythmia because they include both the repolarization prolongation data and the conductance abnormality of cardiomyocyte networks, and can be analyzed through simple waveform peak detection and pattern matching.

17.5.4 On-Chip Quasi *In Vivo* Predictive Cardiotoxicity Measurement (III): Temporal Fluctuation Measurement of Force Generation

One of the harmful side effects of drugs in heart besides electrophysiological phenomenon is mechanophysiological responses in cardiomyocytes such as inhibition of contractile motion. For the precise evaluation of mechanophysiological effects of drugs, we have combined optical image analysis system to the multielectrode system that records the contractile motions of cardiomyocytes with electrical FPD recording simultaneously. As shown in Fig. 17.34, polystyrene microparticles are attached on the surface of cardiomyocytes as displacement point markers and are traced and recorded for analysis of their displacement distances and angles. The measurement index of displacement distance is not always suitable to evaluate fluctuation because the displaced distance itself is usually very small, only less than 10 % of cell size. In contrast, the fluctuation of displacement angle is more sensitive and reflects the beginning of uniformed beating in the cardiomyocyte network.

17.5.5 Importance of Cell Network Size and Spatial Arrangement in On-Chip Cardiomyocyte Network Assay for Precise and Stable Drug Screening Measurement

For drug discovery and toxicology, ICH regulation S7B requests a pair of *in vitro* channel-measurement-based study and *in vivo* animal QT prolongation test, which are called "core battery." Although the *in vitro* patch-clamp analysis of cardiomyocytes is a well-accepted way to investigate the action potential parameters, it is just single-cell-based ion channel study and it is impossible to measure the abnormal cell-to-cell conductivity, which is one of the origins of ventricular tachycardia (VT) including torsade de pointes (TdP), a potentially life-threatening lethal arrhythmia because it

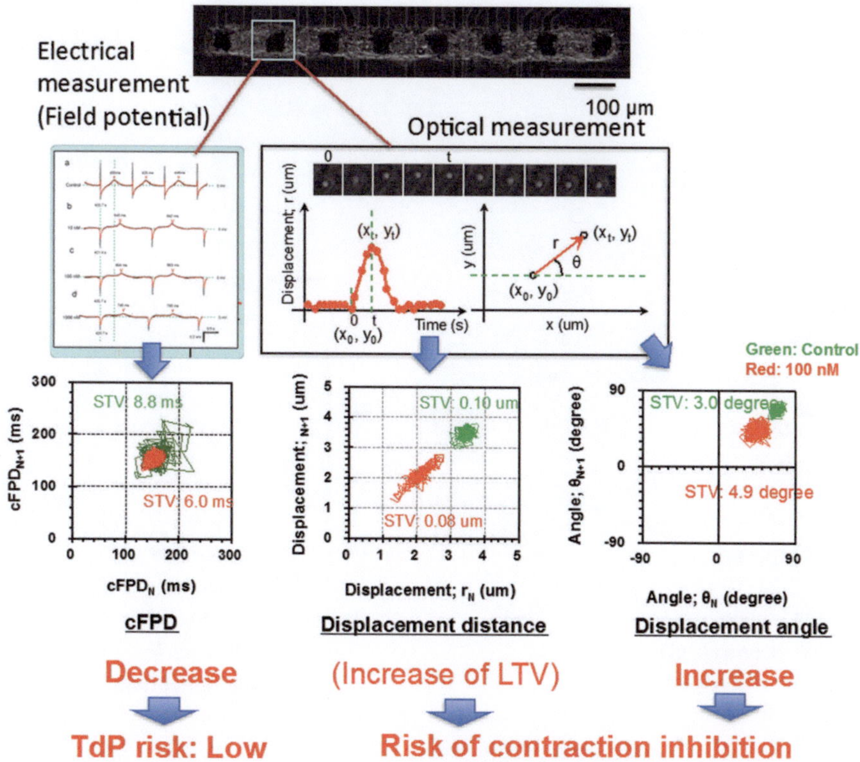

Fig. 17.34 *Simultaneous measurement of tension-Calcium release correlation and FPD*. Optical images of polystyrene microbeads attached on the surface of cardiomyocyte network on the multielectrode array were obtained from bright field microscopy equipped with high-speed CCD video camera with electrical FPD measurement simultaneously. Time-course change of displacement of microbeads on cardiomyocytes and their fluctuation (STV) of displacement distance and angles are recorded and analyzed

may lead to ventricular fibrillation, asystole, and sudden death. Hence, if we can arrange the cardiomyocyte network properly on the chip, the cell-network-based drug screening assay has a potential being suited for drug discovery and toxicology as a minimum model of the ventricle.

Hence, we have investigated the community size and spatial arrangement dependence of cardiomyocyte response against positive compounds using four types of cell network of cardiomyocytes (small cluster, small closed loop, two-dimensional (2D) sheet, and a large closed re-entry loop) in those shaped agarose microchambers fabricated on a multielectrode array chip for simultaneous comparison of their responses with administering Quinidine, a serious ventricular arrhythmiac compound. As shown in Fig. 17.35, we found that the two-dimensional sheet was most durable and stable against administration of compounds even 10 μM high dose, whereas the large closed re-entry loop network was most sensitive and showed lethal arrhythmia from 1 μM low dose administration.

The results indicate the importance of the control of the cell network arrangement because of the different response of cardiomyocytes depending on their

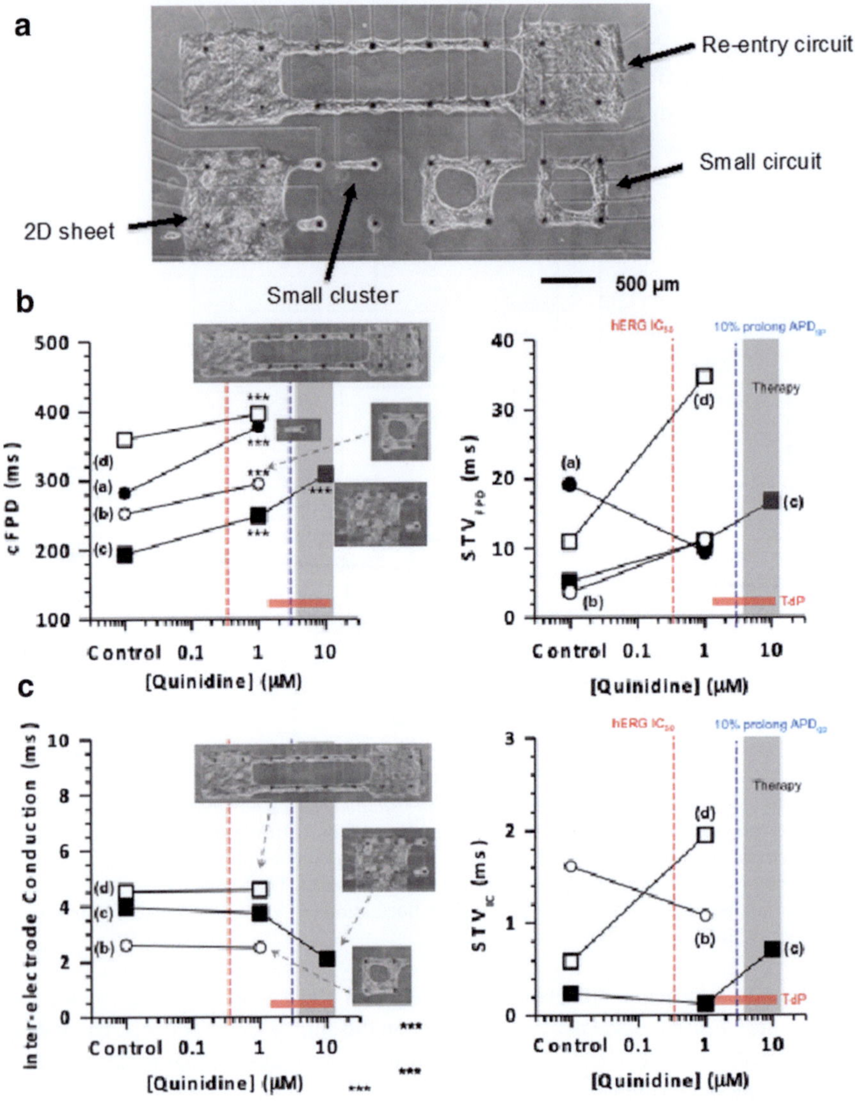

Fig. 17.35 *Two viewpoints of in vitro predictive cardiotoxicity measurement*. (**a**) Micropgraph and schematic drawing of cardiomyocyte network arrangement. Human iPS derived cardiomyocytes are arranged linearly on the microelectrodes on a chip. (**b**) Temporal fluctuation of prolongation of field potential duration (FPD) of single cells (Stratecy I) and temporal fluctuation of conductance between neighboring cells (Strategy II) can be measured simultaneously. (**c**) Acquired temporal fluctuation signals of single cell FPD and cell-to-cell conductance are plotted in the logistic maps and evaluate their uncertainty as a quantitative risk value of lethal arrhythmia of heart

spatial arrangement, and the necessity of the consideration of the proper community size and spatial arrangement pattern of cell networks for representing the proper response of cardiomyocytes against the proper amount of compounds.

17.5.6 Single Cell Shape Control by Agarose Microchamber Shapes: Understanding Correlation of Single Cell Shape and Its Force Generation Direction

As we have described above, agarose microstructures can control the spatial arrangement of cells for the network patterning of neurons, the connection of cardiomyocytes by preventing cell attachment. This photothermally etched agarose microstructure has a potential of not only the control of the network or connection pattern but also the control of the single cell shape by the more precise submicrometer-scaled photothermal etching. Hence, we have examined the control of the shape of single cardiomyocytes with an improved photothermal etching system. And, as an example of its application, we have measured the cell-shape dependence of contractile motion in round- and rectangular-shaped single cardiomyocytes by using an on-chip optical image analysis system.

Using the on-chip submicrometer-scaled photothermal etching technology, single cardiomyocytes successfully controlled the morphological shapes by seeding in 21 μm radius circular- or 20 × 70 μm rectangular-shaped agarose microchambers (Fig. 17.36). To detect the contractility of cardiomyocytes, the cells were labeled with the polystyrene microbeads attached on the surface of target single cardiomyocytes, and the motions of beads were acquired and analyzed using a newly developed wider depth of field optics equipped with 1/100 s high-speed digital camera. Mechanophysiological properties such as displacement and direction of movement were obtained using real-time processing system module at 0.15 μm spatial and 10 ms temporal resolution. Comparison with displacement and direction of contraction in circular- and rectangular-shaped cardiomyocytes were indicated that the rectangular-shaped cardiomyocytes tended to contract along the longitudinal direction like a real heart. This result suggests that the morphological shape of cells affected the function of cells in contractile ability (Kaneko et al. 2011).

17.5.7 What Is Required for the Next Generation of Cardiotoxicity Measurement Importance of Cell Network Size and Spatial Arrangement in On-Chip Cardiomyocyte Network Assay for Precise and Stable Drug Screening Measurement

Lethal arrhythmia has been one of the major safety concerns for the pharmaceutical industry in selecting and developing drug candidates. Integrated assay systems using hERG-transfected HEK-293/CHO-cells (hERG assay), isolated animal tissues (APD or MAPD assay), and conscious and/or anesthetized whole animals (QT or MAPD assay) are currently used to identify QT prolongation, whereas those assay systems are not competent to fully predict the potential lethal arrhythmia such as torsades de pointes (TdP) or ventricular fibrillation (Vf) induced by drugs or candidates. In this context, there is a long-standing and urgent need for a surrogate marker that can distinguish the torsadogenic potential from the QT interval duration. Hence, we have proposed a quasi *in vivo* cardiotoxicity assay, which is a new *in vitro* assay technology platform where human iPS/ES cell-derived cardiomyocytes and on-chip technology

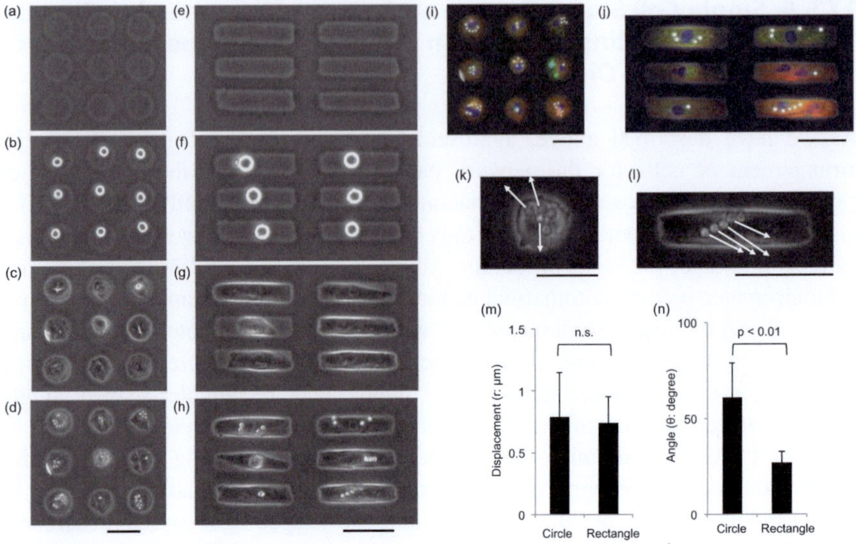

Fig. 17.36 *Shape control of single cardiomycoytes by shape of microchamber structures.* (**a**) and (**e**), phase-contrast images of circular- and rectangular-shaped agarose microchambers, respectively; (**b**) and (**f**), the single cardiomyocytes arranged into agarose microchambers; (**c**) and (**g**), the cardiomyocytes cultured and grown at 2 days in vitro (2 DIV); (**d**) and (**h**), the microbeads (*bright dots*) attached on the surface of cardiomyocytes at 3 DIV. Phase-contrast image superimposed on the fluorescence images of nucleus (*blue*), cardiac myosin (*green*), and actin filament (*red*) in the circular- (**i**) and rectangular- (**j**) shaped single cardiomyocytes. The length and direction of *white arrows* show the displacement and angle of contractile motion of microbeads from a diastolic phase to a systolic phase on the circular- (**k**) and the rectangular- (**l**) shaped single cardiomyocytes. The lengths of *white arrows* show 30-fold real displacement. (**m**) shows the displacement (r) of circular- and rectangular-shaped cardiomyocytes. (**n**) shows the angle (θ) of contractile direction of circular- and rectangular-shaped cardiomyocytes ($n = 7$). The angle (θ) of contractile direction was defined as follows: the angle of the longitudinal direction in rectangular shape defined 0°. The angle 0° of circular-shaped cardiomyocytes was the same longitudinal direction in neighboring rectangular shape in the same plane. The range of angles was defined from 0 to 90° against longitudinal direction. *Error bars* indicate standard deviation. *n.s.* no significance. Bars, 50 μm

are combined and used as an assay tool to bridge the gap between preclinical studies and human clinical settings in terms of cardiotoxicity of new chemical entities for drug development. Potential advantages of the proposed strategy of our quasi *in vivo* assay combining spatiotemporal strategies I and II include (1) using a set of standard human cardiomyocytes prepared from human iPS/ES cells of different races, sexes, and also from patients with various diseases to provide an ideal testing panel platform; (2) to predict lethal arrhythmia (TdP/VT/Vf) by evaluation of temporal fluctuation of single-cell-level ion channels kinetics, and by evaluation of spatial cell-to-cell conductance fluctuation using the on-chip cell network loop which can choose different conductance pathways of human cardiomyocytes among neighboring circulations; and (3) the capacity to quantitatively evaluate the correlation between calcium release and tension generation, and the inhibition on the trafficking pathway of ion-channel proteins by its long-term optical/electrical simultaneous measurement (Fig. 17.37).

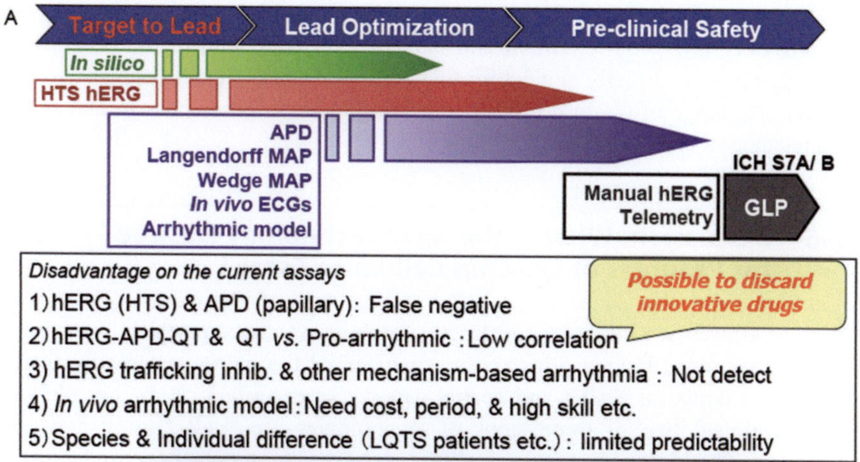

B

	In vitro IKr: HERG, APD	In vivo QT: Dogs, Monkeys	Carlson model: Rabbits	AV-block model: Dogs, Pigs ..	On-chip Q-in vivo model
HERG (IKr)	○	×	×	×	×
Na, Ca	△~○ (false negative/false positive)	△~○	△~○	△~○	△~○
QT (APD/ FPD)	△ (false negative/false positive)	○	○	○	○
TdP (VT/VF)	×	×	△	○	○
Trafficking	△	×~△	×~△	×~△	○
Contractility	×	×	○	○	○
Panel screening	×	×	×	×	○
Compound Amount	○ (~ μg)	× (~ mg)	× (~ mg)	× (~ mg)	○ (~ μg)
Throughput	○	×	×	×	○
Running cost	$	$$$	$$	$$$	$

Fig. 17.37 *Disadvantages of current assays and its potential solution: On-chip quasi in vivo assay*. (**A**) Schematic drawing of drug discovery process and a list of disadvantages of the current assay. (**B**) Comparison of current technologies and on-chip quasi in vivo screening assay

The results in previous subsections also indicate the importance of the community size of cells to acquire reliable and stable results for monitoring of drug effects on them, which we usually neglect for preparation of cells. The results also showed another advantage of on-chip in vitro screening method: quick change and stable maintenance of cells' environment conditions.

17.6 Discussion: What Is the Next Technology Desired for Progress of On-Chip Cellomics Assay?

On-chip cellmics is a kind of a constructive or reconstructive approach from single cells to understand hidden rules and mechanisms in the living systems from two viewpoints, temporal aspect and spatial aspect. As described above, we already have developed two of three steps of technologies, i.e., cell sorting and cell cultivation. The remaining part in our three-step strategy of epigenetic understanding of cell network system is the third step, single cell expression analysis.

We have proposed two major complementary technologies for single-cell-based cell expression analysis: on-chip adaptive electron microscopy detection method combined with a set of 300 nanoparticle probes (Fig. 17.38) (Kim et al. 2007, 2010a, b, c, d), and an ultrafast microdroplet PCR system, in which we can detect the existence of target mRNA or DNA within 3 min ultrafast detection (Fig. 17.39) (Terazono et al. 2010b, c, d).

Especially, the technology for DNA/mRNA analysis of the collected cells has been developed for practically commercialized phase using a newly developed simple ultrahigh-speed real-time μl-sample droplet PCR system featuring rapid water-based heating of sample droplets by quick switching between two hot water circuits (one for denaturation and one for annealing and extension). The module consists of two hot water circuits, a reaction chamber, and a light-fiber fluorescent detector.

As shown in Fig. 17.39A, each hot water circuit has a tank with an internal heater, two diaphragm valves, a fine-adjustment heater, a check valve, and a geared pump and is connected to a PCR reaction chamber made of PDMS and containing a thin-film aluminum chip, on which an array of microindentations are embossed to hold μl-sized water droplets. The water in each circuit is circulated continuously by the pump during reaction to maintain the temperature of the hot water in each tank. It is transferred to the reaction chamber only when the temperature of the reaction chamber needs to be increased to that of the hot water. Fine temperature control of the circulated hot water is achieved by adjusting the appropriate fine-adjustment heater. One circuit maintains the water temperature at 95 °C; this water is used for denaturation of DNA fragments. The other maintains the water temperature at 50–70 °C; this water is used for extension of DNA fragments. Diaphragm valves are used simultaneously to control the circulation of the water and to transfer it when needed to the reaction chamber, where it is circulated under the aluminum chip, resulting in direct transfer of the heat from the water to the chip and then to the droplets. When the water is no longer needed, it is returned to the appropriate hot water tank and again circulated. Using this method, we can switch two temperatures within 0.4 s.

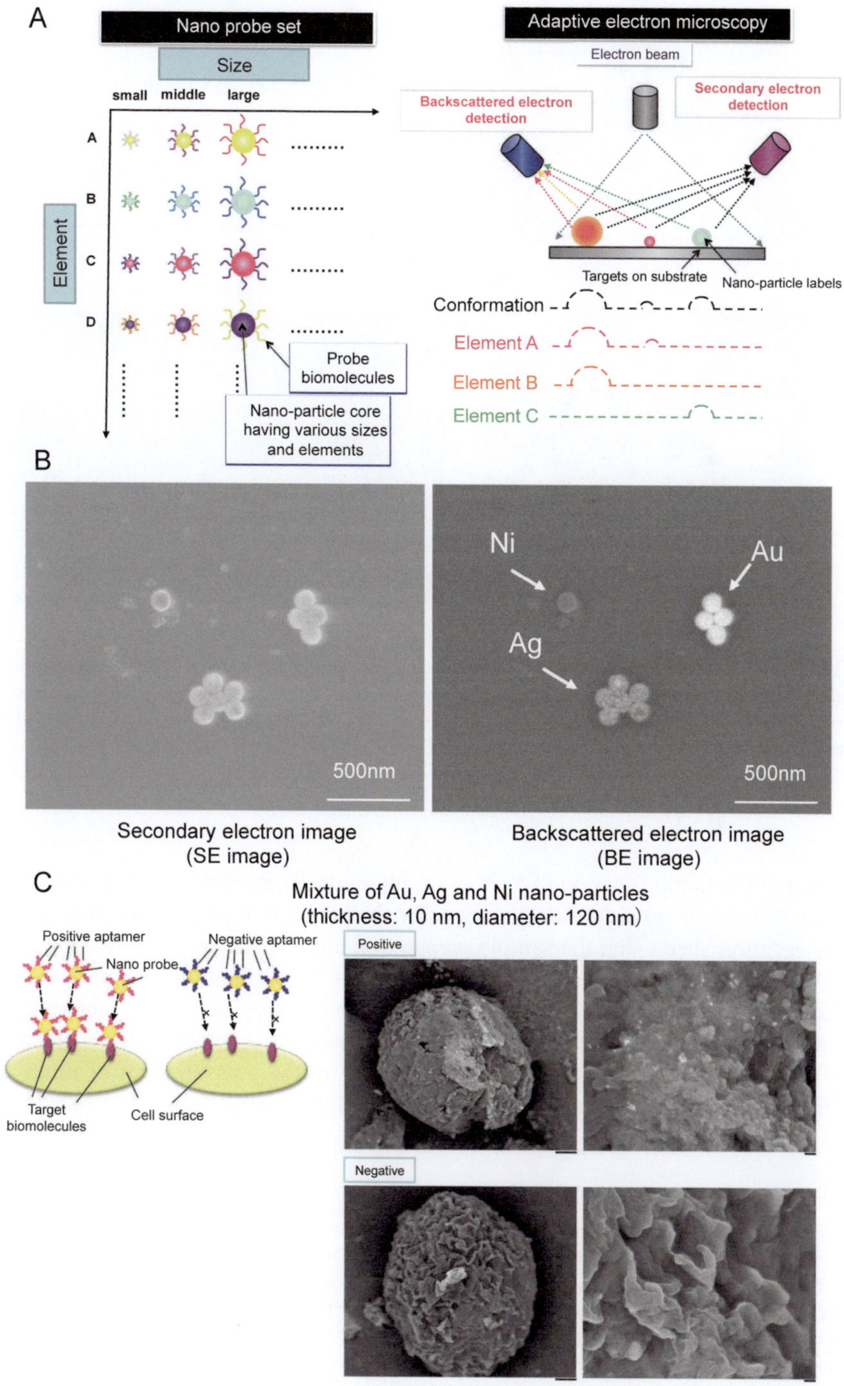

Fig. 17.38 *More than 300 multicolored labeling for single-cell-level expression analysis exploiting multielement alloy nanoparticle probe set with adaptive electron microscopy.* (**A**) Schematic drawing of two core technologies for single-cell-level expression analysis; (1) more than 300 different nanoprobe set, which is made of a two-dimensional combination of 13 different

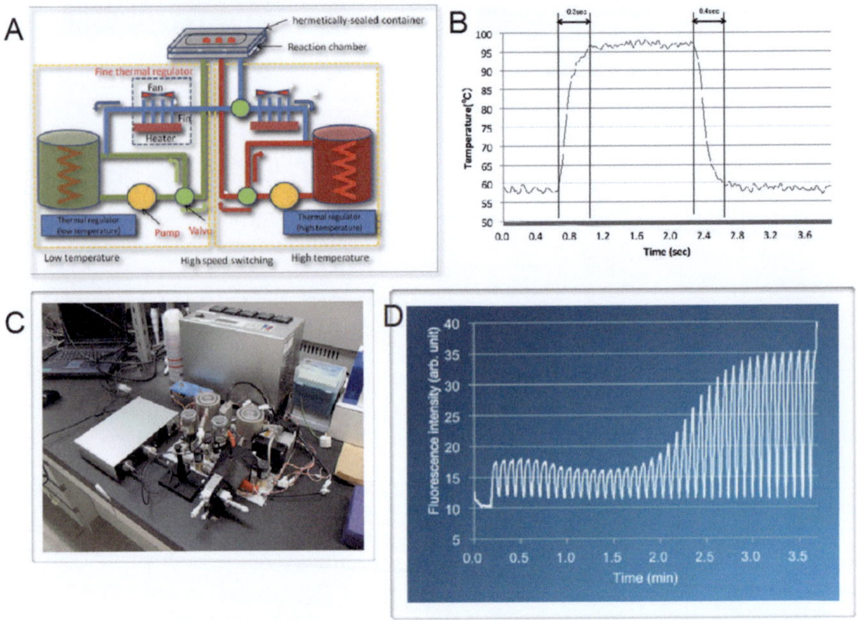

Fig. 17.39 *Ultrafast-microdroplet PCR system.* (**A**) Schematic drawing of principle of ultrafast-microdroplet PCR technology. The system exploited fast heat conduction of strictly temperature-controlled hot water to sample droplets within 0.4 ms. (**B**) Time-course temperature rise and fall of sample droplet. For temperature rise from 60 °C to 95 °C, it took only 0.3 ms, and for fall from 95 °C to 60 °C, only 0.4 ms. (**C**) Photograph of the system. (**D**) Example of PCR reaction, which only needed 3 min for detection

Rapid PCR amplification of a set of droplets containing Bacillus anthracis lined up on the aluminum chip was evidenced by the increase in fluorescent intensity. It was accomplished using the sample preparation module and on-chip imaging flow cell-sorting module within 3.5 min (40 cycles of 1-s denaturation and 3-s extension). As shown in Fig. 17.39D, this is an order of magnitude faster than with a conventional fast PCR system. This integrated system thus enables rapid detection of DNA fragments and is potentially applicable to the measurement of multiple samples simultaneously in a miniaturized microfluidic chip.

Fig. 17.38 (continued) elements and at least three different sizes; (2) adaptive scanning electron microscopy technology for identifying each nanoprobe by its sphere size and its element, which is estimated by comparing its backscattering intensity, which reflect element number, and second electron intensity, which does not reflect element number. (**B**) Example of electron micrographs of second electron images (*left*) and backscattering images of Au, Ag, Ni nanoparticles. (**C**) *Left* schematic drawing explains the mechanism of DNA-aptamer decorated nanoprobes attachment to the target molecules on the surface of target cells. *Right* electron micrographs show the selective attachment of nanoprobes

17.7 Conclusion

We have introduced and discussed a series of on-chip cellomics technologies and their applications as examples for understanding of living systems including epigenetic information from "geometric" viewpoint. We believe the development of new technologies and approaches including these techniques we introduced in this chapter will open the way to the next generation of single-cell-based life science studies and regenerative medicine.

Acknowledgments We thank all the members of Yasuda Laboratory. This work was financially supported by Japan Science and Technology Agency (JST), Grant-in-Aid for Scientific Research of Ministry of Education, Culture, Sports, Science and Technology, New Energy Development Organization (NEDO), On-chip cellomics consortium, Kanagawa Academy of Science and Technology (KAST), and Japan Defense Agency.

References

Anzai Y, Terazono H, Yasuda K (2007) Simple non-invasive cell separation method using magnetic aptamer-conjugated microbeads and nuclease digestion. J Biol Phys Chem 7:83–86

Assenmacher M, Manz R, Miltenyi S, Scheffold A, Radbruch A (1995) Fluorescence-activated cytometry cell sorting based on immunological recognition. Clin Biochem 28(1):39–40

Azuma C, Yasuda K, Tanabe Y, Taniguro H, Kanaya F, Nakayama A, Chen YM, Gong JP, Osada Y (2007) Biodegradation of high-toughness double network hydrogels as potential materials for artificial cartilage. J Biomed Mater Res A 81(2):373–380

Bonner WA, Hulett HR, Sweet RG, Herzenberg LA (1972) Fluorescence activated cell sorting. Rev Sci Instrum 43(3):404–409

Ellington AD, Szostak JW (1990) In vitro selection of RNA molecules that bind specific ligands. Nature 346:818–822

Golan L, Yeheskely-Hayon D, Minai L, Dann EJ, Yelin D (2012) Noninvasive imaging of flowing blood cells using label-free spectrally encoded flow cytometry. Biomed Opt Express 3 (6):1455–1464

Gopinath S (2007) Methods developed for SELEX. Anal Bioanal Chem 387(1):171–182

Gorkov LP (1962) On the forces acting on a small particle in an acoustical field in an ideal fluid. Sov Phys Dokl 6(9):773

Grover SC, Skirtach AG, Gauthier RC, Grover CP (2001) Automated single-cell sorting system based on optical trapping. J Biomed Opt 6(1):14–22

Hatayama T, Yasuda K, Nishiyama E (1994) Characterization of high-molecular-mass heat shock proteins and 42 degrees C-specific heat shock proteins of murine cells. Biochem Biophys Res Commun 204(1):357–365

Hattori A, Yasuda K (2010) Comprehensive study of microgel electrode for on-chip electrophoretic cell sorting. Jpn J Appl Phys 49(6):06GM04

Hattori A, Umehara S, Wakamoto Y, Yasuda K (2003) Measurement of incident angle dependence of swimming bacterium reflection using on-chip single-cell cultivation assay. Jpn J Appl Phys 42(7B): L873

Hattori A, Moriguchi H, Ishiwata S i, Yasuda K (2004) A 1480/1064 nm dual wavelength photothermal etching system for non-contact three-dimensional microstructure generation into agar microculture chip. Sensors Actuators B Chem 100(3):455–462

Hayashi M, Hattori A, Kim H, Terazono H, Kaneko T, Yasuda K (2011) Fully automated on-chip imaging flow cytometry system with disposable contamination-free plastic re-cultivation chip. Int J Mol Sci 12(6):3618–3634

Herr JK, Smith JE, Medley CD, Shangguan D, Tan W (2006) Aptamer-conjugated nanoparticles for selective collection and detection of cancer cells. Anal Chem 78:2918–2924

Herzenberg LA, Sweet RG (1976) Fluorescence-activated cell sorting. Sci Am 234(3):108–117

Hsu CH, Di Carlo D, Chen C, Irimia D, Toner M (2008) Microvortex for focusing, guiding and sorting of particles. Lab Chip 8(12):2128–2134

Hulett HR, Bonner WA, Sweet RG, Herzenberg LA (1973) Development and application of a rapid cell sorter. Clin Chem 19(8):813–816

Imamura M, Aoki H, Eya K, Murakami T, Yasuda K (1995) Balloon angioplasty before Wheat's operation in a patient with Turner's syndrome. Cardiovasc Surg 3(1):70–72

Inoue I, Wakamoto Y, Moriguchi H, Okano K, Yasuda K (2001a) On-chip culture system for observation of isolated individual cells. Lab Chip 1(1):50–55

Inoue I, Wakamoto Y, Yasuda K (2001b) Non-genetic variability of division cycle and growth of isolated individual cells in on-chip culture system. Proc Jpn Acad Ser B Phys Biol Sci 77 (8):145–150

Inoue I, Shiomi D, Kawagishi I, Yasuda K (2004) Simultaneous measurement of sensor-protein dynamics and motility of a single cell by on-chip microcultivation system. J Nanobiotechnol 2 (1):4

Johnson KW, Dooner M, Quesenberry PJ (2007) Fluorescence activated cell sorting: a window on the stem cell. Curr Pharm Biotechnol 8(3):133–139

Kaneko T, Kojima K, Yasuda K (2007) Dependence of the community effect of cultured cardiomyocytes on the cell network pattern. Biochem Biophys Res Commun 356(2):494–498

Kaneko T, Nomura F, Yasuda K (2011) Orientation and community size dependences of pulsatile electrical field stimulation on lined-up and rod-shaped single cardiomyocytes. Jpn J Appl Phys 50(8):080220

Kaneko T, Nomura F, Hamada T, Abe Y, Takamori H, Sakakura T, Takasuna K, Sanbuissho A, Hyllner J, Sartipy P, Yasuda K (2014) On-chip in vitro cell-network pre-clinical cardiac toxicity using spatiotemporal human cardiomyocyte measurement on a chip. Sci Rep 4:4670

Kim H, Oikawa K, Watanabe N, Shigeno M, Shirakawabe Y, Yasuda K (2007) Identification of size differences of gold nanoparticles on cell surface by curvature reconstruction method using atomic force microscopy. Jpn J Appl Phys 46:L184–L186

Kim H, Kira A, Yasuda K (2010a) Non-amplified quantitative detection of nucleic acid sequences using a gold nanoparticle probe set and field-emission scanning electron microscopy. Jpn J Appl Phys 49(6):06GK07

Kim H, Negishi T, Kudo M, Takei H, Yasuda K (2010b) Quantitative backscattered electron imaging of field emission scanning electron microscopy for discrimination of nano-scale elements with nm-order spatial resolution. J Electron Microsc 9(5): 379-385

Kim H, Takei H, Yasuda K (2010c) Production of size-controlled nanoscopic cap-shaped metal shells. Jpn J Appl Phys 49(4):048004

Kim H, Takei H, Yasuda K (2010d) Quantitative evaluation of a gold-nanoparticle labeling method for detecting target DNAs on DNA microarrays. Sensors Actuators B Chem 144 (1):6–10

Kim H, Hayashi M, Terazono H, Takei H, Yasuda K (2011) Production of double-layered metal nanocups for artificial nanospace of biomolecular reaction. Jpn J Appl Phys 50(6):06GJ03

Kim H, Terazono H, Nakamura Y, Sakai K, Hattori A, Odaka M, Girault M, Arao T, Nishio K, Miyagi Y, Yasuda K (2014a) Development of on-chip multi-imaging flow cytometry for identification of imaging biomarkers of clustered circulating tumor cells. PLoS One 9(8): e104372

Kim H, Terazono H, Takei H, Yasuda K (2014b) Cup-shaped superparamagnetic hemispheres for size-selective cell filtration. Sci Rep 4:6362

Kim H, Terazono H, Takei H, Yasuda K (2014c) Fabrication of multilayered superparamagnetic particles based on sequential thermal deposition method. Jpn J Appl Phys 53(6):06JJ01

King LV (1935) On the acoustic radiation pressure on circular discs: inertia and diffraction corrections. Proc R Soc Lond Ser A Math Phys Sci 153(878):1–16

Kojima K, Moriguchi H, Hattori A, Kaneko T, Yasuda K (2003) Two-dimensional network formation of cardiac myocytes in agar microculture chip with 1480 nm infrared laser photo-thermal etching. Lab Chip 3(4):292–296

Kojima K, Kaneko T, Yasuda K (2004) A novel method of cultivating cardiac myocytes in agarose microchamber chips for studying cell synchronization. J Nanobiotechnol 2(1):9

Kojima K, Kaneko T, Yasuda K (2005) Stability of beating frequency in cardiac myocytes by their community effect measured by agarose microchamber chip. J Nanobiotechnol 3(1):4

Kojima K, Kaneko T, Yasuda K (2006) Role of the community effect of cardiomyocyte in the entrainment and reestablishment of stable beating rhythms. Biochem Biophys Res Commun 351(1):209–215

Kozuka T, Tuziuti T, Mitome H (1995) Selective manipulation of micro particles using ultrasound. Tech Rep IEICE US94(82):33

Liu P, Meagher RJ, Light YK, Yilmaz S, Chakraborty R, Arkin AP, Hazen TC, Singh AK (2011) Microfluidic fluorescence in situ hybridization and flow cytometry (muFlowFISH). Lab Chip 11(16):2673–2679

Matsumura K, Yagi T, Yasuda K (2003a) Differential analysis of cell cycle stability in chlamydomonas using on-chip single-cell cultivation system. Jpn J Appl Phys 42(Part 2, No. 7A):L784

Matsumura K, Yagi T, Yasuda K (2003b) Role of timer and sizer in regulation of Chlamydomonas cell cycle. Biochem Biophys Res Commun 306(4):1042–1049

Moriguchi H, Wakamoto Y, Sugio Y, Takahashi K, Inoue I, Yasuda K (2002) An agar-microchamber cell-cultivation system: flexible change of microchamber shapes during cultivation by photo-thermal etching. Lab Chip 2(2):125–132

Moriguchi H, Takahashi K, Sugio Y, Wakamoto Y, Inoue I, Jimbo Y, Yasuda K (2004) On-chip neural cell cultivation using agarose-microchamber array constructed by a photothermal etching method. Electr Eng Jpn 146(2):37–42

Nomura F, Kaneko T, Hattori A, Yasuda K (2011a) Label-free shape-based selection of cardiomyocytes with on-chip imaging cell sorting system. J Bioprocess Biotechnol S3-003:1–6

Nomura F, Kaneko T, Hattori A, Yasuda K (2011b) On-chip constructive cell-network study (II): on-chip quasi-in vivo cardiac toxicity assay for ventricular tachycardia/fibrillation measurement using ring-shaped closed circuit microelectrode with lined-up cardiomyocyte cell network. J Nanobiotechnol 9:39

Nyborg WL (1967) Radiation pressure on a small rigid sphere. J Acoust Soc Am 42(5):947

Ogawa T, Sakata S, Nakamura S, Takuno H, Matsui I, Sarui H, Yasuda K (1994) Thyroid hormone autoantibodies in patients with Graves' disease: effect of anti-thyroid drug treatment. Clin Chim Acta Int J Clin Chem 228(2):113–122

Ohuchi SP, Ohtsu T, Nakamura Y (2006) Selection of RNA aptamers against recombinant transforming growth factor-beta type III receptor displayed on cell surface. Biochimie 88:897–904

Owen CS, Sykes NL (1984) Magnetic labeling and cell sorting. J Immunol Methods 73(1):41–48

Schonbrun E, Gorthi SS, Schaak D (2012) Microfabricated multiple field of view imaging flow cytometry. Lab Chip 12(2):268–273

Shangguan D, Li Y, Tang Z, Cao ZC, Chen HW, Mallikaratchy P, Sefah K, Yang CJ, Tan W (2006) Aptamers evolved from live cells as effective molecular probes for cancer study. Proc Natl Acad Sci U S A 103:11838–11843

Spudich JL, Koshland DE (1976) Non-genetic individuality: chance in the single cell. Nature 262 (5568):467–471

Sugio Y, Kojima K, Moriguchi H, Takahashi K, Kaneko T, Yasuda K (2004) An agar-based on-chip neural-cell-cultivation system for stepwise control of network pattern generation during cultivation. Sensors Actuators B Chem 99(1):156–162

Suzuki I, Yasuda K (2007a) Constructive formation and connection of aligned micropatterned neural networks by stepwise photothermal etching during cultivation. Jpn J Appl Phys 46 (9B):6398–6403

Suzuki I, Yasuda K (2007b) Detection of tetanus-induced effects in linearly lined-up micropatterned neuronal networks: application of a multi-electrode array chip combined with agarose microstructures. Biochem Biophys Res Commun 356(2):470–475

Suzuki I, Sugio Y, Jimbo Y, Yasuda K (2004a) Individual-cell-based electrophysiological measurement of a topographically controlled neuronal network pattern using agarose architecture with a multi-electrode array. Jpn J Appl Phys 43(3B):L403–L406

Suzuki I, Sugio Y, Moriguchi H, Jimbo Y, Yasuda K (2004b) Modification of a neuronal network direction using stepwise photo-thermal etching of an agarose architecture. J Nanobiotechnol 2 (1):7

Suzuki I, Sugio Y, Jimbo Y, Yasuda K (2005) Stepwise pattern modification of neuronal network in photo-thermally-etched agarose architecture on multi-electrode array chip for individual-cell-based electrophysiological measurement. Lab Chip 5(3):241–247

Takahashi K, Orita K, Matsumura K, Yasuda K (2003) On-chip microcultivation chamber for swimming cells using visualized poly(dimethylsiloxane) valves. Jpn J Appl Phys 42(Part 2, No. 9A/B):L1104

Takahashi K, Hattori A, Suzuki I, Ichiki T, Yasuda K (2004) Non-destructive on-chip cell sorting system with real-time microscopic image processing. J Nanobiotechnol 2(1):5

Terazono H, Anzai Y, Soloviev M, Yasuda K (2010a) Labelling of live cells using fluorescent aptamers: binding reversal with DNA nucleases. J Nanobiotechnol 8(1):8

Terazono H, Takei H, Hattori A, Yasuda K (2010b) Development of a high-speed real-time PCR system for rapid and precise nucleotide recognition. Proceeding of SPIE Defense, Security, and Sensing 2010, 76730U

Terazono H, Takei H, Hattori A, Yasuda K (2010c) Development of a high-speed real-time polymerase chain reaction system using a circulating water-based rapid heat-exchange. Jpn J Appl Phys 49(6):06GM05

Terazono H, Takei H, Hayashi M, Hattori A, Yasuda K (2010d) Development of an integrated system for rapid detection of biological agents. Proceeding of SPIE Defense, Security, and Sensing 2010, 766503

Terazono H, Hayashi M, Kim H, Hattori A, Yasuda K (2012) Cell-sorting system with on-chip imaging for label-free shape-based selection of cells. Jpn J Appl Phys 51(6):06FK08

Trickett A, Kwan YL (2003) T cell stimulation and expansion using anti-CD3/CD28 beads. J Immunol Methods 275(1-2):251–255

Tuerk C, Gold L (1990) Systematic evolution of ligands by exponential enrichment: RNA ligands to bacteriophage T4 DNA polymerase. Science 249(4968):505–510

Umehara S, Wakamoto Y, Inoue I, Yasuda K (2003) On-chip single-cell microcultivation assay for monitoring environmental effects on isolated cells. Biochem Biophys Res Commun 305 (3):534–540

Wakamoto Y, Inoue I, Moriguchi H, Yasuda K (2001) Analysis of single-cell differences by use of an on-chip microculture system and optical trapping. Fresenius J Anal Chem 371(2):276–281

Wakamoto Y, Umehara S, Matsumura K, Inoue I, Yasuda K (2003) Development of non-destructive, non-contact single-cell based differential cell assay using on-chip microcultivation and optical tweezers. Sensors Actuators B Chem 96(3):693–700

Yasuda K (1997) Blood concentration by superposition of higher harmonics of ultrasound. Jpn J Appl Phys 36(Part 1, No. 5B):3130

Yasuda K (2000) Non-destructive, non-contact handling method for biomaterials in micro-chamber by ultrasound. Sensors Actuators B Chem 64(1-3):128–135

Yasuda K, Kamakura T (1997) Acoustic radiation force on micrometer-size particles. Appl Phys Lett 71(13):1771–1773

Yasuda K, Kiyama M, Umemura S, Takeda K (1996a) Deoxyribonucleic acid concentration using acoustic radiation force. J Acoust Soc Am 99(2):1248–1251

Yasuda K, Takeda K, Umemura S-i (1996b) Studies on particle separation by acoustic radiation force and electrostatic force. Jpn J Appl Phys 35(Part 1, No. 5B):3295

Yasuda K, Okano K, Ishiwata S (2000) Focal extraction of surface-bound DNA from a microchip using photo-thermal denaturation. Biotechniques 28(5):1006–1011

Yasuda K, Roneker KR, Miller DD, Welch RM, Lei XG (2006) Supplemental dietary inulin affects the bioavailability of iron in corn and soybean meal to young pigs. J Nutr 136 (12):3033–3038

Yasuda K, Hattori A, Kim H, Terazono H, Hayashi M, Takei H, Kaneko T, Nomura F (2013) Non-destructive on-chip imaging flow cell-sorting system for on-chip cellomics. Microfluid Nanofluid 14(6):907–931

Yoon YE, Chang SA, Choi SI, Chun EJ, Cho YS, Youn TJ, Chung WY, Chae IH, Choi DJ, Chang HJ (2012) The absence of coronary artery calcification does not rule out the presence of significant coronary artery disease in Asian patients with acute chest pain. Int J Cardiovasc Imaging 28:389–398

Yoshioka K, Kawashima Y (1955) Acoustic radiation pressure on a compressible sphere. Acustica 5:167–173

Yoshioka M, Mizutani K, Nagai K (1997) Sonic field measurement using light computerized tomography. Jpn J Appl Phys 36:3199

Zborowski M, Chalmers JJ (2005) Magnetic cell sorting. Methods Mol Biol 295:291–300

Zborowski M, Chalmers JJ (2011) Rare cell separation and analysis by magnetic sorting. Anal Chem 83(21):8050–8056

Index

K. Tanishita, K. Yamamoto (eds.), *Vascular Engineering*,
DOI 10.1007/978-4-431-54801-0

D

E

R

S

T

MIX
Papier aus verantwortungsvollen Quellen
Paper from responsible sources
FSC® C105338

If you have any concerns about our products,
you can contact us on
ProductSafety@springernature.com

In case Publisher is established outside the EU,
the EU authorized representative is:
Springer Nature Customer Service Center GmbH
Europaplatz 3, 69115 Heidelberg, Germany

Printed by Libri Plureos GmbH
in Hamburg, Germany